Donated to
SAINT PAUL PUBLIC LIBRARY

OPTICS
SOURCE BOOK

THE McGRAW-HILL SCIENCE REFERENCE SERIES

Acoustics Source Book
Communications Source Book
Computer Science Source Book
Fluid Mechanics Source Book
Meteorology Source Book
Nuclear and Particle Physics Source Book
Physical Chemistry Source Book
Solid-State Physics Source Book
Spectroscopy Source Book

OPTICS SOURCE BOOK

Sybil P. Parker, *Editor in Chief*

McGRAW-HILL BOOK COMPANY

New York   St. Louis   San Francisco
Auckland   Bogotá   Caracas   Colorado Springs   Hamburg
Lisbon   London   Madrid   Mexico   Milan   Montreal
New Delhi   Oklahoma City   Panama   Paris   San Juan
São Paulo   Singapore   Sydney   Tokyo   Toronto

Cover: Infrared thermogram showing thermal effluent entering a slow-moving stream with flow from left to right. (*Barnes Engineering Co.*)

This material has appeared previously in the McGRAW-HILL ENCYCLOPEDIA OF SCIENCE AND TECHNOLOGY, 6th Edition, copyright © 1987 by McGraw-Hill, Inc. All rights reserved.

OPTICS SOURCE BOOK, copyright © 1988 by McGraw-Hill, Inc. All rights reserved. Printed in the United States of America. Except as permitted under the United States Copyright Act of 1976, no part of this publication may be reproduced or distributed in any form or by any means, or stored in a data base or retrieval system, without prior written permission of the publisher.

1 2 3 4 5 6 7 8 9 0   DOC/DOC   8 9 5 4 3 2 1 0 9 8

ISBN 0-07-045506-6

Library of Congress Cataloging in Publication Data:

Optics source book / Sybil P. Parker, editor in chief.
    p. cm. — (McGraw-Hill science reference series)
   "This material has appeared previously in the McGraw-Hill encyclopedia of science and technology, 6th edition"— CIP t.p. verso.
  Bibliography: p.
  Includes index.
  ISBN 0-07-045506-6
  1. Optics.   I. Parker, Sybil P.   II. McGraw-Hill Book Company.   III. Title: McGraw-Hill encyclopedia of science and technology (6th ed.) IV. Series.
QC355.2.069   1988   87-36630
535—dc19

TABLE OF CONTENTS

Introduction	1
Geometrical Optics	3
Optical Imaging Systems	39
Physical Nature of Light	107
Wave Optics	131
Interaction of Light with Matter	175
Interaction of Light with Energy Fields	245
Lasers: Technology and Applications	261
Light Detection and Processing	307
Measurement of Light	341
Human Perception of Light	365
Contributors	385
Index	391

OPTICS
SOURCE BOOK

INTRODUCTION

Narrowly, OPTICS is the science of light and vision; broadly, it is the study of the phenomena associated with the generation, transmission, and detection of electromagnetic radiation in the spectral range extending from the long-wave edge of the x-ray region to the short-wave edge of the radio region. This range, often called the optical region or the optical spectrum, extends in wavelength from about 1 nanometer (4×10^{-8} in.) to about 1 mm (0.04 in.).

In ancient times there was some isolated elementary knowledge of optics, but it was the discoveries of the experimentalists of the early seventeenth century which formed the basis of the science of optics. The statement of the law of refraction by W. Snell, Galileo Galilei's development of the astronomical telescope and his discoveries with it, F. M. Grimaldi's observations of diffraction, and the principles of the propagation of light enunciated by C. Huygens and P. de Fermat all came in this relatively short period. The publication of Isaac Newton's *Opticks* in 1704, with its comprehensive and original studies of refraction, dispersion, interference, diffraction, and polarization, established the science.

So great were the contributions of Newton to optics that a hundred years went by before further outstanding discoveries were made. In the early nineteenth century many productive investigators, foremost among them Thomas Young and A. J. Fresnel, established the transverse-wave nature of light. The relationship between optical and magnetic phenomena, discovered by M. Faraday in the 1840s, led to the crowning achievement of classical optics—the electromagnetic theory of J. C. Maxwell. Maxwell's theory, which holds that light consists of electric and magnetic fields propagated together through space as transverse waves, provided a general basis for the treatment of optical phenomena. In particular, it served as the basis for understanding the interaction of light with matter and, hence, as the basis for treatment of the phenomena of physical optics. In the hands of H. A. Lorentz, this treatment led at the end of the nineteenth century and the beginning of the twentieth to an explanation of many optical phenomena, such as the Zeeman effect, in terms of atomic and molecular structure. The theories of Maxwell and Lorentz are regarded as the culmination of classical optics.

In the twentieth century optics has been in the forefront of the revolution in physical thinking caused by the theory of relativity and especially by the quantum theory. To explain the wavelength dependence of heat radiation, the photoelectric effect, the spectra of monatomic gases, and many other phenomena of physical optics, radical departure from the ideas of Lorentz and Maxwell about the mechanism of the interaction of radiation and matter and about the nature of radiation itself has been found necessary. The chief early quantum theorists were M. Planck, A. Einstein, and N. Bohr; later came L. de Broglie, W. Heisenberg, P. A. M. Dirac, E. Schrödinger, and others.

The science of optics finds itself in a position that is satisfactory for practical purposes but less so from a theoretical standpoint. The theory of Maxwell is sufficiently valid for treating the interaction of high-intensity radiation with systems considerably larger than those of atomic dimensions. The modern quantum theory is adequate for an understanding of the spectra of atoms and molecules and for the interpretation of phenomena involving low-intensity radiation, provided one does not insist on a very detailed description of the process of emission or absorption of radiation. However, a general theory of relativistic quantum electrodynamics valid for all conditions and systems has not been worked out.

The development of the laser has been an outstanding event in the history of optics. The theory of electromagnetic radiation from its beginnings was able to comprehend and treat the properties of coherent radiation, but the controlled generation of coherent monochromatic radiation of high power was not achieved in the optical region until the work of C. H. Townes and A. L. Schawlow in 1958 pointed the way. Many achievements in optics, such as holography and interferometry over long paths, have resulted from the laser.   RICHARD C. LORD

1

GEOMETRICAL OPTICS

Geometrical optics	4
Optical image	10
Magnification	12
Focal length	13
Diopter	13
Aberration	13
Chromatic aberration	20
Ghost image	23
Optical surfaces	24
Mirror optics	26
Optical prism	30
Lens	31
Resolving power	36

GEOMETRICAL OPTICS
ROLAND V. SHACK

The geometry of light rays and their images, through optical systems. Geometrical optics is by far the oldest model proposed for accounting for the behavior of light, going back to classical Greece. It was not until around the beginning of the nineteenth century that the wave nature of light was seriously considered, and in the modern view of the nature of light, geometrical optics as a fundamentally correct model is simply wrong. In spite of this geometrical optics is remarkably robust, remaining as a most practical tool in the solution of optical problems. It has been applied to analyzing laser resonators, to solving problems in interference and diffraction, and even to analyzing the behavior of waveguides, where at first glance it would seem to be totally inappropriate. These developments have been made possible by the generation of "fictitious" rays (all rays are fictitious) or by attributing to rays properties which cannot be accounted for in a strictly geometrical optical model. Nevertheless, the principal application of geometrical optics remains in the field of optical design, where it has been employed since the first optical instruments were developed in the early seventeenth century. This article concentrates on this application. SEE DIFFRACTION; INTERFERENCE OF WAVES; LASER.

Basic concepts. Light is a form of energy which flows from a source to a receiver. It consists of particles (corpuscles) called photons. All photons of a single pure color have the same energy per photon. Different colors have different energies.

For green light there are 2.5×10^{18} photons per joule of energy, or 2.5×10^{18} photons per second per watt of power. This factor increases for red light and diminishes for blue. SEE LIGHT; PHOTON.

Refractive index and dispersion. The speed with which the particles travel depends on the medium. In a vacuum this speed is 3×10^8 m·s^{-1} (1.86×10^5 mi/s) for all colors. In a material medium, whether gas, liquid, or solid, light travels more slowly. Moreover, different colors travel at different rates. This change in speed is a property of the medium, and is measured by the ratio of the speed in a vacuum to the speed in the medium. This ratio is called the refractive index of the medium. The variation in refractive index with color is called dispersion. Generally speaking, blue light travels more slowly than red light, so the medium has a higher refractive index for blue than for red light. In order to deal more precisely and quantitatively with color, and therefore with refractive index and dispersion, it is customary to identify the different colors by their wavelengths, even though waves have no place in geometrical theory. SEE COLOR; DISPERSION (RADIATION); REFRACTION OF WAVES.

Rays. The paths that particles take in going from the source to the receiver are called rays. The particles never interact with each other. The time that it takes a particle to travel from one point to another along a ray is determined by the speed at which the particle travels, which is determined by the refractive index of the medium. The product of the refractive index and the path length is called the optical path length along the ray. The optical path length is equal to the distance that the particle would have traveled in a vacuum in the same time interval.

Sources and wavefronts. A point source is an infinitesimal region of space which emits photons. An extended source is a dense array of point sources. Each point source emits photons along a family of rays associated with it. For each such family of rays there is also a family of surfaces each of which is a surface of constant transit time from the source for all the particles, or alternatively, a surface of constant optical path length from the source. These surfaces are called geometrical wavefronts (**Fig. 1**), although, again, waves have no place in geometrical theory. The reason for this name is that they are often good approximations to the wavefronts predicted by a wave theory. In isotropic media, rays are always normal to the geometrical wavefronts, and the optical path length from one geometrical wavefront to another is the same for all the rays in the family.

Ray paths. The ray path which any particle takes as it propagates is determined by Fermat's principle, which states that the ray path between any two points in space is that path along which the optical path length is stationary (usually a minimum) among all neighboring paths. In a homogeneous medium (one with a constant refractive index) the ray paths are straight lines. In a homogeneous medium containing a point source, the family of rays from the source all radiate outward and the associated geometrical wavefronts are concentric spherical shells.

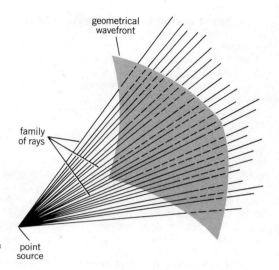

Fig. 1. Geometrical wavefront for a family of rays associated with a point source.

In a system that consists of a sequence of separately homogeneous media with different refractive indices and with smooth boundaries between them (such as a lens system), the ray paths are straight lines in each medium, but the directions of the ray paths will change in passing through a boundary surface. This change in direction is called refraction, and is governed by Snell's law, which states that the product of the refractive index and the sine of the angle between the normal to the surface and the ray is the same on both sides of a surface separating two media. The normal in question is at the point where the ray intersects the surface (**Fig. 2**).

Because this product is the same on both sides of the interface, the angle of the ray in the medium with the lower refractive index is always larger than the corresponding angle on the other side of the surface. When the sine of the larger angle becomes one, the largest value it can have, the angle itself is 90°, but the other angle is smaller. The latter angle is called the critical angle for the two media. If the light is traveling from the denser (higher-index) medium to the less dense medium, there is nothing to prevent the incidence angle from being larger than the critical angle, in which case the light is totally reflected by the surface and does not pass through to the other side. This phenomenon is called total internal reflection. *SEE REFLECTION OF ELECTROMAGNETIC RADIATION.*

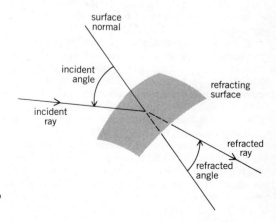

Fig. 2. Refraction of a ray at a smooth surface between two homogeneous media.

Ideal image formation. The primary area of application of geometrical optics is in the analysis and design of image-forming systems. An optical image-forming system consists of one or more optical elements (lenses or mirrors) which when directed at a luminous (light-emitting) object will produce a spatial distribution of the light emerging from it which more or less resembles the object. The latter is called an image. *See Optical image.*

In order to judge the performance of the system, it is first necessary to have a clear idea of what constitutes ideal behavior. Departures from this ideal behavior are called aberrations, and the purpose of optical design is to produce a system in which the aberrations are small enough to be tolerable. *See Aberration.*

First consider the requirements for ideal behavior. In an ideal optical system the rays from every point in the object space pass through the system so that they converge to or diverge from a corresponding point in the image space. This corresponding point is the image of the object point, and the two are said to be conjugate to each other (object and image functions are interchangeable). Another way of expressing the same thing is that an ideal optical system converts every spherical geometrical wavefront in the object space into a spherical geometrical wavefront in the image space with the image point located at its center of curvature.

It is not enough, however, to have a system which forms a perfect point image for every point object. The image points must be in the proper geometrical relationship to constitute a good image.

Since there is a one-to-one correspondence between the conjugate object and image points, the geometry of the object and image spaces must be connected by some mapping transformation. The one generally used to represent ideal behavior is the collinear transformation. The intrinsic properties of the collinear transformation are as follows. If three object points lie on the same straight line, they are said to be collinear. If the corresponding three image points are also collinear, and if this relationship is true for all sets of three conjugate pairs of points, then the two spaces are connected by a collinear transformation. In this case, not only are points conjugate to points, but straight lines and planes are conjugate to corresponding straight lines and planes.

It is attractive to have an ideal behavior in which for every point, straight line, or plane in the object space there is one and only one corresponding point, straight line, or plane in the image space. This does not, however, guarantee that the three-dimensional mapping is distortion-free.

Another feature usually incorporated in the ideal behavior is the assumption that all refracting or reflecting surfaces in the system are figures of revolution about a common axis, and this axis of symmetry applies to the object-image mapping as well. With this axial symmetry, an object line which coincides with the axis has as its conjugate an image line also coinciding with the axis. Therefore every object plane containing the axis, called a meridional plane, has a conjugate which is a meridional plane coinciding with the object plane, and every line in the object-space meridional plane has its conjugate in the same plane. In addition, every object plane perpendicular to the axis must have a conjugate image plane which is also perpendicular to the axis, because of axial symmetry. In the discussion below, the terms object plane and image plane refer to planes perpendicular to the axis unless otherwise modified.

An object line parallel to the axis will have a conjugate line which either intersects the axis in image space or is parallel to it. The first case is called a focal system, and the second an a focal system.

Focal systems. The point of intersection of the image-space conjugate line of a focal system with the axis is called the rear focal point (**Fig. 3**). It is conjugate to an object point on axis at infinity. The image plane passing through the rear focal point is the rear focal plane, and it is conjugate to an object plane at infinity. Every other object plane has a conjugate located at a finite distance from the rear focal point, except for one which will have its image at infinity. This object plane is the front focal plane, and its intersection with the axis is the front focal point. Every object line passing through the front focal point will have its conjugate parallel to the axis in image space.

Now take an arbitrary object plane and its conjugate image plane. Select a point off axis in the object plane and construct a line parallel to the axis passing through the off-axis point. The conjugate line in image space will intersect the axis at the rear focal point and the image plane at some off-axis point. The distance of the object point from the axis is called the object height, and the corresponding distance for the image point is called the image height. The ratio of the

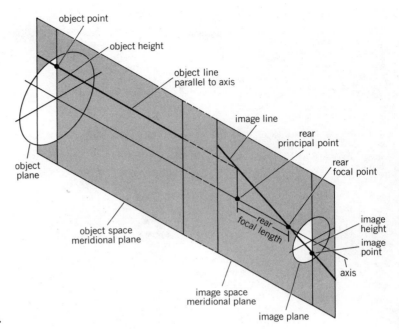

Fig. 3. Focal system.

image height to the object height is called the transverse magnification, and is positive or negative according to whether the image point is on the same side or the opposite side of the axis relative to the object point.

Every pair of conjugate points has associated with it a unique transverse magnification, and a given transverse magnification specifies a unique pair of conjugate planes. The rear focal plane and its conjugate have a transverse magnification of zero, whereas the front focal plane and its conjugate have an infinite transverse magnification.

The conjugate pair which have a transverse magnification of +1 are called the (front and rear) principal planes. The intersections of the principal plane with the axis are called the principal points. The distance from the rear principal point to the rear focal point is called the rear focal length, and likewise for the front focal length. *See Focal length*.

The focal points and principal points are four of the six gaussian cardinal points. The remaining two are a conjugate pair, also on axis, called the nodal points. They are distinguished by the fact that any conjugate pair of lines passing through them make equal angles with the axis. The function of the cardinal points and their associated planes is to simplify the mapping of the object space into the image space.

In addition to the transverse magnification, the concept of longitudinal magnification is useful. If two planes are separated axially, their conjugate planes are also separated axially. The longitudinal magnification is defined as the ratio of the image plane separation to the object plane separation, and is proportional to the product of the transverse magnifications of the two conjugate pairs. In the limit of the separation between the pairs approaching zero, the longitudinal magnification becomes proportional to the square of the transverse magnification.

Only in planes perpendicular to the axis is there an undistorted mapping, because only in such planes is the magnification a constant over the field. An object plane not perpendicular to the axis has a conjugate plane also inclined to the axis, and the magnification varies over the field, resulting in keystone distortion.

Afocal systems. In the case of afocal systems, any line parallel to the axis in object space has a conjugate which is also parallel to the axis (**Fig. 4**). Conjugate planes perpendicular to the axis are still uniquely related, but the transverse magnification is constant for the system, and the

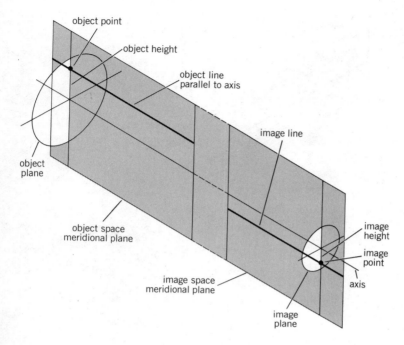

Fig. 4. Afocal system.

same value applied to every pair of conjugate planes. The longitudinal magnification, also constant for the system, is proportional to the square of the transverse magnification regardless of the separation of the pair of conjugate planes. Cardinal points do not exist for afocal systems.

The most common use of an afocal system is as a telescope, where both the object and the image are at infinity. The apparent sizes of the object and image are determined by the angular subtense, which is defined, for finite object or image distance, as the ratio of the height to the distance from the observer, and, for infinite distance, by holding this ratio constant as the distance approaches infinity. The concept of angular magnification is therefore useful; this is defined as the ratio of the transverse magnification to the longitudinal magnification, and is inversely proportional to the transverse magnification. The power of a telescope or a pair of binoculars is the magnitude of the angular magnification. *See* MAGNIFICATION; OPTICAL TELESCOPE.

Paraxial optics. The above discussion of the properties of the ideal image-forming system did not consider the optical properties of the system where the ray paths are determined by Snell's law. Real optical systems do not, in fact, cannot, obey the laws of the collinear transformation, and departures from this ideal behavior are identified as aberrations. However, if the system is examined in a region restricted to the neighborhood of the axis, the so-called paraxial region, where angles and their sines are indistinguishable from their tangents, a behavior is found which is exactly congruent with the collinear transformation. Paraxial ray tracing can therefore be used to determine the ideal collinear properties of the system. If a more extended region than the paraxial is considered, departures from the ideal collinear behavior are observed because additional terms in a series expansion for the trigonometric functions must be taken into account. Nevertheless, even in this extended region the paraxially determined quantities represent the first-order behavior of the system.

Ray tracing is usually done by using a cyclic algorithm, with the coordinate system being transferred from surface to surface through the system as the ray tracing progresses. Two operations are involved, a refraction and a transfer. In refraction, given the incident ray direction and position on the surface, Snell's law is used to determine the emergent direction of the ray. In transfer, the position of the ray on the next surface is determined.

In paraxial ray tracing, which uses the paraxial approximation to Snell's law, the operations of refraction and transfer are simple linear operations, and only two rays need be traced in order to locate the cardinal points of the system. Once the cardinal points have been determined, there are simple equations available to locate any pair of conjugate planes and their associated magnification.

Apertures. The above discussion does not take into account the fact that the sizes of the elements of the optical system are finite, and, in fact, for any optical system, the light that can get through the system to form the image is limited. To determine how much light can get through the system, consider the tracing of a sequence of rays from the axial object point meeting the first surface at increasingly greater heights. Eventually one of these rays and all others beyond it will be blocked by the edge of some aperture in the system. Assuming the aperture is circular and centered on the axis, all the object rays inside the cone determined by the ray which just touches the edge of the aperture will get through the system and participate in the formation of the image. The aperture which limits the cone of rays admitted is called the aperture stop of the system.

An observer who looks into the front of the system from the axial object point sees not the aperture stop itself (unless it is in front of the system), but the image of it formed by the elements preceding it. This image of the aperture stop is called the entrance pupil, and is situated in the object space on the system. The image of the aperture stop formed by the rear elements is the exit pupil, and is situated in the image space of the system. The beam of light which would go from the object point to the image point on axis is spindle-shaped, with a circular cross section everywhere (**Fig. 5**a).

Light from the edge of the object field will also be limited by the same aperture stop and, to first order, will have the same circular cross section at each surface as the axial beam but will be displaced laterally except at the aperture stop and the pupils. The beam will consist of a sequence of sections of skew circular cones (Fig. 5b). The connection between the entrance pupil and the object-space segments of the axial and the off-axis beams is shown in Fig. 5c.

It is also possible that a portion of the beam will be further blocked by the edge of some element which, although large enough to admit the axial beam, will not be large enough to admit all of the off-axis beam. This latter effect is called vignetting.

Marginal ray and chief ray. The characteristics of the axial beam are completely specified by a ray traced from the axial object point toward the edge of the entrance pupil. This ray will graze the edge of the aperture stop and emerge from the edge of the exit pupil proceeding to the axial image point. Such a ray is often called a marginal ray, although other names are also common.

The characteristics of the beam from the edge of the field are the same as those of the axial beam except for the displacement of the center of the beam at each surface. Thus the characteristics of this off-axis beam are described by the marginal ray, determining the cross-sectional radius, and a ray traced from the edge of the object field toward the center of the entrance pupil. This ray will pass through the center of the aperture stop and emerge from the center of the exit pupil proceeding to the image point at the edge of the image field. This ray is commonly called the chief ray.

These two rays, the marginal ray and the chief ray, are all that need be traced to determine the first-order properties of the image-forming system.

Lagrange invariant. At any surface in the system the marginal ray and the chief ray are determined by their heights at the surface and angles they make with respect to the axis. Moreover, they are connected by an interesting invariant relationship, often called the Lagrange invariant. This is obtained by multiplying the height of each ray by the product of the refractive index and the angle of the opposite ray, and taking the difference between the resulting two products.

Because the Lagrange invariant holds at all surfaces throughout the optical system, in particular the object, the image, and the pupils, it is useful in solving many problems in the early stages of laying out an optical system, especially those involving a sequence of subsystems. It also plays a major role in the radiometry of image-forming systems; the total amount of light which can pass through a system is proportional to the square of the Lagrange invariant.

Although the determination of the first-order properties of an optical system are of major

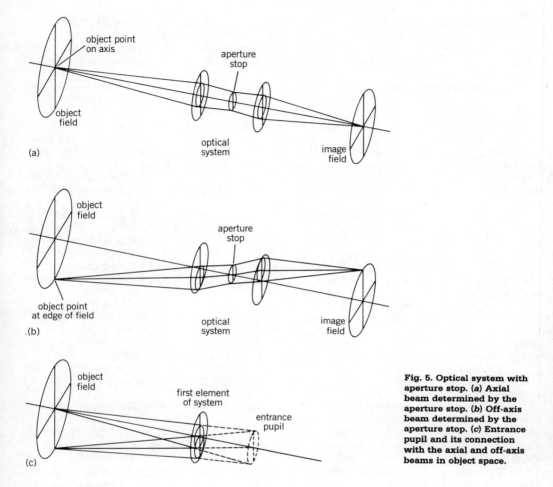

Fig. 5. Optical system with aperture stop. (*a*) Axial beam determined by the aperture stop. (*b*) Off-axis beam determined by the aperture stop. (*c*) Entrance pupil and its connection with the axial and off-axis beams in object space.

importance, this is only the beginning of the task of designing a good optical system. It is then necessary to trace real rays, determine the aberrations, and adjust the system parameters to improve the system's performance while maintaining its first-order behavior. SEE BINOCULARS; CAMERA; EYEPIECE; GUNSIGHTS; LENS; MIRROR OPTICS; OPTICAL MICROSCOPE; OPTICAL PRISM; OPTICAL PROJECTION SYSTEMS; OPTICAL SURFACES; OPTICAL TRACKING INSTRUMENTS; PERISCOPE; RANGEFINDER.

Bibliography. G. J. James, *Geometrical Theory of Diffraction*, 1976, revised 1980; N. S. Kapany and J. J. Burke, *Optical Waveguides*, 1972; R. Kingslake, *Lens Design Fundamentals*, 1978; M. Kline and I. W. Kay, *Electromagnetic Theory and Geometrical Optics*, 1965, reprint 1979; E. Mach, *The Principles of Physical Optics*, 1926; W. Smith, *Modern Optical Engineering*, 1966.

OPTICAL IMAGE
MAX HERZBERGER

The image formed by the light rays from a self-luminous or an illuminated object that traverse an optical system. The image is said to be real if the light rays converge to a focus on the image side and virtual if the rays seem to come from a point within the instrument (see **illus**.).

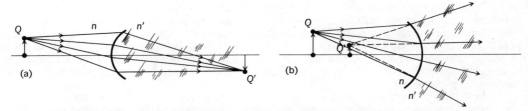

Optical images. (a) Real image. Rays leaving object point Q and passing through the refracting surface separating media n and n' are brought to a focus at the image point Q'. (b) Virtual image. Rays leaving Q and refracted by the concave surface separating n and n' appear to be coming from the virtual image point Q'. As the rays are diverging, they cannot be focused at any point. (*After F. A. Jenkins, Fundamentals of Optics, 4th ed., McGraw-Hill, 1976*)

The optical image of an object is given by the light distribution coming from each point of the object at the image plane of an optical system. The ideal image of a point according to geometrical optics is obtained when all rays from an object point unite in a single image point. However, diffraction theory teaches that even in this case the image is not a point but a minute disk. The diameter of this disk is about 1.22 λ/A, where λ is the wavelength of the light considered and A is the numerical aperture, the sine of the largest cone angle on the image side multiplied by its refractive index (which is usually equal to unity). *See* Focal length; Geometrical optics.

Aberrations. From the standpoint of geometrical optics, if this most desirable type of image formation cannot be achieved, the next best objective is to have the image free from all but aperture errors (spherical aberration). In this case the light distribution in the image plane is still circular, resembling the point image; there is a true coordination of object point and image, although the image may be slightly unsharp. If the aperture errors are small, or if the image is viewed from a distance, such an image formation may be very satisfactory.

Asymmetry and deformation errors may be very disturbing if not held in check, because the light distribution of the image of a point in this case has a decidedly undesirable shape.

When the image of an axis point is considered, the rays through a fixed aperture circle converge to an axis point. For this type of imagery, the term half-sharp image will be used. A small object at the object point is then imaged by a circular stop at the focus of the image bundle with a magnification as given by Eq. (1) where u and u' are the angles of the imaging cone in

$$m = (n \sin u)/(n' \sin u') \tag{1}$$

object and image space, respectively, and n and n' are the corresponding refractive indices.

If the axis point is sharply imaged, an object of finite extent is sharply imaged if, and only if, $m = m_0$ (the gaussian magnification) for all values of u (sine condition).

In the case of aperture errors, the most desirable image formation for an axis point is attained when the different images appear under the same angle from the exit pupil. If k' is the distance of the image point from the exit pupil, $\Delta s'$ is the aperture aberration, and if Eq. (2) gives the magnification error compared with the magnification m_o on the axis, the condition is given by Eq. (3). The fulfillment of this condition gives equal quality for an object near the axis of a system with rotation symmetry.

$$\Delta m = \frac{n \sin u}{n' \sin u'} - m_o \tag{2} \qquad \Delta s'/k' - \Delta m/m_o = \text{constant} \tag{3}$$

Corresponding conditions can be ascertained for the image of an off-axis element if all the asymmetry errors and deformation errors are balanced. *See* Aberration.

Resolution. Two points are resolved by an optical system if the two images lie apart. Photometric analysis of an image may indicate the existence of two object points even if their images overlap, but in such an analysis the illumination of the object, as well as the imagery, plays a role.

In interference experiments, it is found that the image of two self-luminous points (that is, two light sources that are sufficiently separated) is incoherent; that is, the intensities of the two beams simply add. If the two object points are illuminated by the same light source, however, the phase relation of the light at the two points has to be taken into consideration. This is of the greatest importance for microscopes and telescopes, which image very small or distant objects. In this case an artificial change of phase by phase plates and apodization may improve resolution. SEE DIFFRACTION; RESOLVING POWER.

Resolving power is not the only consideration in image formation. The eye recognizes only contrast differences, and therefore objects may not be discerned if the contrast difference is too small. Again, for the image of a point, or an object illuminated by a point light source, means can be found to change the apparent contrast, making it possible to discern biological objects, for example, having small differences of refractive index.

Image analysis. Methods have been suggested for obtaining information about optical images by sine-wave analysis. A sinusoidal test object is imaged by an optical system as a sinusoidal image, but altered in phase and amplitude. A large number of sinusoidal test objects with different frequencies (number of maxima per millimeter) are imaged, and the amplitude and phase are measured.

The curve of amplitude versus frequency gives a measure for resolving power and contrast as a function of the frequency of the test object, whereas the adjusted curve of phase versus frequency describes the lack of symmetry in the image. These amplitude-frequency curves can be measured as well as calculated from the spot diagrams, onto which the effects of diffraction can be superimposed if necessary.

Bibliography. E. U. Condon and H. Odishaw, *Handbook of Physics*, 1967; M. Herzberger, *Modern Geometrical Optics*, 1958, reprint 1978; F. A. Jenkins, *Fundamentals of Optics*, 4th ed., 1976; J. Meyer-Arendt, *Introduction to Classical and Modern Optics*, 2d ed., 1984; Optical Society of America, *Handbook of Optics*, 1978.

MAGNIFICATION
MAX HERZBERGER

A measure of the effectiveness of an optical system in enlarging or reducing an image. For an optical system that forms a real image, such a measure is the lateral magnification m, which is the ratio of the size of the image to the size of the object. If the magnification is greater than unity, it is an enlargement; if less than unity, it is a reduction.

The ratio of the longitudinal (with respect to the optical axis) dimensions of the image to the corresponding dimensions of the object is known as longitudinal magnification, which in first order equals the square of the lateral magnification.

The angular magnification γ is the ratio of the angles formed by the image and the object at the eye. The relation $n'\gamma m = n$ relates angular to lateral magnification. Here n and n' are the refractive indices of the media containing the object and image, respectively. In telescopes the angular magnification (or, better, the ratio of the tangents of the angles under which the object is seen with and without the lens, respectively), can be taken as a measure of the effectiveness of the instrument.

A small off-axis element is imaged with different magnification in both the meridional and sagittal directions. This may be called differential magnification.

Magnifying power is the measure of the effectiveness of an optical system used in connection with the eye. The magnifying power of a spectacle lens is the ratio of the tangents of the angles under which the object is seen with and without the lens, respectively. The magnifying power of a magnifier or an ocular is the ratio of the size under which an object would appear seen through the instrument at a distance of 10 in. (25 cm; the distance of distinct vision) divided by the object size.

E. Abbe suggested defining the magnifying power of an optical system as the ratio of the tangent of the visual angle under which the object appears to the object size. This quantity is

approximately equal to the power of the system, which is the reciprocal of the focal length, $1/f'$.
See Diopter; Lens; Optical image.
Bibliography. M. Born and E. Wolf, *Principles of Optics,* 6th ed., 1980; M. Herzberger, *Modern Geometrical Optics,* 1958, reprint 1978; F. A. Jenkins, *Fundamentals of Optics,* 4th ed., 1976; M Kline and I. W. Kay, *Electromagnetic Theory and Geometrical Optics,* 1965, reprint 1979; J. Strong, *Concepts of Classical Optics,* 1958.

FOCAL LENGTH
Max Herzberger

A measure of the collecting or diverging power of a lens or an optical system. Focal length, usually designated f' in formulas, is measured by the distance of the focal point (the point where the image of a parallel entering bundle of light rays is formed) from the lens, or more exactly by the distance from the principal point to the focal point. See Geometrical optics.

The power of a lens system is equal to n'/f', where n' is the refractive index in the image space (n' is usually equal to unity). A lens of zero power is said to be afocal. Telescopes are afocal lens systems. See Diopter; Lens.

DIOPTER
Max Herzberger

A measure of the power of a lens or a prism. The diopter (also called dioptre) is usually abbreviated D. Its dimension is a reciprocal length, and its unit is the reciprocal of 1 m (39.4 in.). Thus a thin lens of κ diopters has a focal length of $1000/\kappa$ mm or $39.4/\kappa$ in. The lens is collecting for positive κ, diverging for negative κ, and afocal for $\kappa = 0$. See Focal length; Lens.

One can speak of the power of a single surface. The power of a thin lens is then the sum of the powers of its surfaces in diopters. Analogously, the power of a group of (thin) lenses in contact is the sum of the powers of the single lenses.

For a lens which is not rotation-symmetric (having toric or cylindrical surfaces, for instance), two powers—one maximal and one minimal—must be assigned. These correspond to the powers in two perpendicular planes.

The dioptric power of a prism is defined as the measure of the deviation of a ray going through a prism measured at the distance of 1 m (39.4 in.). A prism that deviates a ray by 1 cm (0.394 in.) in a distance of 1 m (39.4 in.) is said to have a power of one prism diopter. See Optical prism.

Spectacle lenses in general consist of thin lenses, which are either spherical, to correct the focus of the eye for near and far distances, or cylindrical or toric, to correct the astigmatism of the eye. An added prism corrects a deviation of the visual axis. The diopter thus gives a simple method for prescribing the necessary spectacles for the human eye. See Eyeglasses.

ABERRATION
Roland V. Shack

A departure of an optical image-forming system from ideal behavior. Ideally, such a system will produce a unique image point corresponding to each object point. In addition, every straight line in the object space will have as its corresponding image a unique straight line. A similar one-to-one correspondence will exist between planes in the two spaces.

This type of mapping of object space into image space is called a collinear transformation. A paraxial ray trace is used to determine the parameters of the transformation, as well as to locate the ideal image points in the ideal image plane for the system. See Geometrical optics; Optical image.

When the conditions for a collinear transformation are not met, the departures from that ideal behavior are termed aberrations. They are classified into two general types, monochromatic aberrations and chromatic aberrations. The monochromatic aberrations apply to a single color, or wavelength, of light. The chromatic aberrations are simply the chromatic variation, or variation with wavelength, of the monochromatic aberrations. SEE CHROMATIC ABERRATION.

Aberration measures. The monochromatic aberrations can be described in several ways. Wave aberrations are departures of the geometrical wavefront from a reference sphere with its vertex at the center of the exit pupil and its center of curvature located at the ideal image point. The wave aberration is measured along the ray and is a function of the field height and the pupil coordinates of the reference sphere (**Fig. 1**).

Transverse ray aberrations are measured by the transverse displacement from the ideal image point to the ray intersection with the ideal image plane.

Some aberrations are also measured by the longitudinal aberration, which is the displacement along the chief ray of the ray intersection with it. The use of this measure will become clear in the discussion of aberration types.

Caustics. Another aberrational feature which occurs when the wavefront in the exit pupil is not spherical is the development of a caustic. For a spherical wavefront, the curvature of the wavefront is constant everywhere, and the center of curvature along each ray is at the center of curvature of the wavefront. If the wavefront is not spherical, the curvature is not constant everywhere, and at each point on the wavefront the curvature will be a function of orientation on the surface as well as position of the point. As a function of orientation, the curvature will fluctuate between two extreme values. These two extreme values are called the principal curvatures of the wavefront at that point. The principal centers of curvature lie on the ray, which is normal to the wavefront, and will be at different locations on the ray.

The caustic refers to the surfaces which contain the principal centers of curvature for the entire wavefront. It consists of two sheets, one for each of the two principal centers of curvature. For a given type of aberration, one or both sheets may degenerate to a line segment. Otherwise they will be surfaces.

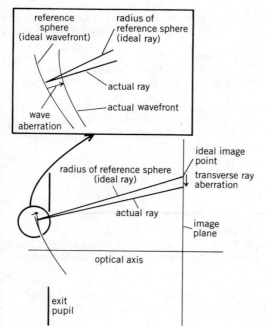

Fig. 1. Diagram of the image space of an optical system, showing aberration measures: the wave aberration and the transverse ray aberration.

GEOMETRICAL OPTICS

The feature which is of greatest interest to the user of the optical system is usually the appearance of the image. If a relatively large number of rays from the same object point and with uniform distribution over the pupil are traced through an optical system, a plot of their intersections with the image plane represents the geometrical image. Such a plot is called a spot diagram, and a set of these are often plotted in planes through focus as well as across the field.

ORDERS OF ABERRATIONS

The monochromatic aberrations can be decomposed into a series of aberration terms which are ordered according to the degree of dependence they have on the variables, namely, the field height and the pupil coordinates. Each order is determined by the sum of the powers to which the variables are raised in describing the aberration terms. Because of axial symmetry, alternate orders of aberrations are missing from the set. For wave aberrations, odd orders are missing, whereas for transverse ray aberrations, even orders are missing. It is customary in the United States to specify the orders according to the transverse ray designation.

Monochromatically, first-order aberrations are identically zero, because they would represent errors in image plane location and magnification, and if the first-order calculation has been properly carried out, these errors do not exist. In any case, they do not destroy the collinear relationship between the object and the image. The lowest order of significance in describing monochromatic aberrations is the third order.

Chromatic variations of the first-order properties of the system, however, are not identically zero. They can be determined by first-order ray tracing for the different colors (wavelengths), where the refractive indices of the media change with color. For a given object plane, the different colors may have conjugate image planes which are separated axially. This is called longitudinal chromatic aberration (**Fig. 2**a). Moreover, for a given point off axis, the conjugate image points may be separated transversely for the different colors. This is called transverse chromatic aberration (Fig. 2b), or sometimes chromatic difference of magnification.

These first-order chromatic aberrations are usually associated with the third-order monochromatic aberrations because they are each the lowest order of aberration of their type requiring correction.

The third-order monochromatic aberrations can be divided into two types, those in which the image of a point source remains a point image but the location of the image is in error, and those in which the point image itself is aberrated. Both can coexist, of course.

Aberrations of geometry. The first type, the aberrations of geometry, consist of field curvature and distortion.

Field curvature. Field curvature is an aberration in which there is a focal shift which varies as a quadratic function of field height, resulting in the in-focus images lying on a curved surface. If this aberration alone were present, the images would be of good quality on this curved surface, but the collinear condition of plane-to-plane correspondence would not be satisfied.

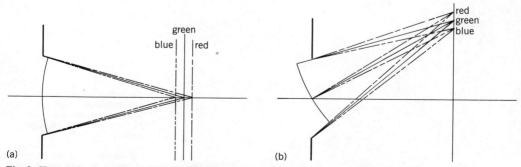

Fig. 2. Chromatic aberration. (a) Longitudinal chromatic aberration. (b) Transverse chromatic aberration.

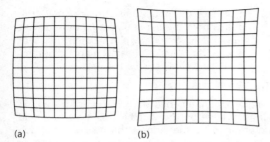

Fig. 3. Distortion. (a) Barrel distortion. (b) Pincushion distortion.

Distortion. Distortion, on the other hand, is an aberration in which the images lie in a plane, but they are displaced radially from their ideal positions in the image plane, and this displacement is a cubic function of the field height. This means that any straight line in the object plane not passing through the center of the field will have an image which is a curved line, thereby violating the condition of straight-line to straight-line correspondence. For example, if the object is a square centered in the field, the points at the corners of the image are disproportionally displaced from their ideal positions in comparison with the midpoints of the sides. If the displacements are toward the center of the field, the sides of the figure are convex; this is called barrel distortion (**Fig. 3**a). If the displacements are away from the center, the sides of the figure are concave; this is called pincushion distortion (Fig. 3b).

Aberrations of point images. There are three third-order aberrations in which the point images themselves are aberrated: spherical aberration, coma, and astigmatism.

Spherical aberration. Spherical aberration is constant over the field. It is the only monochromatic aberration which does not vanish on axis, and it is the axial case which is easiest to understand.

The wave aberration function (**Fig. 4**a) is a figure of revolution which varies as the fourth power of the radius in the pupil. The wavefront itself has this wave aberration function added to

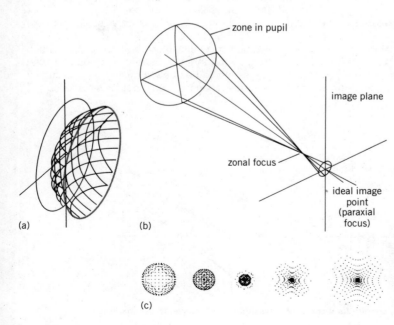

Fig. 4. System with spherical aberration. (a) Wave aberration function. (b) Rays. (c) Spot diagrams through foci showing transverse ray aberration patterns for a square grid of rays in the exit pupil.

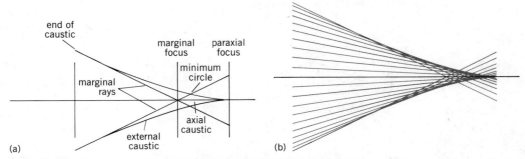

Fig. 5. Caustics in a system with spherical aberration. (*a*) Diagram of caustics and other major features. (*b*) Rays whose envelope forms an external caustic.

the reference sphere centered on the ideal image. The rays (Fig. 4*b*) from any circular zone in the wavefront come to a common zonal focus on the axis, but the position of this focus shifts axially as the zone radius increases. This zonal focal shift increases quadratically with the zone radius to a maximum from the rays from the marginal zone. This axial shift is longitudinal spherical aberration. The magnitude of the spherical aberration can be measured by the distance from the paraxial focus to the marginal focus.

The principal curvatures of any point on the wavefront are oriented tangential and perpendicular to the zone containing the point. The curvatures which are oriented tangential to the zone have their centers on the axis, so the caustic sheet for these consists of the straight line segment extending from the paraxial focus, and is degenerate. The other set of principal centers of curvature lie on a trumpet-shaped surface concentric with the axis (**Fig. 5***a*). This second sheet is the envelope of the rays. They are all tangent to it, and the point of tangency for each ray is at the second principal center of curvature for the ray (Fig. 5*b*).

It is clear that the image formed in the presence of spherical aberration (Fig. 4*c*) does not have a well-defined focus, although the concentration of light outside the caustic region is everywhere worse than it is inside. Moreover, the light distribution in the image is asymmetric with respect to focal position, so the precise selection of the best focus depends on the criterion chosen. The smallest circle which can contain all the rays occurs one-quarter of the distance from the marginal focus to the paraxial focus, the image which has the smallest second moment lies one-third of the way from the marginal focus to the paraxial focus, and the image for which the variance of the wave aberration function is a minimum lies halfway between the marginal and paraxial foci.

Coma. Coma is an aberration which varies as a linear function of field height. It can exist only for off-axis field heights, and as is true for all the aberrations, it is symmetrical with respect to the meridional plane containing the ideal image point. Each zone in its wave aberration function (**Fig. 6***a*) is a circle, but each circle is tilted about an axis perpendicular to the meridional plane, the magnitude of the tilt increasing with the cube of the radius of the zone.

The chief ray (Fig. 6*b*) passes through the ideal image point, but the rays from any zone intersect the image plane in a circle, the center of which is displaced from the ideal image point by an amount equal to the diameter of the circle. The diameter increases as the cube of the radius of the corresponding zone in the pupil. The circles for the various zones are all tangent to two straight lines intersecting at the ideal image point and making a 60° angle with each other. The resulting figure (Fig. 6*c*) resembles an arrowhead which points toward or away from the center of the field, depending on the sign of the aberration.

The upper and lower marginal rays in the meridional plane intersect each other at one point in the circle for the marginal zone. This point is the one most distant from the chief ray intersection with the image plane. The transverse distance between these points is a measure of the magnitude of the coma.

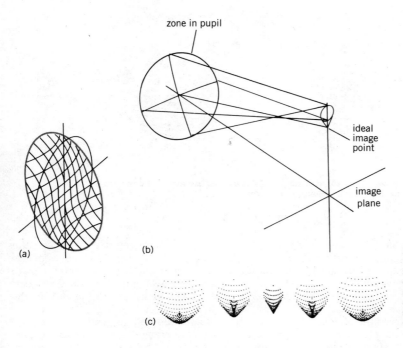

Fig. 6. System with coma. (a) Wave aberration function. (b) Rays. (c) Spot diagrams through foci, showing transverse ray aberration patterns for a square grid of rays in the exit pupil.

Astigmatism. Astigmatism is an aberration which varies as the square of the field height. The wave aberration function (**Fig. 7a**) is a quadratic cylinder which varies only in the direction of the meridional plane and is constant in the direction perpendicular to the meridional plane. When the wave aberration function is added to the reference sphere, it is clear that the principal curvatures for any point in the wavefront are oriented perpendicular and parallel to the meridional plane, and moreover, although they are different from each other, each type is constant over the wavefront. Therefore, the caustic sheets both degenerate to lines perpendicular to and in the meridional plane in the image region, but the two lines are separated along the chief ray.

All of the rays must pass through both of these lines, so they are identified as the astigmatic foci. The astigmatic focus which is in the meridional plane is called the sagittal focus, and the one perpendicular to the meridional plane is called the tangential focus.

For a given zone in the pupil, all of the rays (Fig. 7b) will of course pass through the two astigmatic foci, but in between they will intersect an image plane in an ellipse, and halfway between the foci they will describe a circle (Fig. 7c). Thus, only halfway between the two astigmatic foci will the image be isotropic. It is also here that the second moment is a minimum, and the wave aberration variance is a minimum as well. This image is called the medial image.

Since astigmatism varies as the square of the field height, the separation of the foci varies as the square of the field height as well. Thus, even if one set of foci, say the sagittal, lies in a plane, the medial and tangential foci will lie on curved surfaces. If the field curvature is also present, all three lie on curved surfaces.

The longitudinal distance along the chief ray from the sagittal focus to the tangential focus is a measure of the astigmatism.

The above description of the third-order aberrations applies to each in the absence of the other aberrations. In general, more than one aberration will be present, so that the situation is more complicated. The types of symmetry appropriate to each aberration will disclose its presence in the image.

Higher-order aberrations. The next order of aberration for the chromatic aberrations consists of the chromatic variation of the third-order aberrations. Some of these have been given

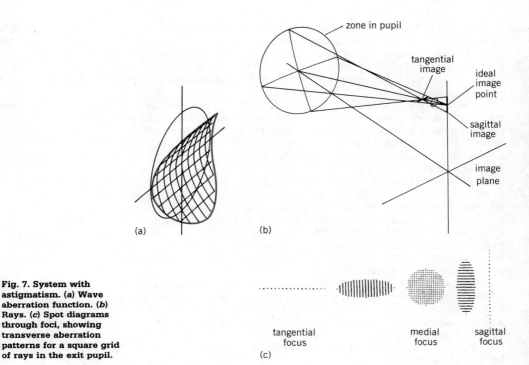

Fig. 7. System with astigmatism. (a) Wave aberration function. (b) Rays. (c) Spot diagrams through foci, showing transverse aberration patterns for a square grid of rays in the exit pupil.

their own names; for example, the chromatic variation of spherical aberration is called spherochromatism.

Monochromatic aberrations of the next order are called fifth-order aberrations. Most of the terms are similar to the third-order aberrations, but with a higher power dependence on the field or on the aperture. Field curvature, distortion, and astigmatism have a higher power dependence on the field, whereas spherical aberration and coma have a higher power dependence on the aperture. In addition, there are two new aberration types, called oblique spherical aberration and elliptical coma. These are not directly related to the third-order terms.

Expansions beyond the fifth order are seldom used, although in principle they are available. In fact, many optical designers use the third order as a guide in developing the early stages of a design, and then go directly to real ray tracing, using the transverse ray aberrations of real rays without decomposition to orders. However, the insights gained by using the fifth-order aberrations can be very useful.

ORIGIN OF ABERRATIONS

Each surface in an optical system introduces aberrations as the beam passes through the system. The aberrations of the entire system consist of the sum of the surface contributions, some of which may be positive and others negative. The challenge of optical design is to balance these contributions so that the total aberrations of the system are tolerably small. In a well-corrected system the individual surface contributions are many times larger than the tolerance value, so that the balance is rather delicate, and the optical system must be made with a high degree of precision.

Insight as to where the aberrations come from can be gained by considering how the aberrations are generated at a single refracting surface. Although the center of curvature of a spherical surface lies on the optical axis of the system, it does not in itself have an axis. If, for the moment, the spherical surface is assumed to be complete, and the fact that the entrance pupil for the surface will limit the beam incident on it is ignored, then for every object point there is a local axis which is the line connecting the object point with the center of curvature. All possible rays from the object point which can be refracted by the surface will be symmetrically disposed about this local axis, and the image will in general suffer from spherical aberration referred to this local axis.

A small pencil of rays about this axis (locally paraxial) will form a first-order image according to the rules of paraxial ray tracing. If the first-order imagery of all the points lying in the object plane is treated in this manner, it is found that the surface containing the images is a curved surface. The ideal image surface is a plane passing through the image on the optical axis of the system, and thus the refracting surface introduces field curvature. The curvature of this field is called the Petzval curvature.

In addition to this monochromatic behavior of the refracting surface, the variation in the index of refraction with color will introduce changes in both the first-order imagery and the spherical aberration. This is where the chromatic aberrations come from.

Thus there are fundamentally only three processes operating in the creation of aberrations by a spherical refracting surface. These result in spherical aberration, field curvature, and longitudinal chromatic aberration referred to the local axis of each image. In fact, if the entrance pupil for the surface is at the center of curvature, these will be the only aberrations that are contributed to the system by this refracting surface.

In general, the pupil will not be located at the center of curvature of the surface. For any off-axis object point, the ray which passes through the center of the pupil will be the chief ray, and it will not coincide with the local axis. The size of the pupil will limit the beam which can actually pass through the surface, and the beam will therefore be an eccentric portion of the otherwise spherically aberrated beam.

Since an aberration expansion decomposes the wave aberration function about an origin located by the chief ray, the eccentric and asymmetric portion of an otherwise purely spherically aberrated wave gives rise to the field-dependent aberrations, because the eccentricity is proportional to the field height of the object point.

In this manner the aberrations which arise at each surface in the optical system, and therefore the total aberrations of the system, can be accounted for. *See* LENS; OPTICAL SURFACES.

Bibliography. M. Born and E. Wolf, *Principles of Optics*, 6th ed., 1980; H. H. Hopkins, *Wave Theory of Aberrations*, 1950; R. Kingslake, *Lens Design Fundamentals*, 1978; W. Smith, *Modern Optical Engineering*, 1966; W. T. Welford, *Aberrations of the Symmetrical Optical System*, 1974.

CHROMATIC ABERRATION
MAX HERZBERGER

The type of error in an optical system in which the formation of a series of colored images occurs, even though only white light enters the system. Chromatic aberrations are caused by the fact that the refraction law determining the path of light through an optical system contains the refractive index n, which is a function of wavelength λ. Thus the image position and the magnification of an optical system are not necessarily the same for all wavelengths, nor are the aberrations the same for all wavelengths.

In this article the chromatic aberrations of a lens system are discussed. For information on other types of aberration SEE ABERRATION. SEE ALSO REFRACTION OF WAVES.

Dispersion formula. When the refractive index of glass or other transparent material is plotted as a function of the square of the wavelength, the result is a set of dispersion curves which appear to have an asymptote in the near-ultraviolet region and a straight portion in the near-infrared (up to 1 micrometer). In the glass catalogs the indices are given for selected wavelengths between 0.365 and 1.01 μm, which are the absorption (emission) bands of certain chemical elements. For an extended discussion of dispersion SEE ABSORPTION OF ELECTROMAGNETIC RADIATION. SEE ALSO OPTICAL MATERIALS.

Values of the four universal functions $a(\lambda)$ for various wavelengths

λ	a_1	a_2	a_3	a_4
1.0140	0.000000	0.000000	0.000000	+1.000000
0.7682	+0.031555	−0.276197	+1.051955	+0.192687
0.6563	0.000000	0.000000	+1.000000	0.000000
0.6438	−0.004774	+0.051075	+0.966511	−0.012813
0.5893	−0.024952	+0.326272	+0.744598	−0.045919
0.5876	−0.025492	+0.336338	+0.735480	−0.046326
0.5461	−0.033080	+0.597795	+0.479049	−0.043764
0.4861	0.000000	+1.000000	0.000000	0.000000
0.4800	+0.009340	+1.036699	−0.052699	+0.006659
0.4358	+0.143980	+1.193553	−0.394569	+0.057036
0.4047	+0.366779	+1.045064	−0.487634	+0.075791
0.3650	+1.000000	0.000000	0.000000	0.000000

For the visible region, the Hartmann formula, Eq. (1), where a and b are constants for a
$$n = a/(\lambda - b) \tag{1}$$
given glass, has been much used. However, if an attempt is made to apply Eq. (1) to the near-ultraviolet or the near-infrared, it proves to be insufficient. Over this more extended range, n can be given as a function of the wavelength by Eq. (2).

$$n = a + b\lambda^2 + \frac{c}{\lambda^2 - 0.028} + \frac{d}{(\lambda^2 - 0.028)^2} \tag{2}$$

This formula, in which c and d are additional constants, enables one to compute n for any wavelength if it is given for four. An equivalent formula is shown by Eq. (3), where $a_1(\lambda)$, $a_2(\lambda)$, . . .

$$n = n_1 a_1(\lambda) + n_2 a_2(\lambda) + n_3 a_3(\lambda) + n_4 a_4(\lambda) \tag{3}$$

are functions of the form of Eq. (2), which assume for $\lambda_1, \lambda_2, \ldots$ the values 1, 0, 0, 0; 0, 1, 0, 0; 0, 0, 1, 0; and 0, 0, 0, 1, respectively. Choosing $\lambda_1 = 0.365 = \lambda^{**}$, $\lambda_2 = 0.4861 = \lambda_F$, $\lambda_3 = 0.6563 = \lambda_C$, $\lambda_4 = 1.014 = \lambda^*$, one finds in the **table** the four universal functions $a(\lambda)$ tabulated for different values of λ. Note that for any wavelength Eq. (4) holds.

$$a_1 + a_2 + a_3 + a_4 = 1 \tag{4}$$

Instead of giving the indices of a glass for the wavelengths specified, one can specify the glass by the refractive index n_F at one wavelength F, the dispersion $\delta = n_F - n_C$, and the partial dispersions in the ultraviolet and infrared, defined respectively by Eqs. (5). The partial dispersion for an arbitrary wavelength λ is then given by Eq. (6).

$$P^{**} = \frac{n^{**} - n_F}{n_F - n_C} \qquad P^* = \frac{n_F - n^*}{n_F - n_C} \tag{5} \qquad P_\lambda = P^{**} a_1(\lambda) + a_2(\lambda) + P^* a_4(\lambda) \tag{6}$$

In the literature the Abbe number is defined by Eq. (7), where n_D is the mean index for the D lines of sodium is frequently used to designate optical glass.

$$\nu = (n_D - 1)/(n_F - n_C) \tag{7}$$

In **Fig. 1**, $(n_F - 1)$ is shown plotted against $(n_F - n_C)$ for various optical materials. One sees that in this kind of plot glasses of the same type lie on a straight line, while on the ordinary plot of n_D versus ν they lie on hyperbolas.

Color correction. A system of thin lenses in contact is corrected for two wavelengths λ_A and λ_B if the power of the combination is the same for both wavelengths as in Eq. (8), where the

$$\phi_A = \phi_B = \Sigma(n_A - 1)K_\kappa = \Sigma(n_B - 1)K_\kappa \tag{8}$$

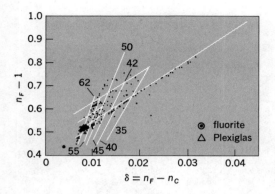

Fig. 1. Plot of $(n_F - 1)$ versus δ for selected glasses and for fluorite and Plexiglas. The numbers on the lines dividing the glasses into groups represent ν values.

K_κ's are the differences between the first and last curvatures of the lenses, and the summation is over all the lenses, each with its particular value of n_A, n_B, and K_κ. Two cemented lenses are corrected for wavelengths C and F, for instance, if Eqs. (9) hold. SEE LENS.

$$\phi_1 + \phi_2 = \phi \qquad \qquad \phi_1/\nu_1 + \phi_2/\nu_2 = \phi_C - \phi_F = 0 \qquad (9)$$

The two lenses are also corrected for a third wavelength if and only if, in addition, one has $P_1 = P_2$, where P_1 and P_2 are given by Eq. (6).

A system corrected for three colors is called an apochromat (in microscopy this term traditionally demands freedom from asymmetry in addition). An apochromat for the ultraviolet portion of the spectrum is possible only if the two glasses in **Fig. 2** have the same P^{**} value. For the infrared, the glasses must have the same P^* value (**Fig. 3**). The $\nu - 1$ values for the two glasses should lie as far apart as possible to give low values for the powers.

Three lenses can be corrected for four wavelengths and therefore practically for the whole spectrum if the glasses lie on the straight line on the plot of P^{**} against P^* (**Fig. 4**). Such a system may be called a superachromat.

In lenses with finite thicknesses and distances, there are in gaussian optics two errors to be corrected. One is a longitudinal aberration, which means that the gaussian images do not lie in the same plane, and the other is a lateral aberration, which means that the images in different colors have different magnifications.

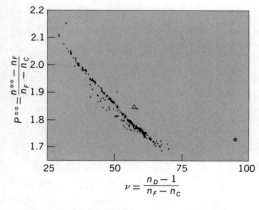

Fig. 2. Plot of P^{**} versus ν for materials of Fig. 1.

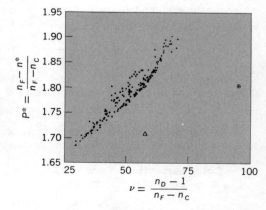

Fig. 3. Plot of P^* versus ν for materials of Fig. 1.

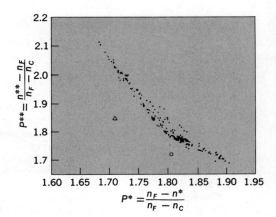

Fig. 4. Plot of P^{**} versus P^* for materials of Fig. 1. $P^* = \dfrac{n_F - n^*}{n_F - n_C}$

In the presence of longitudinal aberration, it is best to balance the lateral aberration so that the apparent sizes of the images as seen from the exit pupil coincide.

A system of two uncorrected lenses, such as a simple ocular, with one finite distance, cannot be corrected for both color errors. Two finite distances are needed to balance both color errors at the same time.

In a general system, all image errors are functions of the wavelength of light. However, in a lens system corrected for color, it is easily possible, with a small adjustment, to balance the correction of the aperture rays (spherical aberration) with respect to color by introducing a small amount of lateral color aberration.

If only two colors can be corrected, the choice of the colors depends on the wavelength sensitivity of the receiving instrument. For visual correction, the values for C and F are frequently brought together. Some optical systems contain filters to permit only a narrow spectral band to pass the instrument. This makes correction for color errors easier.

If light of a large band of wavelengths traverses the instrument, it is frequently desirable to use catadioptric systems, such as the Schmidt camera. The mirror or mirrors of these systems are used to obtain the necessary power without introducing color errors, and an afocal lens system can be added to correct monochromatic aberrations. SEE SCHMIDT CAMERA.

Bibliography. E. U. Condon and H. Odishaw, *Handbook of Physics*, 2d ed., 1967; M. Herzberger, *Modern Geometrical Optics*, 1958, reprint 1978; F. A. Jenkins, *Fundamentals of Optics*, 4th ed., 1976; J. Mayer-Arendt, *Introduction to Classical and Modern Optics*, 2d ed., 1984; Optical Society of America, *Handbook of Optics*, 1978.

GHOST IMAGE
MAX HERZBERGER

An undesired image appearing at the image plane of an optical system. Each surface of an optical system divides the incoming light into two parts: (1) the reflected light, which returns into the first medium, and (2) the refracted light. The reflected light is again divided into two parts when it in turn strikes another dividing surface. The light thus reflected twice forms an image which may be near the plane of the primary image. This may be a false image of the object or an out-of-focus image of a bright source of light in the field of the optical system. Thus a large number of undesired or ghost images may appear. SEE OPTICAL IMAGE; REFLECTION OF ELECTROMAGNETIC RADIATION; REFRACTION OF WAVES.

If the ghost images are far out of focus, they only diminish the contrast in the primary image, a condition known as flare. But if the ghost images are near the focal plane, they are very disturbing. This effect is especially noticeable if there is a bright light source in the field of the instrument, since the ghost image of the light source may have an even greater brightness than the image of the desired object.

This annoying image defect is hardly precalculable, because in a system with N surfaces there are 2^N possible double reflections and 4^N possible quadruple reflections. It has, however, been successfully controlled by antireflection coatings.

Coatings. Lens coatings are films of the proper thickness and refractive index applied to the airglass surfaces of a lens to reduce reflection. For perpendicular incidence on a plane surface bounding a glass of refractive index n, a surface coating with an effective index of $\sqrt{n}$ and a thickness of a quarter of a wavelength causes all light waves of that wavelength returning into the first medium to interfere with each other destructively so that there is no loss by reflection. The coating of lenses with layers of fluorite and other materials has nearly eliminated ghost images from modern optical systems. Different coatings can be combined effectively as color filters for either transmission or reflection. They also can be used as interference filters in interferometry or spectroscopy. SEE COLOR FILTER.

Flare. There are several causes of flare. One is reflection at an even number of lens surfaces which, especially in the case of an illuminated object, may give rise to an out-of-focus image of the light source that will produce background illumination. This type of flare, like ghost images, can be practically eliminated by suitably coating the air surfaces of the lens elements. Flare is also caused by reflection at the mounting of an optical instrument, but this can be eliminated by blackening and suitable baffling. A different type of flare arises from the fact that not all the emergent light is collected within the small nucleus of the image. This flare is due to residual aberrations that exist even in a well-corrected lens.

Flare is especially detrimental if the object has small contrast differences, since it reduces these differences in the image and may bring them below the threshold of recognition. On the other hand, by lightening the shadows of the image of a high-contrast scene, some background illumination, as produced by flare, may enable a photograph to be made with a shorter exposure than would be possible in the absence of it, and may also bring the contrast range of the scene within the range of reproduction of the photographic process.

Bibliography. E. U. Condon and H. Odishaw, *Handbook of Physics*, 3d ed., 1972, reprint, 1982; D. F. Horne, *Optical Production Technology*, 2d ed., 1983; Optical Society of America, *Handbook of Optics*, 1978; W. J. Smith, *Modern Optical Engineering*, 1966; J. Strong, *Concepts of Classical Optics*, 1958.

OPTICAL SURFACES
ROLAND V. SHACK

Interfaces between different optical media at which light is refracted or reflected. From a physical point of view, the basic elements of an optical system are such things as lenses and mirrors. However, from a conceptual point of view, the basic elements of an optical system are the refracting or reflecting surfaces of such components, even though they cannot be separated from the components. Surfaces are the basic elements of an optical system because they are the elements that affect the light passing through the system. Every wavefront has its curvature changed on passing through each surface so that the final set of wavefronts in the image space may converge on the appropriate image points. Also, the aberrations of the system depend on each surface, the total aberrations of the system being the sum of the aberrations generated at the individual surfaces. SEE ABERRATION; REFLECTION OF ELECTROMAGNETIC RADIATION; REFRACTION OF WAVES.

Optical systems are designed by ray tracing, and refraction at an optical surface separating two media of different refractive index is the fundamental operation in the process. The transfer between two surfaces is along a straight line if, as is usually the case, the optical media are homogeneous. The refraction of the ray at a surface results in a change in the direction of the ray. This change is governed by Snell's law.

Spherical and aspheric surfaces. The vast majority of optical surfaces are spherical in form. This is so primarily because spherical surfaces are much easier to generate than nonspherical, or aspheric, surfaces. Not only is the sphere the only self-generating surface, but a number of lenses can be ground and polished on the same machine at the same time if they are mounted together on a common block so that the same spherical surface is produced simultaneously on all of them.

Although aspheric surfaces can potentially improve the performance of a lens system, they are very rarely used. High-quality aspheric surfaces are expensive to produce, requiring the services of a highly skilled master optician. Moreover, lens systems seldom need aspherics because the aberrations can be controlled by changing the shape of the component lenses without changing their function in the system, apart from modifying the aberrations. Also, many lens components can be included in a lens system in order to control the aberrations. S*ee* L*ens*.

On the other hand, mirror systems usually require aspheric surfaces. Unlike lenses, where the shape can be changed to modify the aberrations, mirrors cannot be changed except by introducing aspheric surfaces. Mirror systems are further constrained by the fact that only a few mirrors, usually two, are used in a system because each successive mirror occludes part of the beam going to the mirror preceding it. S*ee* M*irror optics*.

Conics of revolution. The most common form of rotationally symmetric surface is the conic of revolution. This is obtained conceptually by rotating a conic curve (ellipse, parabola, or hyperbola) about its axis of symmetry. There are two forms of ellipsoid, depending on whether the generating ellipse is rotated about its major or its minor axis. In the first case it is a prolate ellipsoid, and in the second it is an oblate ellipsoid. There is only one form of paraboloid, and only the major axis is used in generating the hyperboloid. The departure of conic surfaces from spherical form is shown in the **illustration**.

The classical virtue of the conics of revolution for mirrors is the fact that light from a point located at one focus of the conic is perfectly imaged at the other focus. If these conic foci are located on the axis of revolution, the mirror is free of spherical aberration for such conjugate points. For example, the classical Cassegrain design consists of two mirrors, a paraboloidal primary mirror and a hyperboloidal secondary mirror. The paraboloid forms a perfect image of a point at infinity on its axis. When the light converging to this image is intercepted by the convex hyperboloid with its virtual conic focus at the image point, the final image will be formed at the real conic focus of the hyperboloid. Thus, in the classical Cassegrain design the two mirrors are separately corrected for spherical aberration, and so is the system. However, all other monochromatic aberrations remain uncorrected. S*ee* O*ptical telescope*.

Instead of using the two aspheric terms to correct the same aberration for the two mirrors separately, it is more effective to use the two aspherics to correct two aberrations. A Cassegrain system in which both spherical aberration and coma are corrected by the two aspherics is aplanatic, and is identified as a Ritchey-Chrétien system. The aspheric on the primary is a hyperboloid slightly stronger than the paraboloid of the classical primary, and the aspheric on the secondary is also a hyperboloid, slightly different from that for the classical secondary.

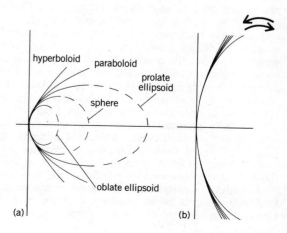

Conics of revolution. (*a*) Cross sections of entire surfaces. (*b*) Cross sections of portions near the optical axis.

General aspherics of revolution. A more general aspheric of revolution is frequently used where terms in even powers of the distance from the axis, usually from the fourth to the twelfth power, are added to the spherical or conic term. This gives greater control over the aberrations, especially if higher-order aberrations are significant. The additional terms are necessary also for describing an aspheric departure from a plane surface where a conic cannot be used, as in the case of a Schmidt corrector plate. *See* SCHMIDT CAMERA.

Nonrotationally symmetric surfaces. Occasionally nonrotationally symmetric surfaces are used in optical systems. The most common varieties are cylindrical and toric surfaces, where the cross sections are circular. They are employed, for example, in the optical systems for taking and projecting wide-screen motion pictures, where the image recorded on the film is compressed in the horizontal direction but not in the vertical. The camera lens compresses the image and the projector lens expands it to fill the wide screen. Noncircular cross sections can also be specified by adding terms in even powers of x and y, the coordinates perpendicular to the optical axis, to the description of the surface, but these surfaces are very difficult to make. *See* CINEMATOGRAPHY.

Eccentric mirrors. Another significant class of aspheric surfaces consists of those which are eccentric portions of rotationally symmetric aspheric surfaces. These are used in mirror systems to eliminate the obscuration of the incoming beam by succeeding mirrors. For small mirrors they are cut out of a large rotationally symmetrical mirror, but for larger mirrors they can be made by computer-controlled generating and polishing machines. *See* GEOMETRICAL OPTICS.

Bibliography. D. F. Horne, *Optical Production Technology*, 2d ed., 1983; D. Malacara, *Optical Shop Testing*, 1978; Optical Society of America, *Handbook of Optics*, 1978.

MIRROR OPTICS
R. R. SHANNON

The use of plane or curved reflecting surfaces for the purpose of reverting, directing, or forming images. The most familiar use of reflecting optical surfaces is for the examination of one's own reflected image in a flat or plane mirror. A single reflection in a flat mirror produces a virtual image which is reverted or reversed in appearance. The use of one or more reflecting surfaces permits light or images to be directed around obstacles, with each successive reflection producing a reversal of the image. A curved mirror, either spherical or conic in form, will produce a real or virtual image in much the same manner as a lens, but generally with reduced aberrations. There will be no chromatic aberrations since the law of reflection is independent of the color or wavelength of the incident light. *See* ABERRATION; OPTICAL IMAGE.

An optical surface which specularly reflects the largest fraction of the incident light is called a reflecting surface. Such surfaces are commonly fabricated by polishing of glass, metal, or plastic substrates, and then coating the surface of the substrate with a thin layer of metal, which may be covered in addition by a single or multiple layers of thin dielectric films. The law of reflection states that the incident and reflected rays will lie in the plane containing the local normal to the reflecting surface and that the angle of the reflected ray from the normal will be equal to the angle of the incident ray from the normal. This law is a special case of the law of refraction in that the angles rather than the sines of the angles of incidence and reflection are equal. Formally, this relation is commonly used in calculations by setting the effective index of refraction prior to incidence on the surface. When this concept is introduced, all of the formulas relating to lenses are applicable to reflective optics. In this article, however, the imaging relations will be described in the most appropriate form for reflecting surfaces. *See* GEOMETRICAL OPTICS.

Plane mirrors. The formation of images in the plane mirrors is easily understood by applying the law of reflection. **Figure 1** illustrates the formation of the image of a point formed by a plane mirror. Each of the reflected rays appears to come from a point image located a distance behind the mirror equal to the distance of the object point in front of the mirror. In Fig. 1, the face of the observer can be considered as a set of points, each of which is imaged by the plane mirror. Since the observer is viewing the facial image from the object side of the mirror, the face will appear to be reversed left for right in the virtual image formed by the mirror. Such a virtual image cannot, of course, be projected on a screen, but can be viewed by a lens, in this

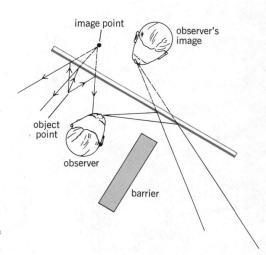

Fig. 1. Formation of images by a plane mirror.

case the eyes of the observer. Figure 1 also indicates the redirection of light by a plane mirror, in that a viewer who cannot observe the object point directly can observe the virtual image of the point formed by the mirror. A simple optical device which is based on this principle is the simple mirror periscope (**Fig. 2**), which uses two mirrors to permit viewing of scenes around an obstacle. In this case two reflections are present and provide an image which is correctly oriented, and not reverted, to the observer. The property of reversion in a complicated mirror system depends upon the location and view direction of the observer, as well as the number of reflections that take place and the orientation of the planes through which the light is directed. *See Periscope.*

Prisms. These are solid-glass optical components that use reflection at the faces to provide redirection of the optical pencils passing through them. The advantage of the use of a prism is that the reflecting surfaces are maintained in accurate location with respect to each other by the integrity of the glass material making up the body of the prism. Difficulties with prisms are that very homogeneous glass is required since the light may make many passes through the prism, and that a prism is optically equivalent to insertion of a long block of glass into the imaging system. The insertion of such a glass block often results in a system which is mechanically shorter

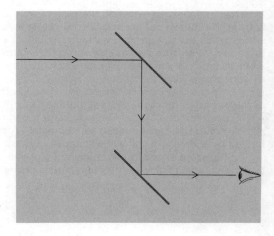

Fig. 2. Simple mirror periscope.

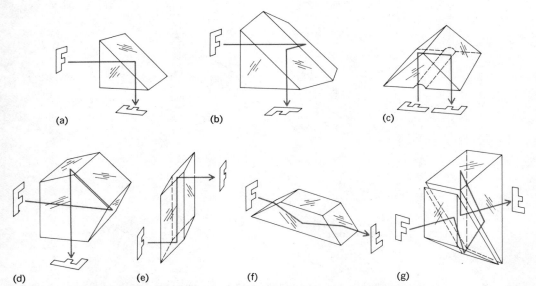

Fig. 3. Prism types. (*a*) Right-angle. (*b*) Amici roof. (*c*) Porro. (*d*) Pentaprism. (*e*) Rhomboid. (*f*) Dove. (*g*) Pechan.

in space, but the aberration balance of the imaging system is changed, frequently requiring a redesign of the associated optical components in order to accommodate the increased glass path. It is obvious that the use of solid glass prisms may introduce more weight into the optical design than the equivalent metal mounts required for a similar arrangement of mirrors in air. In certain cases, the angle of incidence on a reflecting surface within the prism may exceed the critical angle of incidence, and no reflective coating may be required on such a surface. **Figure 3** shows some common types of optical prisms with reflecting surfaces. The applications of such prisms range from simple redirection of light to variable angle of rotation of the image passed through the prism and binocular combination of images. A special case of reflection is the use of a beam splitter which permits the splitting of light, or the combining of two beams by the use of a surface which is partially reflecting and partially transmitting. *See* BINOCULARS; OPTICAL PRISM.

Spherical mirrors. These are reflecting components that are used in forming images. The optics of such mirrors are almost identical to the properties of lenses, with their ability to form real and virtual images. In the case of spherical mirrors, there is a reversal of the direction of light at the mirror so that real images are formed on the same side of the mirror as the object, while virtual images are viewed from the object side but appear to exist on the opposite side of the mirror. Both concave and convex spherical mirrors are commonly encountered. Only a virtual erect image of a real object will be formed by a convex mirror. Such mirrors are commonly used as wide-angle rearview mirrors in automobiles or on trucks. The image formed appears behind the mirror and is greatly compressed in space, with a demagnification dependent on the curvature of the mirror. A concave spherical mirror can form either real or virtual images. The virtual image will appear to the observer as erect and magnified. A common application is the magnifying shaving mirror frequently found in bathrooms. A real image will be inverted, as is the real image formed by a lens, and will actually appear in space between the observer and the mirror.

Figure 4 shows the formation of real and virtual images by a spherical mirror. The equation which applies to all of the image relations is given below. The distances S and S' are mea-

$$\frac{1}{S'} + \frac{1}{S} = \frac{2}{R}$$

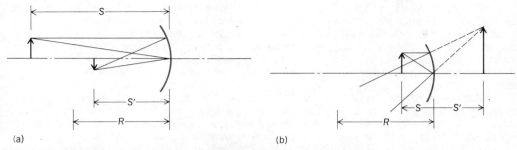

Fig. 4. Formation of images by spherical mirror. (a) Real image. (b) Virtual image.

sured from the surface of the spherical mirror; when either is negative, a virtual image is formed behind the mirror. The constant R is the radius of curvature of the mirror. The magnification of the image is the ratio of the image distance S' to the object distance S.

Conic mirrors. These are a special case of the spherical mirror with improved image quality. A spherical mirror will form an image which is not perfect, except for particular conjugate distances. The use of a mirror which has the shape of a rotated conic section, such as a parabola, ellipsoid, or hyperboloid, will form a perfect image for a particular set of object-image conjugate distances and will have reduced aberrations for some range of conjugate relations. Two of the most familiar applications for conic mirrors are shown in **Fig. 5**. Figure 5a shows the use of a paraboloid of revolution about the optical axis to form the image of an object at an infinite distance. In this drawing the image to be viewed by the observer at the eyepiece is relayed to the side of the telescope tube by a flat folding mirror in what is called a newtonian form of a telescope. This demonstrates one of the difficulties that is found with the use of reflecting optical components to form real images; namely, that the image must often be relayed out of the incident path on the image-forming mirror, otherwise the observer will block some of the light from the object. Not all reflecting systems carry out this relaying in the same manner. The Cassegrain system uses a curved secondary mirror to achieve magnification of the final image while allowing the image to fall outside the telescope barrel.

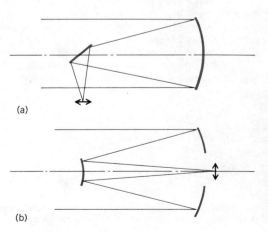

Fig. 5. Applications of conic mirrors. (a) Newtonian telescope. (b) Cassegrain telescope.

Figure 5b shows the use of a paraboloid as the primary mirror with the image relayed to the final image location by a hyperbolic secondary mirror. Such a use of two mirrors permits the construction of a long-focal-length telescope within a relatively short space. This latter form, usually referred to as a Cassegrain telescope, serves as the principal type of modern reflecting astronomical telescope. One of the advantages of this type of telescope design is the freedom from chromatic aberration that would be present in a refracting telescope. SEE OPTICAL SURFACES; OPTICAL TELESCOPE.

Mirror coatings. The reflectivity of a mirror depends on the material used for coating the reflecting surface. The conventional coatings for glass mirror surfaces are silver or aluminum, which are vacuum-deposited or sputtered onto the surface. In some cases, chemical deposition will be used. Most mirrors intended for noncritical uses, such as looking glasses or wall mirrors, will have the reflecting metallic coating placed on the back side of the glass, thus using the glass to protect the coating from oxidation by the atmosphere. Mirrors for most critical or scientific uses require the use of front-surface reflectors, with the reflecting coating on the exposed front surface of the glass. In this case, a hard overcoat of a thin layer of silicon dioxide is frequently deposited over the metal to protect the delicate thin metal surface. The reflectivity of the mirror with respect to wavelength depends on the choice of the metal for the reflector and the material and thickness of material layers in the overcoat. In some cases, a fully dielectric stack will be used as a reflecting coating with special spectral selective properties to form a dichroic beam splitter, as in a color television camera, or as an infrared-transmitting "cold mirror" for a movie projector illumination system. SEE REFLECTION OF ELECTROMAGNETIC RADIATION.

OPTICAL PRISM
MAX HERZBERGER

An optical system consisting of two or more usually plane surfaces of a transparent solid or embedded liquid at an angle with each other. Prisms are used for deviating light. Since the amount of deviation depends on the refractive index of the prism, which varies with wavelength, prisms can also be used for dispersing light. SEE DISPERSION (RADIATION); REFRACTION OF WAVES.

Reflecting prisms. Prisms can be used instead of mirrors for deviating light, with the added advantage that the reflecting surfaces are protected against corrosion. In this case there is at least one internal reflection. When the angles of incidence and emergence are zero, there is no dispersion. The overall dispersion is also zero when the geometry of the prism is such that the dispersion at the entering surface is compensated by dispersion in the opposite sense at the emergent surface. For a detailed discussion of important types of reflecting prisms SEE MIRROR OPTICS.

Dispersing prisms. Dispersing prisms deviate light of different wavelengths by different amounts, and they can therefore be used to separate white light into its monochromatic parts. A parallel beam of light entering the prism leaves the prism as a parallel beam of light but its diameter may be changed. The ratio of its diameter after refraction to its diameter before refraction can be considered as the magnification of the prism.

The prism magnification for a bundle parallel to the prism edge is always equal to unity, whereas it varies in the meridional plane (normal to the edge) as the angle of incidence is varied.

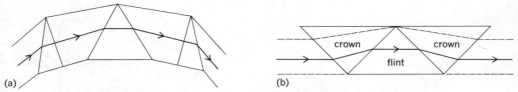

Fig. 1. Two types of dispersing prisms. (a) Rayleigh prism system. (b) Amici direct-vision system consisting of a flint-glass prism and two crown-glass prisms.

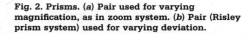

Fig. 2. Prisms. (a) Pair used for varying magnification, as in zoom system. (b) Pair (Risley prism system) used for varying deviation.

It is equal to unity in this plane only if the prism is traversed at minimum deviation. If α is the prism angle, n the refractive index, and δ the deviation, it can be shown that, for minimum deviation, the equation below holds.

$$\sin(\delta + \alpha)/2 = n \sin \alpha/2$$

To increase the dispersion, several prisms with their refracting edges parallel can be used. The Rayleigh prism, shown in **Fig. 1**a, is an example of such a system. By using a prism made of a material, such as flint glass, that has a high dispersion, and adding one or more prisms made of a material having a low dispersion, such as crown glass (Fig. 1b), the deviation can be neutralized without neutralizing the dispersion to give a direct-vision prism system. The arrangement shown in Fig. 1b is known as the Amici prism system. By using a similar arrangement but adjusting the angle so that the dispersion, but not the deviation, is neutralized, it is possible to make a prism system that is achromatic over a small part of the spectrum, like an achromatic lens. SEE LENS.

An achromatic prism in front of an optical system with its refracting edge normal to the meridional plane can be used to change the magnification of the optical system in that plane. This amount can be varied by rotating prism A in **Fig. 2**a about an axis normal to the meridional plane. A second achromatic prism, B, with its edge parallel to the meridional plane, can be used to adjust the sagittal magnification of the optical system. Therefore, an arrangement of two such prisms with their motions linked together can be used for the purpose of forming a variable-focal-length lens system, which is commonly called a zoom lens. SEE ZOOM LENS.

A thin prism is one whose angle is so small that the prism angle expressed in radians is practically equal to the tangent. Such prisms are used in ophthalmology, and their powers are usually expressed in prism diopters. The Risley prism system, used for testing ocular convergence, consists of two thin prisms mounted so that they can be rotated simultaneously in opposite directions, as shown in Fig. 2b. When they are in the orientation sketched at 1, their combined deviation is zero; when both have been rotated by 90° in opposite directions, as shown at 2, their combined deviation is a maximum; at intermediate positions, their combined deviation lies between zero and the maximum, but the plane of deviation is constant. A similar pair of rotating wedges is used in certain types of rangefinders. SEE BINOCULARS; DIOPTER; GEOMETRICAL OPTICS; OPTICAL MATERIALS; PERISCOPE; RANGEFINDER; RESOLVING POWER.

Bibliography. G. A. Boutry, *Instrumental Optics*, 1962; E. U. Condon and H. Odishaw, *Handbook of Physics*, 1967; D. F. Horne, *Optical Instruments and Their Applications*, 1980; F. A. Jenkins, *Fundamentals of Optics*, 4th ed., 1976; Optical Society of America, *Handbook of Optics*, 1978.

LENS
MAX HERZBERGER

A curved piece of ground and polished or molded material, usually glass, used for the refraction of light. Its two surfaces have the same axis. Usually this is an axis of rotation symmetry for both surfaces; however, one or both of the surfaces can be toric, cylindrical, or a general surface with double symmetry. The intersection points of the symmetry axis with the two surfaces are called the front and back vertices and their separation is called the thickness of the lens.

LENS TYPES

There are three lens types, namely, compound, single, and cemented. These are described in the following sections.

Compound lenses. A compound lens is a combination of two or more lenses in which the second surface of one lens has the same radius as the first surface of the following lens and the two lenses are cemented together. Compound lenses are used instead of single lenses for color correction, or to introduce a surface which has no effect on the aperture rays but large effects on the principal rays, or vice versa. Sometimes the term compound lens is applied to any optical system consisting of more than one element, even when they are not in contact.

A group of lenses used together is a lens system. A symmetrical lens is a lens system consisting of two parts, each of which is the mirror image of the other. If one part is a mirror image of the other magnified m times, the system is called hemisymmetric. When $m=1$, the system is often said to be holosymmetric.

Single lenses. The lens diameter is called the linear aperture, and the ratio of this aperture to the focal length is called the relative aperture. This latter quantity is more often specified by its reciprocal, called the f-number. Thus, if the focal length is 2 in. (50 mm) and the linear aperture 1 in. (25 mm), the relative aperture is 0.5 and the f-number is f/2. SEE FOCAL LENGTH.

In precalculation formulas, the lens thicknesses (but not the separations of the lenses) can frequently be neglected. This leads to the convenient fiction of a thin lens.

If ρ_1 and ρ_2 are the front and back curvatures of a lens of refractive index n and thickness d, its power is given by Eq. (1). The curvature of the surface is the reciprocal of its radius.

$$\phi = (n-1)(\rho_1 - \rho_2) + \frac{d(n-1)^2}{n} \rho_1 \rho_2 \qquad (1)$$

The distances from the back vertex to the back nodal point and to the back focal point, respectively, are given by Eq. (2). The last distance is often called the back focus, especially in photographic optics.

$$S'_N \phi = -(d/n)(n-1)\rho_1 \qquad\qquad S'_F \phi = 1 + S'_N \phi \qquad (2)$$

The bending of a lens is a change in the curvature of the two surfaces by the same amount. It does not change the power of a thin lens, which is $(n-1) \cdot (\rho_1 - \rho_2)$. Bending is an important tool of the designer, for it permits the replacing of one lens by another without changing the data of gaussian optics.

When thick lenses are involved, gaussian optics remains constant only if both the powers of the thick lenses and the distance between the back nodal point of the first lens and the front nodal point of the second remain unchanged. Thus a bending of a thick lens should be accompanied by such an adjustment.

The optical center of a thick lens is the image of the nodal point produced inside the lens. All finite rays through the optical center emerge parallel to their respective directions at their entrance.

An optical center exists also in a hemisymmetric system. It is the point of symmetry which divides the separation of the two parts in the ratio $1/m$. If negative values of m are permitted, any single lens is a hemisymmetric system and the point dividing the thickness of the lens (the separation of the two vertices) in a ratio equal to the ratio of the two radii is the optical center of the lens.

A lens is said to be a collecting lens if $\phi > 0$ and a diverging lens if $\phi < 0$. When $\phi = 0$, the lens is afocal. Several types of collecting and diverging lenses are shown in **Fig. 1**.

The surfaces of most lenses are either spherical or planar, but nonspherical surfaces are used on occasion to improve the corrections without changing the power of the lens. SEE OPTICAL SURFACES.

A concentric lens is a lens whose two surfaces have the same center. If the object to be imaged is also at the center, its axis point is sharply imaged upon itself, and since the sine condition is fulfilled, the image is free from asymmetry. Such a lens can be used as an additional system to correct meridional errors.

Another type of lens consists of an aplanatic surface followed by a concentric surface, or

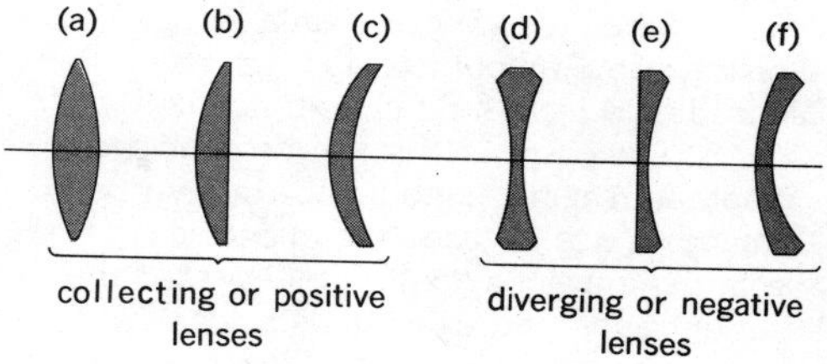

Fig. 1. Common lenses. (a) Biconvex. (b) Plano-convex. (c) Positive meniscus. (d) Biconcave. (e) Plano-concave. (f) Negative meniscus. (*After F. A. Jenkins, Fundamentals of Optics, 4th ed., McGraw-Hill, 1976*)

vice versa. Such a lens divides the focal length of the original lens to which it is attached by n^2, thus increasing the f-number by a factor of n^2 without destroying the axial correction of the preceding system. It does introduce curvature of field which makes a rebalancing of the whole system desirable. *See* Aberration.

Cemented lenses. Consider a compound lens made of two or more simple thin lenses cemented together. Let the power of the κth simple lens be ϕ_K and its Abbe value ν_K. The difference between the powers of the combination for wavelengths corresponding to C and F is given by Eq. (3), where N may be considered to be the effective ν-value of the combination. The ν-

$$\Phi_F - \Phi_C = \Phi/N = \sum \Phi_K/\nu_K \tag{3}$$

values of optical glasses vary between 25 and 70, with the ν-value of fluorite being slightly larger ($\nu = 95.1$). By using compound lenses, effective values of N can be obtained outside this range. Color correction is achieved as N becomes infinite, so that $\Phi_F - \Phi_C = 0$. A lens so corrected is called an achromat. In optical design, it is sometimes desirable to have negative values of N to balance the positive values of the rest of the system containing collecting lenses. Such a lens is said to be hyperchromatic. A cemented lens corrected for more than two colors is said to be apochromatic. A lens corrected for all colors of a sizable wavelength range is called a super-achromatic lens. *See* Chromatic aberration; Optical materials.

LENS SYSTEMS

Optical systems may be divided into four classes: telescopes, oculars (eyepieces), photographic objectives, and enlarging lenses. *See* Eyepiece; Optical microscope.

Telescope systems. A lens system consisting of two systems combined so that the back focal point of the first (the objective) coincides with the front focal point of the second (the ocular) is called a telescope. Parallel entering rays leave the system as parallel rays. The magnification is equal to the ratio of the focal length of the first system to that of the second.

If the second lens has a positive power, the telescope is called a terrestrial or keplerian telescope and the separation of the two parts is equal to the sum of the focal lengths.

If the second lens is negative, the system is called a galilean telescope and the separation of the two parts is the difference of the absolute focal lengths. The galilean telescope has the advantage of shortness (a shorter system enables a larger field to be corrected); the keplerian telescope has a real intermediate image which can be used for introducing a reticle or a scale into the intermediate plane.

Both objective and ocular are in general corrected for certain specific aberrations, while the other abberations are balanced between the two systems.

Photographic objectives. A photographic objective images a distant object onto a photographic plate or film. *See* Photography.

The amount of light reaching the light-sensitive layer depends on the aperture of the optical system, which is equivalent to the ratio of the lens diameter to the focal length. Its reciprocal is called the f-number. The smaller the f-number, the more light strikes the film. In a well-corrected lens (corrected for aperture and asymmetry errors), the f-number cannot be smaller than 0.5.

The larger the aperture (the smaller the f-number), the less adequate may be the scene luminance required to expose the film. Therefore, if pictures of objects in dim light are desired, the f-number must be small. On the other hand, for a lens of given focal length, the depth of field is inversely proportional to the aperture.

Since the exposure time is the same for the center as for the edge of the field, it is desirable for the same amount of light to get to the edge as gets to the center, that is, the photographic lens should have little vignetting.

The camera lens can be considered as an eye looking at an object (or its image), with the diaphragm corresponding to the eye pupil. The gaussian image of the diaphragm in the object (image) space is called the entrance (exit) pupil. The angle under which the object (image) is seen from the entrance (exit) pupil is called the object(image) field angle. For most photographic lenses, the entrance and exit pupils are close to the respective nodal points; for such lenses, the object and the image field angles are equal.

In general, photographic objectives with large fields have small apertures: those with large apertures have small fields. The construction of the two types of systems is quite different. One can say in general that the larger the aperture, the more complex the lens system must be.

There exist cameras (so-called pinhole cameras) that do not contain any lenses. The image is then produced by optical projection. The aperture in this case should be limited to $f/22$.

Other types of lenses. A single meniscus lens, with its concave side toward the object and with its stop in front at its optical center, gives good definition at $f/16$ over a total field of 50° (**Fig. 2**a). The lens can be a cemented doublet for correcting chromatic errors (Fig. 2b). For practical reasons, a reversed meniscus with the stop toward the film is often used.

Combining two meniscus lenses to form a symmetrical lens with central stop makes it possible to correct astigmatic and distortion errors for small apertures as well as large field angles (Fig. 2c).

The basic type of wide-angle objective is the Hypergon, consisting of two meniscus lenses concentric with the regard to stop (Fig. 2d). This type of system can be corrected for astigmatism and field curvature over a total field angle of 180° but it can only be used for a small aperture ($f/12$), since it cannot be corrected for aperture errors. The aperture can be increased to $f/4$ at the expense of field angle by thickening and achromatizing the meniscus lenses and adding symmetrical elements in the center or at the outside of the basic elements.

Two positive achromatic menisci symmetrically arranged around the stop led to the aplanatic type of lens (Fig. 2e). This type was spherically and chromatically corrected. Since the field could not be corrected, a compromise was achieved by balancing out sagittal and meridional field curvature so that one image surface lies in front and the other in back of the film.

Anastigmatic lenses. The discovery of the Petzval condition for field correction led to the construction of anastigmatic lenses, for which astigmatism and curvature of field are corrected. Such lenses must contain negative components.

The Celor (Gauss) type consists of two airspaced achromatic doublets, one on each side of the stop (**Fig. 3**a). The Cooke triplet combines a negative lens at the aperture stop with two

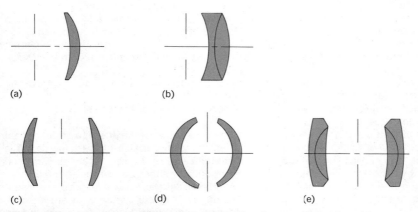

Fig. 2. Older camera lenses. (*a*) Meniscus. (*b*) Simple achromat. (*c*) Periskop. (*d*) Hypergon wide-angle. (*e*) Symmetrical achromat.

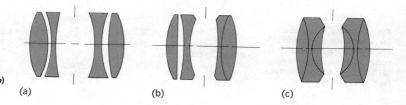

Fig. 3. Types of anastigmats. (*a*) Celor. (*b*) Tessar. (*c*) Dagor.

positive lenses, one in front and the other in back. It is called a Tessar (Fig. 3*b*) if the last positive lens is a cemented doublet, or a Heliar if both positive lenses are cemented. The Dagor type consists of two lens systems that are nearly symmetrical with respect to the stop, each system containing three or more lenses (Fig. 3*c*).

Modern lenses. To increase the aperture, the field, or both, it is frequently advantageous to replace one lens by two separated lenses, since the same power is then achieved with larger radii and this means that the single lenses are used with smaller relative apertures. The replacing of a single lens by a cemented lens changes the color balance, and thus the designer may achieve more favorable conditions. Moreover, the introduction of new types of glass (first the glasses containing barium, later the glasses containing rare earths) led to lens elements which for the same power have weaker surfaces and are of great help to the lens designer, since the errors are reduced.

Of modern designs the most successful are the Sonnar, a modified triplet, one form of which is shown in **Fig. 4***a*; the Biotar (Fig. 4*b*), a modified Gauss objective with a large aperture and a field of about 24°; and the Topogon (Fig. 4*c*), a periscopic lens with supplementary thick menisci to permit the correction of aperture aberrations for a moderate aperture and a large field. One or two plane-parallel plates are sometimes added to correct distortion.

Special objectives. It is frequently desirable to change the focal length of an objective without changing the focus. This can be done by combining a fixed near component behind the stop with an exchangeable set of components in front of the stop. The designer has to be sure that the errors of the two parts are balanced out regardless of which front component is in use. For modern ways to change the magnification SEE ZOOM LENS.

The telephoto objective is a specially contructed objective with the rear nodal point in front of the lens, to combine a long focal length with a short back focus. SEE TELEPHOTO LENS.

The Petzval objective is one of the oldest designs (1840) but one of the most ingenious. It consists in general of four lenses ordered in two pairs widely separated from each other. The first pair is cemented and the second usually has a small air space. For a relatively large aperture, it is excellently corrected for aperture and asymmetry errors, as well as for chromatic errors and distortion. It is frequently used as a portrait lens and as a projection lens because of its sharp central definition. Astigmatism can be balanced but not corrected.

Enlarger lenses and magnifiers. The basic type of enlarger lens is a holosymmetric system consisting of two systems of which one is symmetrical with the first system except that all the data are multiplied by the enlarging factor m. When the object is in the focus of the first system, the combination is free from all lateral errors even before correction. A magnifier in optics is a lens that enables an object to be viewed so that it appears larger than its natural size.

The magnifying power is usually given as equal to one-quarter of the power of the lens expressed in diopters. SEE DIOPTER; MAGNIFICATION.

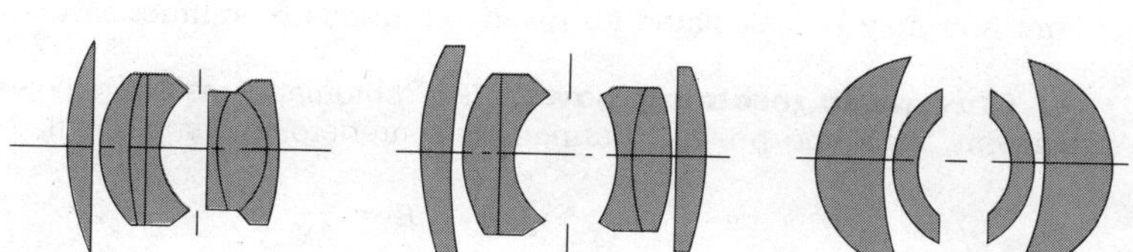

Fig. 4. Modern camera lenses. (*a*) Sonnar. (*b*) Biotar. (*c*) Topogon.

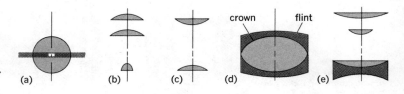

Fig. 5. Typical magnifiers. (a) Sphere with equatorial diaphragm. (b,c) Planoconvex lens combinations. (d) Steinheil triple aplanat. (e) Chevalier type.

Magnifying lenses of low power are called reading glasses. A simple planoconvex lens in which the principal rays are corrected for astigmatism for a position of the eye at a distance of 10 in. (25 cm) is well suited for this purpose, although low-power magnifiers are often made commercially with biconvex lenses. A system called a verant has two lenses corrected for color, astigmatism, and distortion. It is designed for sterosopic vision at low magnification. SEE STEREOSCOPY.

For higher magnifications, many forms of magnifiers exist. One of the basic designs has the form of a full sphere with a diaphragm at the center, as shown in **Fig. 5a**. The sphere may be solid or it may be filled with a refracting liquid. When it is solid, the diaphragm may be formed by a deep groove around the equator. Combinations of thin planoconvex lenses as shown in Fig. 5b and c are much used for moderate powers. Better correction can be attained in the aplanatic magnifier of C. A. Steinheil, in which a biconvex crown lens is cemented between a pair of flint lenses (Fig. 5d).

A design by C. Chevalier (Fig. 5e) aims for a large object distance. It consists of an achromatic negative lens combined with a distant collecting front lens. A magnifying power of up to $10\times$ with an object distance up to 3 in. (75 mm) can be attained.

Bibliography. M. Herzberger, *Modern Geometrical Optics*, 1958, reprint 1973; F. A. Jenkins, *Fundamentals of Optics*, 4th ed., 1976; R. Kingslake et al. (eds.), *Applied Optics and Optical Engineering*, 9 vols., 1965–1983; C. B. Neblette and A. E. Murray, *Photographic Lenses*, 1973; S. F. Ray, *The Photographic Lens*, 1979.

RESOLVING POWER
FRANCIS A. JENKINS AND GEORGE R. HARRISON

A quantitative measure of the ability of an optical instrument to produce separable images. The images to be resolved may differ in position because they represent (1) different points on the object, as in telescopes and microscopes, or (2) images of the same object in light of two different wavelengths, as in prism and grating spectroscopes. For the former class of instruments, the resolving limit is usually quoted as the smallest angular or linear separation of two object points, and for the latter class, as the smallest difference in wavelength or wave number that will produce separate images. Since these quantities are inversely proportional to the power of the instrument to resolve, the term resolving power has generally fallen into disfavor. It is still commonly applied to spectroscopes, however, for which the term chromatic resolving power is used, signifying the ratio of the wavelength itself to the smallest wavelength interval resolved. The figure quoted as the resolving power or resolving limit of an instrument may be the theoretical value that would be obtained if all optical parts were perfect, or it may be the actual value found experimentally. Aberrations of lenses or defects in the ruling of gratings usually cause the actual resolution to fall below the theoretical value, which therefore represents the maximum that could be obtained with the given dimensions of the instrument in question. This maximum is fixed by the wave nature of light and may be calculated for given conditions by diffraction theory. SEE DIFFRACTION; OPTICAL IMAGE.

Chromatic resolving power. The chromatic resolving power R of any spectroscopic instrument, including prisms, gratings, and interferometers, is defined by Eq. (1), where $\delta\lambda$ rep-

$$R = \frac{\lambda}{\delta\lambda} \tag{1}$$

resents the difference in wavelength of two equally strong spectrum lines that can barely be separated by the instrument, and λ the average wavelength of these two lines. It is necessary to specify more precisely the term "barely separated," and for prisms and gratings, in which the width of the lines is determined by diffraction, this is done by use of Rayleigh's criterion. **Illustration** *a* shows the contours of two similar spectrum lines which are at the limit of resolution according to this criterion. The lighter curve represents the line shape due to Fraunhofer diffraction for the wavelength λ, the dashed curve that for $\lambda + \delta\lambda$, and the heavy curve the sum of the two. Rayleigh's criterion specifies that the lines are resolved when the principal maximum of one falls exactly on the first minimum (zero intensity) of the other. Diffraction theory shows that the intensity I of either pattern at the central crossing point is $4/\pi^2$ of that at the maximum, so that the curve representing the sum dips to 81% at the center. The theory also shows that the angular separation $\delta\theta$ of the rays forming the two maxima is λ/a, where a is the linear width of the beam of light emerging from the prism or grating. Hence, quite generally for such an instrument, the resolving power may be defined by Eq. (2).

$$R = \frac{\lambda}{\delta\lambda} = \frac{\lambda}{\delta\theta}\frac{d\theta}{d\lambda} = a\frac{d\theta}{d\lambda} \qquad (2)$$

Expressed in words, Eq. (2) means that

$$\begin{pmatrix}\text{Chromatic}\\\text{resolving power}\end{pmatrix} = \begin{pmatrix}\text{width of}\\\text{emergent beam}\end{pmatrix} \times \begin{pmatrix}\text{angular}\\\text{dispersion}\end{pmatrix}$$

In a given instrument, the calculation of resolving power thus involves finding the last two quantities.

Resolving power of prisms. When a prism is used at minimum deviation, the resolving power depends on the length b of the base of the prism and the slope $dn/d\lambda$ of the dispersion curve giving the wavelength variation of the refractive index n. Thus Eq. (3) holds. Here the

$$R = b\frac{dn}{d\lambda} \qquad (3)$$

assumption is made that the prism is completely filled by the beam of light. If it is not, b must represent the difference in path length between the longest and shortest rays through the prism. S*ee* O*ptical prism*.

Resolving power of gratings. This equals the product of the order of interference m and the total number of rulings N. The order m may be expressed in terms of the grating space s and the angles α and β of incidence and diffraction. Thus Eq. (4) holds. Here w is the width of the ruled area of the grating. For the limiting case of grazing angles of incidence and diffraction,

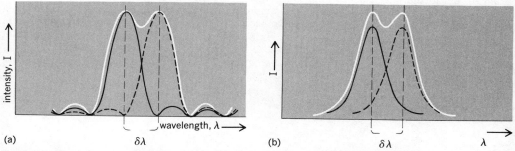

Resolution of two spectrum lines (*a*) when the shape is determined by diffraction (Rayleigh criterion), and (*b*) when the shape follows the Airy formula. The latter is applicable to multiple-beam interferometers.

the maximum possible R is seen to be $2w/\delta$, or the number of wavelengths in twice the width of the grating. SEE DIFFRACTION GRATING.

$$R = mN = \frac{Ns(\sin\alpha + \sin\beta)}{\lambda} = \frac{w(\sin\alpha + \sin\beta)}{\lambda} \quad (4)$$

Resolving power of interferometers. For the type of interferometer most commonly used, the Fabry-Perot interferometer, the resolving power may be expressed as the product of the order of interference $m = 2t/\lambda$, where t is the separation of the interferometer mirrors, and an effective number N_{eff} of interfering beams. For interferometers the line contour of the spectrum lines is not that of Fraunhofer diffraction, but is given by a relations called the Airy formula. This contour has no points of zero intensity but has the general shape shown in illustration b. Therefore, the Rayleigh criterion cannot be applied in the usual way. If, however, the two curves are made to cross at the half-intensity point of each, it is found that there is a dip of approximately 20% in the resultant curve. The value of N_{eff} is therby specified, and the resolving power R is given by Eq. (5), where ρ designates the reflectance of the interferometer plates. SEE INTERFEROMETRY.

$$R = mN_{\text{eff}} = m\left(\frac{\pi\sqrt{\rho}}{1-\rho}\right) \quad (5)$$

Resolving power of telescopes. This depends on the size of the diffraction maximum produced when light from a distant point source passes through a circular aperture of size equal to that of the objective lens or mirror. A graph of the intensity in the diffraction pattern plotted against radial distance closely resembles one of the curves of illustration a, and hence the pattern consists of a central spot surrounded by faint rings. The angular radius of the first dark ring corresponds, by the Rayleigh criterion, to the angular separation of two point sources that are barely resolved. Theory gives this angle, which represents the resolving limit, as defined by Eq. (6) for $\lambda = 560$ nanometers and d, the diameter of the objective lens, in centimeters.

$$\alpha = \frac{1.220\lambda}{d} \text{ radians} = \frac{14.1}{d} \text{ seconds of arc} \quad (6)$$

Resolving power of microscopes. This is determined by diffraction of a circular aperture representing the exit pupil of the microscope objective. There are two important differences between the resolving power of microscopes and that of telescopes. First, the resolving limit of microscopes is expressed in terms of the smallest distance l between two points on the object that are just resolved. Second, this limit depends on the mode of illumination of the object. If the illumination is incoherent, so that there is no constant phase relation between light from adjacent points, the resolving limit is given by Eq. (7), where n, is the refractive index of the material (for

$$l = \frac{0.61\lambda}{n, \sin\alpha} \quad (7)$$

example, oil) in the object space, and α the angle that the extreme ray entering the objective makes with the axis of the instrument. The quantity n, $\sin\alpha$ is called the numerical aperture of the objective. With coherent illumination the resolving limit is given by this formula, with 1.0 in place of 0.61, provided the illumination is central. When the object is illuminated from a point slightly to one side, the factor may be reduced to 0.5. SEE OPTICAL MICROSCOPE; SPECTROSCOPY.

Bibliography. D. F. Horne, *Optical Instruments and Their Applications*, 1980; F. A. Jenkins, *Fundamentals of Optics*, 4th ed., 1976.

OPTICAL IMAGING SYSTEMS

Periscope	40
Eyeglasses	41
Optical projection systems	43
Optical telescope	46
Eyepiece	59
Schmidt camera	60
Gunsights	62
Rangefinder	63
Binoculars	67
Optical tracking instruments	68
Camera	73
Telephoto lens	74
Zoom lens	74
Microscope	75
Optical microscope	77
Reflecting microscope	90
Phase-contrast microscope	92
Interference microscope	96
Fluorescence microscope	101
Adaptive optics	102

PERISCOPE
Edward K. Kaprelian

An optical instrument that permits viewing along a displaced or deflected axis, providing an observer with the view from a position which may be inaccessible or dangerous. Periscopes range in complexity from the simple unit-power tank periscope to the complex multielement submarine periscope.

Tank periscope. This device, intended to protect the user from bullets, employs a pair of plane, parallel, reflecting surfaces (either mirrors or prisms), so arranged in a mount that the path of light through the instrument forms a crude letter Z (**Fig. 1***a*). If powers greater than unity are desired or if the periscope is to be used for sighting, a terrestrial telescope can be added to the periscope, either as a simple, internally contained system (Fig. 1*b*) or entirely in front of or behind the periscope itself, as desired. The reflecting elements of the system which are responsible for deflecting the optical axis are independent of the refracting (telescope) elements which provide the optical power.

It is also possible to arrange the two mirrors of a periscope at right angles to each other, in which case the observer views an inverted image while facing away from the direction from which light enters the instrument. By adding an inverting (astronomical type) telescope to this system, the image is reinverted for the observer.

Periscopes of this type cannot be used for scanning the horizon by rotating the upper mirror because of the image rotation which accompanies such movement. In the panoramic sight (**Fig. 2**), this difficulty is overcome by providing the system with a dove prism which rotates at half the angular speed of the right-angle prism through the action of a differential gear linkage. The combined inversions of the dove prism and the amici prism at the bottom completely compensate for the inversions of the telescope system, while the relative motions of the right-angle prism and the dove prism maintain the image erect during scanning. *See* Mirror optics; Optical prism.

Submarine periscope. In this device, it is necessary to employ a telescope system having a wide field of view and uniform illumination across a field which can be fitted into a long, narrow tube whose length-to-diameter ratio may be 50 or greater. This is achieved by utilizing a plurality of lenses so spaced along the length of the tube as to cause the incoming principal rays from the edge of the field to be deviated from side to side within the tube. In general, the greater the number of lenses, the wider the field of view. One example of the periscopic relay train is shown in **Fig. 3***a* and employs six lenses with three inversions. The typical submarine periscope (Fig. 3*b*) may be considered to be a pair of telescopes facing each other, with such a relay train between them. The usual magnification of submarine periscopes is 6, although some U.S. Navy

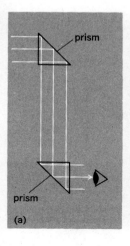 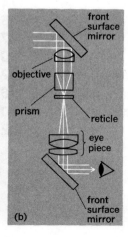

Fig. 1. Tank periscope. (*a*) Simple, with parallel reflecting surfaces. (*b*) With terrestrial telescope.

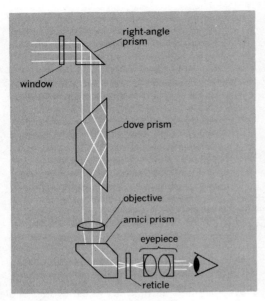

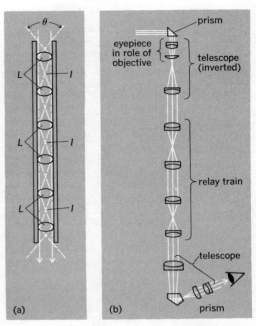

Fig. 2. Panoramic sight, with erect image.

Fig. 3. Periscope relay train. (a) Showing lenses *L*, inversions *I*, and angle of view θ. (b) Between a pair of facing telescopes in a submarine periscope.

periscopes have dual magnifications of 6 and 1.5, the latter being achieved by inserting an inverted galilean telescope into the optical path before the top objective. SEE LENS; MAGNIFICATION.

The submarine periscope can be provided with a built-in rangefinder for fire-control purposes. A conventional coincidence or split-field type of rangefinder may be attached either vertically or horizontally to the upper end of the periscope, the objective of which receives an image from each of the entrance windows of the rangefinder. SEE RANGEFINDER.

Other types. Various modifications of the basic optical systems described here are employed as viewing periscopes in military aircraft and as viewing devices in particle accelerators and nuclear reactors. The cystoscope and endoscope are slender, sometimes mechanically flexible periscopes used for visual examination and photography of body cavities inaccessible to direct observation; an entirely different basis for the design of such instruments is in the use of bundles of optical fibers. SEE OPTICAL FIBERS.

Bibliography. D. F. Horne, *Optical Instruments and Their Applications*, 1980; Optical Society of America, *Handbook of Optics*, 1978.

EYEGLASSES
EDWARD G. STUART AND N. KARLE MOTTET

A general term for optical devices containing corrective lenses for defects in vision or for special purposes. Common visual defects are errors of refraction, which is the bending of light rays so that a sharp retinal image is produced. This deflection is produced when light passes from one material to another of different optical density, as from air to water. Besides optical density, the curvature of the surfaces of the material causes either spreading or focusing of light rays, in accordance with the laws of optics. SEE REFRACTION OF WAVES; VISION.

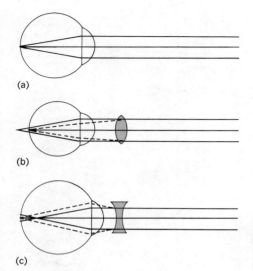

Course of light rays and formation of retinal image in the eye and correction by appropriate lenses. (a) Emmetropia (normal). (b) Hyperopia. (c) Myopia. (*After B. A. Houssay et al., Human Physiology, 2d ed., McGraw-Hill, 1955*)

A normal, or emmetropic, eye focuses light rays from a distance on the retina by means of complementary deflections of the cornea, crystalline lens, and fluids of the eye (**illus**. *a*). Distant light rays, reflected from an object more than 25 ft (7.5 m) from the eye, are considered to be parallel. As an object approaches the eye, however, its reflected rays tend to diverge so that, unless some correction is made, these rays focus behind the retina and vision is blurred. Such correction is known as accommodation and is achieved through alteration of the lens curvature by action of the ciliary muscle so that a retinal focus is obtained. In infants and older persons, failure of accommodation causes blurred near vision but does not affect far vision.

Presbyopia is the progressive loss of ability to accommodate to near vision as a result of aging processes. It usually becomes apparent in the 40–45 age group and may be corrected by convex lenses, usually worn only for close work or reading. Bi-, tri-, or tetrafocal lenses are used under different circumstances to permit rapid shifting from one visual distance to another in failures of accommodation.

Few eyes are completely normal in their refractive ability. Defects in curvatures, densities, and position of the eye structures, as well as variations in the length of the eyeball, or visual axis, are called ametropias. Most are hereditary or developmental defects.

Hyperopia, or farsightedness, results from too short an eyeball so that unaccommodated rays focus behind the retina. Moderate defects may be corrected by accommodation, particularly in young people, but the increased accommodation required for close vision often leads to eyestrain. Convex lenses are used to gain proper focus (illus. *b*).

Myopia, or nearsightedness, results from too long a visual axis so that the focus falls in front of the retina. The ciliary muscles cannot make the lens of the eyeball less curved. Concave lenses are used to permit focusing of far vision (illus. *c*); near vision is often better than that of the normal eye since little accommodation is necessary, less eyestrain is produced with close work, and the retinal image is somewhat larger than in the normal eye.

Astigmatism is a common condition in which the vertical and horizontal meridians of the eye have different curvatures, thereby causing different points of focus. Cylindrical lenses are ground to correct the defect in the abnormal meridian. Most astigmatics also have other defects so that compound lenses are required.

Other lens systems are commonly devised for persons who have unequal vision and to correct for abnormal visual axes and defects in ocular movements caused by muscle imbalance.

Special forms of eyeglasses are made to absorb portions of the light spectrum, such as certain colors and ultraviolet and infrared rays. Aviators, welders, radiologists, furnace workers, and others require lenses of certain physical or optical characteristics. SEE LENS.

OPTICAL PROJECTION SYSTEMS
ARMIN J. HILL

Optical projection is the process whereby a real image of a suitably illuminated object is formed by an optical system in such a manner that it can be viewed, photographed, or otherwise observed. Essential equipment in an optical projection system consists of a light source, a condenser, an object holder, a projection lens, and (usually) a screen on which the image is formed (**Fig. 1**).

The luminance of the image in the direction of observation will depend upon (1) the average luminance of the image of the light source as seen through the projection lens from the image point under consideration, (2) the solid angle subtended by the exit pupil of the projection lens at this image point, and (3) the reflective or transmissive characteristics of the screen. Usually it is desirable to have this luminance as high as possible. Therefore, with a given screen, lens, and projection distance, the best arrangement is to have the light source imaged in the projection lens, with its image filling the exit pupil as completely and as uniformly as possible.

The object is placed between the condenser and the projection lens. If transparent, it can be inserted directly in the light beam; however, it should be positioned, and the optical system should be so designed that it does not vignette (cut off) any of the image of the light source in the projection lens. If the object is opaque, an arrangement known as an epidiascope (**Fig. 2**) is used. A difficulty in the design of this system is to illuminate the object so that all portions will show well in the projected image, without excessive highlights or glare.

If a small uniform source which radiates in accordance with Lambert's law is projected through a well-corrected lens to a screen which is perpendicular to the optic axis of the lens, maximum illuminance of the image will occur on this axis, and illuminance away from the axis will decrease in proportion to the fourth power of the cosine of the angle subtended with the axis at the projection lens. In practice, it is possible to design distortion into the condenser so that the illuminance is somewhat lower on the axis, and considerably higher away from the axis than is given by this fourth-power law. Acceptable illumination for most visual purposes can allow a falloff from center to side in the image of as much as 50%, particularly if the illuminance within a circle occupying one-half of the image area does not drop below 80% of the maximum value. SEE PHOTOMETRY.

Light source. Usually, either an incandescent or an arc lamp is used as the light source. To keep luminance high, incandescent projection lamps operate at such high temperatures that their life is comparatively short. Also, they must be well cooled; all except the smallest sizes require cooling fans for this purpose. Filaments are finely coiled and accurately supported, usually

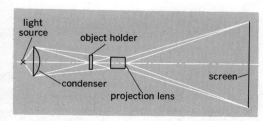

Fig. 1. Simple optical projection system.

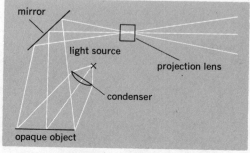

Fig. 2. An epidiascope, or system for projecting an image of an opaque object.

being carefully aligned in a prefocus base so that they will be precisely positioned in the optical system. Spacing between coils is such that a small spherical mirror can be used in back of the lamp to image the coils in the spaces between the coils, thus increasing usable light output nearly twofold.

When a highly uniform field is required, a lamp consisting of a small disk of ceramic material, heated to incandescence by radiofrequency induction, is available. With this, it is possible to maintain illuminance of a projection field of several square inches with a variation of only 2–3%.

Arc lamps are used when incandescent lamps cannot provide sufficient flux to illuminate the screen area satisfactorily. Carbon electrodes with special feed control, magnetic arc stabilization, and other devices are usually used to keep the arc as accurately positioned and as flicker-free as possible. The high-pressure xenon arc lamp is also used as a projection light source. It has a color which more accurately duplicates sunlight than the carbon arc. Its intensity is considerably higher than that of any incandescent lamp, and it avoids most of the problems attendant on the burning of carbons. A special power supply, which ordinarily incorporates a striking circuit, is needed.

A shield between the arc and the object, called a douser, is used to protect the object while the arc is ignited and to provide a quick shutoff for the light beam. Often water cells and other heat-filtering devices are inserted in the beam to keep the heat on the object as low as possible.

Condenser. The condenser system is used to gather as much of the light from the source as possible and to redirect it through the projection lens. Both reflective and refractive systems are used. Reflectors can be of aluminum, although the better ones are of glass with aluminized coatings. They are usually elliptical in shape, with the light source at one focus and the image position in the projection lens at the other.

Refractive systems may be of heat-resistant glass or fused quartz. With arc lamps particularly, the condenser lens is very close to the source in order to provide the high magnification required if the image is to fill the projection lens. Usually, therefore, the condenser requires special cooling. In larger projectors, several elements are used to give the required magnification; these are often aspherical in shape to give the required light distribution.

A well-designed condenser can pick up light in a cone having a half-angle in excess of 50°. This means that, with a well-designed arc lamp or with an incandescent lamp using an auxiliary spherical mirror, more than one-third of the total luminous flux radiated by the source can be directed through the projector.

To obtain the high magnification required with arc sources and large-aperture lenses, a relay type of condenser (**Fig. 3**) may be used. This images the source beyond the first condenser system, and then uses a second lens to relay this image to the projection lens. This arrangement allows for better light-distribution control in the screen image with less waste of light at the object position. Also, an Inconel "cat's-eye" diaphragm at the first image point gives a convenient intensity control for the light beam. For additional information on condensers SEE OPTICAL MICROSCOPE.

Object holder. The function of the object holder is to position the object exactly at the focal point of the projection lens. Slight motion or vibration will reduce image sharpness. There-

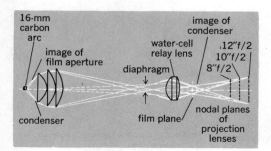

Fig. 3. Relay condenser system having water cell incorporated in second stage. 1″ = 25 mm; 16 mm = 0.63 in.

fore, proper mechanical design of this part of the system is most important. When a succession of objects is to be projected, the holder must be able to position these objects precisely and clamp them firmly in a minimum time. Some designs have been suggested and a few tried which allow the object to move while the system projects a fixed image. However, this requires a motion in some portion of the optical system, which invariably reduces the quality of the projected image. Because of this, such systems have not found wide favor.

The object holder must also provide for protection of the object from unwanted portions of the beam, for cooling the object, where this may be required, and for accurately adjusting its position laterally with respect to the beam. In most systems, focusing of the projection lens is provided by moving the lens itself rather than by disturbing the position of the object longitudinally in the system.

Projection lens. The function of the projection lens is to produce the image of the object on the screen. Its design depends upon the use to be made of the projected image. As examples, a profile or contour projector requires a lens which is well corrected for the aberrations known as distortion and field curvature; an optical printer (or enlarger) requires a lens carefully designed to work at comparatively short conjugate focal distances; and picture projectors must preserve the quality which has been captured by the camera lens. SEE ABERRATION (OPTICS).

Since projection lenses usually transmit relatively intense light beams, they must be designed to be heat-resistant. Their surfaces are usually not cemented. Optical surfaces are coated to reduce reflection, but these coatings must be able to withstand heat without deterioration. Because camera lenses are not designed for this type of operation, they should not be employed as projection lenses unless the projector is specifically designed to use them.

An ideal position at which to control intensity is at the projector lens. Large-aperture lenses containing iris diaphragms are used in the large process projectors of motion picture studios (**Fig. 4**).

Projection screen. Usually a projection screen is used to redirect the light in the image for convenient observation. Exceptions are systems which project the image directly into other optical systems for further processing, or which form an aerial image which is to be viewed only from a localized position.

Screens may be either reflective or translucent, the latter type being used when the image is to be observed or photographed from the side away from the projector. An example is the so-called self-contained projector, which has the screen on the housing holding the other elements.

Fig. 4. Triple-head projector with three 250-A arc lamps and 12-in. (30-mm) $f/2$ projection lenses, incorporating iris diaphragms for intensity control. (*Paramount Pictures*)

Reflective screens may be matte, having characteristics approaching Lambert reflection; directional, producing an image which will appear brighter in the direction of specular reflection; or reflexive, directing most of the light back toward the projector and giving an image of relatively high luminance with low projection intensity, but with a very confined viewing angle. SEE CINEMATOGRAPHY.

OPTICAL TELESCOPE
ROBERT D. CHAPMAN AND BENNY L. KLOCK
R. D. Chapman wrote the section Large Telescopes.

An instrument that collects light energy from a distant source and focuses it into an image that can then be studied by a number of different techniques. This definition of an optical telescope must be narrowed somewhat. Satellite-borne telescopes operated to study celestial x-rays or ultraviolet radiation and ground-based telescopes used to study radio radiation can be called optical telescopes in the sense that they operate according to the principles of geometrical optics. The following discussion stresses ground-based telescopes which are used to study radiation from celestial objects in the wavelength range from the Earth's atmospheric cutoff in the near ultraviolet to the infrared, that is, from 300 to 1000 nanometers. Such instruments may be classified as (1) large astronomical telescopes, used to study the nature of astronomical objects themselves, and (2) astronomical transit instruments, used to study positions and motions of astronomical objects and in the accurate determination of time.

LARGE TELESCOPES

A large astronomical telescope is used by astronomers to study the fundamental problems in the field. The size required for a telescope to be considered large depends on its type, as discussed below.

Types of telescopes. There are basically three types of optical systems in use in astronomical telescopes: refracting systems whose main optical elements are lenses which focus light by refraction; reflecting systems, whose main imaging elements are mirrors which focus light by reflection; and catadioptric systems, whose main elements are a combination of a lens and a mirror. The most notable example of the last type is the Schmidt camera.

In each case, the main optical element, or objective, collects the light from a distant object and focuses it into an image that can then be examined by some means. Specific types of tools that are frequently employed to study astronomical objects are discussed below.

Refracting telescopes. The main optical element, or objective, of a refracting telescope is usually a long-focal-length lens. The objective lens is typically compound; that is, it is made up of two or more pieces of glass, of different types, designed to correct for aberrations such as chromatic aberration. **Figure 1** shows a refractor lens imaging the light of two stars onto a photographic plate. To construct a visual refractor, a lens is placed beyond the images and viewed with the eye. To construct a photographic refractor or simply a camera, a photographic plate is placed at the position of the image. The characteristics of the components of a photographic lens may differ from those of a visual lens.

Generally, refracting telescopes are used in applications where great magnification is required, namely, in planetary studies and in astrometry, the measurement of star positions and

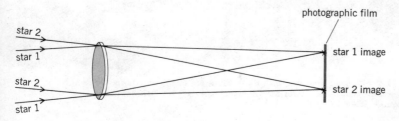

Fig. 1. Refracting optical system used to photograph a star field.

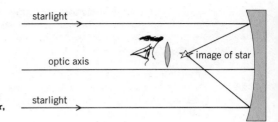

Fig. 2. Viewing a star with a reflecting telescope. In this configuration the observer may block the mirror, unless it is a very large telescope.

motions. For example, most stellar parallaxes have been measured with refractors such as the Sproul Observatory 24-in. (61-cm) $f/18$ refractor. However, this practice is changing, and the traditional roles of refractors are being carried out effectively by a few new reflecting telescopes. This changing role has come about in part because of effective limitations on the size of refracting telescopes. The largest refractor, located at Yerkes Observatory at Williams Bay, Wisconsin, has a 40-in. (1.02-m) objective lens.

A refractor lens must be relatively thin to avoid excessive absorption of light in the glass. On the other hand, the lens can be supported only around its edge and thus is subject to sagging distortions which change as the telescope is pointed from the horizon to the zenith; thus its thickness must be great enough to give it mechanical rigidity. An effective compromise between these two demands is extremely difficult, if not impossible, for a lens over 40 in. (1 m) in diameter, making larger refractors unfeasible.

Reflecting telescopes. The principal optical element, or objective, of a reflecting telescope is a mirror. The mirror forms an image of a celestial object (**Fig. 2**) which is then examined with an eyepiece, photographed, or studied in some other manner.

Reflecting telescopes generally do not suffer from the size limitations of refracting telescopes. The mirrors in these telescopes can be as thick as necessary and can be supported by mechanisms which prevent sagging and thus inhibit excessive distortion. In addition, mirror materials having vanishingly small expansion coefficients (Cer-Vit, ultralow-expansion-fused silica, and others), together with ribbing techniques which allow rapid equalization of thermal gradients in a mirror, have eliminated the major thermal problems plaguing telescope mirrors. Telescopes with mirrors up to 236 in. (6 m) in diameter have been built.

The reflecting telescope has other advantages which make it an attractive system. By using a second mirror (and even a third one, in some telescopes), the optical path in a reflector can be folded back on itself (**Fig. 3***a*), permitting a long focal length to be attained with an instrument housed in a short tube. A short tube can be held by a smaller mounting system and can be housed in a smaller dome than a long-tube refractor, thus decreasing costs.

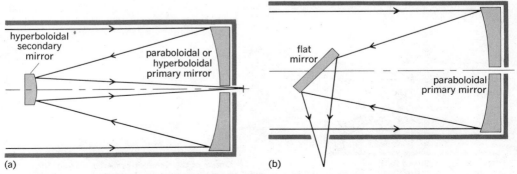

Fig. 3. Diagrams of reflecting telescopes. (*a*) Cassegrain telescope, with either classical or Ritchey-Chrétien optics. (*b*) Newtonian telescope.

Finally, a large mirror has only one surface to be figured—that is, to be ground and polished to the desired optical shape—whereas a refractor lens usually consists of two pieces of glass that must be laboriously figured. The secondary mirror of a reflector is a second surface that needs to be figured, but it is much smaller than the primary and thus is much easier and cheaper to fabricate.

A variety of optical arrangements are possible in large reflecting telescopes, including the prime focus, the newtonian focus, the Cassegrain focus, and the coudé focus.

The newtonian focus is probably most widely used by amateur astronomers in reflectors having apertures on the order of 6 in. (15 cm; Fig. 3b). A flat mirror placed at 45° to the optical axis of the primary mirror diverts the focused beam to the side of the telescope, the image being formed by the paraboloidal primary mirror alone. An eyepiece, camera, or other accessories can be attached to the side of the telescope tube to study the image. In the largest telescopes, with apertures over 100 in. (2.5 m), provision is not usually made for a newtonian focus. Instead, an observing cage is placed inside the tube, where the observer can take accessories to observe the image formed by the primary mirror. This prime focus is identical to the newtonian focus optically, since the newtonian flat does nothing more than divert the light beam. The modern reflectors have fast primary mirrors, so that the focal ratio at the prime focus is $f/2.5$ to $f/6$. Lower focal ratios permit shorter exposures on extended objects such as comets and nebulae.

A Cassegrain system consists of a primary mirror with a hole bored through its center, and a convex secondary mirror which reflects the light beam back through the central hole to be observed behind the primary mirror (Fig. 3a). Since the secondary mirror is convex, it decreases the convergence of the light beam and increases the focal length of the system as a whole. The higher focal ratios ($f/8$ to $f/13$) of Cassegrain systems permit the astronomer to observe extended objects, like planets, at higher spatial resolution and to isolate individual stars from their neighbors for detailed studies.

The classical Cassegrain system consists of a paraboloidal primary mirror and a hyperboloidal secondary mirror. The newtonian focus, prime focus, and Cassegrain focus are not affected by spherical aberration. However, all of the systems are plagued by coma, an optical aberration of the paraboloidal primary. Coma causes a point source off the center of the field of view to be spread out into a comet-shaped image. To correct for the effect of coma, a corrector lens is often used in front of the photographic plate. The design of prime-focus corrector lenses is a major consideration in large telescope design, since the corrector lens itself can introduce additional aberrations. *See* Aberration.

To avoid complicated corrector lenses at the Cassegrain focus, the Ritchey-Chrétien system, an alternate design with both a hyperboloidal primary and a hyperboloidal secondary, is used in modern telescopes. This arrangement is not affected by coma or spherical aberration, so it has a wider field of view than the classical Cassegrain.

Schmidt camera. This is an optical system used almost exclusively for photographic applications such as sky surveys, monitoring of galaxies for supernova explosions, and studies of comet tails. The primary mirror of a Schmidt camera has a spherical shape, and therefore suffers from spherical aberration. To correct for this problem, the light passes through a thin corrector plate as it enters the tube, as illustrated in **Fig. 4**. Schmidt camera correcting plates are among the largest lenses made for astronomical applications. The special features of this system combine

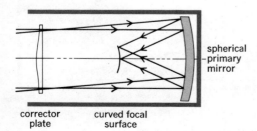

Fig. 4. Optics of the Schmidt system showing the axial and the extra-axial light paths.

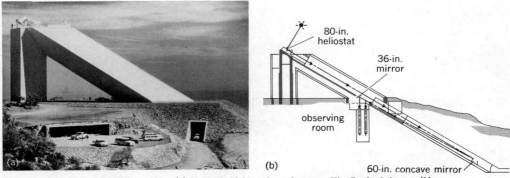

Fig. 5. Unconventional configuration. (a) McMath 60-in. solar telescope, Kitt Peak, Arizona. (b) Diagram of optical elements in the telescope. 1 in. = 2.5 cm.

to produce good images over a far larger angular field than can be obtained in a Cassegrain. The Schmidt camera at the California Institute of Technology/Palomar Observatory, which carried out the National Geographic Society–Palomar Observatory Sky Survey, can photograph a 6° by 6° field. *See Schmidt camera.*

Solar telescopes. Solar instrumentation differs from that designed to study other celestial objects, since the Sun emits great amounts of light energy. One solar instrument, the Robert R. McMath Solar Telescope located at Kitt Peak National Observatory near Tucson, Arizona (**Fig. 5**), consists of an 80-in. (2.03-m) heliostat that reflects sunlight down the fixed telescope tube to a spectrograph. This spectrograph is evacuated to avoid problems that would be created by hot air currents. An alternate design has been used at the Sacramento Peak Observatory, in which the heliostat feeds light into a vertical telescope that is evacuated.

Site selection. Efficient use of a telescope requires a site with a clear, steady atmosphere. Site selection for a large telescope is given considerable attention, every effort being made to locate the instrument in an area that is climatologically and geologically favorable—preferably an area at high elevation, on solid footing, and having a proved record for number of clear nights per year and maximum mean atmospheric stability. The site should be far from the lights and polluted air of large population centers. Today, it is difficult to find excellent sites in the continental United States. The great observatories in California, for instance, all feel the impact of civilization. The most sought-after sites are in the high mountains of Chile and Hawaii (see the **table**).

Tools. Astronomers seldom use large telescopes for visual observations. Instead, they record their data for future study. Modern developments in photoelectric imaging devices are supplanting photographic techniques for many applications. The great advantages of detectors such as charge-coupled devices is their fine sensitivity, and the images can be read out onto a computer-compatible magnetic tape or disk for immediate analysis.

Light received from most astronomical objects is made up of radiation of all wavelengths. The spectral characteristics of the radiation emitted from or reflected by a body may be extracted by special instruments called spectrographs. Wide field coverage is not critical in spectroscopy of stellar objects, so spectrographs are mounted at the Cassegrain and coudé foci. The coudé position has the advantage that its focal point is fixed in position, regardless of where the telescope points in the sky (**Fig. 6**). Thus a spectrograph or other instrument that is too heavy or too delicate to be mounted on the moving telescope tube can be placed at the coudé focus.

Photoelectric imaging devices may be used in conjunction with spectrographs to record spectral information. Photoelectric detectors, usually photomultiplier tubes, are useful tools for classifying stars, monitoring variable stars, and quantitatively measuring the light flux from any source at which the telescope is pointing. The photocathode of the detector is placed just behind the focal plane of the telescope, preceded by a small diaphragm which permits a view of only a very small area of the sky. The phototube converts the incident light energy into an electrical signal, which is subsequently amplified and recorded. Photometry carried out with different filters

Large telescopes of the world

Mirror diameter		Observatory	Year completed
Meters	Inches		
		Some of the largest reflecting telescopes	
6.0	236	Special Astrophysical Observatory, Zelenchukskaya, Crimea, Soviet Union	1976
5.1	200	California Institute of Technology/Palomar Observatory, Palomar Mountain, California	1950
4.0	158	Kitt Peak National Observatory, Arizona	1973
4.0	158	Cerro Tololo Inter-American Observatory, Chile	1976
3.9	153	Anglo-Australian Telescope, Siding Spring Observatory, Australia	1975
3.8	150	United Kingdom Infrared Telescope, Mauna Kea Observatory, Hawaii	1978
3.6	144	Canada-France-Hawaii Telescope, Mauna Kea Observatory, Hawaii	1979
3.6	142	Cerro La Silla European Southern Observatory, Chile	1976
3.2	126	NASA Infrared Telescope, Mauna Kea Observatory, Hawaii	1979
3.0	120	Lick Observatory, Mount Hamilton, California	1959
2.7	107	McDonald Observatory, Fort Davis, Texas	1968
2.6	102	Crimean Astrophysical Observatory, Soviet Union	1960
2.6	102	Byurakan Observatory, Yerevan, Soviet Union	1976
2.5	100	Mount Wilson and Las Campanas, Mount Wilson, California	1917
2.5	100	Cerro Las Campanas, Carnegie Southern Observatory, Chile	1976
2.5	98	Royal Greenwich Observatory, United Kingdom	1967
2.4	94	University of Michigan–Dartmouth College–Massachusetts Institute of Technology, Kitt Peak, Arizona	1986
6 × 1.8	6 × 71	Multi-Mirror Telescope, Mount Hopkins, Arizona	1979
		Largest refracting telescopes	
1.02	40	Yerkes Observatory, Williams Bay, Wisconsin	1897
0.91	36	Lick Observatory, Mount Hamilton, California	1888
0.83	33	Observatoire de Paris, Meudon, France	1893
0.80	32	Astrophysikalisches Observatory, Potsdam, Germany	1899
0.76	30	Allegheny Observatory, Pittsburgh, Pennsylvania	1914

yields basic information about the source with shorter observing time than that required for a complete spectroscopic analysis. S*EE* P*HOTOMETRY*.

Limitations. The largest telescope in operation is the 236-in. (6-m) reflector in the Caucasus Mountains in the Soviet Union. For many applications the Earth's atmosphere limits the effectiveness of larger telescopes.

The most obvious deleterious effect of the Earth's atmosphere is image scintillation and motion, collectively known as "poor seeing." Atmospheric turbulence produces an extremely rapid motion of the image resulting in a smearing of the image. On the very best nights at ideal observing sites, the image of a star will be spread out over a 0.25-arc-second seeing disk; on an average night, the seeing disk may be between 0.5 and 2.0 arc-second. The theoretical resolving power of a 20-in. (50-cm) telescope is 0.25 arc-second, so that any telescope with much over 20-in. aperture is limited by the Earth's atmosphere as far as the finest detail it can resolve. S*EE* R*ESOLVING POWER*.

Telescopes larger than about 20 in. (50 cm) are built, not for resolution, but for light-gathering power, which depends on the area of the primary mirror or objective lens. One of the chief uses of large telescopes is to study phenomena in quasars and galaxies in distant regions of the observable universe. Once again, however, the atmosphere limits what can be observed. The upper atmosphere glows faintly because of the constant influx of charged particles from the Sun. This airglow is a phenomenon similar to the aurora borealis, although it is typically fainter than a visible aurora display. Airglow adds a background exposure or fog to photographic plates that depends on the length of the exposure and the speed (*f*-ratio) of the telescope. The combination of the finite size of the seeing disk of stars and the presence of airglow means that a 33-ft (10-m) telescope could not see an object $(10/6)^2 = 2.8$ times fainter than the faintest object seen by a 20-ft (6-m) telescope. In fact, the gain is much less than that figure. On the other hand, a 33-ft telescope might cost as much as 10 times more than a 20-ft telescope. One solution is placing a large telescope in orbit above the atmosphere.

In practice, the effects of air pollution and light pollution from large cities outweigh the effect of airglow at most observatories in the United States. There are few unspoiled observatory sites left in the continental United States.

Notable telescopes. The definition of a large telescope depends on its type. For a refracting telescope to be considered a large telescope, its objective lens must be larger than about 24 in. (0.6 m), whereas a reflecting telescope will have to exceed 79 in. (2 m) to be considered large. The table lists a few of the largest telescopes. It is of interest to compare the dates of the reflecting telescopes and refracting telescopes. The limitations of refractors were recognized very early. A few telescopes that are notable for their historical importance, large size, or innovative design will be discussed.

Although Galileo's telescopes were small in size, they were notable because of the tremendously important discoveries made with them.

In 1790, William Herschel built a telescope whose mirror was 48 in. (1.2 m) in diameter. Also, William Parsons, the third Earl of Rosse, built a telescope with a 72-in. (1.8-m) mirror about 1840. However, these telescopes were difficult to operate and led to relatively few discoveries. Both were built with alt-azimuth mountings which made it very difficult to follow celestial objects as the Earth rotated. In an equatorial mounting, found in almost all modern telescopes, rotation about one axis, the polar axis, compensates for the Earth's rotation. However, in an alt-azimuth mounting, motion in both altitude and azimuth is required to follow an object.

The 40-in. (1.02-m) refractor at the Yerkes Observatory was completed in 1897. The 40-in. objective lens has a focal length of 62 ft (19 m) and is housed in a 90-ft (23-m) dome. With its delicately balanced equatorial mounting and clock drives, the Yerkes refractor has been a major contributor to astronomy.

The 200-in. (5-m) Hale telescope at Palomar Mountain, California, was completed in 1950. The primary mirror is 200 in. in diameter with a 40-in. (1.02-m) hole in the center. Its focal length is 660 in. (14 m) and it has a paraboloidal figure. The focal ratio of the prime focus is $f/3.3$, and of the Cassegrain focus $f/16$. The Hale telescope was the first that was large enough to have a prime-focus cage where the observer could sit inside the main tube.

Since the completion in 1950 of the Hale reflector, the number of telescopes over 100 in. (2.5 m) in aperture has steadily grown, and a dozen such instruments have been built since 1960.

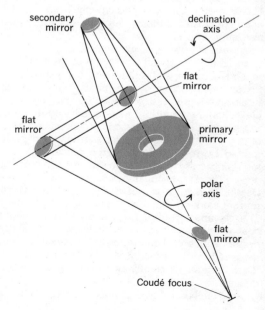

Fig. 6. Optical diagram for a common coudé configuration, showing axes of rotation of the telescope, viewed in isometric projection for clarity.

Fig. 7. The 158-in. (4-m) Mayall reflector of the Kitt Peak National Observatory. The dark tube at the upper end of the telescope is the prime focus cage. The large horseshoe bearing floats on oil pads, seen near each end of the mounting walkway. (*Kitt Peak National Observatory Photograph*)

It is not feasible to describe each of these instruments in detail, although it is worth mentioning the modern instruments which represent the most advanced systems in operation.

The 158-in. (4-m) Mayall reflector at the Kitt Peak National Observatory was dedicated in 1973 (**Fig. 7**). The 158-in. mirror is made from a 24-in.-thick (61-cm) fused quartz disk which is supported in an advanced design mirror cell. It took three years to grind and polish the mirror to its f/2.7 hyperboloidal shape. The prime focus has a field of view six times greater than that of the Hale reflector. The first photographs with the 158-in. mirror showed its outstanding optical characteristics. An identical telescope was subsequently installed at Cerro Tololo Inter-American Observatory, in Chile.

Another notable instrument is the 236-in. (6-m) Soviet reflector. It is supported in an alt-azimuth mounting and like the great telescopes of Herschel and the Earl of Rosse, must be moved in both altitude and azimuth to compensate for the Earth's rotation. This delicately balanced instrument is driven by a computer which calculates the relative rates of motion in azimuth and in altitude required to follow an object, and controls the motors which provide the motion. However, there is yet a third motion required. In an alt-azimuth system, the sky rotates relative to the telescope tube as the telescope tracks a celestial object. This rotation would smear a photograph of the object. To avoid this smearing, the photographic plate (or other observing device) must be rotated by the computer at the proper rate. The Soviet astronomers estimate the alt-azimuth mounting cost half that of an equatorial mounting.

The Smithsonian Astrophysical Observatory/University of Arizona Multi-Mirror Telescope (MMT; **Fig. 8**) is one of the most innovative of the new telescopes. It uses six 71-in. (1.8-m) mirrors working together to provide light-gathering power equivalent to a 177-in. (4.5-m) mirror. The six images from the telescopes are brought to a common focus by a mirror system. The coalignment of the six optical systems is maintained by means of an active laser and computer system which continually moves the secondary mirrors to maintain the alignment.

Observations with the Multi-Mirror Telescope were begun in early 1979, and it quickly demonstrated its capabilities by producing a number of interesting discoveries about stars and quasars.

Two space-based telescopes have been highly successful. The *International Ultraviolet Ex-*

Fig. 8. Multi-Mirror Telescope. (*Multi-Mirror Telescope Observatory, Smithsonian Institution/ University of Arizona*)

plorer (*IUE*) was launched into geosynchronous orbit in early 1978. It is not a giant telescope; its mirror is only 18 in. (45 cm) in diameter. The satellite hangs over the Atlantic Ocean, where it can be operated in real time from both the United States and Europe to obtain high-dispersion ultraviolet spectra of solar system objects, stars, nebulae, and extragalactic objects. The *Einstein Observatory* (*High Energy Astronomy Observatory 2*), launched in November 1978, carries a telescope which images soft x-rays in the 0.3–5-nm wavelength range. The data that scientists have obtained for stars from the *IUE* and the *Einstein* are revolutionizing understanding of stellar physics.

Future developments. The Space Telescope (ST), which is to be launched into Earth orbit by the space shuttle, is a 2.4-m (94.5-in.) Cassegrain reflector, which will feed light into a number of scientific instruments. The instruments on the space telescope can see objects many times fainter than achieved so far. If the results from the small telescope on the *IUE* combined with the *Einstein* are any clue, the Space Telescope should begin a major revolution in understanding the universe. The scientific instruments flown on the Space Telescope include a widefield and planetary camera, a faint-object camera, a faint-object spectrograph, a high-resolution spectrograph, and a high-speed photometer. After several years of operation the instruments will be changed out in orbit and replaced by innovative instruments. The second round of instruments will probably stress some infrared science. NASA plans to operate the Space Telescope through the remainder of the twentieth century.

Astronomers are also considering the future of ground-based telescopes. The concepts for the "next-generation telescope" look substantially different than the traditional designs of large reflectors. The Multi-Mirror Telescope on Mount Hopkins is clearly the first of a generation of multiple-telescope systems that may approach the light-gathering power of a 1000-in. (25-m) telescope. These telescopes will probably be able to sense the distortion in the light waves from celestial objects caused by the Earth's atmosphere, and move the various mirrors relative to one another to correct for the distortions. In such a system, very high-resolution imaging will also be possible.

ASTRONOMICAL TRANSIT INSTRUMENTS

Astronomical transit instruments are telescopic instruments adapted to the observation of the passage, or transit, of an astronomical object across the meridian of the observer. The astronomical transit instrument is the classic instrument of positional astronomy, which is the study of the positions and motions of astronomical objects and the related determination of positions by observation of these astronomical bodies from the Earth (the specific categories of astronomy

Fig. 9. Large transit instrument, Pulkovo Observatory, Soviet Union. (*Courtesy of B. L. Klock*)

concerned with these investigations are astrometry and celestial mechanics). The chief variants of the classic design include the meridian circle, the vertical circle, the horizontal transit circle, the broken or prism transit, and the photographic zenith tube.

The astronomical transit instrument was first developed by the Danish astronomer Ole Roemer in 1689. The modern transit instrument has a telescopic objective with a diameter of 6–10 in. (15–25 cm) and a focal length of 72–90 in. (180–230 cm). The instrument consists of a telescope mounted on a single fixed horizontal axis of rotation. The horizontal axis has a central hollow cube (sometimes a sphere) and two conical semiaxes ending in cylindrical pivots. The objective and eyepiece halves of the telescope are also fastened to the cube of the instrument, perpendicular to the horizontal axis. Rotation of the instrument in its bearings, or wyes, permits the optical axis to sweep only in the plane of the meridian. An accurate clock is the essential ancillary scale by which the transits of the astronomical objects are observed.

The large transit instrument of Pulkovo Observatory, Soviet Union, in use since 1838, is seen in **Fig. 9**.

Applications. The astronomical transit instrument has three interrelated uses. (1) From a known position on the Earth, observations of transits of stars lead to the determination of their right ascension with respect to the astronomical coordinate system. (2) The determination of corrections to the clock may be made by the observation of stars of known position with an instrument situated at a known longitude. (3) Finally, with a knowledge of the positions of the stars observed and of Greenwich time, the longitude of the observer can be computed from observations of the time of transit of a star.

The astronomical transit instrument takes advantage of a special case of the astronomical triangle which is composed of arcs of great circles on the celestial sphere. Its vertices are, respectively, the north celestial pole, the zenith point of the observer, and the celestial object under observation. The angle at the north celestial pole represents the hour angle of the object; hence when the object is on the meridian, the hour angle is zero, and the triangle degenerates to a single arc, a segment of the meridian. At that instant the local sidereal time equals the right ascension of the celestial object.

A divided circle, graduated into fractions of a degree, is seated on the horizontal axis of the transit instrument and is used to set the instrument at the required zenith distance. The instrument has a fastening clamp also situated on the horizontal axis. The clamp may have a fine-motion adjustment mechanism to improve alignment for an observation.

The micrometer, or eyepiece part of the instrument, contains a movable wire, or pair of wires, and a stationary grid of vertical wires for use in registering the transit. At the Bordeaux, Perth, and Pulkovo observatories, photoelectric observations have been made, thereby replacing the observer and yielding higher-quality observations.

An accurate quartz crystal or atomic clock, with a rate of less than $0\overset{s}{.}001$ per day, is used as the scale for recording the transit data in conjunction with some form of data storage, such as punched tape, a printing chronograph, or a computer.

Corrections. It is extremely difficult to adjust the instrument to the point of perfection, where the mean wire will trace the true meridian as the instrument is rotated on its pivots; therefore corrections must be determined and applied to the observational data. The three principal instrument errors that require correction are azimuth, collimation, and level. These errors do not remain constant even though the instrument may be well constructed and mounted. The principal cause for their change is attributed to variations of temperature in the environment of the pavilion housing the instrument. This temperature is ambient since the roof of the pavilion must be open to make observations. Modern design of transit instruments attempts to utilize new advances in metallurgy to minimize the thermal influence on the instrument.

The azimuth correction is the horizontal angle between the axis of rotation and the true east-west direction. This correction is usually less than a few seconds of arc. It may be determined via observational data made on circumpolar stars in conjunction with artificial stars (marks) located several hundred feet to the north or to the south of the instrument pavilion.

The collimation correction is the angle between the line from the optical center of the telescope objective to the mean wire in the micrometer and the plane perpendicular to the horizontal axis of rotation. If this line intersects the horizon to the west of the south point, the stars appear to transit late and a negative correction is required. The reverse is true for stars which appear to transit early because of this error. The magnitude of this correction is generally no larger than $0\overset{s}{.}010$–$0\overset{s}{.}020$. There are several methods of determining the collimation error. With large, permanently mounted instruments two horizontal collimating telescopes are placed 15–20 ft (4.5–6 m) to the north and to the south of the main instrument. These two telescopes have cross hairs in their focal planes, with a diffuse illumination source behind them. The transit instrument generally has a large hole in its central cube section that can be opened to permit the collimating telescopes to sight through it. The cross hairs of the collimating telescopes are then aligned on each other so that they lie in a plane passing approximately north and south of the transit instrument. The transit instrument is then pointed first toward one telescope and then the other. The position of the cross-hair image is observed in each case, and half the sum of their displacements is the collimation error. The collimation may also be determined by making observations on a distant mark and then reversing the telescope on its pivots and repeating the readings.

The level correction is the angle that the axis of rotation makes with the plane of the horizon. The value of this constant may be determined by observing the reflected images of the cross wires in a mercury horizon, when the telescope is pointed toward the nadir. The mercury basin is placed under the instrument only while the level correction is being determined. The level can also be determined through a striding level supported on a special frame designed to ride on, or hang from, each end of the horizontal axis. The magnitude of this correction is normally no greater than a few seconds of arc.

The clock correction of the transit instrument represents the error between the true sidereal time that the star should transit the local meridian and the time of transit recorded by the local sidereal clock. At one time the clocks had rates which were significant enough to warrant a correction; however, today they have sufficient accuracy so that this correction may be generally regarded as a constant for each night's work. This correction is evaluated through observations on bright stars whose positions are already well known.

Meridian circle. The major astrometrical observatories of the world have astronomical transit instruments called meridian or transit circles. These instruments are similar to the transit instrument previously described, except they have a micrometer eyepiece which has an extra pair of moving wires perpendicular to the vertical set. These wires are used to measure the zenith distance or declination of the celestial object in conjunction with readings taken from a large,

Fig. 10. Six-in. (150-mm) transit circle, U.S. Naval Observatory. (*Official U.S. Naval Observatory photograph*)

accurately calibrated circle attached to the horizontal axis. The circle may be read photographically, photoelectrically, or electronically.

The 6-in. (15-cm) transit circle of the U.S. Naval Observatory (**Fig. 10**) was designed and constructed at the end of the nineteenth century, but it has been improved continuously with the latest technological developments. It is the first transit circle in the world to have an electronic circle and the first to have the micrometer data read directly into an electronic computer. In addition to the electronic circle data, the divided glass circle is scanned photoelectrically with six scanning micrometers. The data from the scanners are entered into the computer for real-time processing. The probable error of a single set of circle readings is $0''07$. However, the declination of a star may contain uncertainties of $0''25-0''50$ due to errors from other sources, such as atmospheric refraction, mechanical flexure of the instrument, and residual errors in the divided circle.

The 6-in. (15-cm) transit circle is used not only to observe the brighter stars (as faint as ninth magnitude) but also the Sun, Moon, planets, and several of the brighter asteroids. These observations are made visually, with the observer seated on a couch with an adjustable back to lend head support during an observation. The probable error of a single observation of the right ascension data of an equatorial star is about $0^s.12$ with the use of the motor-driven micrometer. A computer presets the basic speed of the right-ascension wires and then the observer adds or subtracts to this speed through pushbuttons on a hand keyboard as the celestial object transits the field of view.

The U.S. Naval Observatory has refurbished its 7-in. (18-cm) transit circle with new instrumentation featuring an automatic micrometer, automatic setting, and a circle scanning system similar to the 6-in. (15-cm) transit circle system. The 7-in. transit circle was sent to the southern hemisphere in 1984 in order to improve knowledge of the positions and motions of celestial objects in that part of the sky.

The Tokyo Astronomical Observatory at Mitaka has constructed a new transit circle with automatic tracking and automatic reading of the circle. The circle reading system features a unique charge-coupled-device linear array.

Another configuration for the meridian circle is that of the Carlsbad meridian circle, originally located in Denmark, in which the light from a star on the meridian is reflected by a 45° flat mirror to a horizontal reflecting telescope oriented east-west. A photoelectric micrometer and circle scanning system will complement the configuration. A joint Anglo-Danish meridian circle program sent the modernized Carlsberg Meridian Circle to the Canary Islands in 1984.

The meridian (transit) circle is the instrument used to determine fundamental star positions in most of the world except the Soviet Union. There the philosophy is to separately measure each coordinate, right ascension, and declination, with its own particular kind of telescope. Right ascensions are measured with the transit instrument, and declinations are measured with a vertical circle. The vertical circle is similar to the meridian circle, except that its micrometer contains wires which permit measurement only in the vertical plane. A photographic vertical circle has been constructed at Pulkovo Observatory in Leningrad. With this instrument the observations are made photographically rather than visually. The Pulkovo observatory also has a horizontal mirror transit circle.

Photographic zenith tube. This specially designed telescope is used for the accurate determination of time. The world's largest instrument of this kind (26 in. or 65 cm) operates at the U.S. Naval Observatory (**Fig. 11**). The telescope tube is mounted in a permanent vertical position (**Fig. 12**). Stars passing close to the zenith of the observatory form the selected list of observed objects.

The optical axis of this instrument is folded back upon itself by a pool of mercury, so that the image is formed just below the center of the objective. The objective is designed so that its second nodal point is slightly behind the last glass surface so the focal plane can be brought to this point by raising or lowering the pool of mercury. When the adjustment is accomplished, the time of passage of a star image through the nodal point becomes independent of the tilt of the instrument as the plate carriage is reversed through 180° midway through the observation. This reversal cancels the instrumental errors introduced into the images recorded before and after the reversal. The photographic plate is mounted in the carriage, which is motor-driven across the field at the exact speed of the star image. As the carriage proceeds, the plate is marked at several positions at known clock times. Thus, the position of the star image on the plate, with respect to the time marks, gives directly the correction to the clock.

Most of the major time services in the world, with the exception of the Soviet Union, use the photographic zenith tube (PZT). The time services at the Sternberg Astronomical Institute in Moscow and the Pulkovo Observatory make use of the broken-back, or prism, transit instrument.

Fig. 11. Exterior view of the world's largest photographic zenith tube (26 in. or 65 cm), at the U.S. Naval Observatory. (*Courtesy of T. J. Rafferty*)

Fig. 12. Photographic zenith tube, U.S. Naval Observatory. (*Official U.S. Naval Observatory photograph*)

Fig. 13. Broken-back transit, Sternberg Astronomical Institute, Moscow, Soviet Union. (*Courtesy of B. L. Klock*)

The U.S. Naval Observatory 26-in. (65-cm) photographic zenith tube has a wide field to complement its natural ability to reach to fainter magnitudes.

Broken-back transit. In the broken-back, or prism, transit (**Fig. 13**) the telescope is bent at the axis of rotation by the insertion of a prism at the intersection of the optical and rotational axes. This places the eyepiece at one end of the rotation axis, where it remains in a stationary position except when the instrument is reversed on its bearings. The reversal is usually performed midway through an observation. This aids in the partial elimination of some of the Soviet Union have adopted a special photoelectric micrometer to their broken-back transits which enables them to obtain clock corrections of the order of $0^s.005$ ("s" denotes seconds of time). This is the same order of precision achieved with the photographic zenith tube.

Bibliography. A. N. Adams and D. K. Scott, *Publ. U.S. Nav. Observ.*, 2d ser., vol. 19, pt. 2, 1968; G. Burbidge and A. Hewitt (eds.), *Telescopes for the Nineteen Eighties*, 1981; J. Cornell, Six new eyes peer from Mount Hopkins, *Sky Telesc.*, 58:23–24, 1979; J. Cornell and J. Carr (eds.), *Infinite Vistas: New Tools for Astronomy*, 1985; D. L. Crawford (ed.), *The Construction of Large Telescopes*, 1966; H. Eichorn, *Astronomy of Star Positions*, 1974; First 4-meter photographs from Kit Peak, *Sky Telesc.*, 46(1):10–13; W. C. Gliese et al. (eds.), *New Problems in Astrometry*, 1974; P. D. Hemenway, Washington 6-inch transit circle, *Sky Telesc.*, February 1966; H. C. King, *The History of the Telescope*, 1955, reprint 1979; G. P. Kuiper and B. M. Middlehurst (eds.), *Telescopes*, 1960, reprint 1977; J. B. Oke, Palomar's Hale telescope: The first 50 years, *Sky Telesc.*, 58:505–509, 1979; T. Page and L. W. Page, *Telescopes*, 1966; Photographic report from Kit Peak, *Sky Telesc.*, 45(1):10–17, 1973; A. G. Davis Philips, A visit to the Soviet Union's 6-meter reflector, *Sky Telesc.*, 47(5):290–295, 1974; V. V. Podobed, *Fundamental Astronomy*, 1965; W. T. Powers and R. S. Aikens, Image orthicon astronomy at the Dearborn Observatory, *Appl. Opt.*, 2(2):157–163, 1963; G. M. Sanger and R. R. Shannon, Optical fabrication techniques for the MMT, *Sky Telesc.*, 46(5):280–284, 1973; D. H. Schulte, Auxiliary optical systems for the Kitt Peak telescopes, *Appl. Opt.*, 2(2):141–151, 1963; C. L. Tichenor, Notes on modern telescope mountings, *Sky Telesc.*, 35(5):290–295, 1968; A. B. Underhill, The international ultraviolet explorer satellite, *Sky Telesc.*, 46(6): 377–379, 1973; P. van de Kamp, *Principles of Astrometry*, 1967; E. W. Woolard and G. M. Clemence, *Spherical Astronomy*, 1966; The x-ray eyes of Einstein, *Sky Telesc.*, 57:527–534, 1979.

EYEPIECE
MAX HERZBERGER

A lens or optical system which offers to the eye the image originating from another system (the objective), at a suitable viewing distance. The image can be virtual. *See* OPTICAL IMAGE.

In modern instruments, most eyepieces (also called oculars) are not independently corrected for all errors. They are designed to balance out certain residual aberrations of the objective or (as in the microscope) of a group of objectives, for instance, chromatic difference of magnification. *See* ABERRATION.

The Ramsden eyepiece consists of two planoconvex lenses, the field lens and the eye lens, with their plane sides out. Both of these lenses have the same power and focal length; their separation is equal to their common focal length. The field lens is near the image formed by the objective. The Ramsden acts as a field stop. The Kellner eyepiece is a Ramsden eyepiece with an achromatic eye lens. *See* GEOMETRICAL OPTICS; LENS.

The Huygens eyepiece also consists of two planoconvex lenses, but the plane sides of both lenses face the eye. The focal length of the field lens is in general three times that of the eye lens, and the separation is twice the focal length of the eye lens. The Huygens eyepiece is usually corrected for astigmatism and distortion. A type of Huygens eyepiece in which the eye lens is achromatized to compensate for the color errors of the objective is called a compensating eyepiece. The field lens may also be achromatized.

Huygens and Ramsden eyepieces (see **illus**.) are in general used for low magnifying powers. For higher powers, eyepieces assume more complex forms to correct astigmatism, field curvature, and distortion for larger field angles.

The orthoscopic eyepiece consists of a single lens, made up of three cemented elements, to which a planoconvex lens is added.

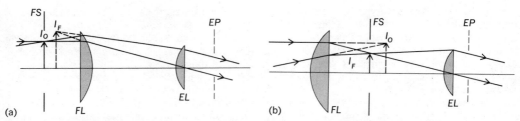

Eyepieces: (a) Ramsden and (b) Huygens. FL = field lens; EL = eye lens; FS = field stop; EP = exit pupil or eye point; I_O = image (real or virtual) formed by the preceding system; I_F = image (virtual or real) formed by the preceding system and the field lens.

Complex types of oculars enable corrections to be made over a field angle of 70° or more on the image side. They are used especially in military instruments.

Bibliography. P. F. Carlson (ed.), *Introduction to Applied Optics for Engineers*, 1977; D. F. Horne, *Optical Instruments and Their Applications*, 1980; R. Kingslake (ed.), *Applied Optics and Optical Engineering*, vol. 3: *Optical Components*, 1966; Optical Society of America, *Handbook of Optics*, 1978; M. Young, *Applied Optics*, 1977.

SCHMIDT CAMERA
Jay M. Pasachoff

A wide-field telescope that uses a thin aspheric front lens and a larger concave spherical mirror to focus the image **(Fig. 1)**; it is also known as a Schmidt telescope. The German optician Bernhard Schmidt devised the scheme in 1931. The field of best focus is located midway between the lens and the mirror and is curved convexly toward the mirror, with a radius of curvature equal to the focal length. Usually film or photographic plates are bent to match this curved focus. With shorter focal lengths, a field-flattening lens may be used. These telescopes are known as Schmidt cameras because they are always used photographically, and no focus accessible to the eye is provided in basic Schmidts. *See* Geometrical optics; Lens; Mirror optics.

Schmidt telescopes are very fast, some with focal ratios in the vicinity of f/1. Thus they are sensitive to objects of low surface brightness. Schmidt telescopes have no coma; because the only lens element is so thin, they suffer only slightly from chromatic aberration and astigmatism. The front element is sometimes known as a corrector plate. Often the outer surface of the corrector plate is plane, with the inner surface bearing the figure. This corrector plate reduces the spherical aberration severely, giving extremely sharp images. The Schmidt design became widely used after 1936, when the secret of the fabrication of the corrector plate was released. Schmidt is credited as much with his skill in figuring the fourth-degree curves on the corrector plate as with the design itself. The largest Schmidt cameras **(Fig. 2)** are listed in the **table**. *See* Aberration; Optical surfaces.

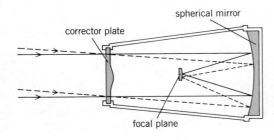

Fig. 1. Cross section of Schmidt camera with aspherical corrector plate. (After J. M. Pasachoff, *Astronomy: From the Earth to the Universe*, 2d ed., Saunders College Publishing, 1983)

OPTICAL IMAGING SYSTEMS

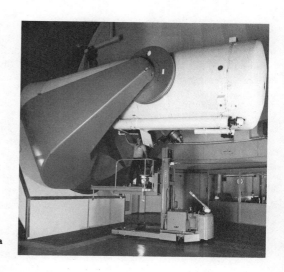

Fig. 2. European Southern Observatory Schmidt in La Silla, Chile. (*European Southern Observatory*)

Sky surveys. The Schmidt camera at the Palomar Observatory in California was used in the 1950s to survey the northern two-thirds of the sky in the Palomar Observatory–National Geographic Society Sky Survey (POSS). The plate pairs—one in the red and one in the blue—provide a first-epoch coverage on which many nebulae, galaxies, clusters of galaxies, and other objects were discovered. Each plate, about 14 in. (35 cm) square, covers an area about 6° square. The POSS is a basic reference in most observatories. It can be obtained as glass, film, or paper copies, often as negatives to preserve detail. A smaller Schmidt telescope (18 in. or 46 cm) at the Palomar Observatory has discovered many asteroids with orbits that cross that of the Earth.

The European Southern Observatory (ESO) Schmidt in Chile and the United Kingdom Schmidt in Australia have been used to continue the survey to the southern hemisphere, with improved emulsions to survey from declination −20° to the south pole. The project is jointly conducted by the European Southern Observatory and the Science Research Council of the United Kingdom. The European Southern Observatory telescope carries out the blue survey, and the

The largest Schmidt cameras

Telescope or institution	Location	Diameter		Focal ratio	Date completed
		Corrector plate, in. (m)	Spherical mirror, in. (m)		
Karl Schwarzschild Observatory	Tautenberg (near Jena), East Germany	53 (1.3) (removable)	79 (2.0)	f/2	1960
Palomar Observatory	Palomar Mountain, California	48 (1.2)	72 (1.8)	f/2.5	1948
United Kingdom Schmidt, Siding Spring Observatory	Warrumbungle National Park, New South Wales, Australia	48 (1.2)	72 (1.8)	f/2.5	1973
Tokyo Astronomical Observatory	Kiso Mountains, Japan	41 (1.1)	60 (1.5)	f/3.1	1975
European Southern Observatory Schmidt	La Silla, Chile	39 (1.0)	64 (1.6)	f/3	1972

United Kingdom telescope carries out the red survey. The survey includes 606 blue and 606 red plates to cover the one-quarter of the sky that cannot be reached by the Palomar Schmidt. The United Kingdom Schmidt has also completed an infrared survey of the Milky Way and Magellanic Clouds, and has undertaken a blue-red study of the region just south of the celestial equator to provide images showing stars about 1.5 magnitudes fainter than those on the Palomar survey. In addition to mapping, the plates are used to help make optical identifications of southern radio and x-ray sources. Study of the plates has been speeded by the COSMOS automatic plate-measuring machine.

A new corrector plate designed for the Palomar Schmidt has allowed a second-epoch sky survey to be undertaken in the blue, the red, and the infrared. Advances in film technology will allow a fainter limiting magnitude to be reached.

Schmidt-Cassegrain design. Many amateur astronomers use telescopes of the Schmidt-Cassegrain design, in which a small mirror attached to the rear of the corrector plate reflects and refocuses the image through a hole in the center of the primary mirror. These telescopes, often in sizes of 5 in. (12.5 cm), 8 in. (20 cm), and 14 in. (35 cm), are portable and relatively inexpensive for the quality of image. Their fields are much narrower than a standard Schmidt design, however. Schmidt cameras are also available for amateurs.

Related systems. The success of the Schmidt design has led to many other types of catadioptric systems, with a combination of lenses and mirrors. The desire for a wide field has led to most modern large telescopes to be built to the Ritchey-Chrétien design instead of the traditional paraboloid, though these fields are perhaps one-third the diameter of those of Schmidts.

Schmidt optics are often used in microscopes and in projection televisions. SEE OPTICAL TELESCOPE.

Bibliography. British astronomers look south, *Sky Telesc.*, 64:543, December 1982; G. B. Kuiper and B. Middlehurst (eds.), *Stars and Stellar Systems*, vol. 1: *Telescopes*, 1960; J. J. Labrecque, Testing a Schmidt corrector at a finite distance, *Sky Telesc.*, 37:250–251, April 1969; The largest Schmidt's first 20 years, *Sky Telesc.*, 62:554–557, December 1981; D. Overbye, Exploring the southern sky, *Sky Telesc.*, 58:30–36, July 1979; J. M. Pasachoff, *Contemporary Astronomy*, 3d ed., 1985; R. M. West, ESO Schmidt photographs of the southern sky, *Sky Telesc.*, 48:225–229, October 1974.

GUNSIGHTS
EDWARD K. KAPRELIAN

Optical instruments which establish an optical line or axis for the purpose of aiming a weapon. The axis includes the observer's eye, a suitable mark in the instrument, and the target. Most gunsights employ as their basis either a telescope or a partially reflective mirror.

Rifle sights. A typical rifle sight (**Fig. 1**) consists of a terrestrial telescopic system having an objective, an eyepiece, an erector lens, and a reticle. Sometimes a field lens is employed to ensure uniform illumination. The magnifying powers customarily range from unity to 12 diameters, magnifications of 2½ to 9 being most common. Instruments are available in which the power may be varied optically over a wide range to suit existing conditions. Large eyepieces are used to provide a wide field of view and a substantial eye relief, the latter so that the eyepiece will not strike the user's eye in recoil. In order to adjust for elevation or windage, either the reticle may be moved, usually by means of screws, or the entire sight may be tilted relative to the gun barrel. Machine-gun sights are also telescopic in nature and customarily employ a prism for erecting the

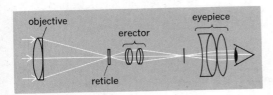

Fig. 1. Rifle sight.

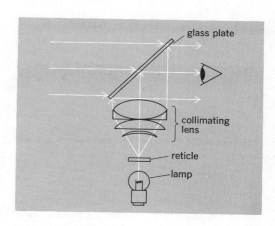

Fig. 2. Reflex sight.

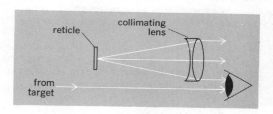

Fig. 3. Collimator sight.

image. Magnifying powers range between 1½ and 3, and the reticle may be illuminated by a small lamp to permit night use. See EYEPIECE.

Aircraft gunsights. These are usually of the reflector type (also known as reflex sights) and employ in their simplest form a lamp, a reticle, a collimating lens, and a glass plate or partially reflecting mirror (**Fig. 2**). The collimator images the reticle pattern at infinity, and the mirror superimposes this image over the target area. The collimator's pupil is so large that the eye can be positioned at any point within a large cylindrical volume in back of the sight. The collimating system may employ a simple spherical mirror, a Schmidt lens, or a Mangin mirror. A Mangin mirror consists of a negative meniscus lens, the shallower surface of which is silvered to act as a spherical mirror, while the other surface corrects for the spherical aberration of the reflecting surface. The magnification of this type of sight is unity. See MIRROR OPTICS.

Artillery sights. These can assume various forms, the simplest of which is the collimator sight (**Fig. 3**), consisting of an objective having a reticle at its focus. When the eye is so placed as to receive light simultaneously from the target and the reticle, the latter appears superimposed on the former and a line of sight is established. Other forms of artillery sights include the elbow telescope, which is a conventional terrestrial telescope containing a prism for producing a right-angle bend, and the panoramic sight. For a discussion of panoramic sights SEE PERISCOPE. SEE ALSO LENS.

Bibliography. D. F. Horne, *Optical Instruments and their Applications*, 1980; Optical Society of America, *Handbook of Optics*, 1978; W. J. Smith, *Modern Optical Engineering*, 1966.

RANGEFINDER

EDWARD K. KAPRELIAN

An optical instrument for measuring distance, usually from its position to a target point. Light from the target enters the optical system through two windows spaced apart, the distance between the windows being termed the base length of the rangefinder. The rangefinder operates as

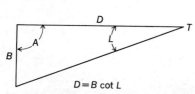

Fig. 1. Range triangle; symbols explained in the text.

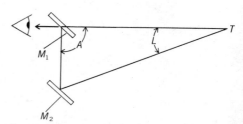

Fig. 2. Simple coincidence rangefinder; symbols explained in the text.

an angle-measuring device for solving the triangle comprising the rangefinder base length and the line from each window to the target point. Rangefinders can be classified in general as being of either the coincidence or the stereoscopic type.

Coincidence rangefinders. In these types, one-eyed viewing through a single eyepiece provides the basis for manipulation of the rangefinder adjustment to cause two images or parts of each to match or coincide. This type of device is used, in its simpler forms, in photographic cameras. The range triangle for such a rangefinder is shown in **Fig. 1**, where B is the base, angle A a right angle, D the range, and L the convergence angle at the target T. The relationship is given by Eq. (1). The basic optical arrangement is shown in **Fig. 2**, where M_1 and M_2 are a

$$D = B \cot L \tag{1}$$

semitransparent mirror and a reflecting mirror, respectively. When coincidence is obtained, that is, when the target T is seen in the same apparent position along either path, the rangefinder equation is satisfied. In many small rangefinders coincidence is achieved by rotating mirror M_2 through a small angle while viewing the images in mirror M_1; other deviating means which do not require rotating of mirror M_2 are also employed. **Figure 3**a and b shows the appearance of the images in a superposed field rangefinder. If mirror M_1 is made fully reflecting and arranged to fill only the lower half of the field of view, allowing the eye to view target T directly through the upper half, a split-field type of rangefinder is produced; the images appear as shown in **Fig. 4**a and b.

Camera rangefinders. These usually have base lengths varying from less than 1 in. (25 mm) to more than 3 in. (75 mm), with magnifications between 0.5 and 2.5. The lower powers are found in arrangements where the rangefinder and viewfinder are combined for viewing through a single eyepiece; here a magnification of unity or less is necessary to cover the usual 42–52° angle of view.

Military types. Coincidence rangefinders of the military type ordinarily employ a pentaprism, or its two-mirror equivalent, at each window. These prisms, permanently fixed in place,

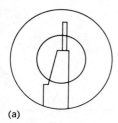

(a)   (b)

Fig. 3. View in superposed field rangefinder. (a) Images in coincidence. (b) Images unmatched.

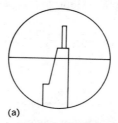 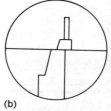

(a)   (b)

Fig. 4. View in split-field rangefinder. (a) Images matched. (b) Images unmatched.

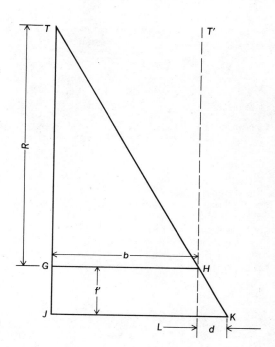

Fig. 5. Range triangle for military coincidence rangefinder; symbols explained in text.

reflect the light rays inward at an angle exactly 90° to the angle of incidence. The light from each prism passes through a telescope objective, and an image of the target formed by each objective is combined and directed by a prism assembly in a single eyepiece for simultaneous viewing. **Figure 5** shows the geometry to such a rangefinder. Here b is the base length of the rangefinder, and f' is the focal length of its objective lenses. Light from an infinitely distant target would arrive along lines TG and $T'H$, while light from a target distance R would arrive along lines TG and TK. The objectives, located at G and H and with their axes on lines TG and TH, respectively, form images of the target T at J and at K, respectively. With an infinitely distant target, objective H would form its image at L rather than at K. This displacement d, equal to LK, is called the parallactic displacement and, since triangles TGH and HLK are similar, is the measure of range. The relationship is given by Eq. (2).

$$R = \frac{bf'}{d} \qquad (2)$$

The optical arrangement of a typical military coincidence rangefinder is shown in **Fig. 6**. The rangefinder is adjusted so that if the object being viewed is at infinity (dotted line path through objective H), both images of the object are seen exactly superimposed. If the object is closer than infinity, the axis of the image which is formed by objective H is displaced by a deviation means, such as prism D, in the manner and amount described by Eq. (2), so that superimposition is achieved. The amount of such displacement is shown on the scale as the distance reading. Other means for obtaining deviation include oppositely rotating wedges (diasporometer), two equal coaxial oppositely arranged prisms with variable spacing, and slidable positive and negative lenses (swing wedge).

Stereoscopic rangefinders. These are entirely different, although externally they resemble coincidence rangefinders except for the fact that they possess two eyepieces. Both eyes are used with this instrument, rangefinding being accomplished by stereoscopic vision. It is essen-

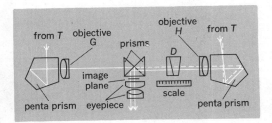

Fig. 6. Military coincidence rangefinder.

tially a large stereobinocular fitted with special reticles which allow a skilled user to superimpose the stereo image formed by the pair of reticles over the images of the target seen in the eyepieces, so that the reticle marks appear to be suspended over the target and at the same apparent distance. In the typical stereoscopic rangefinder (**Fig. 7**) the objectives form their images in the plane of their respective reticles. A diasporometer in one half of the instrument is used to vary the angle between the beams reaching the eyes from the target until it equals that between the beams reaching the eyes from the two reticles, at which point the reticle image appears to be at the same distance as the target. SEE STEREOSCOPY.

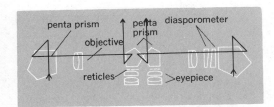

Fig. 7. Stereoscopic rangefinder.

Rangefinder errors. The accuracy of a rangefinder depends upon the accuracy of the eye in judging the coincidence of two lines, as well as upon the characteristics of the instrument and the range distance. Rangefinder errors are proportional to the square of the range and inversely proportional to the base length and to the magnification. A 1.1-yd (1-m) base and a 3.3-yd (3-m) base rangefinder of 15 power would have errors of 7.3 and 2.7 yd (6.7 and 2.5 m) at 1000-yd (914-m) range and 470 yd (430 m) and 170 yd (155 m) at 8000-yd (7315-m) range, respectively.

Heightfinders. These are modified forms of rangefinders originally used in the directing of antiaircraft fire. A heightfinder depends for its operation upon the knowledge of the range R to the aircraft and its angle of elevation θ. The height is given by Eq. (3).

$$H = R \sin \theta \tag{3}$$

In practice, one way that the height may be measured is by multiplying range and elevation angle in a gear train computer. Another method employs an additional pair of deviation prisms within the rangefinder, geared so as to produce a deviation which varies as $\sin \theta$.

Depression rangefinders. These are used in coast defense positions, where such instruments can be installed on a cliff top, and this elevation is used as the rangefinder base. The device measures the depression angle of the target, which it converts directly to range after including necessary corrections for tide, curvature of the Earth, and refraction of the atmosphere.

Stadia methods of rangefinding. These are suitable for some applications. If the size of the object is known or can be guessed accurately, its distance can be determined by measuring its size in the eyepiece reticle of a telescopic system. SEE LENS; OPTICAL PRISM.

Bibliography. D. F. Horne, *Optical Instruments and Their Applications*, 1980; Optical Society of America, *Handbook of Optics*, 1978.

BINOCULARS
Harold E. Rosenberger

Optical instruments designed for use with both eyes to give enhanced views of distant objects. The distinguishing performance feature of binoculars, compared to monocular instruments, is the depth perception obtainable.

When an object is viewed with both eyes, the distance of approximately 2.5 m (65 mm) between the eyes requires each to view the object from different angles. To bring the images onto the fovea centralis of each eye, the eyes rotate inward until the lines of sight converge at the object. This convergence is the signal to the brain which stimulates the sense of depth, and its magnitude is the parameter from which judgment of depth is made. When two objects fall within an observer's view but are located at different distances, their respective convergence angles differ, and this difference makes the observer aware of the difference in distances. *See* Stereoscopy; Vision.

The minimum difference between convergence angles which the average adult can detect is about 30 seconds of arc, which is approximately the value corresponding to an object at 500 yd (450 m). The convergence angles for more distant objects are less, and therefore objects beyond 500 yd (450 m) cannot be resolved stereoscopically.

With binoculars this range can be greatly extended. Binoculars employing prism-erecting systems (see **illus**.) can be so constructed that the distance between objectives is twice that

Modern prism binocular. (*Bausch and Lomb Optical Co.*)

between the eyepieces. Hence, if the convergence angle at an object subtended by the unaided eyes is α, the angle subtended by the binocular objectives is 2α. This angle in the real field of the binocular becomes $2M\alpha$ in the apparent field, M being the magnification of the binocular. Thus the minimum convergence angle in the real field becomes 30 seconds/$2M$, and the maximum range at which objects can be resolved stereoscopically, using, for example, a seven-power binocular, is extended to about 7000 yd (6300 m). *See* Optical prism.

To ensure freedom from fatigue and eyestrain for the observer, the optical axes of the binocular halves must be aligned parallel to each other within a few minutes of arc. A complicating factor in the maintenance of this parallelism is the hinge mechanism joining the two halves, which is provided to allow adjustment of the intereyepiece distance to match the distance between the observer's eyes. The accuracy to which the parts must be made and the requirement of absolute stability of parts after assembly to provide and maintain parallelism largely account for the higher cost of quality binoculars compared to monocular telescopes. *See* Eyepiece; Magnification.

Bibliography. American Institute of Physics, *The Binoculars*, 1975; D. F. Horne, *Optical Instruments and Their Applications*, 1980.

OPTICAL TRACKING INSTRUMENTS
GEORGE A. ECONOMOU

A family of optical instruments used for precise time-correlated observation of distant airplanes, missiles, and artificial satellites, all of which travel at apparent velocities much greater than those of most astronomical objects. The instruments supply permanent engineering records for the determination of spatial position, missile attitude, structural behavior, and performance of specific mechanisms. These observations enable engineers to correct design, improve performance, and collect scientific data from missiles and satellites at extreme distances and altitudes.

Optical tracking instruments were used initially for photographic recording of distant engineering events and for the determination of their spatial position. Tracking telescopes were the basic engineering event-recording systems, while cinetheodolites and ballistic cameras were used for precise spatial position. Electrooptical sensors and computer developments have expanded the family of optical tracking instruments to include television and laser detection and tracking of objects out to geosynchronous orbits. The proliferation of sensors has led to the development of flexible optical tracking instruments (**Fig. 1**) capable of accommodating a variety of photographic and electrooptical sensors.

These sensors have been combined with computer controllers to develop automatic tracking instruments offering both event and position functions in one instrument. The further addition of laser ranging capability permits single-station position solutions along with tracking telescope engineering event recording.

Spatial position determination. This can be considered the determination of the position of a moving target using dynamic adaptions of the methods of civil surveying. The classical techniques utilize a minimum of four instruments on precisely measured baselines to locate a

Fig. 1. KINETO Model 433 Tracking Mount shown with 100- and 200-in. focal-length (2.5- and 5.1-m) motion picture cameras, both with 12-in. (30-cm) aperture, and a 40–240-in. focal-length (1.0–6.1-m), 6-in.-aperture (15-cm) television camera. The fourth platform is reserved for either a laser ranger or a thermal imaging system. (*Contraves Goerz Corp.*)

Fig. 2. Contraves Skytrack cinetheodolite with 45- and 120-in. (1.2- and 3.0-m) focal-length, 12-in.-aperture (30-cm) objective. (*Contraves AG*)

moving target. Each instrument records the data for computing the direction of the line of sight to the target for each instant of time. This information, in the form of analog or digital elevation and azimuth angles, the times of observation, and the known location of each instrument, is used to triangulate for the location of the missile as a function of time.

Most of the spatial position work is performed by cinetheodolites (**Fig. 2**), which are surveying theodolites having 35-mm motion picture cameras with 45–120-in. focal-length (1.2–3-m) lenses substituted for the surveyor's eye and telescope. The system of cameras is synchronized up to a maximum of 30 frames per second from a master control station for simultaneous exposure as the cinetheodolites follow the moving missile. Each photograph records the elevation angle, the azimuth angle, the missile image, and the reticle lines which define the instrumental axis (**Fig. 3**). The angles are measured with optically graduated circles accurate to 2 seconds of arc. The tracking error (the difference between the instrumental line of sight and the line of sight to the missile) is determined from the missile photograph. The tracking error is then used to determine the missile line of sight from the recorded instrumental angles. Photography of known targets and precise leveling are used to establish the relationship between the instrument angles and the test coordinate system.

During the procedure, the operator follows the missile visually through a 5–20-power sighting telescope, guiding the cinetheodolite by a servo control stick as required to keep the cinetheodolite axis pointed at the missle. Tracking can also be controlled by slaving to a remote radar, by a preprogrammed computer, or by a television automatic tracker. Laser ranging is being added to the instrument for single-station laser-radar operation.

A minimum of three cinetheodolites is used to ensure optimum triangulation as the missile instrument geometry changes with missile motion. The accuracy of cinetheodolite systems under

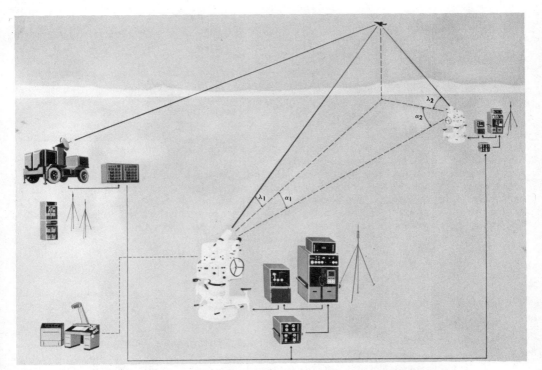

Fig. 3. Cinetheodolite triangulation. The cinetheodolites are on a precisely surveyed baseline. The missile is not precisely on the optical axis. The angles α and λ, therefore, must be corrected for this offset before triangulating for position of missile.

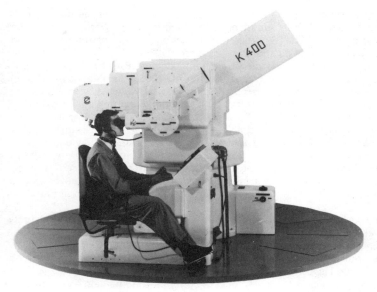

Fig. 4. K-400 cinetheodolite. (*Contraves AG*)

field conditions is 5–15 seconds of arc. The degradation of the 2-second laboratory accuracy is the result of thermal and dynamic deformation under field conditions.

The necessity for higher servo performance to allow computer-controlled and automatic tracking led to the development of the Contraves 12-in.-aperture (300-mm) Skytrack and the 16-in.-aperture (400-mm) K-400 cinetheodolites (**Fig. 4**). Smoother and more precise control is achieved with the direct-drive torque motors. This permits infrared or television automatic tracking in addition to computer or one-person tracking control. The 1-second accuracy and the aperture of the K-400 make it the most accurate telescope-theodolite in existence.

Ballistic cameras. These are fixed-axis, wide-angle, photographic-plate cameras capable of more precise spatial position determination by recording on one plate multiple exposures of the missile against a stellar background. Use of a static system and precisely cataloged star positions decreases necessity for long-term mechanical stability and accuracy, allowing ballistic cameras to achieve 2–5 seconds' angular accuracy.

Pyrotechnic flares of electronic stroboscopic lamps at the missile are used to indicate the missile positions against the night sky. The image of each flare is measured with respect to the surrounding stars. The use of lenses of less than 10-micrometer distortion allows the lines of sight from camera to missile to be determined with 25 reference stars. The lines of sight are used to determine the missile positions by methods similar to those used with cinetheodolites.

The high accuracy of ballistic cameras is the result of the static mode of operation. The camera shutter remains open for the entire time of passage of the missile across the field of view. The photographing on a single plate of more than 100 missile points against the star field permits precise position velocity and acceleration measurements without degradation by the flexure, mislevel, and vibration of tracking motion. The location of both the stars and missile outside the Earth's atmosphere decreases the effect of geometric distortion caused by the uncertainty of atmospheric refraction corrections.

The 20-in. focal-length (500-mm) F/1 Baker-Nunn camera, designed to photograph satellites against background reference, represents the ultimate in photographic ballistic camera design.

For satellite tracking, the ballistic cameras have been replaced by the electrooptical GEODSS (Ground Based Electro-Optical Deep Space Surveillance) system (**Fig. 5**). This system, with a 40-in.-aperture (1-m) F/2 telescope, uses a Silicon Intensified Target Vidicon to detect satellite motion relative to the stellar background. With 10-s exposures, the system can detect

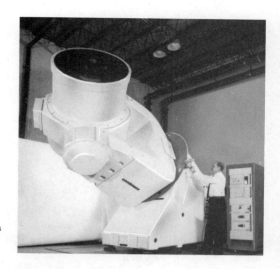

Fig. 5. Ground-Based Electro-Optical Deep Space Surveillance System (GEODSS) tracking mount with 40-in.-aperture (1-m), f/2 optical system. (*Contraves Goerz Corp.*)

18.5-magnitude objects within its 2° field. This is equivalent to a 40-in. (1-m) object in a geosynchronous orbit (at an altitude of 22,000 mi or 36,000 km). The system operates in the sidereal mode for surveillance, but changes to the track mode after acquisition. The GEODSS system has the obvious advantage of real-time output and more rapid coverage of the sky than would be possible with photographic techniques.

Tracking telescopes. These are long-focal-length telescopes mounted to track missiles in flight precisely while collecting missile performance data. The first systems were crude attempts to track manually with 35-mm cameras of 12–24-in. (30–60-cm) focal length. Increased focal length led to the use of geared, manually driven naval gun mounts and variable-speed, belt-driven machine gun mounts with the telescopes substituted for the armament. In all such systems, the tracking operator observes the missile through an optical sight while controlling the orientation of the telescope to ensure that the missile remains within its field.

The requirements for precise tracking by heavy, long-focal-length telescopes led to the development of complex telescope mounts such as the ROTI MK II, a 24-in.-aperture (60-cm) 100–500-in. focal-length (2.5–13-m) telescope with automatic focus, automatic exposure control, and a 10–60 frame-per-second Photosonics 70-mm camera for missile photography. The 3000-lb (1360-kg) telescope, carried by a 9000-lb (4080-kg) mount, is driven by a geared electrohydraulic servo system. In the normal aided-tracking mode of operation, the operator observes the missile through a 20-, 30-, or 40-power tracking sight, and controls the motion of the telescope by exerting light finger pressure on a control knob. The smooth control is necessary if the system resolution of 0.5 second is not to be lost because of relative motion of the image during the exposure.

The telescope and mount are shielded from the thermally disturbing sunlight by a dome. The assembled mount is located on a 20-ft (6-m) tower to prevent degradation of the resolution by the thermally disturbed air immediately over the terrain under the line of sight. A Thor missile photographed at 35 mi (56 km) with this telescope is shown in **Fig. 6**.

The advent of computer control and electrooptical automatic tracking systems has resulted in the adaption of direct-drive torque motor systems to replace the electrohydraulic systems. The dual 24-in. (60-cm) telescope DOAMS (Distant Object Attitude Measurement System; **Fig. 7**) is the most advanced instrument of this type. This combination of 100- and 200-in. focal-length (2.5- and 5-m) systems with 0.6 arc-second digital angle encoders delivers both engineering event recording and spatial position determination at the White Sands Missile Range.

Satellite optical tracking. Satellites equipped with retroflectors have been placed in orbits at altitudes which minimize gravitational and atmospheric anomalies. Satellites, such as Lageos with a 3700-mi (6000-km) orbit, are the basis of laser ranging measurements which require precise optical tracking to be effective.

Fig. 6. Thor missile photographed at 35 mi (56 km) using 24-in.-diameter (61-cm), 500-in.-focal-length (12.7-m) tracking telescope, driven by electrohydraulic servo system. (*U.S. Air Force*)

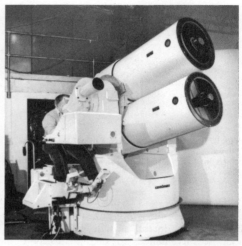

Fig. 7. Distant Object Attitude Measuring System (DOAMS), with 24-in.-aperture (61-cm), 100- and 200-in. focal-length (2.5- and 5.1-m) telescopes for 70-mm high-speed photography. (*Contraves Goerz Corp.*)

Fig. 8. Mobile Optical Measuring System (MOMS), a transportable 30-in.-aperture (76-cm) laser ranging system. The laser is housed in the clean room with a series of mirrors relaying the light to the transmitting optics. (*Contraves Goerz Corp.*)

Fig. 9. The 60-in.-aperture (1.5-m) lunar and satellite ranging system. (*Australian National Map Service*)

The orbit of the satellite becomes the basis of measurements because of its regularity. Operation requires that the optical tracking systems must direct a 10-arc-second laser beam along the predicated laser path. Hundreds of laser ranging measurements are then made from two locations. The satellite ephemeris then becomes the yardstick for measuring the location relative to other measured points.

This technique is used by systems, such as MOMS (Mobile Optical Measurement System; **Fig. 8**), in crucial dynamic studies. The five MOMS units are transported to seismically active areas for the measurements of the motion of tectonic plates which are believed to be one of the causes of earthquakes. The tectonic plates move between 2 and 8 in. (5 and 20 cm) per year. The same technique of laser ranging to retroreflectors on satellites is also used for geodetic and geophysical studies of polar motion, Earth rotation, gravimetric and tide models, and precise geoid determination.

A 60-in.-aperture (1.5-m) system (**Fig. 9**) is used by the Australian National Map Service for both satellite and lunar ranging. The main telescope is used as a common receiver and transmitter. Lunar ranging requires that a laser beam of 2-arc-second divergence be used, which requires corresponding precision in the optical tracking. SEE CAMERA; LASER; LENS; OPTICAL TELESCOPE.

CAMERA
JEROME P. O'NEILL, JR.

The basic instrument of photography: in its simplest form, a light-tight box with a lens at one end, which forms an image of the subject on light-sensitive film at the other end. A pinhole may be substituted for the lens, but forms an unsharp image. Nearly all modern cameras contain a glass or plastic lens to focus a sharp image of the subject onto the film; a shutter to control the length of time the film is exposed, normally a fraction of a second; a viewfinder, to show the subject being photographed, and a mechanism to hold the film and change film between exposures. Many cameras include a lens aperture control, to regulate image brightness on the film. Advanced cameras may accept a variety of lenses for different photographic effects. Many accessories are commonly used: flash units, exposure meters to measure light, filters to control color rendition, tripods, and close-up focusing attachments.

Focusing. To make a sharply focused photograph, the camera lens must be of good optical quality, producing a sharp image. Equally important, the image must be accurately focused on the film.

Even when sharply focused, no lens is perfect. The image of a tiny point of light, such as a star, will be not a point but a small, blurred disk called a circle of confusion. But the human eye is also imperfect; if the blurred circle is small enough, it will be seen as a point. For most people, a circle of $1/100$ in. or less, seen from 10 in. (25 cm) away, will appear to be a sharp point. Thus if a lens can produce photographs with this degree of resolution, it is sharp.

Focusing is done by moving the lens toward or away from the film. The lens is closest to the film when focused on distant subjects, for instance, landscapes, which are said to be at optical infinity. For most normal focal-length lenses, infinity may mean anything farther than 100 ft (30 m) away. Focusing on subjects closer than infinity requires moving the lens away from the film, either by means of a helical-thread lens mount (almost universal for 35-mm cameras), or by moving the entire camera front with the lens (as with view cameras). SEE FOCAL LENGTH.

Focusing is very straightforward with view and reflex cameras: the lens is adjusted for maximum sharpness of the ground-glass image. For cameras with no ground-glass screen, indirect means must be used. The simplest is a focusing scale on the camera body or lens mount, listing distances in feet or meters. The photographer measures or estimates the distance to the subject, then sets it on the scale, which focuses the lens for that distance. Some amateur cameras use pictorial symbols instead of numbers: a head-and-shoulders symbol for close-ups at 3 ft (1 m); mountains, for landscapes at infinity.

The most accurate indirect focusing system is the optical rangefinder. This is usually built into the top of the camera, with an eyepiece in the back. The photographer looks through this while adjusting the focus and sees the subject split into a double image by a prism system. The prisms are adjusted so that when the two images merge, the camera lens is focused sharply. SEE RANGEFINDER.

Depth of focus is sometimes used as a synonym for depth of field. However, the correct optical definition for depth of focus is different: it is the range of sharp focus at the film plane, *within* the camera. This range may be only a few thousandths of an inch, which is why the film must be positioned accurately and held flat. SEE GEOMETRICAL OPTICS.

Finders. Cameras need a viewing device to show what is being photographed. With view and reflex cameras, the image is shown on the ground-glass screen. For other cameras, a viewfinder must be added. Some early cameras simply used a V-shaped line on the top of the camera for the photographer to sight along. Almost as simple is the wire-frame finder or peep sight, still sometimes used for sports photography (and called a sportsfinder) because it gives a clear view of action outside the actual picture area. Most often, optical viewfinders are used. They provide a large, bright image when the photographer looks through an eyepiece in the back of the camera. Rangefinder cameras usually combine rangefinder and viewfinder; the double image for focusing appears in the center. Since any nonreflex viewfinder must be slightly separated from the camera lens, there is always some parallax error—not usually a problem except for close-ups. Some viewfinders correct for parallax.

Shutters. Any device that allows light to reach the film for a controllable length of time can be a shutter. Early photographers simply removed the lens cap, then replaced it, to time the long exposures they needed. To make shorter exposures, many shutters have been devised which open briefly. Some have been operated by pneumatic bulb, some by rubber bands; later types usually work by spring tension.

Two types of shutters are manufactured. Cameras with noninterchangeable (fixed) lenses usually have shutters located between the lens elements; these are known as central or between-the-lens shutters. They may have speeds as fast as $1/500$ s. To avoid having a shutter in every lens, cameras with interchangeable lenses often build the shutter into the camera body instead. Placed directly in front of the film plane, these focal-plane shutters often have speeds to $1/1000$ or $1/2000$ s. Most shutters permit manually timed long exposures with a T (time) or a B (bulb) setting. All current shutters have flash synchronization contacts built in.

Conventional shutters work mechanically. Setting the speed adjusts a tiny clockwork timing mechanism powered by springs; winding the film advance lever cocks the springs (some cameras have separate cocking levers). Pressing the release button trips the shutter open; it stays open the set time, then closes. Later designs replace the clockwork with miniature electronic timing circuits. These can easily be connected to a built-in light meter for automatic exposure control. Electronic shutters can also make long time exposures, if desired.

TELEPHOTO LENS
MAX HERZBERGER

A photographic lens system specially designed to give a large image of a distant object in a camera of relatively short focal length. A telephoto lens generally consists of a positive lens system and a negative lens system, separated by a considerable distance. If color correction is desired, each of the partial systems must be color-corrected. It is usually not easy to correct distortion in a teleobjective, but occasionally it has been achieved. SEE FOCAL LENGTH.

Teleobjectives of very long focal length with infrared correction are important for aerial reconnaissance, where reduction of the back focus is necessary because of space. SEE CAMERA; LENS.

ZOOM LENS
MAX HERZBERGER

A system of lenses in which two or more parts are moved with respect to each other to obtain a continuously variable focal length and hence magnification, while the image is kept in the same image plane. SEE FOCAL LENGTH.

OPTICAL IMAGING SYSTEMS

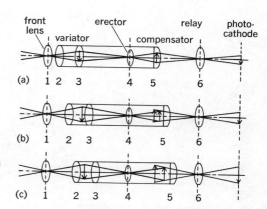

Zoomar lens in three operating positions. (*a*) Wide-angle. (*b*) Medium-angle. (*c*) Telephoto. Lens elements 2, 3, and 5 are mounted in movable barrel connected to zoom handle. Lens elements 1, 4, and 6 are stationary, mounted in lens housing. (*After D. G. Fink, ed., Television Engineering Handbook, McGraw-Hill, 1957*)

If the diaphragm is opened at the same time so that its linear opening increases with the focal length, it is possible to keep the relative aperture of the whole system constant while the focal length varies. In any case, the system must be constructed so that the errors do not vary too much in the shifting. Thus the designer must strive for a design in which the image errors are small, at least for the beginning and end positions of the "zooming" procedure, and do not become too large for intermediate positions.

In general, a complicated cam is needed to control the motions of the parts of the system. However, it is possible to simplify the mechanism if the focal plane is not kept precisely constant, but only required to coincide exactly with a given plane for several focal lengths while approximately coinciding for others. Such a system is called an optically compensated varifocal system.

Some early variable-focal-length lenses contained 15 or more separated elements, but this number has been reduced to as few as 4. The zoom ratio (the ratio of maximum to minimum power) is in general 3:1, but it has been possible in some designs to increase the range to 4:1 or even more. Moving sets of achromatic prisms are sometimes used to achieve the zooming effect.

The Zoomar lens, a variable-focal-length lens, is shown in the **illustration** in three operating positions. The focal length is determined by the television camera operator by means of a zoom handle which is moved forward or backward. *See* Lens; Telephoto lens.

Bibliography. L. Gaunt, *Zoom and Special Lenses*, 1981; W. Hawken, *Zoom Lens Photography*, 1981; C. B. Neblette and A. E. Murray, *Photographic Lenses*, 2d ed., 1973.

MICROSCOPE
Oscar W. Richards

An instrument used to obtain an enlarged image of a small object. The image may be seen, photographed, or sensed by photocells or other receivers, depending upon the nature of the image and the use to be made of the information of the image. Microscopes are classified as simple or compound according to the kind of radiation used to form the image, and the use for which they are designed.

Simple microscope. A simple microscope, hand lens, or magnifier usually is a round piece of transparent material, ground thinner at the edge than at the center, which can form an enlarged image of a small object. Commonly, simple microscopes are double convex or planoconvex lenses, or systems of lenses acting together to form the image (**Fig. 1**). The lens can be mounted in a simple holder, in a folding case for hand use, or with a support which has a mechanical focusing mechanism, stage, and mirror to make a dissecting microscope. *See* Lens.

Compound microscope. The compound microscope utilizes two lenses or lens systems. One lens system forms an enlarged image of the object and the second magnifies the image

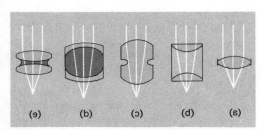

Fig. 1. Some common types of magnifier. (*a*) Double convex. (*b*) Doublet. (*c*) Coddington. (*d*) Hastings triplet. (*e*) Achromat. (*After F. A. Jenkins, Fundamentals of Optics, 4th ed., McGraw-Hill, 1976*)

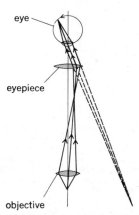

Fig. 2. Compound microscope. (*After F. A. Jenkins, Fundamentals of Optics, 4th ed., McGraw-Hill, 1976*)

formed by the first. The total magnification is then the product of the magnifications of both lens systems. Theoretically, several simple microscopes could be used in line, each to magnify the image of the one before it. Practically, the losses from aberrations, reflections, and other defects limit the compound microscope to two such systems (**Fig. 2**).

The typical compound microscope consists of a stand, a stage to hold the specimen, a movable body-tube containing the two lens systems, and mechanical controls for easy movement of the body and the specimen. The lens system nearest the specimen is called the objective; the one nearest the eye is called the eyepiece or ocular. A mirror is placed under the stage to reflect light into the instrument when the illumination is not built into the stand. For objectives of higher numerical aperture than 0.4, a condenser is provided under the stage to increase the illumination of the specimen. Various optical and mechanical attachments may be added to facilitate the analysis of the information in the doubly enlarged image.

Special compound microscopes have two image-forming systems to give an enlarged image of an image. These instruments utilize electrons, x-rays, sound, or other forms of radiation for image formation, and electromagnetic or electrostatic fields or mirrors to form the enlarged images. Because these images are not visible, photography, television, and special receivers must be used to record and analyze the image. Special microscopes are usually named according to the kind of radiation used, such as electron, x-ray, ion, and ultrasonic.

The compound microscopes which employ a lens system are essentially similar in their working principles. Basically, they are modifications of the ordinary laboratory microscope (bright-field microscope) for specialized or specific purposes. Some of the types are as follows.

Light or photon microscopes utilize light of wavelengths from 380 to 760 nanometers for image formation. Such microscopes include the laboratory or bright-field microscope, and modifications of it such as the capillary, centrifuge, chemical, comparison, crystallographic, dark-field, dissecting, fluorescence, integrating, interference, inverted microprojection, museum, nuclear track, petrographic, phase, phosphorescence, and profile microscopes.

Reflecting microscopes utilize a mirror rather than a lens system. The infrared microscope uses radiation of wavelengths greater than 700 nm, and the ultraviolet employs light of 180–400 nm. The ultraviolet microscope requires reflecting optics or special quartz and crystal objectives. Color-translating microscopes employ three different wavelengths of light to reveal details produced by ultraviolet or other nonvisible radiation.

In the electron, proton, x-ray, and beta-ray microscopes the image is usually recorded on a fluorescent screen or is photographed.

Mechanical vibrations, generated into an elastic system, are the basis for the ultrasonic microscope, for locating foreign bodies or analyzing reflecting surfaces. SEE FLUORESCENCE MICROSCOPE; INTERFERENCE MICROSCOPE; OPTICAL MICROSCOPE; PHASE-CONTRAST MICROSCOPE; REFLECTING MICROSCOPE.

OPTICAL MICROSCOPE

MAX HERZBERGER, OSCAR W. RICHARDS, AND LAWRENCE A. HARRIS

M. Herzberger wrote the sections Lenses and Condensers. L. A. Harris prepared the section Near-Infrared Microscopy.

An instrument used to obtain an enlarged image of a small object. In general, a compound microscope consists of a light source, a condenser, an objective, and an ocular or eyepiece, which can be replaced by a recording device such as a photoelectric tube or a photographic plate. The optical microscope is limited by the wavelengths of the light used and by the materials available for manufacturing the lenses. Some optical errors, for instance, curvature of field and lateral color, are not generally corrected in the objective but neutralized in the ocular. This article includes general discussions of microscope lenses and condensers, and specific discussions of the ordinary microscope, inverted microscope, comparison microscope, dissecting microscope, and metallurgical microscope, and a discussion of microscopy. SEE ABERRATION.

LENSES

The quality and design of the lens system determines the magnifying power, details of image formation, and color correcting capabilities of a light microscope.

Magnifying power. The magnifying power of a compound microscope is the product of the magnification of the objective and the magnifying power of the eyepiece. The latter is computed like that of any magnifier. The magnification of the objective is equal to the distance from the second focal point to the image formed by the objective, divided by the focal length. An objective of 18-mm (0.7-in.) focal length thus has a power of 10×. It is customary to specify objectives in terms of magnifying power instead of focal length. The distance mentioned is called the optical tube length (generally 180 mm or 7 in.), and is to be distinguished from the mechanical tube length, which is the length of the mechanical tube itself. SEE MAGNIFICATION.

Objectives. A microscope objective consists of a set of achromatic lenses which are partially or wholly corrected for longitudinal color, aperture errors, and asymmetry errors. The numerical aperture (NA) of the system is given by $n \sin u$, where n is the refractive index of the object space and u is the angle made by the ray of largest aperture and the axis.

A microscope objective generally consists of a collection of positive lenses comparatively close together. Color magnification and curvature of field in the objective are frequently balanced out at least partly by the ocular. An objective of 16-mm (0.63-in.) focal length is shown in **Fig. 1**a.

For high magnifications, the first lens is planoconvex, with the convex surface either concentric or aplanatic to the aperture rays. An objective of this construction is shown in Fig. 1b. To increase the aperture, the object can be embedded in an immersion liquid (a special oil) having a refractive index equal or nearly equal to the index of the first lens. Such an arrangement is termed a homogeneous immersion system. A cover glass of fixed but small thickness separates the object from the immersion liquid for biological investigations. Figure 1c shows an immersion objective of 2-mm (0.08 in.) focal length and 1.40 NA.

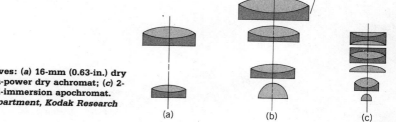

Fig. 1. Microscope objectives: (a) 16-mm (0.63-in.) dry achromat; (b) typical high-power dry achromat; (c) 2-mm (0.08-in.), NA 1.40, oil-immersion apochromat. (*Photographic Service Department, Kodak Research Laboratory*)

Color correction. Achromatic color correction is not good enough for lenses of high aperture. The use of fluorite led to objectives corrected for more than two colors, and most achromatic lenses of high aperture are now made with fluorite. An objective whose chromatic errors are corrected for three colors and whose aperture and asymmetry errors are also corrected was termed apochromatic by E. Abbe. The lateral chromatism of such an objective is great and must be removed by a special compensating eyepiece. *See Eyepiece.*

Experiments have been made to design high-aperture microscopes for a field larger than the usual 3°. This has been achieved by adding either a negative lens or a very thick positive lens having a negative Petzval sum. Such objectives are termed plane-achromats. Attempts have also been made to correct field curvature in the eyepiece.

Because resolving power depends upon wavelength, objectives which are corrected for ultraviolet radiation have been designed. Such objectives are generally corrected for a single wavelength and are termed monochromats. They are made of quartz.

Catadioptric systems. Catadioptric systems have been developed for microscopes. Their great advantage is their comparatively small chromatic aberration. A type consisting of a single element is shown in **Fig. 2**a. Pure mirror systems have no color aberrations. In catadioptric systems, therefore, it is customary to assign all the power to the mirror or mirrors, keeping the refracting system nearly afocal. The chromatic errors of the entire system remain small, and the refracting part can be used to correct the remaining monochromatic errors. However, in catadioptric systems part of the aperture is obscured by the mirror and the ensuing diffraction may damage the fine detail in the image. All microscopic work in the ultraviolet region is done with catadioptric systems. A type designed for this purpose is sketched in Fig. 2b. *See Schmidt camera.*

Image formation. Geometrical optics is not sufficient to explain all the details of image formation at high magnifications. According to geometrical optics, a point should be imaged by a perfect objective as a point, but there is diffraction at the aperture, and there may also be diffraction at the object. Diffraction theory applied to the aperture shows that, because of the finite aperture of the optical system, a spherical concentric wave bounded by the exit pupil produces a light pattern in the image plane in which the light is distributed over a disk of diameter $1.22\lambda/NA$ with faint rings outside the disk. If the magnifying power of the microscope becomes so large that the rings are visible, the image will contain details which are not in the object. This is undesirable. Thus, the aperture determines the useful magnification of the microscope; it is important that the aperture be enlarged when the magnification has to be increased.

Because the microscope is used to study very small objects at usually high magnifications, diffraction at the aperture is more noticeable in microscopes than in other optical systems. Moreover, most objects viewed with microscopes are so small that there is a significant amount of diffraction at the object, as is seen in the case of dark-field illumination and phase microscopy.

This is of special importance if the object has a periodic structure since even a point light source will then give rise to an imagelike structure. Interference of the rays coming from the light source in phase and being diffracted at the structured object causes this effect and gives rise to sets of images in different planes. This theory (Abbe's theory of image formation in the microscope) has been successfully carried further in F. Zernike's theory, which led to the construction of the phase microscope. *See Point source.*

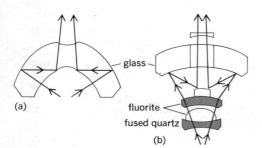

Fig. 2. Two types of catadioptric objective. (a) Maksutov type. (b) 53X, NA 0.72, ultraviolet objective, designed by Gray. Glass elements in the latter serve purely as reflectors. (*Photographic Service Department, Kodak Research Laboratory*)

In the case of illumination with a large cone of light, these interference patterns fall together at the gaussian focus, so that there is only one image plane. The results in this case, however, can also be derived from Abbe's theory, but only if integration over the aperture is carried out. SEE INTERFERENCE MICROSCOPE; PHASE-CONTRAST MICROSCOPE.

CONDENSERS

An external auxiliary lens is used to condense the light from a light source so that the object is brightly and uniformly illuminated. The usual purpose of a condenser system is to make sure that as much light as possible coming from the object goes through an optical system.

Condensers are used in macroscopic projection, in which an illuminated film or slide is imaged with the help of a projection objective or magnifier. In microscope systems, they are used to direct the light from a light source so that the rays from any object point fill most of the entrance pupil.

Condenser system. A condenser system is usually arranged to image the light source onto the entrance pupil of the optical system (Köhler illumination). The condenser is generally corrected for spherical aberration, color, and sine condition, although the requirements are slightly different than in an image-forming system. Condensers frequently consist of a number of planoconvex lenses with the plane side toward the objective. Sometimes one surface is made aspheric to improve the light concentration. Condensers for projection optics are rarely achromatized, but the effect of color magnification is decreased by vignetting the colored borders. For microscope substage condensers, however, achromatism is a necessary requirement.

Aperture. The aperture of the condenser must be at least as large as that of the objective with which it is used. Because microscope objectives are generally designed in such a way that they are excellently corrected for color, aperture, and asymmetry errors for only about seven-eighths of their aperture, the condenser need fill only this much of the entrance pupil. A good test of whether the condenser of a microscope is well adjusted is to remove the object and ocular and see whether the exit pupil of the objective is filled uniformly with light up to seven-eighths of its aperture.

Dark field. Thus far, only condensers for giving a bright field have been discussed. In microscope practice, it is sometimes advantageous to increase the contrast of small objects by making them appear as bright objects on a dark background. This is achieved by arranging the condenser so that direct light is cut off and only light diffracted or dispersed from the object enters the microscope. The cardioid condenser in **Fig. 3** is an example of such a dark-field condenser.

The principles outlined here have been elaborated by the development of phase microscopy. In the phase microscope, an annular diaphragm is placed at the front focus of the substage condenser so that the object is illuminated by a hollow cone of parallel light. A phase plate is put

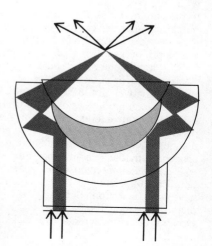

Fig. 3. Cardioid condenser. Shaded meniscus area is air space; other unshaded areas are portions of condenser through which light passes. (*Photographic Service Department, Kodak Research Laboratory*)

at the image of the diaphragm formed by the objective. This phase plate usually consists of a transparent annular layer evaporated on a transparent plate, the added layer corresponding in size and shape to the image of the diaphragm and having a thickness equal to one-fourth of a selected wavelength of light. A metallic layer is generally superimposed to diminish the transmittance of the annulus. When diffraction takes place at the object, the central maximum has a phase shift of $\lambda/2$ with respect to the diffracted light. This phase shift results in an intensity difference in the image which increases the contrast. SEE MICROSCOPE.

TYPES OF MICROSCOPE

The types of microscope discussed in this section are variations of the light- or bright-field microscope.

Light microscope. The mirror, condenser, oculars, and body tube of the light microscope are frequently known as the optical train. The stand, stage, and adjustments comprise the mechanical part of the microscope (**Fig. 4**).

A mirror is usually attached to the substage of the microscope to reflect light along the optic axis of the microscope. When no condenser is used, the concave mirror is used because it concentrates more light on the specimen; a plane mirror is used with a condenser.

The condenser concentrates light onto the specimen at an angle to fill the objective. Laboratory microscopes are usually supplied with an uncorrected two-lens Abbe condenser with NA 1.25, consisting of a double convex lower and a hyperhemisphere upper lens. A 1.40 NA Abbe is useful for concentrating radiation for fluorescence microscopy. Well-corrected objectives require aplanatic condensers corrected for chromatic aberration for efficient observation.

The objective is the basic part of the microscope; it forms the image that is again enlarged by the eyepiece. Objectives vary from a simple doublet lens to complex corrected lens systems. Achromatic objectives are corrected for spherical aberration in one color and for chromatic aberrations in two colors. Apochromatic objectives are corrected to focus three colors together and the spherical aberration is minimized for two colors. Some semiapochromatic and fluorite objectives are of intermediate correction. Fluorite objectives include a lens of crystal fluorite. SEE CHROMATIC ABERRATION.

As previously indicated, one important measure of an objective is the numerical aperture. With air between the specimen and the objective the maximum NA is about 0.92. Water-immersion objectives have a greater NA, and with immersion oil, an NA of 1.4 is available. Objectives with an NA of 1.6 have been made but require special immersion and mounting media. The resolving power of an objective, the least distance at which two objects can be seen to be separate, is equal to the wavelength of light λ divided by the sum of the numerical apertures of the condenser and objective used. The larger the numerical aperture, the greater is the resolving power. The depth of field seen in focus at one time and the working distance of objective decrease with an increase in the numerical aperture. The light passed through a microscope is proportional to the square of the numerical aperture and to the inverse of the square of the magnification. The numerical aperture is engraved on the objective and can be measured with an apertometer. Objectives are described also by the equivalent focal length. Objectives of shorter focal length have less depth of field, less working distance, and greater magnification.

Photomicrographic objectives are designed to produce a flat image with little distortion. Some objectives obtain a flatter field by means of a concave rather than a flat front lens. Apochromatic objectives are undercorrected and compensating oculars must be used with them to complete the correction for color and for best resolution.

For convenience, two to five objectives can be mounted on a revolving nosepiece to be parfocal and parcentric, so that the specimen remains almost in focus at the center of the field as the objectives are changed. For more critical work, utilizing interference and polarizing microscopes, individually adjustable quick-change nosepieces are employed.

The commonly used Huygenian ocular has a fairly flat field with marked pincushion distortion (**Fig. 5**). Compensating oculars complete the color correction for apochromatic objectives and have less distortion, but they do have curvature of field. To obtain a flat field with minimum distortion for microprojection or photomicrography special projection oculars are designed, one type with a color-corrected minus lens called a negative amplifier, and the other a positive-projec-

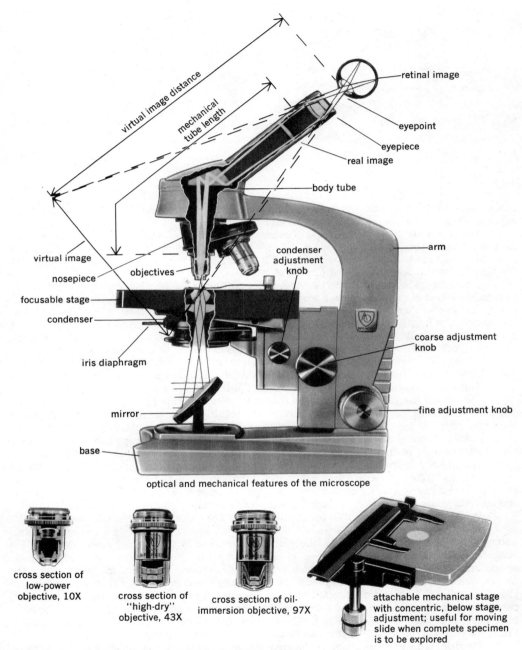

Fig. 4. Diagram of light microscope. (*American Optical Corp.*)

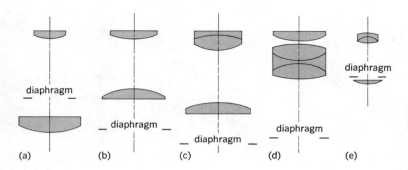

Fig. 5. Types of ocular. (*a*) Huygenian. (*b*) Ramsden. (*c*) Wide-field. (*d*) Compensating (high eyepoint). (*e*) Projection.

tion ocular with a focusable eye lens. Oculars with a high exit pupil are useful for spectacle wearers. Other oculars are designed to give a wide field of view.

The monocular body tube may be of adjustable length. American microscopes are designed for a mechanical tube length of 160 mm (6.3 in.) and a cover-glass thickness of 0.18 mm (0.007 in.). The draw tube is lengthened for thinner and shortened for thicker cover glasses to correct for the spherical aberrations from cover glasses of incorrect thickness.

Binocular bodies are designed for the use of both eyes. Most binocular bodies use prisms to reflect one-half of the light to each eye. Because each eye sees the same field, these binocular bodies do not give stereoscopic vision. The binocular is often longer than the monocular body and the proper tube length is maintained with a compensating lens. Because the tube length is fixed, it is essential for critical microscopy with binocular bodies to use cover glasses 0.18 mm thick. Binocular bodies can be made with stops, or polarizing materials, so that each eye sees with the corresponding half-aperture of the objective and stereoscopic vision is possible, but at one-half the resolving power that would have been obtained with full aperture.

Trinocular bodies are binocular bodies with a third tube for a camera.

The lowest magnification which will resolve the detail required should be used. High magnification gives images that are less bright and therefore difficult to see. Magnification greater than that required for complete visible detail is called empty magnification and is useless for seeing, although sometimes helpful for measurement. For visual microscopy the total magnification (magnification of the objective times that of the ocular) should be about 1000 times the numerical aperture in use. Apochromatic objectives and compensating oculars are desirable for best vision and color photomicrography. The research microscope has a larger, heavier stand with more precise adjustments for convenience and measurements. For special applications, microscopes with phase, interference, polarizing, and other special equipment are available.

Inverted microscope. The inverted microscope has the body of the microscope, including the objective and the ocular, below the stage and the illumination above the stage for transmitted light. With opaque materials, the vertical illuminator is used under the stage near the objective. The inverted microscope is especially useful for the examination of surfaces (**Fig. 6**). Specimens placed on the stage can be held substantially in focus. Large and awkward specimens can be moved over the stage more readily than with the usual microscope. The inverted microscope is also useful for microdissection and the observation of hanging-drop preparations and is convenient for observing chemical reactions, melting-point determinations, and photomicrography. The camera can be included in the base, as in the metallograph microscope, for stability. Either monocular or binocular bodies can be used with the inverted microscope.

Comparison microscope. The comparison microscope is an arrangement of two microscopes connected by a special viewing ocular so that the field of one microscope is seen at one side of a vertical dividing line and the field of the other microscope on the opposite side of the dividing line; or it may be a projection type of microscope in which the image is compared with a template or known pattern.

When two microscopes and a comparison ocular are used, the magnifications of each must be matched. The specimens are placed on the microscopes and usually require separate lighting.

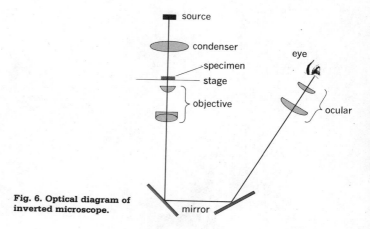

Fig. 6. Optical diagram of inverted microscope.

A common application is the examination of bullets. A test bullet fired from the suspected gun is placed under one microscope and the bullet recovered from the scene of the crime is placed under the other microscope. Holders which allow rotation of the bullets are turned, and if the grooves on the two bullets match, it is evident that they came from the same gun. Comparison microscopes can be used to compare grain size, structure, distribution of elements, color, and various other characteristics of any two specimens. Either transmitted light, vertical illumination, or a combination of lighting types can be used.

A projection microscope with a built-in screen can also be used as a comparison microscope. The image of the part is focused on the screen, and the contours of the image are measured with scales or compared with a template of the desired form, for example, a profile of a gear or a screw thread. The stage is usually modified to have a proper holder for the kind of specimen to be examined.

Dissecting microscope. Dissecting microscopes are of two types. The simplest is a magnifying glass mounted on a support above a glass plate, used for the dissection of materials. A mirror may be present to reflect light to the specimen.

The more usual dissecting microscope, often called a Greenough microscope, is a stereoscopic microscope composed of two separate microscopes fastened together and used as a single unit on one stand (**Fig. 7**).

This is a truly stereoscopic instrument because the right eye sees the specimen from the right side and the left eye from the left side. Prisms are usually included in the body tube to erect the image; thus movements of the specimen are direct and are not reversed as with the monobjective microscope. Because two objectives are required, the mechanical difficulty of placing the lenses close together limits the numerical aperture to approximately 0.12. There is no advantage in using more than 120 diameters magnification because further magnification would be empty or useless.

The binocular microscope was developed for biological dissecting, but with the need for small parts in industry it is now used extensively for the examination and assembly of small parts such as transistors. The body and focusing adjustment can be mounted on lathes and other machinery when the work must be controlled visually to close tolerances; they can be mounted on a stand with a long arm for examination of large specimens or may be used to assist the surgeon in the operating room.

When the angle of the objectives and oculars is the same, the specimen is seen in true depth. By changing these angles the perception of depth can be increased or decreased. Hyperstereoscopy is useful for biological dissection.

Examination with the stereoscopic microscope is helpful in orienting the microscopist to the specimen before instruments of greater magnification are used. By proper illumination both opaque and transmitted materials can be examined.

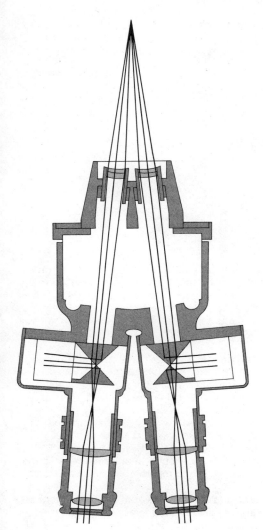

Fig. 7. Diagram of light rays as they pass through a binocular biobjective microscope.

Metallurgical microscope. The metallurgical microscope is a laboratory microscope with a focusing stage and a vertical illuminator, used primarily for the examination of metal surfaces (**Fig. 8**).

The specimens are usually embedded in molded plastic so that the surface is at a definite position, surfaced with a series of increasingly finer abrasives, and polished. To differentiate the constituents of the metal, the surface is etched lightly by chemical treatment before examination with the microscope.

The metallurgical microscope shows the structure of the metal, including grain size, as well as the nature and distribution of the components. The roughness or polishability of the metal can be studied. Because many of the metals used commercially are mixtures, or alloys, rather than pure metals, the metallurgical microscope is important for analysis and for assessing the effects of heat treatment and surface changes.

OPTICAL IMAGING SYSTEMS

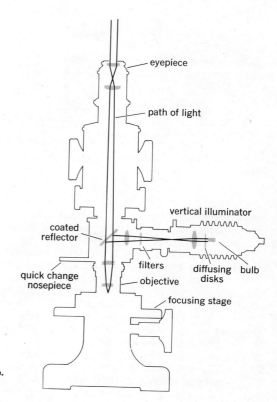

Fig. 8. Light path of metallurgical microscope. (*American Optical Corp.*)

MICROSCOPY

This discussion of microscopy considers the following with relation to the optical microscope: illumination, calibration and measurement, immersion fluids, mountants, and the hanging-drop method.

Illumination. Illumination of the microscope is obtained from a bright surface or from a luminous source concentrated with the aid of condensing lenses. Sources commonly used are tungsten-filament lamps and carbon, mercury, and xenon arcs. With an illuminated-surface light source the lamp is positioned and the microscope condenser focused so that the field and aperture of the microscope are lighted as uniformly as possible. Illuminators with focusing condensers and iris diaphragms, called research lamps, are used for critical, Köhler, and other more efficient methods.

Critical illumination. In critical illumination an image of the light source is focused on the specimen from a uniform source, for example, a ribbon-filament lamp. This method is no more and no less critical than other methods, but the name has evolved and is so used (**Fig. 9**). Parallel light from an illuminator focused to infinity is directed into the microscope condenser which is then focused so that the image of the source is in the plane of the specimen.

Köhler's method. This method is used with coiled filaments or other sources of irregular form or brightness to obtain a uniformly illuminated field. An image of the filament large enough to fill the opening of the iris is focused on the microscope condenser. The microscope condenser then is focused so that the image of the iris diaphragm on the lamp is in focus with the specimen and the lamp iris is opened only enough to fill the field of view. The iris of the microscope condenser is opened only enough to illuminate the back aperture of the objective. No ground glass is

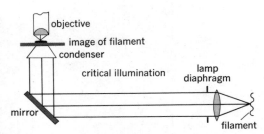

Fig. 9. Light path in critical illumination. (*American Optical Corp.*)

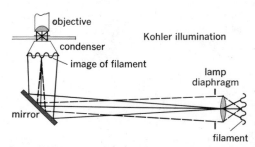

Fig. 10. Light path in Köhler illumination. (*American Optical Corp.*)

used. The condensing lens of the illuminator becomes the effective source and the field is uniformly illuminated, even though the source itself is not uniform. Loss of contrast from misplaced glare light is minimized (**Fig. 10**).

Shillaber's type 3. This method is useful when the lens systems of the condensers are inadequate to illuminate the field and aperture of the microscope.

A ground glass is placed close to the condensing lens of the illuminator between the lens and the lamp bulb, or the surface of the lamp is ground on the side facing the lens. Because the ground surface does not diffuse all the light, the condensing lens can concentrate more light into the microscope than could be obtained from the same area of a surface-type illuminator. An effectively larger source results which is useful with low-power (searcher) objectives for a large field of view and when the lamp filament is not large enough for the condensing system—for example, some sources built into the base of the microscope. This method is preferable when a diffuser must be used, but more stray or glare light is produced than with the Köhler method.

Oblique illumination. Oblique illumination occurs when the mirror is moved to one side of the optical axis of the microscope, the nearly closed iris diaphragm is moved away from the axis of the condenser, a diaphragm is moved across the condenser aperture, or part of the exit pupil (Ramsden disk) is covered. A plastic, pseudo–three-dimensional appearance may occur, and parallel detail in the specimen is emphasized with this one-sided lighting. Interpretation of what is seen is difficult.

Vertical illumination. Vertical illumination uses a partly silvered mirror, or a prism, to reflect light through the objective onto the specimen. The light reflected back through the objective from the specimen forms the image which is seen.

Epi-illumination is vertical illumination with the illuminating light passing around a special objective to the specimen. Only the light reflected from the specimen passes through the objective. Better definition results from the separation of the paths of the illuminating and viewing light.

Metals, ores, minerals, and opaque materials with adequate reflecting surfaces are examined with vertical illumination.

Dark-field illumination. In dark-field illumination the specimen appears bright, or self-luminous, against a black background. The illuminating beam is a hollow cone of light formed by an opaque stop at the center of the condenser, large enough to prevent any direct light from entering the objective. A specimen placed at the concentration of the light cone is seen with the light scattered or diffracted by it, and the smallest particle revealable depends upon the intensity of the available light, even though the particle may be too small for resolution as to size and shape. Size can sometimes be inferred from the number of particles found in a given volume of the specimen (**Fig. 11**). Ultramicroscopy and Rheinberg illumination are modifications of dark-field illumination.

Ultramicroscopy, used for the examination of colloids and smokes, is a dark field obtained with an intense narrow beam of light directed through the specimen at right angles to the optical axis of the microscope.

Rheinberg illumination or optical staining is a modification of the dark-field method. The

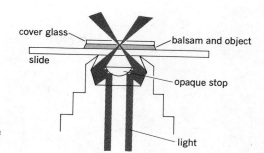

Fig. 11. Dark-field illumination. (*American Optical Corp.*)

central disk is transparent and colored, rather than opaque, and an annulus of a complementary color fills the remaining condenser aperture. The specimen is seen in the color of the annulus against a background of the color of the central disk; for example, when the annulus is yellow and the background blue, the specimen appears yellow against a blue background.

Modification by filters. Filters are used between the microscope illuminator and the microscope to control the intensity or quality of illumination. Filters can be liquids in a flat-sided container or solids. Solid filters are made of colored glass, gelatin, or other materials. The type of filter used depends on the requirements of the examination.

Clear filters of water or of heat-absorbing glass are used to remove the excess heat from the lamp beam when delicate specimens must be protected. Blue filters remove excess yellow from tungsten light so that it resembles daylight. Neutral filters of glass or metal (Inconel) deposited on glass are used to reduce the amount of light without altering its color.

Colored filters change the quality of the light by selectively absorbing certain wavelengths and are grouped into broadband and narrow-band types. Broadband filters are used to increase visibility by modifying the color contrast of the specimen; for example, a yellow (minus blue) filter transmits all of the spectrum except the blue. Complementary filters increase contrast, and filters of similar color decrease color contrast. Polychromatic filters, transmitting two or more spectral colors (for example, a minus green filter passing blue and red) are useful with some stained specimens. Smaller regions of the spectrum are isolated with narrow-band-pass filters, interference filters, and by combinations of filters. Monochromatic light usually is obtained from a single spectral line of an arc source rather than with filters.

Calibration and measurement. The size of the object under examination is frequently a desirable criterion for identification. A reference reticule is placed on the diaphragm in the focal plane of the ocular and seen in focus with the specimen from measurement in microscopy. A cross-hair reticule (**Fig. 12**) is satisfactory for position; scales (**Fig. 13**) are used for measuring

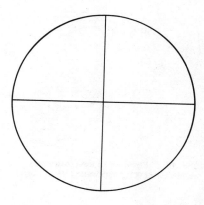

Fig. 12. Cross-hair reticule. (*American Optical Corp.*)

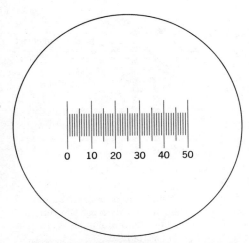

Fig. 13. Scale reticule, nine times normal size. (*American Optical Corp.*)

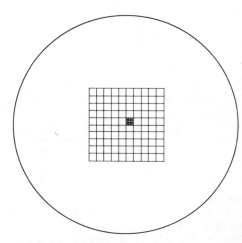

Fig. 14. Net reticule, used for counting and drawing. (*American Optical Corp.*)

distance and a net reticule (**Fig. 14**) for counting and drawing. Oculars with a focusing eye lens permit focusing the scale to the microscopist's eye.

Angular measurements are made by orienting one side of the specimen to the cross hair, reading the position of the graduated stage of the microscope, and rotating the stage until the other side is in line with the cross hair. The difference between the first and second readings of the scale gives the angle. Another method uses a goniometer ocular and measures the angle with the scale of the goniometer.

Linear measurements are made by placing one edge of the specimen at the cross hair, reading the position of the graduated mechanical stage, moving the specimen to the other side,

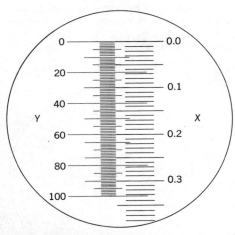

Fig. 15. Calibration of eyepiece scale Y, with a stage micrometer X. (*American Optical Corp.*)

Fig. 16. Screw micrometer ocular. (*American Optical Corp.*)

and reading the position of the mechanical stage. The difference of the readings is the distance. This method is useful for larger specimens because the mechanical stages are ordinarily not graduated closer than 0.1 mm (0.004 in.). For measurements of smaller specimens a scale is used in the eyepiece and the size of the specimen obtained from the scale (Fig. 13). The actual or true distance of the ocular scale depends upon the magnification of the microscope and is obtained by calibration with the aid of a stage micrometer. The true distance x seen on the stage micrometer which corresponds to the number of divisions y of the eyepiece micrometer is noted. Dividing this true distance by the number of divisions in the eyepiece micrometer gives the distance $c = x/y$ each one subtends. The number of eyepiece divisions covered by the specimen multiplied by the calibration constant c gives the actual length of the specimen (**Fig. 15**). Change of eyepiece, objective, magnification, or tube length of the microscope necessitates recalibration.

More precise measurements can be made with a screw micrometer eyepiece, a filar micrometer, or a step micrometer eyepiece. The movable scales with graduated controls facilitate measurement, but must also be calibrated (**Fig. 16**).

NEAR-INFRARED MICROSCOPY

Near-infrared microscopy is an optical method that can be used for studying a variety of materials that are opaque in transmitted visible light (400–700 nm) yet translucent in the near infrared (700–1200 nm). The method utilizes the near-infrared optical microscope, a device for the conversion of the near-infrared image to a visible image.

The heart of the microscope is the image converter tube which is positioned above the microscope objective in the plane of the primary image (**Fig. 17**). The near-infrared to visible conversion is achieved within the image tube, a cathode tube containing a positive terminal

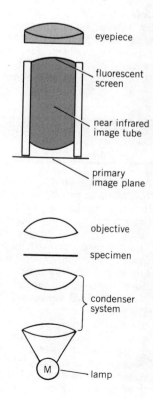

Fig. 17. Arrangement of near-infrared image tube in light microscope.

coated with fluorescent material. The impingement of the near-infrared light image onto the negative terminal of the cathode tube releases electrons which are focused by means of an electrostatic lens onto the fluorescent screen.

The image formed on the fluorescent screen either can be viewed directly by means of an eyepiece positioned above the screen or can be photographed with standard cameras using ordinary film. Sharpening of the image is achieved by use of near-infrared filters located in front of the image tube. These filters serve to suppress the visible-light image and limit the wavelength range of the transmitted beam. However, such filters also tend to reduce the percentage of transmitted light so that a need arises for a light source that has a sufficiently intense near-infrared component (such as xenon and tungsten filament lamps).

The transmission of near-infrared light, like that of visible light, follows Lambert's law, $I_T = I_0 e^{-\mu x}$, where I_0 = the intensity of the light entering the specimen, I_T = the intensity after passing through a specimen of thickness x, and μ = the absorption coefficient of the particular specimen, resulting in a linear decrease in transmission with increasing thickness.

Transmission near-infrared microscopy can be used in the study of all near-infrared transmitting materials in a manner identical to visible-light microscopy. All optical measurements commonly made using visible light are attainable in near-infrared light.

Bibliography. R. D. Allen, *Optical Microscopy*, 1985; R. Barer, A. Bouwers, and D. S. Gray, Reflecting microscope, *Proc. London Conf. Opt. Instrum.*, pp. 43–76, 1951; H. Boegehold, *Das Optische System des Mikroskops*, 1958; A. Bouwers, *Achievements in Optics*, 1946; S. Bradbury, *An Introduction to the Optical Microscope*, 1984; C. W. Mason, *Handbook of Chemical Microscopy*, vol. 1, 4th ed., 1983; Eastman Kodak Co., *Photomicrography*, 14th ed., 1944; R. D. Enoch and R. M. Lambert, An infrared polarizing microscope for observation of domains in thick samples of magnetic oxides, *J. Phys.*, E. 3, 1970; P. Gray, *The Microtomist's Formulary and Guide*, 1954, reprint 1975; L. A. Harris, The application of near-infrared microscopy to materials science, *Microstruc. Sci.*, 6:119–129, 1978; M. Hertzberger, *J. Opt. Soc. Amer.*, 1936; R. Kingslake, *Applied Optics and Optical Engineering*, 7 vols., 1965–1979; O. Lummer and F. Reiche, *Die Lehre von der Bildentstehung in Mikroskop*, von E. Abbe, 1910; C. E. McClung and R. M. Jones (eds.), *Handbook of Microscopical Technique for Workers in Animal and Plant Tissues*, 3d rev. ed., 1950, reprint 1964; C. P. Shillaber, *Photomicrography in Theory and Practice*, 1944; M. Spencer, *Fundamentals of Light Microscopy*, 1982; J. Strong, *Concepts of Classical Optics*, 1958; V. K. Zworykin and G. A. Morton, Applied electron optics, *J. Opt. Soc. Amer.*, 16:181–189, 1936.

REFLECTING MICROSCOPE
DAVID S. GREY AND OSCAR W. RICHARDS
O. W. Richards is author of the section Uses.

A microscope whose objective is composed of two mirrors, one convex and the other concave. The imaging properties are independent of the wavelength of light, and this freedom from chromatic aberration allows the objective to be used even for infrared and ultraviolet radiation. Although the reflecting microscope is simple in appearance, the construction tolerances are so small and so difficult to achieve that the system is used only when refracting objectives are unsuitable. The distance from the objective to the specimen can be made very large; this large working distance is useful in special applications, such as examining objects situated within metallurgical furnaces. Reflecting microscopes have been mainly used for microspectrometry in the infrared and the ultraviolet, and for ultraviolet microphotography.

Optical system. The concave mirror collects light from the specimen under examination. The convex mirror intercepts this light just above the point where the rays would be focused by the concave mirror, and redirects the light up the microscope tube to the plane where the primary image is formed (see **illus**.). The curvatures and spacing of the mirrors can be so chosen that the optical aberrations of the mirrors are mutually compensating, provided that the numerical aperture does not exceed about 0.60 (visible light). The limiting aperture is somewhat smaller than this for ultraviolet radiation and somewhat larger for infrared, and is smaller for large working distances than for small working distances.

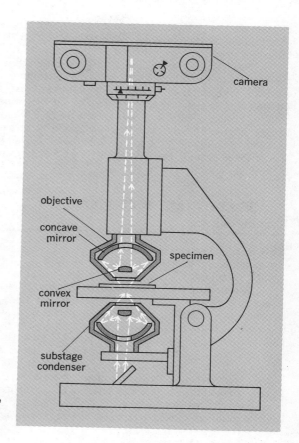

Reflecting microscope arranged for photomicrography.

The optical designer, in attempting to achieve numerical apertures larger than about 0.60, may resort to mirrors that depart slightly from spherical curvature; alternatively, he may add one or more lens elements. By such means, objectives of numerical aperture greater than unity and free of chromatic aberration can be realized.

The convex mirror unavoidably obscures a small central portion of the aperture. This alters the diffraction of light (upon which the ultimate resolving power of the objective depends) but fortunately this effect is negligible if proper care is used in the design and construction.

Substage illumination of the specimen can sometimes be achieved by a conventional refracting condenser if only visible light is used. Usually, however, the chromatic aberration of the condenser is troublesome and a condenser similar to the objective is used. Chromatic aberration of the ocular is usually not serious; hence the ocular may be of conventional form, although usually constructed of fused silica, fluorite, or other materials transparent to the infrared or ultraviolet. The ocular may be omitted for special applications in which the primary image formed by the objective is examined directly.

Uses. The reflecting microscope is used for examination with nonvisible radiation, or when it is desirable to focus the microscope with visible light and examine with nonvisible ultraviolet radiation. Because reflecting objectives lack chromatic aberration, the focus is the same for different wavelengths. However, in catadioptric objectives with both refracting and reflecting surfaces, spherical aberration occurs.

The reflecting microscope is primarily used in the form of a microspectrophotometer for examination with ultraviolet radiation. Different proteins and nucleic acids have different absorption curves, and the ultraviolet instrument can be used to estimate the amounts of these and other materials. The resolution is better with ultraviolet radiation than with visible light, and this increased resolving power is helpful.

The infrared microscope also requires reflecting optics and can reveal detail in materials that are opaque to light; for example, molybdenum is opaque to light but shows a distinct fiber structure with infrared. Wood, corals, and many red-dyed materials can be examined with the infrared microscope. It is useful in the detection of forgeries. One disadvantage of infrared microscopy in biology comes from the absorption of infrared radiation by water. Dried specimens are usually different in form from living ones. When materials can be dried, or when the water absorption can be subtracted, the infrared microscope gives information concerning the chemical structure of the specimen. Resolving power becomes proportionately less as the longer wavelengths of infrared are used.

PHASE-CONTRAST MICROSCOPE
Robert Barer

A microscope used for making visible differences in phase or optical path in transparent or reflecting specimens. It is an important instrument for studying living cells and is used in biological and medical research.

Microscopy of transparent objects. When a light wave passes through an absorbing object, it is reduced in amplitude and intensity. Since the human eye and the photographic plate are sensitive to variations in intensity, such an object will give a visible image when viewed through an ordinary microscope. A perfectly transparent object does not absorb light, so that the intensity remains unaltered, and such an object is essentially invisible in an ordinary microscope. However, the light that has passed through the transparent object is slowed down by it and arrives at the eye a minute fraction of a second later than it would otherwise have done. Such delays, or in technical language, phase changes or differences in optical path, are not detected by the eye, and the fundamental problem in the microscopy of transparent objects, of which living cells are important examples, is to convert them into visible intensity changes. For further discussions of basic principles see Absorption of electromagnetic radiation; Diffraction; Interference of waves; Light.

Simple theory. A consideration follows of what happens when a light wave passes through a partially absorbing object, such as a stained biological specimen. In **Fig. 1**, let A be the incident wave. When the wave passes through the object, energy is absorbed so that the transmitted wave B is reduced in height or amplitude. The intensity is proportional to the square of the amplitude. Wave B can also be represented as the sum of the incident wave A and another wave shown by a dotted line. This wave has a real physical existence and represents the light scattered or diffracted by the object. The German physicist E. Abbe was the first to stress the importance of diffraction in image formation and showed that the final image was formed by interference or addition of the incident and diffracted light. If the incident wave and the diffracted wave are added together algebraically, that is, upward displacements being treated as positive and downward ones as negative, the resultant is wave B. In the case of a partially absorbing

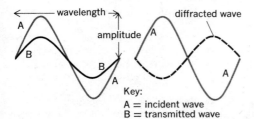

Fig. 1. Effect of a partially absorbing object on a light wave. The diffracted wave is half a wavelength out of phase with the incident wave.

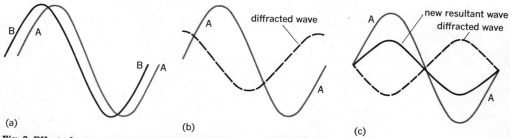

Fig. 2. Effect of a transparent object on a light wave. (*a*) Incident wave A is unaltered in amplitude but delayed in phase and is transmitted as wave B. (*b*) B can again be represented as the sum of the incident wave A and a diffracted wave. The diffracted wave is no longer half a wavelength out of phase with A, but if the phase difference can be changed to half a wavelength (*c*) the new resultant wave will resemble that produced by a partially absorbing object.

object, therefore, the trough of the diffracted wave coincides with the crest of the incident wave, that is, the two waves differ in phase by half a wavelength.

Consideration of a perfectly transparent refractile object follows. Since no light is absorbed, the height of the wave is unchanged, but if the object has a higher refractive index than its surroundings, the wave will be delayed and can be represented by wave B in **Fig. 2**. Once again, wave B can be represented as the sum of the incident wave A and a diffracted wave (Fig. 2*b*). This diffracted wave is no longer half a wavelength out of phase with wave A. If wave B is only slightly delayed, the diffracted wave is approximately one-quarter of a wavelength out of phase with A. Suppose now that the incident wave A and the diffracted wave could somehow be separated and their phase relationships altered; in particular, suppose the phase difference between them could be increased from about one-quarter of a wavelength to half a wavelength, as in Fig. 2*c*. The crest of one wave would now coincide with the trough of the other, and the new resultant wave would be indistinguishable from wave B in Fig. 1. In other words, the invisible transparent object would be indistinguishable from a partially absorbing object. This was first realized by Frits Zernike, who was awarded the 1953 Nobel Prize in physics for his invention of phase contrast.

Practical realization. The essential features of a phase-contrast microscope are shown in **Fig. 3**. The practical problem is to find some way of separating the incident or direct light from that diffracted by the object. This is done by placing a diaphragm D of easily recognizable shape, such as an annulus, at the front focal plane of the substage condenser C. Light from each point of the focal plane passes as a parallel pencil of rays through the specimen S and is brought to a focus at the rear focal plane P of the objective O. Thus, on removing the eyepiece, an image of

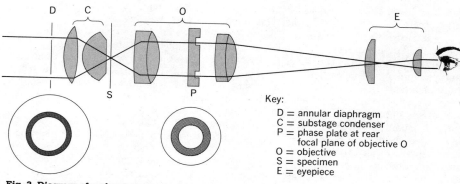

Key:
D = annular diaphragm
C = substage condenser
P = phase plate at rear focal plane of objective O
O = objective
S = specimen
E = eyepiece

Fig. 3. Diagram of a phase-contrast microscope.

the annulus will be seen at the back of the objective lens. This image corresponds to the incident light. In addition, when a specimen is present, some light is diffracted by it and spreads out to fill the whole of the back lens of the objective. Thus, apart from the small area of overlap over the image of the annulus, the direct and diffracted waves are essentially separated at the plane P. A phase plate is now inserted at this level. This can be a transparent disk with an annular groove of such dimensions that it coincides exactly with the image of the diaphragm D. All the direct light now passes through the groove in the phase plate, whereas the diffracted light passes mainly outside the groove. Since the diffracted light has to pass through a greater thickness of transparent material than the direct light, a phase difference, depending on the refractive index of the phase-plate material and on the thickness of the groove, is introduced between them. If this phase difference is about one-quarter of a wavelength, the basic conditions for phase contrast will have been achieved (Fig. 2). If the phase plate is made to retard the incident wave by a quarter of a wavelength, the crests and troughs of the two waves will coincide, giving a resultant of greater amplitude. Refractile details will appear bright (negative contrast) instead of dark (positive contrast). The commercial Anoptral system is a type of negative phase contrast.

In practice, a partially absorbing layer of metal or dye is usually deposited over the phase-plate annulus. Figure 2c shows that the amplitude of the direct wave is usually greater than that of the diffracted wave so that, although the resultant wave is reduced in amplitude, it is still far from zero, and a transparent object will not appear perfectly black, but gray. If the amplitude of the direct wave is reduced after it has passed through the object, it can be made equal to that of the diffracted wave so that the resultant is zero, and the object appears perfectly black. Unfortunately, in order to match the amplitude of all possible diffracted waves, it would be necessary to make a variable-absorption phase plate. Such a device is expensive, and in practice most manufacturers use fixed phase plates with absorptions of between 50 and 80%. Each objective must have a corresponding substage diaphragm. The commonest arrangement is a rotatable wheel in the substage condenser containing three or four diaphragms of different sizes and also a clear circular aperture which enables the instrument to be used with conventional illumination.

Image interpretation. The phase change or optical path difference ϕ introduced by an object is defined by the relationship $\phi = (n_o - n_m)t$, where n_o is the refractive index of the object, n_m that of the surrounding medium, and t the object thickness; ϕ is usually expressed either in terms of the wavelength, in which case t must also be expressed in the same unit, or in angular measure. In the latter notation, one wavelength is equal to 360° or 2π radians. Although the phase-contrast microscope converts variations in ϕ into variations in intensity, the relationship between these quantities is not a simple one. In general, as ϕ increases, the image becomes darker; but beyond a certain value of ϕ, it becomes lighter again, and finally, for very large values of ϕ, the contrast actually becomes reversed, so that the image becomes brighter than the background. Thus, it cannot be assumed that because one part of the image appears darker than another, it will necessarily correspond to a region of greater optical path. Even these basic relationships are disturbed because the theoretical requirement of complete separation between the direct and diffracted light can never be achieved in practice. This results in the appearance of a bright halo around every dark detail and an accentuation of edges and sharp discontinuities in the object. Although these effects do not allow the instrument to be used for quantitative measurements of optical path, they are not really disadvantageous for purely observational work.

Biological applications. The phase-contrast microscope is a routine instrument for the examination of living cells. It made it possible to study the structure of living cells under excellent optical conditions and with no loss in resolving power. Accurate observations thus became much easier to make, and in particular, the use of phase-contrast cinemicrography made it possible to study changes in cell structure during such processes as the movement and division of cells with great clarity. The method is also useful for the study of unstained tissue sections and for the comparison of material in the electron and optical microscopes.

A quantitative application is microrefractometry. It follows from the basic definition of phase change, namely $\phi = (n_o - n_m)t$, that if the refractive index n_m of the mounting medium is made equal to that of the object n_o, ϕ becomes zero irrespective of the object thickness t, and since the intensity is a function of ϕ, the object will become invisible. The phase-contrast microscope can thus be used as a very sensitive null indicator for measuring refractive indices. The principle is to immerse the object in a series of media of graded refractive index until one is found

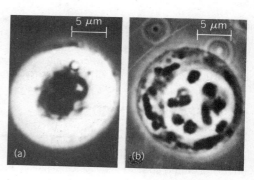

Fig. 4. Phase-contrast photograph of living locust spermatocyte. (*a*) In a saline medium. The refractive index of the cell is so much higher than the index of the medium that all internal detail is obscured. (*b*) Similar cell immersed in 9% bovine plasma albumin solution. The refractive index difference is much less, and the internal nucleoplasm and chromosomes can be clearly seen.

that makes the object invisible. This is one of the most sensitive methods of microrefractometry available, and it can be extended to the quantitative refractometry of living cells (**Fig. 4**). In this case, a suitable nontoxic medium must be used. The most suitable medium is a concentrated protein solution such as Armour's bovine plasma albumin fraction V, with added salts. The importance of cell refractometry rests on the fact that there is a linear relationship between the refractive index n of a solution and the concentration C of the dissolved substance. Thus, $n = n_o + \alpha C$ where n_o is the refractive index of the solvent, for example, water, and α is a characteristic constant known as the specific refraction increment. The mean value of α for most cellular constituents may be taken as 0.0018, and the formula can be used to calculate C, the total solid concentration in the cell. The water concentration can also be deduced, and if the cell volume is known, the dry and wet mass of the cell is also known. *See* INTERFERENCE MICROSCOPE.

Nonbiological applications. The study of transparent specimens such as crystals and fibers calls for little comment except to point out that, if the specimen is birefringent, a single polarizer must be used in order to avoid a confused image.

Industrial applications include the examination of chemicals, oils, waxes, soaps, paints, foods, plastics, rubber, resins, emulsions, and textiles.

Surface structure can often be studied by phase contrast with vertical illumination. Either the specimen must be naturally reflecting, or a thin film of reflecting material must be deposited on it or on a surface replica. A typical arrangement is shown in **Fig. 5**. A condenser lens forms an image of a lamp on an annular diaphragm. A semireflecting cover slip is placed just above the objective. A projector lens and the objective combine to form an image of the annulus at a plane just above the semireflector, and a phase plate is inserted here (compare Fig. 3). Normally, it is desirable to place the phase plate at the rear focal plane of the objective; but if this were done here, the incident light would have to pass through the metalized part of the phase plate and would lose intensity. Stray light and glare would also occur because of reflections at the surface

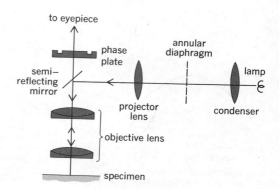

Fig. 5. Phase-contrast microscope with vertical illumination for reflecting specimens.

of the phase plate. Apart from these modifications, the system is similar to that for phase contrast with transmitted light. Differences in height in the reflecting specimen will produce differences in optical path length, and these will be converted into intensity differences.

Bibliography. W. Burrells, *Microscope Technique*, 1980; G. Wade (ed.), *Acoustic Imaging: Cameras, Microscopes, Phased Arrays, and Holographic Systems*, 1976.

INTERFERENCE MICROSCOPE
ROBERT BARER

A microscope used for visualizing and measuring differences in phase or optical path in transparent or reflecting specimens. It is closely allied to the phase-contrast microscope. SEE PHASE-CONTRAST MICROSCOPE.

Simple theory. In the phase-contrast microscope, the final image is produced by separating the incident wave and the wave diffracted by the object, introducing a phase difference between them, and then recombining the altered waves. The final resultant wave has a lower amplitude than the incident wave, and a transparent object thus becomes visible. The same result is achieved in interference microscopy by a different method. The incident and diffracted waves are not separated, but interference is produced between the transmitted wave and another wave which originates from the same source. In **Fig. 1**, A represents the incident wave, B the wave transmitted by a perfectly transparent object. Since no energy is absorbed, B has the same amplitude as A, but is slightly delayed or altered in phase relative to A. The interfering wave is represented by the broken line. According to the Abbe theory the final image is formed by the interference or summation of all waves that pass through the optical system. If the interfering wave and the transmitted wave B are added together algebraically, the resultant wave in Fig. 1 will have a lower amplitude than the incident wave A, so that the transparent object will now appear to absorb light. The resultant wave can be changed by altering the phase and amplitude of the interfering wave, thus enabling the appearance of the image to be varied. Since two waves can only interfere if they have the same wavelength and are derived from a common source, it is usual to divide the light from the source into two parts by means of a beam-splitting device, such as a semireflecting mirror. One beam is made to traverse the object, the other travels along a similar path, but does not pass through the object. When the two beams are recombined under suitable conditions, there is interference, and the differences in optical path introduced by various parts of the object can be seen as variations in intensity or color. SEE INTERFERENCE OF WAVES.

Practical realization. The basic principles may be illustrated by reference to the Linnik microscope, which has been used for studying the surface structure of reflecting specimens (**Fig. 2**). If M_2 is tilted slightly, relative to M_1, interference fringes appear, and irregularities in M_1 are seen as displacements in the fringe system. If the two mirrors are exactly parallel, the fringe will be of infinite width so that irregularities in M_1 will appear as intensity variations against a uniformly illuminated background, as in phase contrast.

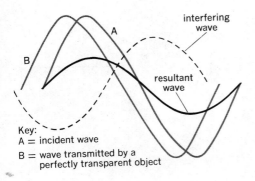

Fig. 1. Basic principle of interference microscopy. Interference is produced between B and another wave, shown by the broken line, derived from the same source. The resultant differs in amplitude from wave A so that the transparent object becomes visible.

Key:
A = incident wave
B = wave transmitted by a perfectly transparent object

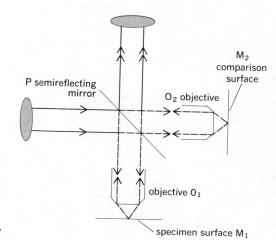

Fig. 2. Linnik interference microscope for reflecting specimens. The light from the source is divided by the semireflecting mirror P into two beams, one of which is focused through the objective O_1 onto the specimen surface M_1, the other through the objective O_2 onto the comparison surface M_2. Interference occurs between the reflected beams which are reunited by P and then viewed through the eyepiece.

Designs for use with transmitted light are more complex. A theoretically ideal system is shown in **Fig. 3**. By introducing suitable filters and transparent plates in the comparison path, the interfering beam can be varied in phase and amplitude without affecting the beam through the object. An instrument based on these principles is produced by Leitz but in view of the complexity and expense of such systems, designs have been developed using only one microscope.

The Dyson microscope (**Fig. 4**), one of the first commercial interference instruments, was manufactured by Vickers but is now obsolete. A light ray from the substage condenser (not shown) is split at the semireflecting upper surface of a glass plate P_1 into two beams b and c. Beam b passes through the object O on the slide S, which is oil immersed to another glass plate P_2 with semireflecting surfaces. Plate P_2 is cemented to a glass hemispherical block B, the upper surface of which is metallized, except near the optical axis A. Beam c is reflected downward to a reflecting spot R, then up through the system to B and also to A. Thus, at A, interference can take place between beam b, which has passed through the object, and beam c, which has passed lateral to it. The image at A is magnified by an ordinary microscope objective L, to which block B is attached by a collar. The relative amplitude of the two beams is fixed, but the phase difference between them can be varied by making P_1 and P_2 very slightly wedge-shaped and by moving P_2 by means of a calibrated screw S_1. Interference fringes can be made to appear by rotating P_1 and P_2 relative to each other; when their wedge axes are parallel, the fringes are of infinite width, giving a uniform background as in phase contrast.

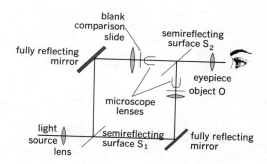

Fig. 3. A theoretical interference system for transparent specimens. The semireflecting mirror S_1 divides the light into two beams which follow similar paths. One beam passes through a microscope containing the object O, the other through a microscope containing a blank comparison slide. The beams are reunited at S_2 and are viewed through the eyepiece.

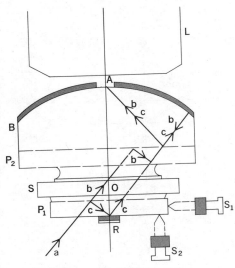

Fig. 4. Dyson interference microscope.

Key:
- O = object
- S = slide
- B = glass hemispherical block
- A = optical axis
- L = microscope objective
- S_1, S_2 = calibrated screws
- b, c = split light beams from substage condenser
- P_1, P_2 = semireflecting glass plates
- R = reflecting spot

The Smith-Baker microscope, manufactured by Vickers, uses birefringent crystals as beam splitters. The essential components are shown in **Fig. 5**. Birefringent calcite plates P_c, P_o are cemented to the front lenses of the condenser L_c and objective L_o, and the microscope is illuminated with plane polarized light from a substage polarizer. Each ray is split at P_c into an ordinary ray o and an extraordinary ray e. The extraordinary rays are focused on the object O; the ordinary rays pass to one side of it at S. The two sets of rays are recombined by the birefringent plate P and pass through another sheet of Polaroid (not shown), which acts as an analyzer and brings the rays into the correct state of polarization for interference to occur. Thus, once again the possibility occurs of having interference between waves which have passed through the object and others which have not. The relative phase and amplitude of the interfering waves can be varied easily by rotating the polarizer and analyzer and by introducing calibrated birefringent plates known as compensators. *See* BIREFRINGENCE; POLARIZED LIGHT.

Differential interference contrast (DIC) microscope. Some interference microscopes, such as the Smith-Baker shown in Fig. 5, produce two images of an object sheared laterally with respect to each other. It is often advantageous to make this shear quite large so that the object, or part of it, can be seen against a clear background. If, however, the shear is made very small (less than the resolving power), only one image will appear and the optical path through a specimen point will be compared not with that through a clear or fixed reference area, but with that through a neighboring point of the specimen. The contrast will then depend on the rate of change or gradient of optical path (in the direction of the shear). Consider a globule of uniform refractive index but thicker at its center than at the periphery. The optical path gradient will be maximal at one edge, fall to zero at the center where the globule is virtually flat, and then change sign, rising to another maximum at the opposite edge. Since the contrast depends both on steepness and sign, the globule will appear bright on one side of the center and dark on the opposite side. Shearing occurs only in one direction depending on the orientation of the beam-splitting element. This leads to an asymmetrical distribution of light and dark areas on either side of the globule and gives it a shadow-cast appearance with an impression of depth. It is important to remember that the appearance is related to optical path differences and not to true height or thickness. The temptation to regard the image as revealing surface structure (as in a scanning electron microscope) must be resisted. Thus an aqueous vacuole in a cell may produce an actual surface bulge but might appear as a depression or concavity. Despite such problems of interpre-

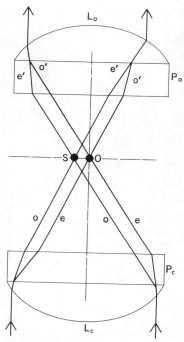

Fig. 5. Smith-Baker microscope.

Key:
P_o, P_c = birefringent calcite plates
L_c = condenser front lenses
L_o = objective front lenses
O = object
o = ordinary ray
e = extraordinary ray
S = point at which ordinary rays pass object
o′ = ordinary ray in birefringent
e′ = extraordinary ray in birefringent calcite plate

tation, the differential interference contrast microscope provides high sensitivity for the detection of transparent structures and can give high resolution.

Measurement of phase changes. The basic principle of one method of measurement may be seen from Fig. 1. If a single uniformly refractile object occupies part of the field, the background light will be produced by wave A and the light through the object by wave B. Now, if the interfering wave is made equal in amplitude to A, but exactly out of phase with it, destructive interference will occur so that there will be no background light. The object will be seen by virtue of the light diffracted by it. In other words, it will be seen by dark ground illumination. Now suppose the interfering wave is changed in phase by means of the appropriate control until it interferes destructively with B. The object will now appear dark against a lighter background. It is clear from the diagram that the phase change that has to be introduced to achieve this will be equal to that produced by the transparent object itself, that is, to the phase difference between waves A and B.

Comparison with phase-contrast type. The phase-contrast microscope can be regarded as an imperfect interference microscope in which there is incomplete separation between the two interfering beams. This results in the halo effect and the accentuation of edges. These effects can be reduced or eliminated in some interference microscopes so that accurate measurements of phase change can be made. This is not necessarily advantageous for purely observational work, however, as the optical deficiencies of phase contrast actually increase the contrast of internal details, particularly in living cells. The relative simplicity, cheapness, and ease of manipulation of the phase-contrast microscope make it the instrument of choice for routine observational work and the measurement of refractive indices, but the interference microscope is essential when it is necessary to measure a phase change.

Biological applications. Apart from the considerations already mentioned, the use of the interference microscope for observing transparent specimens such as living cells calls for little comment.

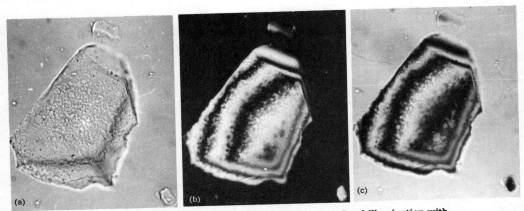

Fig. 6. Fragment of glass mounted in oil. (a) Photographed with conventional illumination with reduced substage aperture. (b) Photographed in the Baker interference microscope under darkground conditions. (c) Photographed with reversed contrast. The interference fringes present a contour map of variations in optical path through the object. Magnification ×800.

The main use of the interference microscope in biology is for the determination of dry mass. There is a linear relationship between the refractive index n of a solution and the concentration C of dissolved substances, expressed by $n = n_o + \alpha C$. The term α is a characteristic constant known as the specific refraction increment whose mean value for most cellular constituents may be taken as 0.0018. If a cell is regarded as a protoplasmic jelly of refractive index n surrounded by a medium of refractive index n_o, the optical path difference ϕ introduced by a region of the cell of thickness t is defined by $\phi = (n - n_o)t$; hence by combining these relationships, $\phi = \alpha C t$. Multiplying throughout by A, the projected area of the cell, $\phi A = \alpha A c t$ is obtained. However, $Act/100$ is simply the dry mass M of the cell, so that $\phi A = 100\alpha M$ or finally $M/A = \phi/100\alpha$. In other words, the phase change ϕ is proportional to the mass per unit area of the cell. The value of the interference microscope is at present severely limited by the fact that in order to determine the total mass of

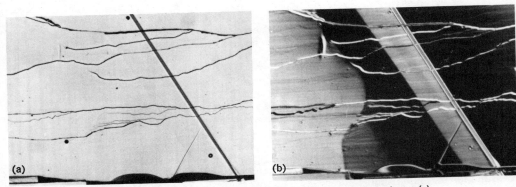

Fig. 7. Application of the Nomarski interference microscope to a metallurgical specimen. (a) Cleavage surface of a zinc crystal, showing mechanical twins and irregular cleavage steps; vertical illumination. (b) As a, with slight surface tilts (~1°) clearly shown and much fine detail revealed between the large cleavage steps. Magnification × 100. (Courtesy of A. R. Entwisle)

a heterogeneous cell, it is necessary to make numerous measurements at every point of the cell and to integrate the result. A number of automatic integrating devices have been developed for this purpose. Since the phase change ϕ depends on the product of refractive index and thickness, one of these quantities can be calculated if the other is measured. Thus, the combination of interference microscopy and microrefractometry is capable of providing much useful quantitative information about the cell. In some cases, it is possible to determine the concentration of solids and of water, the total dry and wet mass, the cell volume and thickness. Interference microscopy can also be used for studying changes in dry mass produced by enzymic digestion or specific extraction procedures.

Nonbiological applications. Like the phase-contrast microscope, the interference microscope can be applied to the study of any transparent specimen, such as fibers, crystals, and so on. A most important application is the study of reflecting specimens, such as metallic surfaces or metallized replicas of surfaces. These are of special importance in metallurgy and engineering. For this type of work, it is usual to operate the instrument with fringes in the field of view; the shape of the fringe gives a contour map of the surface (**Fig. 6**). The sensitivity is such that irregularities of the order of 0.01 of a wavelength or even less can be observed. The method thus gives valuable information about the quality of ground, polished, or etched surfaces. The differential (Nomarski) type of interference microscope has proved particularly useful for studying surface structure of reflecting specimens since it gives a true relief image uncomplicated by variations in refractive index (**Fig. 7**). SEE DIFFRACTION; RESOLVING POWER.

Bibliography. R. D. Allen, G. B. David, and G. Nomarski, Z. Wiss. Mikrosk., 69:193–221, 1969; A. W. Pollister (ed.), Physical Techniques in Biological Research, vol. 3A, 2d ed., 1966; J. Padawer, J. Roy. Microsc. Soc., 88:305–349, 1967; K. F. A. Ross, Phase Contrast and Interference Microscopy for Cell Biologists, 1967.

FLUORESCENCE MICROSCOPE
OSCAR W. RICHARDS

A variation of the compound laboratory light microscope which is arranged to transmit ultraviolet, violet, and sometimes blue radiations to a specimen. The specimen then fluoresces, that is, appears to be self-luminous and often colored. The phenomenon of fluorescence is thought to involve an electronic rearrangement in the irradiated substance. SEE OPTICAL MICROSCOPE.

Materials with absorption-spectra maxima below 320 nanometers require a quartz condenser in place of the usual glass condenser of the microscope. The microscope should have an aluminized front-surface mirror because silver is a less efficient reflector of ultraviolet.

The fluorescent image looks bright and has good contrast, although the amount of light is small. With weak fluorescence a monocular microscope must be used and the work is done in a darkened room. With more intense fluorescence a binocular microscope can be used. The air-glass surfaces of the microscope should be coated to decrease loss of light. Although all nonfluorescing objectives can be used, the 8-mm and 20× ocular is often the most useful combination.

An Abbe condenser of 1.40 numerical aperture concentrates more radiation on the specimen than condensers of smaller aperture and is used for the bright-field cross-filter method. Achromatic-corrected condensers may produce glare from autofluorescence of the cemented surfaces and are avoided, as are objectives with fluorite elements. Other investigators prefer the dark-field condenser in place of the bright-field condenser.

The absorbing filter is usually in the microscope between the objective and the observer's eye to remove any other exciting radiation not absorbed by the specimen. This may be a nearly colorless filter absorbing only ultraviolet when the specimen is to be observed in full color, or it may be of a color complementary to the radiation used. For example, with materials absorbing blue and ultraviolet radiation a yellow filter absorbs the blue radiation beyond the specimen and passes the yellow fluorescence of the specimen to the eye. The cross-filter combination is chosen to give the best visibility for the specimen.

ADAPTIVE OPTICS
James C. Wyant

The science of optical systems that measure and correct wavefront aberrations in real time, that is, simultaneously with the operation of the system. This possibility of improving the performance of an optical system has many applications for both passive imaging systems and laser systems. For example, adaptive optics offers the possibility of removing the effect of atmospheric turbulence on the images of objects viewed through the air; of allowing diffraction-limited imaging of astronomical objects from the ground; of compensating for the degrading effects of the atmosphere in transmitting laser radiation; and of correcting for static or dynamic figure errors in optical surfaces or aberrations in the gain medium of lasers. Active correction and control of optical surfaces allows new methods of construction of large optical telescopes, such as the phasing together of numerous smaller elements or the effective use of thinner, less precise optical surfaces. These techniques may significantly reduce the cost and time to fabricate large telescopes. They also may make feasible the employment of very large optical elements for ground or space applications. SEE LASER.

To appreciate the usefulness of an adaptive optics system, it is useful to look at the effect of the atmosphere upon the performance of any Earth-based passive imaging system. Neglecting atmospheric effects, a telescope with a perfect 40-in.-diameter (1-m) aperture operating at a wavelength of 0.5 micrometer has a resolution capability that extends out to 10 cycles per arc-second. However, for typical atmospheric conditions the resolution is limited to about 1 cycle per arc-second, that is, the effective aperture diameter is on the order of 4 in. (0.1 m). Adaptive optics has the potential for making the effects of the atmosphere nearly negligible for many situations. While H. W. Babcock noted this fact as early as 1953, it was only in the mid-1970s that the necessary technology became available to make atmospheric compensation a reality.

Basic techniques. Adaptive optics systems consist of three basic components (**Fig. 1**). All presently working techniques (other than nonlinear phase conjugation techniques) use a deformable mirror as the wavefront corrector. The main difference between different techniques is the wave-front sensor. Three main adaptive optics techniques have been investigated: multidither COAT (coherent optical adaptive techniques); sharpness-function maximization; and phase-conjugate.

Multidither COAT system. This is used in optical systems that transmit laser radiation and must concentrate this radiation into as small an area as possible. Three particular optical systems in which multidither COAT techniques are especially useful for improving the performance are optical communications systems, laser fusion systems, and laser weapons. To understand the operation of a multidither COAT system, consider a two-element system as shown in **Fig. 2**. One element has a fixed phase, while the optical phase of the second element can be adjusted. At the target, the far-field diffraction pattern for a two-element array varies sinusoidally with distance along the direction of the two-element array. A low-amplitude phase modulation or dither of approximately ±20° is applied to the adjustable array element. This modulation causes the diffraction pattern to move back and forth. If the target contains a small region, commonly called a glint, that strongly reflects light back toward the receiver, and if this highly reflecting region cannot be resolved by the optics used to transmit the laser radiation, the light reflected back toward the receiver is amplitude-modulated at the dither frequency. This amplitude modulation is sensed by a single intensity detector. The signal is fed back through electronics to the element phase shifter that adjusts the array element phase to position the diffraction pattern maximum intensity on the target glint. This maximization will occur even if there are time-varying phase aberrations between the transmitter and the target. If the optical system has more than two

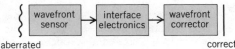

Fig. 1. Adaptive wavefront compensation system.

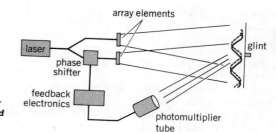

Fig. 2. Two-element dithered COAT system. (*After R. R. Shannon and J. C. Wyant, Applied Optics and Optical Engineering, Academic Press, 1979*)

elements, the maximum intensity can be further increased, and thus the radiation can be confined to a smaller area. *See Diffraction*.

Three characteristics of a multidither COAT system should be remembered: active illumination is required; a glint is necessary; and although a means of modulating the phase is required, only a single detector is needed.

Sharpness-function technique. For this technique, the correction is dithered to maximize a given function, such as the integral of the square of the irradiance in the image plane or the amount of energy within a certain region. While the sharpness-function technique has limited value in low-energy astronomical applications, it can be a powerful technique in laser systems. For example, in a laser fusion system where the goal is to maximize the amount of energy within a given-diameter area, the sharpness-function technique can be very useful. In this instance a mirror in the system would be deformed to maximize the amount of radiation passing through a given-size aperture placed in the focal plane of the system, as shown in **Fig. 3**. This system has the particular advantage that only a single detector is required, and the quantity of primary interest, namely the amount of energy within a given area, is measured directly.

Phase-conjugate system. In operating phase-conjugate systems, the wavefront to be corrected is measured directly by using either a geometric or interferometric test, as described below. Generally the system requires on the order of one detector per correcting element. For many systems, active illumination is not required, although it can be used if a glint is available. The biggest advantage of many phase-conjugate techniques is that they can work with a wide variety of light sources: a laser source is not required.

1. *Geometric wavefront sensors.* A common geometric wavefront sensor for an adaptive optics system is a Hartmann sensor. In the Hartmann sensor the wavefront disturbance being measured is imaged onto a lens array. If the incoming wavefront were perfect, each lens would focus the light at its own null position. Local wavefront slope variations, however, displace the foci from their ideal positions. This image displacement, which can be measured, is directly related to the slope error in the wavefront.

2. *Interferometric wavefront sensors.* Although there are several interferometric wavefront sensors that are applicable to adaptive optics, all the common ones use heterodyne interferometry

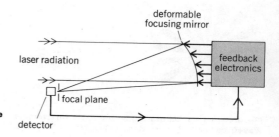

Fig. 3. Sharpness-function technique. The focusing mirror is deformed to maximize the amount of radiation passing through the aperture placed in the focal plane.

techniques for the following four reasons: high phase measurement accuracy; phase measurement independent of irradiance distribution; good measurements with small signals; and real-time measurement.

The basic feature of a heterodyne interferometer is that a frequency shift is introduced between the two interfering beams to cause the irradiance of the interference pattern to vary sinusoidally with time at a rate equal to the frequency difference between the two interfering beams. Detectors such as silicon diodes or photomultiplier tubes are used to measure this sinusoidally varying irradiance. The phase of the sinusoidal signal coming out of each detector, which is the information that needs to be determined, can be measured by using any one of several well-developed electronic techniques.

The most popular interferometer in adaptive optics is the radial Ronchi grating shown in **Fig. 4**, which like the Hartmann sensor measures wavefront slope. In the Ronchi grating interferometer the light is focused onto the grating. The interference of the 0 and +1 diffraction orders and the 0 and −1 diffraction orders gives an interferogram. If the grating is rotated, the different diffracted orders are frequency-shifted an amount equal to the number of grating lines passing a given point times the order number. The irradiance of the interference pattern produced by any two orders will vary sinusoidally at a frequency equal to the difference frequency between the two diffracted orders. The phase of this signal, which gives the wavefront slope, can be measured electronically.

Geometrically, one can think of the Ronchi grating interferometer as a chopper. If a wavefront slope error is present, the focused spot will shift sideways a small amount. The detector simply measures the light passing through the transparent positions of the Ronchi grating. If the spot shifts sideways, the time at which the light is transmitted through the rotating grating will change. This time change will be proportional to the spot displacement and thus the slope of the wavefront. By measuring this time delay as a function of pupil position, the wavefront slope variation across the aperture is obtained. *See* INTERFEROMETRY.

Wavefront correctors. The requirements for a wavefront corrector depend upon the application. As an example, consider the situation for correcting atmospheric turbulence. For turbulence correction, the response must be 100 to 1000 Hz, and the resolution should be at least 300 points for a 40-in.-diameter (1-m) aperture. Between correction points, linear interpolation is necessary. A correction accuracy of at least 1/20 of the wavelength of the light of interest is required with a correction amplitude of 1 to 2 micrometers, plus overall tilt.

In most applications, the wavefront corrector used in an adaptive optics system is a deformable mirror. Often the deformation is produced by a piezoelectric transducer exerting a force upon the mirror surface. The mirror may be the large primary mirror used in the telescope; more often it is a small mirror that is conjugate to the phase-disturbing medium or element. The mirror may be segmented or may be monolithic. Available mirrors can produce a few micrometers of motion. In a multidither system, two mirrors are generally used; one mirror has a low-amplitude, high-frequency (10–100 Hz) response, while the second has a large-amplitude, low-frequency response. In working with high-energy laser systems, it is often necessary to use a cooled mirror.

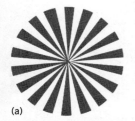
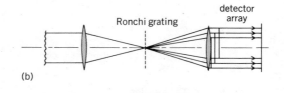

Fig. 4. Radial Ronchi grating lateral shear interferometer. (*a*) Radial Ronchi grating. (*b*) Wavefront imaged on grating. (*c*) Lateral sheared images showing 0, +1, and −1 diffraction orders.

Nonlinear phase conjugation. Nonlinear phase conjugation offers the potential to perform adaptive optics by use of a single component to do both the measuring of the wavefront aberration and the actual correction. By use of a nonlinear material, such as a crystal, the phase reversal of an aberrated wavefront is performed at very high rates, much higher than could be performed by using deformable mirrors. This phase reversal completely compensates for any aberration in the beam passing through the nonlinear material. SEE OPTICAL PHASE CONJUGATION.

Operating systems. Since many of the applications of adaptive optics are military-related, much of the work performed has not been discussed in the open literature. Four systems that have been discussed will be briefly described.

An 18-element multidither COAT system that has demonstrated a convergence time of 1.2 millisecond has been constructed. The optical system uses a 0.488-μm argon laser and an array of beam splitters, phase shifters, and beam combiners. The system has demonstrated moving glint tracking and multiple glint discrimination. Nearly diffraction-limited phasing performance has been achieved both in the laboratory and over a 330-ft (100-m) outdoor propagation range with real atmospheric turbulence.

The correction of atmospheric turbulence has been demonstrated with an image sharpness-function technique. This system consists of a telescope that has a 12 in. × 2 in. (30 cm × 5 cm) aperture and used six movable mirrors to compensate for atmospherically induced phase deformation. A feedback system adjusts the mirrors in real time to maximize the intensity of light passing through a narrow slit in the image plane. The image-sharpening system was installed and operated on equatorial mounts at both the Leuschner Observatory and Lick Observatory. As **Fig. 5** shows, the feedback system greatly improves the sharpness of the image.

A 14-in.-aperture (36-cm) adaptive optics system was produced and used with a 36-in. (91-cm) telescope at the Princeton University Observatory to reduce the photographic seeing disk of Sirius from several arc-seconds to less than an arc-second. The adaptive mirror was controlled by use of piezoelectric crystals. A sharpness-function wavefront sensor was used; the light passing

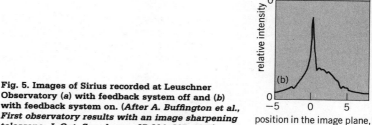

Fig. 5. Images of Sirius recorded at Leuschner Observatory (*a*) with feedback system off and (*b*) with feedback system on. (*After A. Buffington et al., First observatory results with an image sharpening telescope, J. Opt. Soc. Amer., 67:304–305, 1977*)

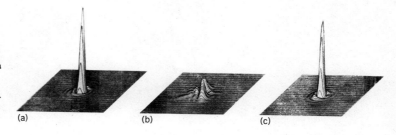

Fig. 6. Image of point source. (*a*) No aberration present. (*b*) Aberration present. (*c*) Aberration nearly removed with 21-element adaptive optics system.

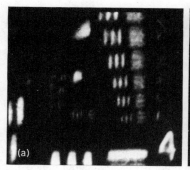 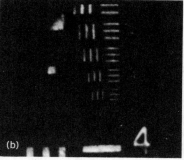

Fig. 7. Image of 3-bar-resolution target. (a) Aberration present. (b) Aberration nearly removed with 21-element adaptive optics system. (*After J. W. Hardy et al., Real-time atmospheric compensation, J. Opt. Soc. Amer., 67:360–369, 1977*)

through a hole corresponding to 1.30 arc-seconds was compared to the light that did not pass through the hole.

A phase-conjugate adaptive optics system has been described for correcting atmospheric turbulence. The instrument uses the radial Ronchi grating interferometer and a 21-element monolithic PZT (lead zirconate titanate) mirror. **Figure 6** gives a representation of the theoretical improvement that can be obtained with the 21-element phase-conjugate correction system. Figure 6a shows the intensity distribution of the image of a point if no aberration is introduced; Fig. 6b shows the intensity distribution of the image of a point if a rather large amount of aberration is present: Fig. 6c shows how well the image in Fig. 6b can be improved if the 21-element adaptive optic system removed as much aberration as is theoretically possible. **Figure 7a** shows an actual image obtained with aberration if the 21-element adaptive optics system is not used, while Fig. 7b shows the improved image obtained by using the 21-element correction system. These two photos clearly show the image improvement that can be obtained by the use of adaptive optics. SEE ABERRATION.

An adaptive optics system was designed for a 5-ft-diameter (1.5-m) mirror. The effective spacing of the detectors and correcting elements in this system is approximately 4.3 in. (11 cm). A radial Ronchi grating lateral shear interferometer is used as the wavefront sensor. The system is designed to work with sources as faint as ninth to eleventh magnitude. This system makes it possible to demonstrate the operating limits of an atmospheric compensating correction system using state-of-the-art technology.

Bibliography. A. Buffington et al., *J. Opt. Soc. Amer.*, 67:304–305, 1977; J. W. Hardy, J. E. Lefebvre, and C. L. Koliopoulous, *J. Opt. Soc. Amer.*, 67:360–369, 1977; Nonlinear optical phase conjugation, *Opt. Eng.*, 21:156–283, 1982; Optical phase conjugation, *J. Opt. Soc. Amer.*, 73:525–660, 1983; R. R. Shannon and J. C. Wyant, *Applied Optics and Optical Engineering*, vol. 7, 1979.

3
PHYSICAL NATURE OF LIGHT

Light	108
Electromagnetic wave	117
Electromagnetic radiation	118
Poynting's vector	121
Radiation pressure	122
Infrared radiation	122
Ultraviolet radiation	126
Incandescence	128
Photon	129

LIGHT

Kenneth M. Evenson and George W. Stroke

K. M. Evenson is author of the two sections Principal Effects: Speed of Light and Stabilized Lasers.

The term light, as commonly used, refers to the kind of radiant electromagnetic energy that is associated with vision. In a broader sense, light includes the entire range of radiation known as the electromagnetic spectrum. The branch of science dealing with light, its origin and propagation, its effects, and other phenomena associated with it is called optics. Spectroscopy is the branch of optics that pertains to the production and investigation of spectra.

Any acceptable theory of the nature of light must stem from observations regarding its behavior under various circumstances. Therefore, this article begins with a brief account of the principal facts known about visible light, including the relation of visible light to the electromagnetic spectrum as a whole, and then describes the apparent dual nature of light. The remainder of the article discusses various experimental and theoretical considerations pertinent to the study of this problem.

PRINCIPAL EFFECTS

The electromagnetic spectrum is a broad band of radiant energy which extends over a range of wavelengths running from trillionths of inches to hundreds of miles; wavelengths of visible light are measured in hundreds of thousandths of an inch. Arranged in order of increasing wavelength, the radiation making up the electromagnetic spectrum is termed gamma rays, x-rays, ultraviolet rays, visible light, infrared waves, microwaves, radio waves, and very long electromagnetic waves. SEE ELECTROMAGNETIC RADIATION.

Speed of light. The fact that light travels at a finite speed or velocity is well established. In round numbers, the speed of light in vacuum or air may be said to be 186,000 mi/s or 300,000 km/s.

Measurements of the speed of light, c, which had attracted physicists for 308 years, came to an end in 1983 when the new definition of the meter fixed the value of the speed of light. The speed of light is one of the most interesting and important of the fundamental physical constants. It is used to convert light travel times to distance, as in laser and radio measurements of the distance to the Moon and planets. It relates mass m to energy E in Einstein's equation, $E = mc^2$. To fix the value of c in the new definition, highly precise values of c were obtained by extending absolute frequency measurements into a region of the electromagnetic spectrum where wavelengths can be most accurately measured. These advances were facilitated by the use of stabilized lasers and high-speed tungsten-nickel diodes which were used to measure the lasers' frequencies. The measurements of the speed of light and of the frequency of lasers yielded a value of the speed of light limited only by the standard of length which was then in use. This permitted a redefinition of the meter in which the value of the speed of light assumed an exact value, 299,792,458 m/s. The meter is defined as the length of the path traveled by light in vacuum during a time interval of 1/299 792 458 of a second. SEE LASER.

Prior to the observations of O. Roemer in 1675, the speed of light was thought to be infinite. Roemer noted a variation of the orbiting periods of the moons of Jupiter that depended on the annual variation in the distance between Earth and Jupiter. He correctly ascribed the variation to the time it takes light to travel the varying distance between the two planets. The accuracy of Roemer's value of c was limited by a 30% error in the knowledge of the size of the planetary orbits at that time.

The first terrestrial measurement of c was performed by H. L. Fizeau in 1849. His measurement of the time it took light to travel to a distant mirror and return resulted in a value accurate to 15%.

J. C. Maxwell's theory of electromagnetic radiation showed that both light and radio waves were electromagnetic and hence traveled at the same speed in vacuum. This discovery soon led to another method of measuring c: it was the product of the frequency and wavelength of an electromagnetic wave. In 1891 R. Blondlot first used this method to determine a value of c by measuring both the wavelength and the frequency of an electromagnetic, radio-frequency wave. His measurement demonstrated that c was the same for radio and light waves. It is this method which now exhibits the greatest accuracy, and it is used in the most accurate measurements of c using a laser's frequency and wavelength.

In 1958 K. D. Froome reported the speed of light c to be 299,792,500 m/s, with an uncertainty of plus or minus 100 m/s. He measured both the frequency and the wavelength of millimeter waves from klystron oscillators to obtain this result. The major uncertainty lay in measuring the wavelength of the radiation. Since short wavelengths can be measured much more accurately than long wavelengths, a shorter-wavelength source was needed to improve the accuracy further; the stabilized laser soon provided such a source. However, a means of measuring its incredibly high frequency was needed. This problem, too, was soon overcome with the discovery of the tungsten-nickel point-contact diode.

Stabilized lasers. Before the advent of the laser, the most spectrally pure light came from the emission of radiation by atoms in electric discharges. The spectral purity of such radiation was about 1 part per million. Lasers, in contrast, have exhibited short-period spectral purities some 10^8 times greater than this. However, the frequency of this laser radiation was free to wander over the entire emission line, and a means of stabilizing and measuring the frequency was necessary before it could be used in a measurement of c. The technique of sub-Doppler saturated absorption spectroscopy permitted the "locking" of the frequency of the radiation to very narrow spectral features so that the frequency (and, of course, the wavelength) remained fixed. Several different lasers at different wavelengths have been stabilized, and they serve as precise frequency and wavelength sources. Three of the most common are: the helium-neon laser at a wavelength of 3.39 micrometers stabilized with a saturated absorption in methane, the 10-micrometer carbon dioxide (CO_2) laser stabilized to a saturated fluorescence in carbon dioxide, and the common red helium-neon laser stabilized to an iodine-saturated absorption. The frequencies and wavelengths of these three lasers have been measured to yield values of c.

Measurement of wavelength. Precise measurements of wavelengths are commonly made in Fabry-Perot interferometers, in which two wavelengths are compared by the observation of interference fringes of waves reflecting between two mirrors. A bright fringe occurs when the optical path length between the high-reflectivity mirrors is a multiple of a half-wavelength.

With the use of special Fabry-Perot interferometers, wavelength measurements of stabilized lasers were made with accuracies limited by the length standard then in use, the 605.8-nm orange radiation from the krypton atom. This limitation affected all speed-of-light measurements, with a resulting uncertainty of about 4 parts in 10^9. *See* INTERFEROMETRY.

Frequency measurement. The measurement of the frequency of an electromagnetic wave in the laser region is performed by a heterodyne technique in which harmonics are generated in high-speed nonlinear devices. For the most accurate measurements of c, the accuracy of the frequency measurements were 10 times more accurate than the wavelength measurements; hence, the uncertainties were dominated by the wavelength measurements.

Results. The values of c obtained from the frequency and wavelength measurements of various lasers are shown in **Fig. 1**. Because of the stability and reproducibility of the stabilized lasers, separate frequency and wavelength measurements were sometimes combined to give independent values of c. The first 1972 measurement did not involve an absolute counting of the laser frequency and was somewhat less accurate. For the first time in history, the various values of the speed of light were in agreement. Prior to 1958, the measured values of c often varied outside the limits of error quoted by the experimenters, and even prompted some observers to think that c might be changing with time. The 1974 meeting of the Consultative Committee for the Definition of the Meter (CCDM) recommended that 299,792,458 m/s be the value of c to be used for converting wavelength to frequency and vice versa, and in all other precise applications involving c. This number was arrived at by the consideration of the first four values of c in Fig. 1. The subsequent measurements of c have shown that this was a good choice, and it is the value used in the redefinition of the meter. The 1983 measurement is the only entry for visible light, and is in excellent agreement with the others. In 1983 the General Conference on Weights and Measures (CCPM) used exactly 299,792,458 m/s in the redefinition of the meter. With this new definition, fixing the value of the speed of light, the era of speed-of-light measurements was at an end.

Diffraction and reflection. One of the most easily observed facts about light is its tendency to travel in straight lines. Careful observation shows, however, that a light ray spreads slightly when passing the edges of an obstacle. This phenomenon is called diffraction. The reflec-

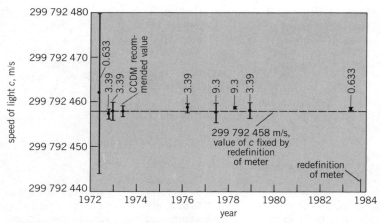

Fig. 1. Laser speed-of-light measurement made from 1972 to 1984. Separate frequency and wavelength measurements were combined to obtain the 1976, 1977, and 1983 measurements. Numbers accompanying data bars indicate the wavelengths in micrometers at which the measurements were carried out.

tion of light is also well known. The Moon, as well as all other satellites and planets in the solar system, are visible only by reflected light; they are not self-luminous like the Sun. Reflection of light from smooth optical surfaces occurs so that the angle of reflection equals the angle of incidence, a fact that is most readily observed with a plane mirror. When light is reflected irregularly and diffusely, the phenomenon is termed scattering. The scattering of light by gas particles in the atmosphere causes the blue color of the sky. When a change in frequency (or wavelength) of the light occurs during scattering, the scattering is referred to as the Raman effect. SEE DIFFRACTION; RAMAN EFFECT; REFLECTION OF ELECTROMAGNETIC RADIATION; SCATTERING OF ELECTROMAGNETIC RADIATION.

Refraction. The type of bending of light rays called refraction is caused by the fact that light travels at different speeds in different media—faster, for example, in air than in either glass or water. Refraction occurs when light passes from one medium to another in which it moves at a different speed. Familiar examples include the change in direction of light rays in going through a prism, and the bent appearance of a stick partially immersed in water. SEE REFRACTION OF WAVES.

Interference and polarization. In the phenomenon called interference, rays of light emerging from two parallel slits combine on a screen to produce alternating light and dark bands. This effect can be obtained quite easily in the laboratory, and is observed in the colors produced by a thin film of oil on the surface of a pool of water. SEE INTERFERENCE OF WAVES.

Polarization of light is usually shown with Polaroid disks. Such disks are quite transparent individually. When two of them are placed together, however, the degree of transparency of the combination depends upon the relative orientation of the disks. It can be varied from ready transmission of light to almost total opacity, simply by rotating one disk with respect to the other. SEE POLARIZED LIGHT.

Chemical effects. When light is absorbed by certain substances, chemical changes take place. This fact forms the basis for the science of photochemistry. Rapid progress has been made toward an understanding of photosynthesis, the process by which plants produce relatively complex substances such as sugars in the presence of sunlight. This is but one example of the all-important response of plant and animal life to light.

HISTORICAL APPROACH

Early in the eighteenth century, light was generally believed to consist of tiny particles. Of the phenomena mentioned above, reflection, refraction, and the sharp shadows caused by the straight path of light were well known, and the characteristic of finite velocity was suspected. All of these phenomena except refraction clearly could be expected of streams of particles, and Isaac

PHYSICAL NATURE OF LIGHT

Newton showed that refraction would occur if the velocity of light increased with the density of the medium through which it traveled.

This theory of the nature of light seemed to be completely upset, however, in the first half of the nineteenth century. During that time, Thomas Young studied the phenomena of interference, and could see no way to account for them unless light were a wave motion. Diffraction and polarization had also been investigated by that time. Both were easily understandable on the basis of a wave theory of light, and diffraction eliminated the "sharp-shadow" argument for particles. Reflection and finite velocity were consistent with either picture. The final blow to the particle theory seemed to have been struck in 1849, when the speed of light was measured in different media and found to vary in just the opposite manner to that assumed by Newton. Therefore, in 1850, it seemed finally to be settled that light consisted of waves.

Even then, however, there was the problem of the medium in which light waves traveled. All other kinds of waves required a physical medium, but light traveled through a vacuum—faster, in fact, than through air or water. The term ether was proposed by James Clerk Maxwell and his contemporaries as a name for the unknown medium, but this scarcely solved the problem because no ether was ever actually found. Then, near the beginning of the twentieth century, came certain work on the emission and absorption of energy that seemed to be understandable only if light were assumed to have a particle or corpuscular nature. The external photoelectric effect, the emission of electrons from the surfaces of solids when light is incident on the surfaces, was one of these. At that time, then, science found itself in the uncomfortable position of knowing a considerable number of experimental facts about light, of which some were understandable regardless of whether light consisted of waves or particles, others appeared to make sense only if light were wavelike, and still others seemed to require it to have a particle nature.

THEORY

The study of light deals with some of the most fundamental properties of the physical world and is intimately linked with the study of the properties of submicroscopic particles on the one hand and with the properties of the entire universe on the other. The creation of electromagnetic radiation from matter and the creation of matter from radiation, both of which have been achieved, provide a fascinating insight into the unity of physics. The same is true of the deflection of light beams by strong gravitational fields, such as the bending of starlight passing near the Sun.

A classification of phenomena involving light according to their theoretical interpretation provides the clearest insight into the nature of light. When a detailed accounting of experimental facts is required, two groups of theories appear which, in the majority of cases, account separately for the wave and the corpuscular character of light. The quantum theories seem to obviate questions concerning this dual character of light, and make the classical wave theory and the simple corpuscular theory appear as two very useful limiting theories. It happens that the wave theories of light can cope with a considerable part of the phenomena involving electromagnetic radiation. Geometrical optics, based on the wave theory of light, can solve many of the more common problems of the propagation of light, such as refraction, provided that the limitations of the underlying theory are not disregarded. SEE GEOMETRICAL OPTICS.

Phenomena involving light may be classed into three groups: electromagnetic wave phenomena, corpuscular or quantum phenomena, and relativistic effects. The relativistic effects appear to influence similarly the observation of both corpuscular and wave phenomena. The major developments in the theory of light closely parallel the rise of modern physics.

Wave phenomena. Interference and diffraction, mentioned earlier, are the most striking manifestations of the wave character of light. Their fundamental similarity can be demonstrated in a number of experiments. The wave aspect of the entire spectrum of electromagnetic radiation is most convincingly shown by the similarity of diffraction pictures produced on a photographic plate, placed at some distance behind a diffraction grating, by radiations of different frequencies, such as x-rays and visible light. The interference phenomena of light are, moreover, very similar to interference of electronically produced microwaves and radio waves.

Polarization demonstrates the transverse character of light waves. Further proof of the electromagnetic character of light is found in the possibility of inducing, in a transparent body that is being traversed by a beam of plane-polarized light, the property of rotating the plane of polarization of the beam when the body is placed in a magnetic field. SEE FARADAY EFFECT.

The fact that the velocity of light had been calculated from electric and magnetic parameters (permittivity and permeability) was at the root of Maxwell's conclusion in 1865 that "light, including heat and other radiations if any, is a disturbance in the form of waves propagated . . . according to electromagnetic laws." Finally, the observation that electrons and neutrons can give rise to diffraction patterns quite similar to those produced by visible light has made it necessary to ascribe a wave character to particles.

Electromagnetic-wave propagation. Electromagnetic waves can be propagated through free space, devoid of matter and fields, and with a constant gravitational potential; through space with a varying gravitational potential; and through more or less absorbing material media which may be solids, liquids, or gases. Radiation can be transmitted through waveguides with cylindrical, rectangular, or other boundaries, the insides of which can be either evacuated or filled with a dielectric medium.

From electromagnetic theory, and especially from the well-known equations formulated by Maxwell, a plane wave disturbance of a single frequency f is propagated in the x direction with a phase velocity $v = \lambda f = \lambda \omega / 2\pi$, where λ is the free-space wavelength and $\omega = 2\pi f$. The wave can be described by the equation $y = A \cos \omega(t - x/v)$. Two disturbances of same amplitude A, of respective angular frequencies ω_1 and ω_2 and of velocities v_1 and v_2, propagated in the same direction, yield the resulting disturbance y', defined in Eq. (1). Here $\Delta\omega = \omega_1 - \omega_2$ and $\omega =$

$$y' = y_1 + y_2 = 2A \cos \tfrac{1}{2}[(\Delta\omega)t - x\Delta(\omega/v)] \times \cos(\omega t - \omega x/v) \tag{1}$$

$\tfrac{1}{2}(\omega_1 + \omega_2)$. The ratio $u = \Delta\omega/\Delta(\omega/v)$ is defined as the group velocity, which is the speed of a light signal, just as the ratio $\omega/(\omega/v)$ is identical with the phase velocity, which is the speed of a wavefront. In the limit for small $\Delta\omega$, $u = d\omega/d(\omega/v)$. Noting that $\omega = 2\pi v/\lambda$, $d\omega = 2\pi(\lambda\, dv - v\, d\lambda)/\lambda^2$, and $d(\omega/v) = -2\pi\, d\lambda/\lambda^2$, an important relation, Eq. (2), between group and phase

$$u = v - \lambda\, dv/d\lambda \tag{2}$$

velocity is obtained. This shows that the group velocity u is different from the phase velocity v in a medium with dispersion $dv/d\lambda$. In a vacuum, $u = v = c$. With the help of Fourier theorems, the preceding expression for u can be shown to apply to the propagation of a wave group of infinite length, but with frequencies extending over a finite small domain. Furthermore, even if the wave train were emitted with an infinite length, modulation or "chopping" would result in a degrading of the monochromacity by introduction of new frequencies, and hence in the appearance of a group velocity. Considerations of this nature were not trivial in earlier measurements of the velocity of light, but were quite fundamental to the conversion of instrumental readings to a value of c. Similar considerations apply to the incorporation of the effects of the medium and the boundaries involved in the experiments. Complications arise in the regions of anomalous dispersion (absorption regions), where the phase velocity can exceed c and $dv/d\lambda$ is positive. SEE DISPERSION (RADIATION).

Refractive index. A plane wavefront, in going from a medium in which its phase velocity is v into a second medium where the velocity is v', changes direction at the interface. By geometry, it can be shown that $\sin i/\sin i' = v/v'$, where i and i' are the angles which the light path forms with a normal to the interface in the two media. It can also be shown that the path between any two points in this system is that which minimizes the time for the light to travel between the points (Fermat's principle). This path would not be a straight line unless $i = 90°$ or $v = v'$. Snell's law states that $n \sin i = n' \sin i'$, where n is the index of refraction of the medium. It follows that the refractive index of a medium is $n = c/v'$, because the refractive index of a vacuum, where $v = c$, has the value 1.

Dispersion. The dispersion $dv/d\lambda$ of a medium can easily be obtained from measurements of the refractive index for different wavelengths of light. The weight of experimental evidence, based on astronomical observations, is against a dispersion in a vacuum. Observation of the light reaching the Earth from the eclipsing binary star Algol, 120 light-years distant (1 light-year $\cong$ 6 × 10^{12} mi or 9.5 × 10^{12} km is the distance traversed in vacuum by a beam of light in 1 year), shows that the light for all colors is received simultaneously. The eclipsing occurs every 68 h 49 min. Were there a difference of velocity for red light and for blue light in interstellar space as great as 1 part in 10^6, this star would show a measurable time difference in the occurrence of the eclipses in these two colors.

Corpuscular phenomena. In its interactions with matter, light exchanges energy only in discrete amounts, called quanta. This fact is difficult to reconcile with the idea that light energy is spread out in a wave, but is easily visualized in terms of corpuscles, or photons, of light. SEE PHOTON.

Blackbody (heat) radiation. The radiation from theoretically perfect heat radiators, called blackbodies, involves the exchange of energy between radiation and matter in an enclosed cavity. The observed frequency distribution of the radiation emitted by the enclosure at a given temperature of the cavity can be correctly described by theory only if it is assumed that light of frequency ν is absorbed in integral multiples of a quantum of energy equal to $h\nu$, where h is a fundamental physical constant called Planck's constant. This startling departure from classical physics was made by Max Planck early in the twentieth century.

Photoelectric effect. When a monochromatic beam of electromagnetic radiation illuminates the surface of a solid (or less commonly, a liquid), electrons are ejected from the surface in the phenomenon known as photoemission, or the external photoelectric effect. The kinetic energies of the electrons can be measured electronically by means of a collector which is negatively charged with respect to the emitting surface. It is found that the emission of these photoelectrons, as they are called, is immediate, and independent of the intensity of the light beam, even at very low light intensities. This fact excludes the possibility of accumulation of energy from the light beam until an amount corresponding to the kinetic energy of the ejected electron has been reached. The number of electrons is proportional to the intensity of the incident beam. The velocities of the electrons ejected by light at varying frequencies agree with Eq. (3), where m is the

$$\tfrac{1}{2} m v_{max}^2 = h(\nu - \nu_0) \tag{3}$$

mass of the electron, v_{max} the maximum observed velocity, ν the frequency of the illuminating light beam, and ν_0 a threshold frequency characteristic of the emitting solid.

In 1905 Albert Einstein showed that the photoelectric effect could be accounted for by assuming that, if the incident light is composed of photons of energy $h\nu$, part of this energy, $h\nu_0$, is used to overcome the forces binding the electron to the surface. The rest of the energy appears as kinetic energy of the ejected electron.

Compton effect. The scattering of x-rays of frequency ν_0 by the lighter elements is caused by the collision of x-ray photons with electrons. Under such circumstances, both a scattered x-ray photon and a scattered electron are observed, and the scattered x-ray has a lower frequency than the impinging x-ray. The kinetic energy of the impinging x-ray, the scattered x-ray, and the scattered electron, as well as their relative directions, are in agreement with calculations involving the conservation of energy and momentum.

Quantum theories. The need for reconciling Maxwell's theory of the electromagnetic field, which describes the electromagnetic wave character of light, with the particle nature of photons, which demonstrates the equally important corpuscular character of light, has resulted in the formulation of several theories which go a long way toward giving a satisfactory unified treatment of the wave and the corpuscular picture. These theories incorporate, on one hand, the theory of quantum electrodynamics, first set forth by P. A. M. Dirac, P. Jordan, W. Heisenberg, and W. Pauli, and on the other, the earlier quantum mechanics of L. de Broglie, Heisenberg, and E. Schrödinger. Unresolved theoretical difficulties persist, however, in the higher-than-first approximations of the interactions between light and elementary particles. The incorporation of a theory of the nucleus into a theory of light is bound to call for additional formulation.

Dirac's synthesis of the wave and corpuscular theories of light is based on rewriting Maxwell's equations in a hamiltonian form resembling the hamiltonian equations of classical mechanics. Using the same formalism involved in the transformation of classical into wave-mechanical equations by the introduction of the quantum of action $h\nu$, Dirac obtained a new equation of the electromagnetic field. The solutions of this equation require quantized waves, corresponding to photons. The superposition of these solutions represents the electromagnetic field. The quantized waves are subject to Heisenberg's uncertainty principle. The quantized description of radiation cannot be taken literally in terms of either photons or waves, but rather is a description of the probability of occurrence in a given region of a given interaction or observation.

Relativistic effects. The measured magnitudes of such characteristics as wavelength and frequency, velocity, and the direction of the radiation in a light beam are affected by a relative motion of the source with respect to the observer, occurring during the emission of the signal-carrying electromagnetic wave trains.

A difference in gravitational potential also affects these quantities. Several important observations of this nature are listed in this section, followed by a discussion of several important results of general relativity theory involving light.

Velocity in moving media. In 1818 A. Fresnel suggested that it should be possible to determine the velocity of light in a moving medium, for example, to determine the velocity of a beam of light traversing a column of liquid of length d and of refractive index n, flowing with a velocity v relative to the observer, by measuring the optical thickness nd. The experiment was carried out by Fizeau in a modified Rayleigh interferometer, shown in **Fig. 2**, by measuring the fringe displacement in O corresponding to the reversing of the direction of flow. If v' is the phase velocity of light in the medium (deduced from the refractive index by the relation $v = c/n$), it is found that the measured velocity v_m in the moving medium can be expressed as $v_m = v' + v(1 - 1/n^2)$ rather than $v_m = v' + v$, as would be the case with a newtonian velocity addition.

Aberration. J. Bradley discovered in 1725 a yearly variation in the angular position of stars, the total variation being 41 seconds of arc. This effect is in addition to the well-known parallax effect, and was properly ascribed to the combination of the velocity of the Earth in its orbit and the speed of light. Bradley used the amplitude of the variation to arrive at a value of the velocity of light. George Airy compared the angle of aberration in a telescope before and after filling it with water, and discovered, contrary to his expectation, that there was no difference in angle.

Michelson-Morley experiment. The famous Michelson-Morley experiment, one of the most significant experiments of all time, was performed in 1887 to measure the relative velocity of the Earth through inertial space. Inertial space is space in which Newton's laws of motion hold. Dynamically, an inertial frame of reference is one in which the observed accelerations are zero if no forces act. A point in an orbit is the center of such a frame.

The rotation of the Earth about its axis, with tangential velocities never exceeding 0.3 mi/s (0.5 km/s) is easily demonstrated mechanically (Foucault's pendulum, precession of gyroscopes) and optically (Michelson's rectangular interferometer). The surface of the Earth is not an inertial frame. In its orbit around the Sun, on the other hand, the Earth has translational velocities of the order of 18 mi/s (30 km/s), but this motion cannot be detected by mechanical experiments because of its orbital nature. The hope existed, however, that optical experiments would permit the detection (and measurement) of the relative motion of the Earth through inertial space by comparing the times of travel of two light beams, one traveling in the direction of the translation through inertial space, and the other at right angles to it. The hope was based on the now dis-

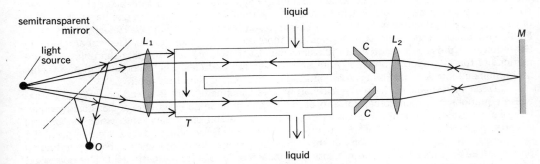

Fig. 2. Fizeau's experiment. C, compensator plates; M, mirror; L_1, L_2, lenses; T, tube; O, interference fringes.

proven notion that the velocity of a light would be equal to the constant c only when measured with respect to the inertial space, but would be measured as smaller $(c - v)$ or greater $(c + v)$ with respect to a reference frame, such as the Earth, moving with a velocity v in inertial space if a light beam were projected respectively in the direction and in the opposite direction of translation of this frame. According to classical velocity addition theorems (which, as is now known, do not apply to light), a velocity difference of $2v$ would be detected under such circumstances.

Not only does the Earth move in an orbit around the Sun, but it is carried with the Sun in the galactic rotation toward Cygnus with a velocity of several hundreds of miles per second, and the Galaxy itself is moving with a high speed in its local spiral group. Speeds of hundreds and possibly thousands of miles per second should be detectable by measurements on Earth in two orthogonal directions, assuming of course that the Earth motion is itself with respect to inertial space, or indeed that such a space has the physical meaning ascribed to it. The unexpected result of the experiment was that no such velocity difference could be detected, that is, no relative motion could be detected by optical means.

The Michelson-Morley apparatus (**Fig. 3**) consists of a horizontal Michelson interferometer with its two arms at right angles. The mirrors are adjusted so that the central white-light fringe falls on the cross hair of the observing telescope. This indicates equality of optical phase, and therefore an equality of the times taken by the light beams to travel from the beam-splitting surface to each of the two mirrors and back. Rotation of the entire system by 90°, or indeed by any angle, as well as repetition of the experiment at various times of the year all are found to leave the central white-light fringe and associated fringe system undisplaced, indicating no change in the time required by the light to traverse the two arms of the interferometer when their directions relative to the direction of the Earth's motion are varied. Had there been a difference in the velocity of light in the two directions OM and ON, the two arms would be of unequal length in the initial adjustment. For example, if the light traveled faster in the direction OM (on the average, going back and forth), then the corresponding arm would have to be longer so as to make the time of travel equal in both arms. If the apparatus were turned through 90°, the shorter arm would take the place of the longer arm, and the "faster" light would now travel in the shorter arm, and the "slower" light in the longer arm; a noticeable fringe displacement would, but actually does not, take place.

Einstein's theory accounts for the null result by the simple explanation that no relative motion between the apparatus and the observer exists in the experiment. No change in measured length has occurred in either direction, and because the propagation of light is isotropic, no velocity difference or detection of relative motion should be expected.

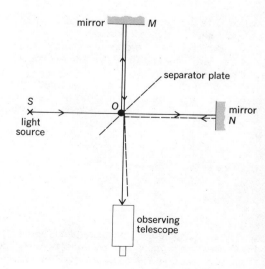

Fig. 3. Michelson-Morley experiment.

Gravitational and cosmological redshifts. Two different kinds of shifts, or displacements of spectral lines toward the red end of the spectrum, are observed in spectrograms taken with starlight or light from nebulae. One is the rare, but extremely significant, gravitational redshift or Einstein shift, which has been measured in some spectra from white dwarf stars. The other, much more widely encountered, is the redshift in spectra from external galaxies usually described as being caused by a radial Doppler effect characterizing an "expansion of the universe."

The most famous example of the gravitational redshift is that observed in the spectrum of the so-called dark companion of Sirius. The existence of this companion was predicted by F. W. Bessel in 1844 on the basis of a perturbation in the motion of Sirius, but because of its weak luminosity (1/360 that of the Sun, 1/11,000 that of Sirius), it was not observed until 1861. This companion is a white dwarf of a mass comparable to that of the Sun (0.95), but having a relatively small radius of 11,000 mi (18,000 km) and the fantastically high density of 61,000 times that of water. This companion shows a shift in its spectral lines relative to the ones emitted by Sirius itself; the shift of 0.03 nm for the Balmer β-line of hydrogen was reliably determined by W. S. Adams. In 1960, R. V. Pound and G. A. Rebka, Jr., measured the gravitational redshift in a laboratory experiment involving the Mössbauer effect.

The cosmological redshift is a systematic shift observed in the spectra of all galaxies, best measured with the calcium H and K absorption lines. Distances of galaxies are determined photometrically by measurements of their intensities, and it is found that the wavelength shift toward the red increases with the distance of the galaxies from the Earth. (The Earth does not have a privileged position if expansion of the universe is indeed involved. Rather its position is somewhat like that of a person in a crowd which is dispersing—each individual experiencing an ever-increasing distance from every other one.) The change of wavelength with distance d is given by Eq. (4), where $1/H$ is $1-2 \times 10^{10}$ light-years.

$$\Delta\lambda/\lambda = Hd \tag{4}$$

Results of general relativity. The propagation of light is influenced by gravitation. This is one of the fundamental results of Einstein's general theory of relativity which has been subjected to experimental tests and found to be verified. Three important results involving light need to be singled out.

1. The velocity of light, measured by the same magnitude c independently of the state of motion of the frame in which the measurement is being carried out, depends on the gravitational potential Φ of the field in which it is being measured according to Eq. (5). Here $\Phi = -GM/R$,

$$c = c_0\left(1 + \frac{\Phi}{c^2}\right) \tag{5}$$

where G is the universal constant of gravitation (6.670×10^{-11} in SI units), M the mass of the celestial body, R the radius of the body, and c_0 the velocity of light in a vacuum devoid of fields.

For example, the absolute value of the term Φ/c^2 is about 3000 times greater on the Sun than on Earth, making the measurements of c smaller by two parts in 10^6 on the Sun as compared to measurements on Earth.

2. The frequency ν of light emitted from a source in a gravitational field with the gravitational potential Φ is different from the frequency ν_0 emitted by an identical source (atomic, nuclear, molecular) in a field-free region, according to Eq. (6). Spectral lines in sunlight should be

$$\nu = \nu_0\left(1 + \frac{\Phi}{c^2}\right) \tag{6}$$

displaced toward the red by two parts in 10^6 when compared to light from terrestrial sources.

3. Light rays are deflected when passing near a heavenly body according to Eq. (7), where

$$\alpha = \frac{4GM}{c^2R} \tag{7}$$

α is the angular deflection in radians, and R the distance of the beam from the center of the heavenly body of mass M. The deflection is directed so as to increase the apparent angular distance of a star from the center of the Sun when starlight is passing near the edge of the Sun. The deflection according to this equation should be 1.75 seconds of arc, a value which compares favorably with eclipse measurements of the star field around the Sun in 1931. These measurements indicated values up to 2.2 seconds of arc when compared with photographs of the same field 6 months earlier. Measurements of the deflection of radio waves from extremely small-diameter celestial radio sources in 1974–1975 agreed with Einstein's theory to within 1%. This prediction of Einstein's theory might seem less surprising today when the corpus-cular-photon character of light is widely known, and when a newtonian M/R^2 attraction might be considered to be involved in the motion of a corpuscle with the velocity c past the Sun. However, application of Newton's law predicts a deviation only half as great as the well verified relativistic prediction.

Matter and radiation. The possibility of creating a pair of electrons—a positively charged one (positron) and a negatively charged one (negatron)—by a rapidly varying electromagnetic field (gamma rays of high frequency) was predicted as a consequence of Dirac's wave equation for a free electron and has been experimentally verified. I. Curie and F. Joliot, as well as J. Chadwick, P. M. S. Blackett, G. P. Occhialini, and others have compared the number of positrons and negatrons ejected by gamma rays passing through a thin sheet of lead (and other materials) and have found them to be the same, after accounting for two other groups of electrons also appearing in the experiment (photoelectrons and recoil electrons). Other examples of negatron-positron pair production include the collision of two heavy particles, a fast electron passing through the field of a nucleus, the direct collision of two electrons, the collision of two light quanta in a vacuum, and the action of a nuclear field on a gamma ray emitted by the nucleus involved in the action.

Evidence of the creation of matter from radiation, as well as that of radiation from matter, substantiates Einstein's equation, Eq. (8), which was first expressed in the following words: "If a

$$E = mc^2 \tag{8}$$

body [of mass m] gives off the energy E in the form of radiation, its mass diminishes by E/c^2."

In regard to exchanges of energy and momentum, electromagnetic waves behave like a group of particles with energy as in Eq. (9) and momentum as in Eq. (10).

$$E = mc^2 = h\nu \tag{9} \qquad p = h\nu/c = h/\lambda \tag{10}$$

Finally, many experiments with photons show that they also possess an intrinsic angular momentum, as do particles. Circularly polarized light, for example, carries an experimentally observable angular momentum, and it can be shown that, under certain circumstances, an angular momentum can be imparted to unpolarized or plane-polarized light (plane wave passing through a finite circular aperture). In any case, the angular momentum will be quantized in units of $h/2\pi$.

The inverse process to the creation of electron pairs is the annihilation of a positron and a negatron, resulting in the production of two gamma-ray quanta (two-quantum annihilation). Nuclear chain reactions are known to involve similar processes.

Bibliography. P. G. Bergmann, *Introduction to the Theory of Relativity*, 1976; M. Born and E. Wolf, *Principles of Optics*, 6th ed., 1980; S. Fluegge (ed.), *Handbuch der Physik*, vol. 24, 1956; K. D. Froome and L. Essen, *The Velocity of Light and Radio Waves*, 1969; J. L. Hall and J. Carlsten (eds.), *Laser Spectroscopy III*, 1977; E. Hecht and A. Zajac, *Optics*, 2d ed., 1987; F. A. Jenkins, *Fundamentals of Optics*, 4th ed., 1976; M. V. Klein and T. E. Furtak, *Optics*, 2d ed., 1986; D. J. E. Knight and W. R. C. Rowley, *Survey Review XXIV*, 185:131–134, 1977; R. Loudon, *The Quantum Theory of Light*, 2d ed., 1983; J. Meyer-Arendt, *Introduction to Classical and Modern Optics*, 1984; F. L. Pedrotti and L. S. Pedrotti, *Introduction to Optics*, 1987; H. H. Skilling, *Fundamentals of Electric Waves*, 2d ed., 1948, reprint 1974; A. J. W. Sommerfeld, *Lectures on Theoretical Physics*, vol. 4, 1954; J. Strong, *Concepts of Classical Optics*, 1958.

ELECTROMAGNETIC WAVE
WILLIAM R. SMYTHE

A disturbance, produced by the acceleration or oscillation of an electric charge, which has the characteristic time and spatial relations associated with progressive wave motion. A system of electric and magnetic fields moves outward from a region where electric charges are accelerated, such as an oscillating circuit or the target of an x-ray tube. The wide wavelength range over which such waves are observed is shown by the electromagnetic spectrum. The term electric wave, or Hertzian wave, is often applied to electromagnetic waves in the radar and radio range. Electromagnetic waves may be confined in tubes, such as wave guides, or guided by transmission lines. They were predicted by J. C. Maxwell in 1864 and verified experimentally by H. Hertz in 1887. *See* ELECTROMAGNETIC RADIATION.

ELECTROMAGNETIC RADIATION
William R. Smythe

Energy transmitted through space or through a material medium in the form of electromagnetic waves. The term can also refer to the emission and propagation of such energy. Whenever an electric charge oscillates or is accelerated, a disturbance characterized by the existence of electric and magnetic fields propagates outward from it. This disturbance is called an electromagnetic wave. The frequency range of such waves is tremendous, as is shown by the electromagnetic spectrum in the **table**. The sources given are typical, but not mutually exclusive, as is shown by the fact that the atomic interstellar hydrogen radiation whose wavelength is 0.210614 m falls in the radar region. The other monochromatic radiation listed is that from positron-electron annihilation whose wavelength is 2.42626×10^{-12} m.

Detection of radiation. In theory, any electromagnetic radiation can be detected by its heating effect. This method has actually been used over the range from x-rays to radio. Ionization effects measured by cloud chambers, photographic emulsions, ionization chambers, and Geiger counters have been used in the gamma- and x-ray regions. Direct photography can be used from the gamma-ray to the infrared region. Fluorescence is effective in the x-ray and ultraviolet ranges. Bolometers, thermocouples, and other heat-measuring devices are used chiefly in the infrared

Electromagnetic spectrum

Frequency, Hz	Wavelength, m	Nomenclature	Typical source
10^{23}	3×10^{-15}	Cosmic photons	Astronomical
10^{22}	3×10^{-14}	γ-rays	Radioactive nuclei
10^{21}	3×10^{-13}	γ-rays, x-rays	
10^{20}	3×10^{-12}	x-rays Positron-electron annihilation	Atomic inner shell
10^{19}	3×10^{-11}	Soft x-rays	Electron impact on a solid
10^{18}	3×10^{-10}	Ultraviolet, x-rays	Atoms in sparks
10^{17}	3×10^{-9}	Ultraviolet	Atoms in sparks and arcs
10^{16}	3×10^{-8}	Ultraviolet	Atoms in sparks and arcs
10^{15}	3×10^{-7}	Visible spectrum	Atoms, hot bodies, molecules
10^{14}	3×10^{-6}	Infrared	Hot bodies, molecules
10^{13}	3×10^{-5}	Infrared	Hot bodies, molecules
10^{12}	3×10^{-4}	Far-infrared	Hot bodies, molecules
10^{11}	3×10^{-3}	Microwaves	Electronic devices
10^{10}	3×10^{-2}	Microwaves, radar	Electronic devices
10^{9}	3×10^{-1}	Radar, interstellar hydrogen	Electronic devices
10^{8}	3	Television, FM radio	Electronic devices
10^{7}	30	Short-wave radio	Electronic devices
10^{6}	300	AM radio	Electronic devices
10^{5}	3000	Long-wave radio	Electronic devices
10^{4}	3×10^{4}	Induction heating	Electronic devices
10^{3}	3×10^{5}		Electronic devices
100	3×10^{6}	Power	Rotating machinery
10	3×10^{7}	Power	Rotating machinery
1	3×10^{8}		Commutated direct current
0	Infinity	Direct current	Batteries

and microwave regions. Crystal detectors, vacuum tubes, and transistors cover the microwave and radio frequency ranges.

Free-space waves. A charge in simple harmonic (linear sinusoidal) motion in a vacuum generates a simple wave which becomes spherical at distances from the source much larger than the amplitude of the motion and so great that many oscillations have occurred before the disturbance arrives. The wave is plane when the dimensions of the area observed are very small compared with the radius of spherical curvature. In this case the choice of the rectangular coordinates x and z as the directions of the oscillation and of the observation or field point, respectively, permits the electric intensity **E** and the magnetic flux density **B** to be written as Eq. (1). The field

$$E_x = vB_y = E_0 \cos[\omega(t - v^{-1}z)] \tag{1}$$

amplitude E_0 is constant over the specified area and not dependent on z if the z-range is small compared with the source distance, as in stellar radiation. The angular frequency of the source is ω radians per second, which is the frequency ν in hertz multiplied by 2π. The velocity of the wave is v, the direction of propagation z, and the time t. The wavelength λ is $2\pi v/\omega$. If t is in seconds and z in meters, then v is in meters per second, and λ in meters. It is found that in a lossless, isotropic, homogeneous medium Eq. (2) holds; here μ is the permeability, and ϵ the

$$v = (\mu\epsilon)^{-1/2} \tag{2}$$

capacitivity, or dielectric constant. This wave is transverse because **E** and **B** are normal to z. It is plane-polarized because E_x and B_y are parallel to fixed axes. The plane of polarization is taken as that defined by the electric vector and the direction of propagation.

Plane waves. An electromagnetic disturbance is a plane wave when the instantaneous values of any field element such as **E** and **B** are constant in phase over any plane parallel to a fixed plane. These planes are called wavefronts. In empty unbounded space, **E** and **B** lie in the wavefront normal to each other; if the wave is unpolarized, their direction fluctuates in this plane in random fashion. If the plane waves are bounded, as on transmission lines and in wave guides, the amplitudes may vary over the wavefront, and in the case of wave guides and crystals some of the elements will not in general lie in the wavefront. The equation for an undamped plane wave whose front is normal to z is Eq. (3), where F is one of the field elements such as **E** or **B**. Note

$$F = \Phi_1(x,y) f_1(z - vt) + \Phi_2(x,y) f_2(z + vt) \tag{3}$$

that if an observer sees a certain value of $\Phi_1(x,y)$ at z and then jumps instantaneously in the z direction to a point $z + \Delta z$, the observer will, after waiting a time $\Delta z/v$, see the same value $\Phi_1(x,y)$ because Eq. (4) is valid. Thus, the first term represents a wave moving in the z direction with a

$$f(z - vt) = f[z + \Delta z - v(t + \Delta z/v)] \tag{4}$$

velocity v. The sound term represents a wave in the negative z direction. The form of $\Phi_1(x,y)$ and $\Phi_2(x,y)$ depends on the boundary conditions.

Spherical waves. A wave is spherical when the instantaneous value of any field element such as **E** or **B** is constant in phase over a sphere. The radiation from any source of finite dimensions becomes spherical at great distances in an unbounded, isotropic, homogeneous medium. The equation for an undamped spherical wave is Eq. (5). The first term represents a diverging

$$F = r^{-1}\Phi_1(\theta,\varphi) f(r - vt) + r^{-1}\Phi_2(\theta,\varphi) f(r + vt) \tag{5}$$

and the second a converging wave. Again, the form of $\Phi_1(\theta,\varphi)$ and $\Phi_2(\theta,\varphi)$ depends on the nature of the source and other boundary conditions.

Damped waves. If there are energy losses which are proportional to the square of the amplitude, as in the case of a medium of conductivity γ which obeys Ohm's law, then the wave is exponentially damped, and Eq. (1) becomes Eq. (6). The symbol α is called the attenuation

$$E_x = E_0 e^{-\alpha z} \cos(\omega t - \beta z) \tag{6}$$

constant, and β the wave number or phase constant which equals ω/v', where v' is the damped-wave velocity. The electric wave amplitude at the origin has been taken as E_0. The ratio of E_0 to B_0, as well as that of α to β, depends on the permeability μ, the capacitivity ϵ, and the conductivity γ of the medium. In terms of the phasor $\check{E}_x$, Eq. (6) may be written as the real part of Eq. (7). This is exactly the form for the current on a transmission line. (Phasors are complex numbers

$$E_x = \check{E}_x e^{j\omega t} = E_0 e^{-(\alpha + j\beta)z} e^{j\omega t} \tag{7}$$

of form such that, when multiplied by $e^{j\omega t}$, the real part of the product gives the amplitude, phase, and time dependence.)

Wave impedance. Those trained in transmission line theory find it useful to apply the same techniques to wave theory. Consider an isolated tubular section of the wave in Eq. (1) bounded by $x = 0$, $x = 1$, and $y = 0$, $y = 1$ as a transmission line. The potential across the line between $x = 0$ and $x = 1$ is **E**. The line integral of **B** around the $x = 0$ boundary from $y = 0$ to $y = 1$ is μI by Ampère's law and equals **B** because **B** is zero on the negative side. Thus, the impedance of the line is, making use of Eqs. (1) and (2), given by Eq. (8). This depends only on

$$\check{Z}_k = \frac{V}{I} = \frac{\mu E}{B} = \frac{E}{H} = \left(\frac{\mu}{\epsilon}\right)^{1/2} = \eta \tag{8}$$

the properties of the medium and is known as the wave impedance. In transmission line theory the ratio μ/ϵ would be replaced by the ratio of the series impedance $\check{Z}_L = j\omega L$ to the shunt

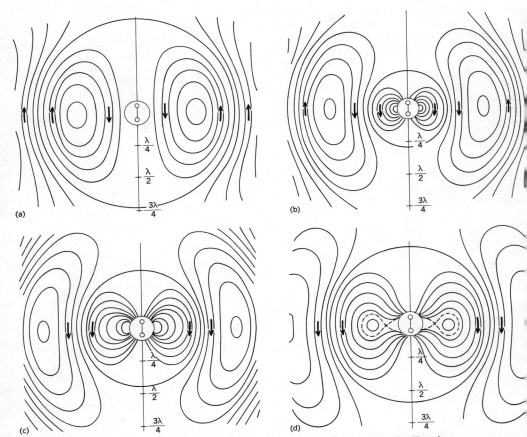

Diagrams of electric dipole. The outward moving electric field lines generated by the Hertzian oscillator are shown at successive eighth-period intervals. (a) $t = 0$; (b) $t = T/8$; (c) $t = T/4$; (d) $t = 3T/8$.

admittance $\check{Y} = j\omega C$, where L is the inductance per unit length, and C the capacitance per unit length across the line. If there is a resistance R per unit length across the line, then $1/R$ must be added to Y. This resistance is $1/\gamma$ for the tubular section. Thus, for a conducting medium, Eq. (8) becomes Eq. (9). The last term is a common transmission line form. The reflection and refraction

$$\check{Z}_k = \left(\frac{j\omega\mu}{\gamma + j\omega\epsilon}\right)^{1/2} = \frac{j\omega\mu}{\alpha + j\beta} \tag{9}$$

of plane waves at plane boundaries separating different mediums may be calculated by transmission line formulas with the aid of Eqs. (8) and (9).

Electric dipole. A charge undergoing simple harmonic motion in free space is a dipole source when the amplitude of the motion is small compared with the wavelength. The term is loosely applied to the hertzian oscillator, usually pictured as a dumbbell-shaped conductor in which the electrons oscillate from one end to the other, leaving the opposite end periodically positive. An electric dipole of moment M is defined as the product qa when two large, equal and opposite charges, $+q$ and $-q$, are placed a small distance a apart. A dipole is oscillating when M is periodic in time and is the simplest source of spherical waves. Much can be learned by a study of H. Hertz's picture of the outward moving electric field lines at successive time intervals of one-eighth period in a plane which passes through the hertzian oscillator axis, shown in the **illustration**. The most striking feature of the pictures is that, after breaking loose from the dipole, all electric field lines are closed, which means that the divergence of **E** is zero. This is true of all unbounded waves. It is also noteworthy that the waves become truly spherical with a fixed wavelength λ only in a direction perpendicular to the dipole and at a distance which greatly exceeds the dipole dimensions. This distance is beyond the edges of the picture. Lengths $\lambda/4$, $\lambda/2$, and $3\lambda/4$ are marked off on the axis for comparison. The magnetic field lines are circles coaxial with the oscillator, so they intersect the plane of the diagram normally. They are most dense where the electric lines are closely spaced. The radiant energy emitted by atoms and molecules is essentially radiation of the dipole type. SEE ABSORPTION OF ELECTROMAGNETIC RADIATION; DIFFRACTION; INFRARED RADIATION; INTERFERENCE OF WAVES; LIGHT; POLARIZATION OF WAVES; REFLECTION OF ELECTROMAGNETIC RADIATION; REFRACTION OF WAVES; SCATTERING OF ELECTROMAGNETIC RADIATION; ULTRAVIOLET RADIATION.

Bibliography. M. Born and E. Wolf, *Principles of Optics*, 6th ed., 1980; A. G. Brown, *X-Rays and Their Applications*, 1975; F. A. Jenkins, *Fundamentals of Optics*, 4th ed., 1976; H. P. Neff, *Basic Electromagnetic Fields*, 1981; C. R. Paul and S. A. Nasar, *Introduction to Electromagnetic Fields*, 1982; S. Ramo and J. R. Whinnery, *Fields and Waves in Communications Electronics*, 2d ed., 1984; F. H. Read, *Electromagnetic Radiation*, 1980; V. Rojansky, *Electromagnetic Fields and Waves*, 1980; W. R. Smythe, *Static and Dynamic Electricity*, 3d ed., 1968; R. K. Wangsness, *The Electromagnetic Field*, 1979.

POYNTING'S VECTOR
WILLIAM R. SMYTHE

A vector, the outward normal component of which, when integrated over a closed surface in an electromagnetic field, represents the outward flow of energy through that surface. It is given by Eq. (1), where **E** is the electric field strength, **H** the magnetic field strength, **B** the magnetic flux

$$\Pi = \mathbf{E} \times \mathbf{H} = \mu^{-1}\mathbf{E} \times \mathbf{B} \tag{1}$$

density, and μ the permeability. This can be shown with the aid of Maxwell's equations, Eqs. (2),

$$\mathbf{H} \cdot (\nabla \times \mathbf{E}) - \mathbf{E} \cdot (\nabla \times \mathbf{H}) = \nabla \cdot (\mathbf{E} \times \mathbf{H}) = -\mathbf{i} \cdot \mathbf{E} - \mathbf{E} \cdot \frac{\delta \mathbf{D}}{\delta t} - \mathbf{H} \cdot \frac{\delta \mathbf{B}}{\delta t} \tag{2}$$

where **D** is the electric displacement and i the current density. Integration over any volume v and use of the divergence theorem to replace one volume integral by a surface integral give Eq. (3),

$$-\int (\mathbf{E} \times \mathbf{H}) \cdot \mathbf{n}\, dS = \int_v \left[\frac{\delta}{\delta t}(\tfrac{1}{2}\mathbf{B} \cdot \mathbf{H}) + \frac{\delta}{\delta t}(\tfrac{1}{2}\mathbf{D} \cdot \mathbf{E}) + \mathbf{E} \cdot \mathbf{i}\right] dv \tag{3}$$

where **n** is a unit vector normal to dS. In the volume integral, $\frac{1}{2}\mathbf{B} \cdot \mathbf{H}$ is the magnetostatic energy density, and $\frac{1}{2}\mathbf{D} \cdot \mathbf{E}$ is the electrostatic energy density, so the integral of the first two terms represents the rate of increase of energy stored in the magnetic and electric fields in v. The product of $\mathbf{E} \cdot \mathbf{i}$ is the rate of energy dissipation per unit volume as heat; or, if there is a motion of free charges so that $\mathbf{i}$ is replaced by ρv, ρ being the charge density, it is the energy per unit volume used in accelerating these charges. The net energy change must be supplied through the surface, which explains the interpretation of Poynting's vector.

It should be noted that this proof permits an interpretation of Poynting's vector only when it is integrated over a closed surface. In quantum theory, where the photons are localized, it could be interpreted as representing the statistical distribution of photons over the surface. Perhaps this justifies the common practice of using Poynting's vector to calculate the energy flow through a portion of a surface.

When an electromagnetic wave is incident on a conducting or absorbing surface, theory predicts that it should exert a force on the surface in the direction of the difference between the incident and the reflected Poynting's vector. SEE ELECTROMAGNETIC RADIATION; RADIATION PRESSURE.

RADIATION PRESSURE
WILLIAM R. SMYTHE

Pressure exerted by electromagnetic radiation on objects on which it impinges. This pressure is caused by the fact that electromagnetic radiation transmits energy and possesses momentum. In the case of a plane electromagnetic wave incident normally on a plane absorbing sheet, the mean pressure is ϵE_0^2, where E_0 is the amplitude of the electric field and ϵ is the dielectric constant of the medium. If the wave impinges normally on a perfectly reflecting, plane conducting sheet, then standing waves are formed, and the average pressure is twice that on the absorbing sheet. These pressures are very small (about 10^{-9} pascal or 10^{-14} atm if E is a few volts per meter), but were measured successfully by E. F. Nichols and G. F. Hull in 1903. The effect is conspicuous in the case of a comet near the Sun, where the radiation pressure from the Sun forces the lighter cometary constituents away from the Sun. SEE ELECTROMAGNETIC RADIATION.

INFRARED RADIATION
WILLIAM L. WOLFE

Electromagnetic radiation in which wavelengths lie in the range from about 1 micrometer to 1 millimeter. This radiation therefore has wavelengths just a little longer than those of visible light and cannot be seen with the unaided eye. The radiation was discovered in 1800 by William Herschel, who used a prism to refract the light of the Sun onto mercury-in-glass thermometers placed just past the red end of the visible spectrum generated by the prism. Because the techniques and materials used to collect, focus, detect, and display infrared radiation are different from those of the visible, and because many of the applications of infrared radiation are also quite different, a technology has arisen, and many scientists and engineers have specialized in its application. SEE ELECTROMAGNETIC RADIATION.

Sources. The source can be described by the spectral distribution of power emitted by an ideal radiating body. This distribution is characteristic of the temperature of the body. A real body is related to it by a radiation efficiency factor also called emissivity. It is the ratio at every wavelength of the emission of a real body to that of the ideal under identical conditions. **Figure 1** shows curves for these ideal bodies radiating at a number of different temperatures. The higher the temperature, the greater the total amount of radiation. The total number of watts per square meter is given by $5.67 \times 10^{-8} T^4$, where T is the absolute temperature in kelvins (K). Higher temperatures also provide more radiation at shorter wavelengths. This is evidenced by the maxima of these curves moving to shorter wavelengths with higher temperatures. Ideal radiators are also called blackbodies.

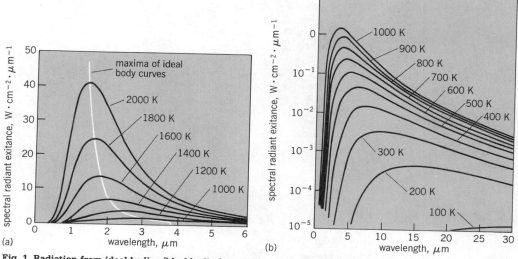

Fig. 1. Radiation from ideal bodies (blackbodies) at different temperatures, shown on (a) a linear scale and (b) a logarithmic scale. °F = (K × 1.8) − 460.

The sources can be either cooperative or uncooperative. Some examples of the former include tungsten bulbs (sometimes with special envelopes), globars, and Nernst glowers. These are made of rare-earth oxides and carbons. They closely approximate blackbodies and are used mostly for spectroscopy. Lasers have been used in special applications. Although they provide very intense, monochromatic, and coherent radiation, they are limited in their spectral coverage. The principal infrared lasers have been carbon dioxide (CO_2) and carbon monoxide (CO) gas lasers and lead-tin-tellurium (PbSnTe) diodes. *See* LASER.

Transmitting medium. The radiation of one of these sources propagates from the source to the optical collector. This path may be through the vacuum of outer space, 3 ft (1 m) of laboratory air, or some arbitrary path through the atmosphere. **Figure 2** shows a low-resolution transmission spectrum of the atmosphere and the transmissions of the different atmospheric constituents. Two of the main features are the broad, high transmission regions between 3 and 5 micrometers and between 8 and 12 micrometers. These are the spectral regions chosen for most terrestrial applications. **Figure 3** shows a small section of this spectrum, illustrating its complexity. The radiation from the source is filtered by the atmosphere in its propagation so that the flux on the optical system is related to both the source spectrum and the transmission spectrum of the atmosphere.

Optical system. A lens, mirror, or a combination of them is used to focus the radiation onto a detector. Since glass is opaque in any reasonable thickness for radiation of wavelengths longer than 2 μm, special materials must be used. The **table** lists the properties of some of the most useful materials for infrared instrumentation. In general, these are not as effective as glass is in the visible, so many infrared optical systems use mirrors instead. The mirrors are characterized by the blank and by its coating. Blanks are usually made of aluminum, beryllium, or special silica materials. The choices are based on high strength, light weight, and good thermal and mechanical stability. They are coated with thin evaporated layers of aluminum, silver, or gold. The reflectivities of thin-film metallic coatings increase with wavelength, and the requirements for surface smoothness are also less stringent with increasing wavelength. *See* MIRROR OPTICS; OPTICAL MATERIALS; OPTICAL TELESCOPE.

Detectors. Photographic film is not useful for most of the infrared spectrum. The response of the silver halide in the film, even when specially processed, is only sensitive to about 1.2 μm.

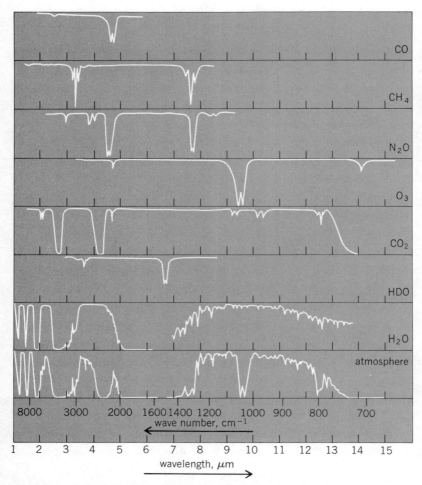

Fig. 2. Low-resolution transmission spectra of the atmosphere and of constituent gases.

It cannot respond to radiation in the important 3–5-μm and 8–12-μm atmospheric transmission bands. If there were a film for the infrared, it would have to be kept cold and dark before and after exposure. Accordingly, infrared systems have used point (elemental) detectors or arrays of them. These detectors are based either on the generation of a change in voltage due to a change in the detector temperature resulting from the power focused on it, or on the generation of a change in voltage due to some photon-electron interaction in the detector material. This latter effect is sometimes called the internal photoelectric effect. Electrons which are bound to the different atomic sites in the crystal lattice receive a quantum of photon energy. They are freed from their bound lattice positions and can contribute to current flow. The energy in electronvolts required to do this is $1.24/\lambda$ (where λ is the wavelength in micrometers). Thus only a very small binding energy, about 0.1 eV, is permitted in photon detectors. The thermal agitation of the lattice could cause spontaneous "emission," so most infrared photodetectors are cooled to temperatures from 10 to 100 K (-424 to $-280°F$). This does not affect the speed of response of photodetectors, which depends upon photon-electron interactions, but it does slow down thermal detectors. These are sometimes cooled because they generally respond to a relative change in temperature (and for a

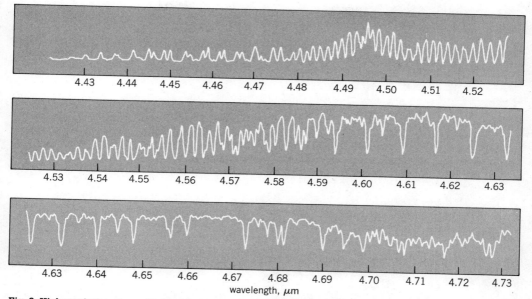

Fig. 3. High-resolution atmospheric transmission spectrum between 4.43 and 4.73 μm.

lower temperature, a small absolute change gives a larger relative change), and thermal noise is also reduced at low temperatures. *See* BOLOMETER; RADIOMETRY.

Electronic circuitry. The voltage or current from the detector or detector array is amplified by special circuitry, usually consisting of metal oxide semiconductor field-effect transistors (MOSFETs) which are designed for low-temperature operations. The amplified signals are then handled very much like television signals. One important feature of most of these systems is that the system does not yield a direct response; only changes are recorded. Thus a "dc restore" or absolute level must be established with a thermal calibration source. The black level of the display can then be chosen by the operator.

Reticle. A reticle or chopper is an important feature of nonimaging systems. A typical reticle and its use are shown in **Figure 4**. An image of the scene is portrayed on a relatively large fraction of the reticle—anywhere from 10 to 100%. The lens just behind it collects all the light that passes through the reticle and focuses it on a detector. The reticle is rotated. If the scene is uniform, there will be no change in the detector output. However, if the scene has a point source (like the image of the hot exhaust of an aircraft engine), then a point image is formed on the reticle and a periodic detector output is generated. The phase and frequency of this output can

Properties of materials for infrared instrumentation

Material	Transmission region, μm	Approximate refractive index	Comment
Germanium	2–20	4	Opaque when heated
Silicon	2–15	3.5	Opaque when heated
Fused silica	0.3–2.5	2.2	Strong, hard
Zinc selenide	0.7–15	2.4	Expensive
Magnesium fluoride	0.7–14	1.6	Not very strong
Arsenic sulfur glass	0.7–12	2.2	Not always homogeneous
Diamond	0.3–50	1.7	Small sizes only
Salt	0.4–15	1.5	Attacked by moisture

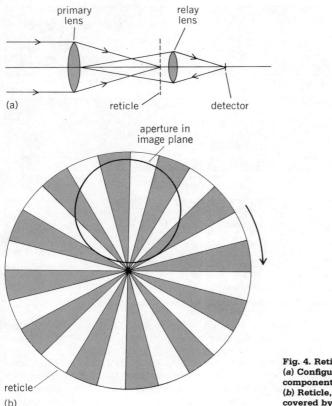

Fig. 4. Reticle system. (*a*) Configuration of components of system. (*b*) Reticle, showing area covered by image.

be used with properly designed reticles to obtain the angular coordinates of the point source. The reticle pattern can also be used to reduce the modulation obtained from almost uniform scenes which radiate from large areas. Reticles are used in most infrared tracking systems, although other schemes are sometimes employed.

Bibliography. S. S. Ballard (ed.), *Proceedings of the Institute of Radio Engineers: Infrared*, special issue, September 1959; R. D. Hudson, Jr., *Infrared System Engineering*, 1969; R. B. Johnson and W. L. Wolfe (eds.), *Selected Papers on Infrared Design*, SPIE vol. 513, 1984; R. Vanzetti, *Practical Applications of Infrared Techniques*, 1972; W. L. Wolfe and G. J. Zissis, *The Infrared Handbook*, 1978.

ULTRAVIOLET RADIATION
Fred W. Billmeyer

Electromagnetic radiation in the wavelength range 4–400 nanometers. The ultraviolet region begins at the short wavelength (violet) limit of visibility and extends to the wavelength of long x-rays. It is loosely divided into the near (400–300 nm), far (300–200 nm), and extreme (below 200 nm) ultraviolet regions (see **illus**.). In the extreme ultraviolet, strong absorption of the radiation by air requires the use of evacuated apparatus; hence this region is called the vacuum ultraviolet. Important phenomena associated with ultraviolet radiation include biological effects and applications, the generation of fluorescence, and chemical analysis through characteristic absorption or fluorescence.

Biological effects of ultraviolet radiation include erythema or sunburn, pigmentation or tanning, and germicidal action. The wavelength regions responsible for these effects are indicated in the figure. Important biological uses of ultraviolet radiation include therapy, the production of vitamin D, the prevention and cure of rickets, and the disinfection of air, water, and other substances.

Fluorescence and phosphorescence are phenomena often generated as a result of the absorption of ultraviolet radiation. These phenomena are utilized in fluorescent lamps, in fluorescent dyes and pigments, in ultraviolet photography, and in phosphors. The effectiveness of ultraviolet radiation in generating fluorescence is shown in the illustration. *See* Photography.

Chemical analysis may be based on characteristic absorption of ultraviolet radiation. Alternatively, the fluorescence arising from absorption in the ultraviolet region may itself be analyzed or observed. *See* Fluorescence microscope.

Sources of ultraviolet radiation include the Sun (although much solar ultraviolet radiation is absorbed in the atmosphere); arcs of elements such as carbon, hydrogen, and mercury; and incandescent bodies. The wavelengths produced by some sources of ultraviolet radiation are indicated in the illustration.

Artificial sources of ultraviolet light are often used to simulate the effects of solar ultraviolet radiation in the study of the deterioration of materials on exposure to sunlight. Trace amounts of chemicals which strongly absorb ultraviolet radiation may effectively stabilize materials against such degradation.

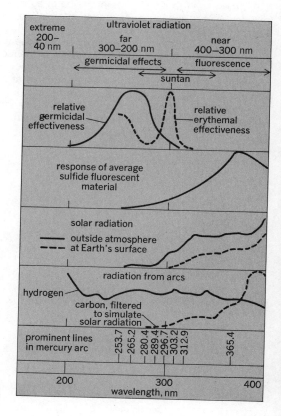

Phenomena associated with ultraviolet radiation.

Detectors of ultraviolet radiation include biological and chemical systems (the skin, the eye of an infant, or eye without a lens, and photographic materials are sensitive to this radiation), but more useful are physical detectors such as phototubes, photovoltaic and photoconductive cells, or radiometric devices.

Bibliography. W. Harm, *Biological Effects of Ultraviolet Radiation*, 1980; C. N. Rao, *Ultraviolet and Visible Spectroscopy; Chemical applications*, 3d ed., 1975.

INCANDESCENCE
HOWARD W. RUSSELL AND GEORGE R. HARRISON

The emission of visible radiation by a hot body. A theoretically perfect radiator, called a blackbody, will emit radiant energy according to Planck's radiation law at any temperature. Prediction of the visual brightness requires additional consideration of the sensitivity of the eye, and the radiation will be visible only for temperatures of the blackbody which are above some minimum. The relation between brightness and temperature is plotted in the **illustration**. As shown, the minimum temperature for incandescence for the dark-adapted eye is about 730°F (390°C). Under these ideal observing conditions, the incandescence appears as a colorless glow. The dull red light commonly associated with incandescence of objects in a lighted room requires a temperature of about 930°F (500°C). *SEE VISION.*

Not all sources of light are incandescent. A cold gas under electrical excitation may emit light, as in the so-called neon tube or the low-pressure mercury-vapor lamp. Ultraviolet light from mercury vapor may excite visible light from a cold solid, as in the fluorescent lamp. Luminescence is the term used to refer to the emission of light due to causes other than high temperature, and includes thermoluminescence, in which emission of previously trapped energy occurs on moderate heating.

Flames are made luminous by incandescent particles of carbon. Gas flames can be made to produce intense light by the use of a gas mantle of thoria containing a small amount of ceria. This mantle is a good emitter of visible light, but a poor emitter of infrared radiation. As less heat is lost in the long waves, the mantle operates at a higher temperature than a blackbody would and, hence, produces more intense visible light.

A useful criterion of an incandescent source is its color temperature, the temperature at which a blackbody has the same color, although not necessarily the same brightness. The color

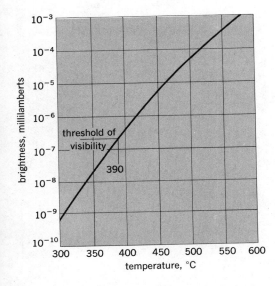

Graph showing the relation between the brightness of a blackbody and temperature. °F = (°C × 1.8) + 32.

Approximate color temperatures of common light sources

Source	Color temperature, °F (K)
Candle	3005 (1925)
Kerosine lamp	3140 (2000)
Common tungsten-filament 100-watt electric light bulb	4600 (2800)
Carbon arc	6700 (4000)
Sun	10,000 (5800)

temperature of common light sources depends upon operating conditions. Approximate values are given in the accompanying **table**.

Bibliography. M. M. Benarie, Optical pyrometry below red heat, *J. Opt. Soc. Amer.*, 47:1005–1009, 1957; R. W. Ditchburn, *Light*, 3 vols., 1977; F. A. Jenkins, *Fundamentals of Optics*, 4th ed., 1976; Optical Society of America, *Handbook of Optics*, 1978.

PHOTON
Murray Sargent III

A quantum of a single mode (that is, single wavelength, direction, and polarization) of the electromagnetic field. There are also two other definitions of photon in use, not entirely consistent with the first definition or each other: an elementary light particle or "fuzzy ball," and an informal unit of light energy. The fuzzy-ball definition emphasizes a particle character of light suggested, for example, by momentum exhibited in the Compton effect and light levitation phenomena. Although this definition is often justified by the random arrivals of counts in photoelectron detection, light waves incident on a quantum-mechanical detector yield the same behavior. More critically, the fuzzy-ball picture lacks a rigorous foundation and is not required for the explanation of any fundamental phenomenon. As an informal unit of energy, the photon equals $h\nu$, where h is Planck's constant ($= 6.626 \times 10^{-34}$ joule-second), and ν is the frequency of the light in hertz.

The definition as a single-mode light quantum has rigorous foundation in quantum electrodynamics, and contradicts the fuzzy-ball definition in that, according to Fourier analysis, light of a single wavelength must be spread out. Other theories, typified by "neoclassical" theory, attempt to explain the interaction between light and matter by quantizing only the matter's response, that is, without using the photon. However, quantum electrodynamics remains the only theory capable of quantitatively explaining spontaneous emission, the Lamb shift, and the anomalous magnetic moment of the electron.

Bibliography. C. Cohen-Tannoudji, B. Diu, and F. Laloë, *Quantum Mechanics*, 1977; M. O. Scully and M. Sargent III, The concept of the photon, *Phys. Today*, 25(3):38–47, March 1972.

WAVE OPTICS

Wave optics	132
Point source	132
Inverse-square law	132
Huygens' principle	133
Interference of waves	133
Fringe	142
Interferometry	142
Interference filter	151
Diffraction	153
Diffraction grating	162
Polarization of waves	165
Polarized light	166

WAVE OPTICS
Richard C. Lord

The branch of optics which treats of light (or electromagnetic radiation in general) with explicit recognition of its wave nature. The counterpart to wave optics is ray optics or geometrical optics, which does not assume any wave character but treats the propagation of light as a straight-line phenomenon except for changes of direction induced by reflection or refraction. See GEOMETRICAL OPTICS.

Any optical phenomenon which is correctly describable in terms of geometrical optics can also be correctly described in terms of wave optics. However, the many phenomena of interference, diffraction, and polarization are incontrovertible evidence of the wave nature of light, and geometrical optics often gives an incomplete or incorrect description of the behavior of light in an optical system. This is especially true if changes of refractive index occur within a space which is of the order of several wavelengths of the light. See DIFFRACTION; ELECTROMAGNETIC WAVE; INTERFERENCE OF WAVES; POLARIZED LIGHT.

POINT SOURCE
McAllister Hull, Jr.

A source having definite position but no extension in space. In discussing radiation, it is convenient to define the concept of point source. If the radiation propagates in radially straight lines (or, which is the same thing, in spherical waves) from the point source, conservation of energy demands that the intensity of the radiation decrease in any direction inversely as the square of the distance from the source. No physical source is actually a mathematical point, but for distances sufficiently large compared to dimensions of the source, the inverse-square law may be a good approximation. See INVERSE-SQUARE LAW.

INVERSE-SQUARE LAW
William R. Smythe

Any law in which a physical quantity varies with distance from a source inversely as the square of that distance. When energy is being radiated by a point source (see **illus**.), such a law holds, provided the space between source and receiver is filled with a nondissipative, homogeneous, isotropic, unbounded medium. All unbounded waves become spherical at distances r, which are large compared with source dimensions so that the angular intensity distribution on the expanding wave surface, whose area is proportional to r^2, is fixed. Hence emerges the inverse-square law. See POINT SOURCE.

Similar reasoning shows that the same law applies to mechanical shear waves in elastic media and to compressional sound waves. It holds statistically for particle sources in a vacuum, such as radioactive atoms, provided there are no electromagnetic fields and no mutual interactions. The term is also used for static field laws such as the law of gravitation and Coulomb's law in electrostatics.

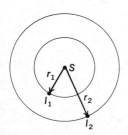

Point source S emitting energy of intensity I. The inverse-square law states that $I_1/I_2 = r_2^2/r_1^2$.

HUYGENS' PRINCIPLE
Francis A. Jenkins and William W. Watson

An assumption regarding the behavior of light waves, originally proposed by C. Huygens in the seventeenth century to explain the fact that light travels in straight lines and casts sharp shadows. Large-scale waves, such as sound waves or water waves, bend appreciably into the shadow. The special behavior of light may be explained by Huygens' principle, which states that "each point on a wavefront may be regarded as a source of secondary waves, and the position of the wavefront at a later time is determined by the envelope of these secondary waves at that time." Thus a wave WW originating at S is shown in **illus.** a at the instant it passes through an aperture. If a large number of circular secondary waves, originating at various points on WW, are drawn with the radius r representing the distance the wave would travel in time t, the envelope of these secondary waves is the heavily drawn circular arc $W'W'$. This represents the wave after t. If, as Huygens' principle requires, the disturbance is confined to the envelope, it will be 0 outside the limits indicated by points W'.

Careful observation shows that there is a small amount of light beyond these points, decreasing rapidly with distance into the geometrical shadow. This is called diffraction. See Diffraction.

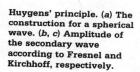
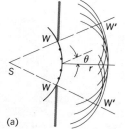
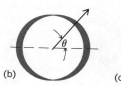
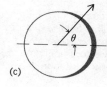

Huygens' principle. (a) The construction for a spherical wave. (b, c) Amplitude of the secondary wave according to Fresnel and Kirchhoff, respectively.

The Huygens-Fresnel principle, a modification of Huygens' original formulation, is capable of explaining diffraction. A. Fresnel in 1814 postulated that the amplitude of any secondary wave decreases in proportion to $\cos \theta$, when θ is the angle between the normal to the original wavefront and any point on the secondary wave (see illus. b, where the thickness of the arc indicates the amplitude). Fresnel then modified Huygens' requirement that the disturbance be confined to the envelope, by specifying that at any point the disturbance was the resultant of all displacements due to secondary waves reaching that point. In this way Fresnel was able to explain the complicated diffraction patterns that are produced by sending light through small apertures. Subsequent theoretical investigations by G. Kirchhoff showed that the correct obliquity factor should be $1 + \cos \theta$ instead of $\cos \theta$ (illus. c). Approximations made by Fresnel had compensated for this. A discrepancy in the phase of the resultant wave of one-quarter period was also explained by Kirchhoff.

INTERFERENCE OF WAVES
Bruce H. Billings

The process whereby two or more waves of the same frequency or wavelength combine to form a wave whose amplitude is the sum of the amplitudes of the interfering waves. The interfering waves can be electromagnetic, acoustic, or water waves, or in fact any periodic disturbance.

The most striking feature of interference is the effect of adding two waves in which the trough of one wave coincides with the peak of another. If the two waves are of equal amplitude, they can cancel each other out so that the resulting amplitude is zero. This is perhaps most dramatic in sound waves; it is possible to generate acoustic waves to arrive at a person's ear so as to cancel out noise that is disturbing him. In optics, this cancellation can occur for particular wavelengths in a situation where white light is a source. The resulting light will appear colored. This gives rise to the iridescent colors of beetles' wings and mother-of-pearl, where the substances involved are actually colorless or transparent.

Two-beam interference. The quantitative features of the phenomenon can be demonstrated most easily by considering two interfering waves. The amplitude of the first wave at a particular point in space can be written as Eq. (1), where A_0 is the peak amplitude, and ω is 2π times the frequency. For the second wave Eq. (2) holds, where $\varphi_1 - \varphi_2$ is the phase

$$A = A_0 \sin(\omega t + \varphi_1) \quad (1) \qquad B = B_0 \sin(\omega t + \varphi_2) \quad (2)$$

difference between the two waves. In interference, the two waves are superimposed, and the resulting wave can be written as Eq. (3). Equation (3) can be expanded to give Eq. (4). By writing

$$A + B = A_0 \sin(\omega t + \varphi_1) + B_0 \sin(\omega t + \varphi_2) \quad (3)$$

$$A + B = (A_0 \sin \varphi_1 + B_0 \sin \varphi_2) \cos \omega t + (A_0 \cos \varphi_1 + B_0 \cos \varphi_2) \sin \omega t \quad (4)$$

Eqs. (5) and (6), Eq. (4) becomes Eq. (7), where C^2 is defined in Eq. (8). When C is less than A or

$$A_0 \sin \varphi_1 + B_0 \sin \varphi_2 = C \sin \varphi_3 \quad (5) \qquad A_0 \cos \varphi_1 + B_0 \cos \varphi_2 = C \cos \varphi_3 \quad (6)$$

$$A + B = C \sin(\omega t + \varphi_3) \quad (7) \qquad C^2 = A_0^2 + B_0^2 + 2A_0 B_0 \cos(\varphi_2 - \varphi_1) \quad (8)$$

B, the interference is called destructive. When it is greater, it is called constructive. For electromagnetic radiation, such as light, the amplitude in Eq. (7) represents an electric field strength. This field is a vector quantity and is associated with a particular direction in space, the direction being generally at right angles to the direction in which the wave is moving. These electric vectors can be added even when they are not parallel. For a discussion of the resulting interference phenomena SEE POLARIZED LIGHT.

In the case of radio waves or microwaves which are generated with vacuum tube or solid-state oscillators, the frequency requirement for interference is easily met. In the case of light waves, it is more difficult. Here the sources are generally radiating atoms. The smallest frequency spread from such a light source will still have a bandwidth of the order of 10^7 Hz. Such a bandwidth occurs in a single spectrum line, and can be considered a result of the existence of wave trains no longer than 10^{-8} s. The frequency spread associated with such a pulse can be written as notation (9), where t is the pulse length. This means that the amplitude and phase of the wave

$$\Delta f \approx \frac{1}{2\pi t} \quad (9)$$

which is the sum of the waves from two such sources will shift at random in times shorter than 10^{-8} s. In addition, the direction of the electric vector will shift in these same time intervals. Light which has such a random direction for the electric vector is termed unpolarized. When the phase shifts and direction changes of the light vectors from two sources are identical, the sources are termed coherent.

Splitting of light sources. To observe interference with waves generated by atomic or molecular transitions, it is necessary to use a single source and to split the light from the source into parts which can then be recombined. In this case, the amplitude and phase changes occur simultaneously in each of the parts at the same time.

Young's two-slit experiment. The simplest technique for producing a splitting from a single source was done by T. Young in 1801 and was one of the first demonstrations of the wave nature of light. In this experiment, a narrow slit is illuminated by a source, and the light from this slit is caused to illuminate two adjacent slits. The light from these two parallel slits can interfere, and the interference can be seen by letting the light from the two slits fall on a white screen. The screen will be covered with a series of parallel fringes. The location of these fringes can be derived approximately as follows: In **Fig. 1**, S_1 and S_2 are the two slits separated by a distance d. Their plane is a distance l from the screen. Since the slit S_0 is equidistant from S_1 and S_2, the intensity and phase of the light at each slit will be the same. The light falling on P from slit S_1 can be

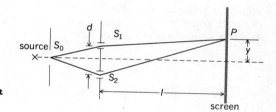

Fig. 1. Young's two-slit interference.

represented by Eq. (10) and from S_2 by Eq. (11), where f is the frequency, t the time, c the velocity

$$A = A_0 \sin 2\pi f \left(t - \frac{x_1}{c} \right) \quad (10) \qquad B = A_0 \sin 2\pi f \left(t - \frac{x_2}{c} \right) \quad (11)$$

of light; x_1 and x_2 are the distances of P from S_1 and S_2, and A_0 is the amplitude. This amplitude is assumed to be the same for each wave since the slits are close together, and x_1 and x_2 are thus nearly the same. These equations are the same as Eq. (1) and (2), with $\varphi_1 = x_1/c$ and $\varphi_2 = x_2/c$. Accordingly, the square of the amplitude or the intensity at P can be written as Eq. (12). In general, l is very much larger than y so that Eq. (12) can be simplified to Eq. (13). Equation (13) is

$$I = 4A_0^2 \cos^2 \frac{2\pi f}{c} (x_1 - x_2) \quad (12) \qquad I = 4A_0^2 \cos^2 \pi \left(\frac{yd}{l\lambda} \right) \quad (13)$$

a maximum when Eq. (14) holds and a minimum when Eq. (15) holds, where n is an integer.

$$y = n\lambda \frac{l}{d} \quad (14) \qquad y = (n + 1/2) \lambda \frac{l}{d} \quad (15)$$

Accordingly, the screen is covered with a series of light and dark bands called interference fringes. If the source behind slit S_0 is white light and thus has wavelengths varying perhaps from 400 to 700 nm, the fringes are visible only where $x_1 - x_2$ is a few wavelengths, that is, where n is small. At large values of n, the position of the nth fringe for red light will be very different from the position for blue light, and the fringes will blend together and be washed out. With monochromatic light, the fringes will be visible out of values of n which are determined by the diffraction pattern of the slits. For an explanation of this SEE DIFFRACTION.

The energy carried by a wave is measured by the intensity, which is equal to the square of the amplitude. In the preceding example of the superposition of two waves, the intensity of the individual waves in Eqs. (1) and (2) is A^2 and B^2, respectively. When the phase shift between them is zero, the intensity of the resulting wave is given by Eq. (16).

$$(A + B)^2 = A^2 + 2AB + B^2 \quad (16)$$

This would seem to be a violation of the first law of thermodynamics, since this is greater than the sum of the individual intensities. In any specific experiment, however, it turns out that the energy from the source is merely redistributed in space. The excess energy which appears where the interference is constructive will disappear in those places where the energy is destructive. This is illustrated by the fringe pattern in the Young two-slit experiment. The energy on the screen from each slit alone is given by Eq. (17), where A_0^2 is the intensity of the light from each

$$E_1 = \int_0^\infty A_0^2 \, dy \quad (17)$$

slit as given by Eq. (10). The intensity from the two slits without interference would be twice this value. The intensity with interference is given by Eq. (18). The comparison between $2E_1$ and E_3

$$E_3 = \int_0^\infty 4A^2 \cos^2 \left[2\pi \left(\frac{yd}{l\lambda} \right) \right] dy \quad (18)$$

need be made only over a range corresponding to one full cycle of fringes. This means that the argument of the cosine in Eq. (18) need be taken only from zero to π. This corresponds to a section of screen going from the center to a distance $y = l\lambda/2d$. From the two slits individually, the energy in this section of screen can be written as Eq. (19).

$$2E_1 = 2 \int_0^{l\lambda/2d} A_0^2 \, dy = \frac{A_0^2 l\lambda}{d} \tag{19}$$

With interference, the energy is given by Eq. (20). Equation (20) can be written as Eq. (21).

$$E_3 = \int_0^{l\lambda/2d} 4A_0^2 \cos^2\left[2\pi\left(\frac{yd}{l\lambda}\right)\right] dy \quad (20) \qquad E_3 = \frac{l\lambda}{2\pi d}\int_0^\pi 4A_0^2 \cos^2\varphi \, d\varphi = \frac{A_0^2 l\lambda}{d} \tag{21}$$

Thus, the total energy falling on the screen is not changed by the presence of interference. The energy density at a particular point is, however, drastically changed. This fact is most important for those waves of the electromagnetic spectrum which can be generated by vacuum-tube oscillators. The sources of radiation or antennas can be made to emit coherent waves which will undergo interference. This makes possible a redistribution of the radiated energy. Quite narrow beams of radiation can be produced by the proper phasing of a linear antenna array.

The double-slit experiment also provides a good illustration of Niels Bohr's principle of complementarity.

Fresnel double mirror. Another way of splitting the light from the source is the Fresnel double mirror (**Fig. 2**). Light from the slit S_0 falls on two mirrors M_1 and M_2 which are inclined to each other at an angle of the order of a degree. On a screen where the illumination from the two mirrors overlaps, there will appear a set of interference fringes. These are the same as the fringes produced in the two-slit experiment, since the light on the screen comes from the images of the slits S_1' and S_2' formed by the two mirrors, and these two images are the equivalent of two slits.

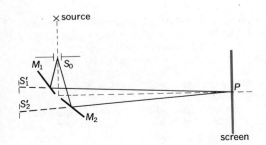

Fig. 2. Fresnel's double-mirror interference.

Fresnel biprism. A third way of splitting the source is the Fresnel biprism. A sketch of a cross section of this device is shown in **Fig. 3**. The light from the slit at S_0 is transmitted through the two halves of the prism to the screen. The beam from each half will strike the screen at a different angle and will appear to come from a source which is slightly displaced from the original slit. These two virtual slits are shown in the sketch at S_1' and S_2'. Their separation will depend on the distance of the prism from the slit S_0 and on the angle θ and index of refraction of the prism material. In Fig. 3, a is the distance of the slit from the biprism, and l the distance of the biprism

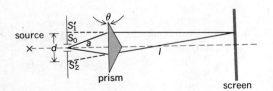

Fig. 3. Fresnel biprism interference.

WAVE OPTICS   **137**

Fig. 4. Equipment for demonstrating Fresnel biprism interference.

from the screen. The distance of the two virtual slits from the screen is thus $a + l$. The separation of the two virtual slits is given by Eq. (22), where μ is the refractive index of the prism material.

$$d = 2a(\mu - 1)\theta \tag{22}$$

This can be put in Eq. (14) for the two-slit interference pattern to give Eq. (23) for the position of a bright fringe.

$$y = n\lambda \frac{a + l}{2a(\mu - 1)\theta} \tag{23}$$

A photograph of the experimental equipment for demonstrating interference with the Fresnel biprism is shown in **Fig. 4**. A typical fringe pattern is shown in **Fig. 5**. This pattern was

Fig. 5. Interference fringes formed with Fresnel biprism and mercury-arc light source.

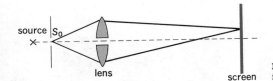

Fig. 6. Billet split-lens interference.

obtained with a mercury-arc source, which has several strong spectrum lines, accounting in part for the intensity variation in the pattern. The pattern is also modified by diffraction at the apex of the prism.

Billet split lens. The source can also be split with the Billet split lens (**Fig. 6**). Here a simple lens is sawed into two parts which are slightly separated.

Lloyd's mirror. An important technique of splitting the source is with Lloyd's mirror (**Fig. 7**). The slit S_1 and its virtual image S'_2 constitute the double source. Part of the light falls directly on the screen, and part is reflected at grazing incidence from a plane mirror. This experiment differs from the previously discussed experiments in that the two beams are no longer identical. If the screen is moved to a point where it is nearly in contact with the mirror, the fringe of zero path difference will lie on the intersection of the mirror plane with the screen. This fringe turns out to be dark rather than light, as in the case of the previous interference experiments. The only explanation for this result is that light experiences a 180° phase shift on reflection from a material of higher refractive index than its surrounding medium. The equation for maximum and minimum light intensity at the screen must thus be interchanged for Lloyd's mirror fringes.

Amplitude splitting. The interference experiments discussed have all been done by splitting the wavefront of the light coming from the source. The energy from the source can also be split in amplitude. With such amplitude-splitting techniques, the light from the source falls on a surface which is partially reflecting. Part of the light is transmitted, part is reflected, and after further manipulation these parts are recombined to give the interference. In one type of experiment, the light transmitted through the surface is reflected from a second surface back through the partially reflecting surface, where it combines with the wave reflected from the first surface (**Fig. 8**). Here the arrows represent the normal to the wavefront of the light passing through surface S_1 to surface S_2. The wave is incident at A and C. The section at A is partially transmitted to B, where it is again partially reflected to C. The wave leaving C now consists of two parts, one of which has traveled a longer distance than the other. These two waves will interfere. Let AD be the perpendicular from the ray at A to the ray going to C. The path difference will be given by Eq. (24), where μ is the refractive index of the medium between the surfaces S_1 and S_2, and AB and CD are defined in Eqs. (25) and (26).

$$\Delta = 2\mu(AB) - (CD) \quad (24) \qquad (AB) = d/\cos r \quad (25) \qquad (CD) = 2(AB)\sin r \cos i \quad (26)$$

From Snell's law, Eq. (27) is obtained and thus Eq. (28) holds.

$$\sin i = \mu \sin r \quad (27) \qquad \Delta = \frac{2\mu d}{\cos r} - \frac{2\mu d}{\cos r}\sin^2 r = 2\mu d \cos r \quad (28)$$

The difference in terms of wavelength and the phase difference are, respectively, given by Eqs. (29) and (30).

$$\Delta' = \frac{2\mu d \cos r}{\lambda} \quad (29) \qquad \Delta\varphi = \frac{4\pi\mu d \cos r}{\lambda} + \pi \quad (30)$$

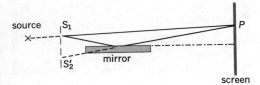

Fig. 7. Lloyd's mirror interference.

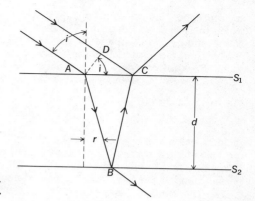

Fig. 8. Dielectric-plate-reflection interference.

The phase difference of π radians is added because of the phase shift experienced by the light reflected at S_1. The experimental proof of this 180° phase shift was shown in the description of interference with Lloyd's mirror. If the plate of material has a lower index than the medium in which it is immersed, the π radians must still be added in Eq. (30), since now the beam reflected at S_2 will experience this extra phase shift. The purely pragmatic necessity of such an additional phase shift can be seen by considering the intensity of the reflected light when the surfaces S_1 and S_2 almost coincide. Without the extra phase shift, the two reflected beams would be in phase and the reflection would be strong. This is certainly not proper for a film of vanishing thickness. Constructive interference will take place at wavelengths for which $\Delta\varphi = 2m\pi$, where m is an integer. If the surfaces S_1 and S_2 are parallel, the fringes will be located optically at infinity. If they are not parallel, d will be a function of position along the surfaces and the fringes will be located near the surface. The intensity of the fringes will depend on the value of the partial reflectivity of the surfaces.

Testing of optical surfaces. Observation of fringes of this type can be used to determine the contour of a surface. The surface to be tested is put close to an optically flat plate. Monochromatic light is reflected from the two surfaces and examined as in Fig. 8. One of the first experiments with fringes of this type was performed by Isaac Newton. A convex lens is pressed against a glass plate and illuminated with monochromatic light. A series of circular interference fringes known as Newton's rings appear around the point of contact. From the separation between the fringes, it is possible to determine the radius of curvature of the lens.

Thin films. Interference fringes of this two-surface type are responsible for the colors which appear in oil films floating on water. Here the two surfaces are the oil-air interface and the oil-water interface. The films are close to a visible light wavelength in thickness. If the thickness is such that, in a particular direction, destructive interference occurs for green light, red and blue will still be reflected and the film will have a strong purple appearance. This same general phenomenon is responsible for the colors of beetles' wings.

Channeled spectrum. Amplitude splitting shows clearly another condition that must be satisfied for interference to take place. The beams from the source must not only come from identical points, but they must also originate from these points at nearly the same time. The light which is reflected from C in Fig. 8 originates from the source later than the light which makes a double traversal between S_1 and S_2. If the surfaces are too far apart, the spectral regions of constructive and destructive interference become so close together that they cannot be resolved. In the case of interference by wavefront splitting, the light from different parts of a source could only be considered coherent if examined over a sufficiently short time interval. In the case of amplitude splitting, the interference when surfaces are widely separated can only be seen if examined over a sufficiently narrow frequency interval. If the two surfaces are illuminated with white light and the eye is used as the analyzer, interference cannot be seen when the separation is more than a few wavelengths. The interval between successive wavelengths of constructive interference becomes so small that each spectral region to which the eye is sensitive is illuminated, and no color is seen. In this case, the interference can again be seen by examining the reflected light with a spectroscope. The spectrum will be crossed with a set of dark fringes at those wavelengths for which there is destructive interference. This is called a channeled spectrum. For large separations of the surfaces, the separation between the wavelengths of destructive interference becomes smaller than the resolution of the spectrometer, and the fringes are no longer visible.

Fresnel coefficient. The amplitude of the light reflected at normal incidence from a dielectric surface is given by the Fresnel coefficient, Eq. (31), where A_0 is the amplitude of the

$$A = A_0 \frac{n_1 - n_2}{n_1 + n_2} \tag{31}$$

incident wave and n_1 and n_2 are the refractive indices of the materials in the order in which they are encountered by the light. In the simple case of a dielectric sheet, the intensity of the light reflected normally will be given by Eq. (32), where B is the amplitude of the wave which has passed through the sheet and is reflected from the second surface and back through the sheet to join A. The value of B is given by Eq. (33), where the approximation is made that the intensity of

$$C^2 = A^2 + B^2 + 2AB \cos \varphi \tag{32}$$

$$B = \frac{n_2 - n_3}{n_2 + n_3} \tag{33}$$

the light is unchanged by passing through the first surface and where n_3 is the index of the material at the boundary of the far side of the sheet.

Nonreflecting film. An interesting application of Eq. (32) is the nonreflecting film. A single dielectric layer is evaporated onto a glass surface to reduce the reflectivity of the surface to the smallest possible value. From Eq. (32) it is clear that this takes place when $\cos \varphi = -1$. If the surface is used in an instrument with a broad spectral range, such as a visual device, the film thickness should be adjusted to put the interference minimum in the first order and in the middle of the desired spectral range. For the eye, this wavelength is approximately in the yellow so that such films reflect in the red and blue and appear purple. The index of the film should be chosen to make $C^2 = 0$. At this point Eqs. (34)–(36) hold. Equation (36) can be reduced to Eq. (37). In the

$$(A - B)^2 = 0 \tag{34}$$

$$\frac{n_1 - n_2}{n_1 + n_2} = \frac{n_2 - n_3}{n_2 + n_3} \tag{35}$$

$$n_1 n_2 - n_2^2 + n_1 n_3 - n_2 n_3 = n_1 n_2 - n_1 n_3 + n_2^2 - n_2 n_3 \tag{36}$$

$$n_2 = \sqrt{n_1 n_3} \tag{37}$$

case of a glass surface in air, $n_1 = 1$ and $n_3 \cong 1.5$. Magnesium fluoride is a substance which is frequently used as a nonreflective coating, since it is hard and approximately satisfies the relationship of Eq. (37). The purpose of reducing the reflection from an optical element is to increase its transmission, since the energy which is not reflected is transmitted. In the case of a single element, this increase is not particularly important. Some optical instruments may have 15–20 air-glass surfaces, however, and the coating of these surfaces gives a tremendous increase in transmission.

Haidinger fringes. When the second surface in two-surface interference is partially reflecting, interference can also be observed in the wave transmitted through both surfaces. The interference fringes will be complementary to those appearing in reflection. Their location will depend on the parallelism of the surfaces. For plane parallel surfaces, the fringes will appear at infinity and will be concentric rings. These were first observed by W. K. Haidinger and are called Haidinger fringes.

Multiple-beam interference. If the surfaces S_1 and S_2 are strongly reflecting, it is necessary to consider multiple reflections between them. For air-glass surfaces, this does not apply since the reflectivity is of the order of 4%, and the twice-reflected beam is much reduced in intensity.

In **Fig. 9** the situation in which the surfaces S_1 and S_2 have reflectivities r_1 and r_2 is shown. The space between the surfaces has an index n_2 and thickness d. An incident light beam of amplitude A is partially reflected at the first surface. The transmitted component is reflected at S_2 and is reflected back to S_1 where a second splitting takes place. This is repeated. Each successive

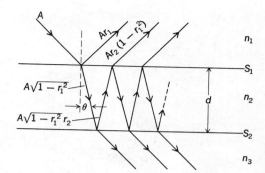

Fig. 9. Multiple reflection of wave between two surfaces.

component of the waves leaving S_1 is retarded with respect to the next. The amount of each retardation is given by Eq. (38).

$$\varphi = \frac{4\pi n d}{\lambda} \cos \theta \tag{38}$$

Equation (7) was derived for the superposition of two waves. It is possible to derive a similar expression for the superposition of many waves. From Fig. 9, the different waves at a plane somewhere above S_1 can be represented by the following expressions:

Incoming wave $= A \sin \omega t$
First reflected wave $= A r_1 \sin \omega t$
Second reflected wave $= A(1 - r_1^2) r_2 \sin (\omega t + \varphi)$
Third reflected wave $= -A(1 - r_1^2)^2 r_1 r_2^2 \times \sin (\omega t + 2\varphi)$

By inspection of these terms, one can write down the complete series. As in Eq. (3), the sine terms can be broken down and coefficients collected. A simpler method is to multiply each term by $i = \sqrt{-1}$ and add a cosine term with the same coefficient and argument. The individual terms then are all of the form of expression (39), where m is an integer.

$$B e^{-i\omega t} e^{-im\varphi} \tag{39}$$

The individual terms of expression (39) can be easily summed. For the reflected wave one obtains Eq. (40). Again, as in the two-beam case, the minimum in the reflectivity R is obtained

$$R = \frac{r_1 + r_2 e^{-i\varphi}}{1 + r_1 r_2 e^{-i\varphi}} \tag{40}$$

when $\varphi = N\pi$, where N is an odd integer and $r_1 = r_2$. The fringe shape, however, can be quite different from the earlier case, depending on the values of the reflectives r_1 and r_2. The greater these values, the sharper become the fringes.

It was shown earlier how two-beam interference could be used to measure the contour of a surface. In this technique, a flat glass test plate was placed over the surface to be examined and monochromatic interference fringes were formed between the test surface and the surface of the plate. These two-beam fringes have intensities which vary as the cosine squared of the path difference. It is very difficult with such fringe to detect variations in fringe straightness or, in other terms, variations of surface planarity that are smaller than 1/20 wavelength. If the surface to be examined is coated with silver and the test surface is also coated with a partially transmitting metallic coat, the reflectivity increases to a point where many beams are involved in the formation of the interference fringes. The shape of the fringes is given by Eq. (40). The shape of fringes for different values of r is shown in **Fig. 10**. With high-reflectivity fringes, the sensitivitity to a departure from planarity is increased far beyond 1/20 wavelength.

It is thus possible with partially silvered surfaces to get a much better picture of small irregularities than with uncoated surfaces. The increase in sensitivity is such that steps in cleaved

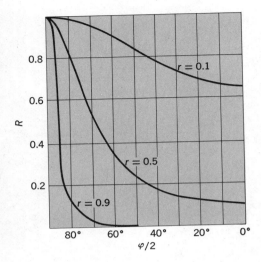

Fig. 10. The shape of multiple-beam fringes for different values of surface reflectivity.

mica as small as 1 nm in height can be seen by examining the monochromatic interference fringes produced between a silvered mica surface and a partially silvered glass flat. SEE INTERFERENCE FILTER; INTERFEROMETRY.

Bibliography. M. Born and E. Wolf, *Principles of Optics*, 6th ed., 1980; O. S. Heavens, *Optical Properties of Thin Solid Films*, 1955; E. Hecht and A. Zajac, *Optics*, 2d ed., 1987; F. A. Jenkins, *Fundamentals of Optics*, 4th ed., 1976; M. V. Klein and T. E. Furtak, *Optics*, 2d ed., 1986; J. Meyer-Arendt, *Introduction to Classical and Modern Optics*, 2d ed., 1984.

FRINGE
FRANCIS A. JENKINS AND WILLIAM W. WATSON

The part of optics using the light or dark bands produced by interference or diffraction of light. Distances between fringes are usually very small, because of the short wavelength of light. Fringes are clearer and more numerous when produced with light of a single color.

Diffraction fringes are formed when light from a point source, or from a narrow slit, passes by an opaque object of any shape. The Fraunhofer diffraction fringes produced when the light approaches and leaves the obstacle in essentially plane waves are especially important in the theory of optical instruments. SEE DIFFRACTION; RESOLVING POWER.

Interference fringes are obtained by bringing together two or more beams of light that have originated from a common source. This is usually accomplished by means of an apparatus especially designed for the purpose called an interferometer, although interference fringes may also be seen in nature. Examples are the colors in a soap film and in an oil film on water. When the fringes are controllable, for example, by changing the paths traversed by the two beams in an interferometer, they are valuable for accurate measurement of small distances and of slight differences in refractive index. SEE INTERFERENCE OF WAVES; INTERFEROMETRY.

If the light from a laser is passed through an interferometer, features such as coherence and phase fluctuations may be checked by observing the sharpness and drift of the fringes. SEE LASER.

INTERFEROMETRY
JAMES C. WYANT

The design and use of optical interferometers. Optical interferometers based on both two-beam interference and multiple-beam interference of light are extremely powerful tools for metrology and

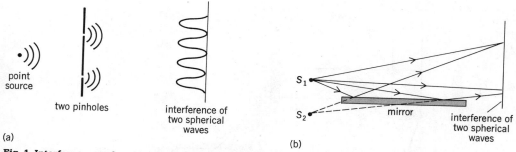

Fig. 1. Interference produced by division of wavefront. (*a*) Young's two-pinhole interferometer. (*b*) Lloyd's mirror.

spectroscopy. A wide variety of measurements can be performed, ranging from determining the shape of a surface to an accuracy of less than a millionth of an inch (25 nanometers) to determining the separation, by millions of miles, of binary stars. In spectroscopy, interferometry can be used to determine the hyperfine structure of spectrum lines. By using lasers in classical interferometers as well as holographic interferometers and speckle interferometers, it is possible to perform deformation, vibration, and contour measurements of diffuse objects that could not previously be performed.

Basic classes of interferometers. There are two basic classes of interferometers: division of wavefront and division of amplitude. **Figure 1** shows two arrangements for obtaining division of wavefront. For the Young's double pinhole interferometer (Fig. 1*a*), the light from a point source illuminates two pinholes. The light diffracted by these pinholes gives the interference of two point sources. For the Lloyd's mirror experiment (Fig. 1*b*), a mirror is used to provide a second image S_2 of the point source S_1, and in the region of overlap of the two beams the interference of two spherical beams can be observed. There are many other ways of obtaining division of wavefront; however, in each case the light leaving the source is spatially split, and then by use of diffraction, mirrors, prisms, or lenses the two spatially separated beams are superimposed.

Figure 2 shows one technique for obtaining division of amplitude. For division-of-amplitude interferometers a beam splitter of some type is used to pick off a portion of the amplitude of the radiation which is then combined with a second portion of the amplitude. The visibility of the resulting interference fringes is a maximum when the amplitudes of the two interfering beams are equal. *See* INTERFERENCE OF WAVES.

Michelson interferometer. The Michelson interferometer (**Fig. 3**) is based on division of amplitude. Light from an extended source S is incident on a partially reflecting plate (beam splitter) P_1. The light transmitted through P_1 reflects off mirror M_1 back to plate P_1. The light which is reflected proceeds to M_2 which reflects it back to P_1. At P_1, the two waves are again partially reflected and partially transmitted, and a portion of each wave proceeds to the receiver R, which

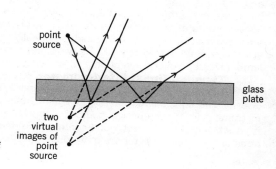

Fig. 2. Division of amplitude.

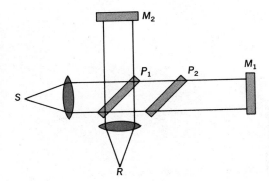

Fig. 3. Michelson interferometer.

may be a screen, a photocell, or a human eye. Depending on the difference between the distances from the beam splitter to the mirrors M_1 and M_2, the two beams will interfere constructively or destructively. Plate P_2 compensates for the thickness of P_1. Often when a quasimonochromatic light source is used with the interferometer, compensating plate P_2 is omitted.

The function of the beam splitter is to superimpose (image) one mirror onto the other. When the mirrors' images are completely parallel, the interference fringes appear circular. If the mirrors are slightly inclined about a vertical axis, vertical fringes are formed across the field of view. These fringes can be formed in white light if the path difference in part of the field of view is made zero. Just as in other interference experiments, only a few fringes will appear in white light, because the difference in path will be different for wavelengths of different colors. Accordingly, the fringes will appear colored close to zero path difference, and will disappear at larger path differences where the fringe maxima and minima for the different wavelengths overlap. If light reflected off the beam splitter experiences a one-half-cycle relative phase shift, the fringe of zero path difference is black, and can be easily distinguished from the neighboring fringes. This makes use of the instrument relatively easy.

The Michelson interferometer can be used as a spectroscope. Consider first the case of two close spectrum lines as a light source for the instrument. As the mirror M_1 is shifted, fringes from each spectral line will cross the field. At certain path differences between M_1 and M_2, the fringes for the two spectral lines will be out of phase and will essentially disappear; at other points they will be in phase and will be reinforced. By measuring the distance between successive maxima in fringe contrast, it is possible to determine the wavelength difference between the lines.

This is a simple illustration of a very broad use for any two-beam interferometer. As the path length L is changed, the variation in intensity $I(L)$ of the light coming from an interferometer gives information on the basis of which the spectrum of the input light can be derived. The equation for the intensity of the emergent energy can be written as Eq. (1), where β is a constant

$$I(L) = \int_0^\infty I(\lambda) \cos^2\left(\frac{\beta L}{\lambda}\right) d\lambda \tag{1}$$

and $I(\lambda)$ is the intensity of the incident light at different wavelengths λ. This equation applies when the mirror M_1 is moved linearly with time from the position where the path difference with M_2 is zero, to a position which depends on the longest wavelength in the spectrum to be examined. From Eq. (1), it is possible mathematically to recover the spectrum $I(\lambda)$. In certain situations such as in the infrared beyond the wavelength region of 1.5 micrometers, this technique offers a large advantage over conventional spectroscopy in that its utilization of light is extremely efficient.

Twyman-Green interferometer. If the Michelson interferometer is used with a point source instead of an extended source, it is called a Twyman-Green interferometer. The use of the laser as the light source for the Twyman-Green interferometer has made it an extremely useful instrument for testing optical components. The great advantage of a laser source is that it makes it possible to obtain bright, good-contrast, interference fringes even if the path lengths for the two arms of the interferometer are quite different. SEE LASER.

Figure 4 shows a Twyman-Green interferometer for testing a flat mirror. The laser beam is expanded to match the size of the sample being tested. Part of the laser light is transmitted to the reference surface, and part is reflected by the beam splitter to the flat surface being tested. Both beams are reflected back to the beam splitter, where they are combined to form interference fringes. An imaging lens projects the surface under test onto the observation plane.

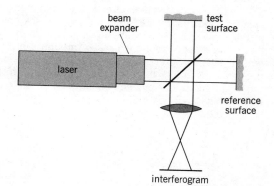

Fig. 4. Twyman-Green interferometer for testing flat surfaces.

Fringes (**Fig. 5**) show defects in the surface being tested. If the surface is perfectly flat, then straight, equally spaced fringes are obtained. Departure from the straight, equally spaced condition shows directly how the surface differs from being perfectly flat. For a given fringe, the difference in optical path between light going from laser to reference surface to observation plane and the light going from laser to test surface to observation plane is a constant. (The optical path is equal to the product of the geometrical path times the refractive index.) Between adjacent fringes (Fig. 5), the optical path difference changes by one wavelength, which for a helium-neon laser corresponds to 633 nm. The number of straight, equally spaced fringes and their orientation depend upon the tip-tilt of the reference mirror. That is, by tipping or tilting the reference mirror the difference in optical path can be made to vary linearly with distance across the laser beam.

Deviations from flatness of the test mirror also cause optical path variations. A height change of half a wavelength will cause an optical path change of one wavelength and a deviation from fringe straightness of one fringe. Thus, the fringes give surface height information, just as a topographical map gives height or contour information.

Fig. 5. Interferogram obtained with the use of a Twyman-Green interferometer to test a flat surface.

The existence of the essentially straight fringes provides a means of measuring surface contours relative to a tilted plane. This tilt is generally introduced to indicate the sign of the surface error, that is, whether the errors correspond to a hill or a valley. One way to get this sign information is to push in on the piece being tested when it is in the interferometer. If the fringes move toward the right when the test piece is pushed toward the beam splitter, then fringe deviations from straightness toward the right correspond to high points (hills) on the test surface and deviations to the left correspond to low points (valleys).

The basic Twyman-Green interferometer (Fig. 4) can be modified (**Fig. 6**) to test concave-spherical mirrors. In the interferometer, the center of curvature of the surface under test is placed at the focus of a high-quality diverger lens so that the wavefront is reflected back onto itself. After this retroflected wavefront passes through the diverger lens, it will be essentially a plane wave, which, when it interferes with the plane reference wave, will give interference fringes similar to those shown in Fig. 5 for testing flat surfaces. In this case it indicates how the concave-spherical mirror differs from the desired shape. Likewise, a convex-spherical mirror can be tested. Also, if a high-quality spherical mirror is used, the high-quality diverger lens can be replaced with the lens to be tested.

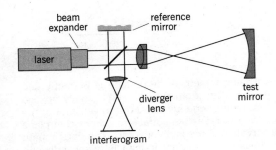

Fig. 6. Twyman-Green interferometer for testing spherical mirrors or lenses.

Fizeau interferometer. One of the most commonly used interferometers in optical metrology is the Fizeau interferometer, which can be thought of as a folded Twyman-Green interoferometer. In the Fizeau, the two surfaces being compared, which can be flat, spherical, or aspherical, are placed in close contact. The light reflected off these two surfaces produces interference fringes. For each fringe, the separation between the two surfaces is a constant. If the two surfaces match, straight, equally spaced fringes result. Surface height variations between the two surfaces cause the fringes to deviate from straightness or equal separations, where one fringe deviation from straightness corresponds to a variation in separation between the two surfaces by an amount equal to one-half of the wavelength of the light source used in the interferometer. The wavelength of a helium source, which is often used in a Fizeau interferometer, is 587.56 nm hence one fringe corresponds to a height variation of approximately 0.3 μm.

Mach-Zehnder interferometer. The Mach-Zehnder interferometer (**Fig. 7**) is a variation of the Michelson interferometer and, like the Michelson interferometer, depends on a amplitude splitting of the wavefront. Light enters the instrument and is reflected and transmitted by the semitransparent mirror M_1. The reflected portion proceeds to M_3, where it is reflected through the cell C_2 to the semitransparent mirror M_4. Here it combines with the light transmitted by M to produce interference. The light transmitted by M_1 passes through a cell C_1, which is similar to C_2 and is used to compensate for the windows of C_1.

The major application of this instrument is in studying airflow around models of aircraft missiles, or projectiles. The object and associated airstream are placed in one arm of the interferometer. Because the air pressure varies as it flows over the model, the index of refraction varies and thus the effective path length of the light in this beam is a function of position. When the variation is an odd number of half-waves, the light will interfere destructively and a dark fringe will appear in the field of view. From a photograph of the fringes, the flow pattern can be mathematically derived.

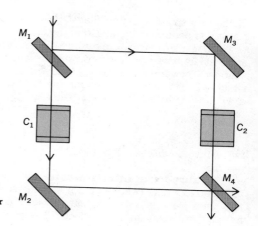

Fig. 7. Mach-Zehnder interferometer.

A major difference between the Mach-Zehnder and the Michelson interferometer is that in the Mach-Zehnder the light goes through each path in the instrument only once, whereas in the Michelson the light traverses each path twice. This double traversal makes the Michelson interferometer extremely difficult to use in applications where spatial location of index variations is desired. The incoming and outgoing beams tend to travel over slightly different paths, and this lowers the resolution because of the index gradient across the field.

Shearing interferometers. In a lateral-shear interferometer, an example of which is shown in **Fig. 8**, a wavefront is interfered with a shifted version of itself. A bright fringe is obtained at the points where the slope of the wavefront times the shift between the two wavefronts is equal to an integer number of wavelengths. That is, for a given fringe the slope or derivative of the wavefront is a constant. For this reason a lateral-shear interferometer is often called a differential interferometer.

Another type of shearing interferometer is a radial-shear interferometer. Here, a wavefront is interfered with an expanded version of itself. This interferometer is sensitive to radial slopes.

The advantages of shearing interferometers are that they are relatively simple and inexpensive, and since the reference wavefront is self-generated, an external wavefront is not needed. Since an external reference beam is not required, the source requirements are reduced from those of an interferometer such as a Twyman-Green. For this reason, shearing interferometers, in particular lateral-shear interferometers, are finding much use in applications such as adaptive optics systems for correction of atmospheric turbulence where the light source has to be a star, or planet, or perhaps just reflected sunlight. SEE ADAPTIVE OPTICS.

Michelson stellar interferometer. A Michelson stellar interferometer can be used to measure the diameter of stars which are as small as 0.01 second of arc. This task is impossible with a ground-based optical telescope since the atmosphere limits the resolution of the largest telescope to not much better than 1 second of arc.

The Michelson stellar interferometer is a simple adaptation of Young's two-slit experiment. In its first form, two slits were placed over the aperture of a telescope. If the object being observed were a true point source, the image would be crossed with a set of interference bands. A second point source separated by a small angle from the first would produce a second set of fringes. At certain values of this angle, the bright fringes in one set will coincide with the dark fringes in the second set. The smallest angle α at which the coincidence occurs will be that angle subtended at the slits by the separation of the peak of the central bright fringe from the nearest dark fringe. This angle is given by Eq. (2), where d is the separation of the slits, λ the dominant wavelength

$$\frac{\lambda}{2d} = \alpha \tag{2}$$

of the two sources, and α their angular separation. The measurement of the separation of the sources is performed by adjusting the separation d between the slits until the fringes vanish.

Consider now a single source in the shape of a slit of finite width. If the slit subtends an angle at the telescope aperture which is larger than α, the interference fringes will be reduced in contrast. For various line elements at one side of the slit, there will be elements of angle α away which will cancel the fringes from the first element. By induction, it is clear that for a separation d' such that the slit source subtends an angle as given by Eq. (3) the fringes from a single slit

$$\alpha' = \frac{\lambda}{d'} \tag{3}$$

will vanish completely. For additional information on the Michelson stellar interferometer SEE DIFFRACTION.

Fabry-Perot interferometer. All the interferometers discussed above are two-beam interferometers. The Fabry-Perot interferometer (**Fig. 9**) is a multiple-beam interferometer since the two glass plates are partially silvered on the inner surfaces, and the incoming wave is multiply reflected between the two surfaces. The position of the fringe maxima is the same for multiple-beam interference as two-beam interference; however, as the reflectivity of the two surfaces increases and the number of interfering beams increases, the fringes become sharper.

A quantity of particular interest in a Fabry-Perot is the ratio of the separation of adjacent maxima to the half-width of the fringes. It can be shown that this ratio, known as the finesse, is given by Eq. (4), where R is the reflectivity of the silvered surfaces.

$$\mathcal{F} = \frac{\pi\sqrt{R}}{1-R} \tag{4}$$

The multiple-beam Fabry-Perot interferometer is of considerable importance in modern optics for spectroscopy. All the light rays incident on the Fabry-Perot at a given angle will result in a single circular fringe of uniform irradiance. With a broad diffuse source, the interference fringes will be narrow concentric rings, corresponding to the multiple-beam transmission pattern. The position of the fringes depends upon the wavelength. This is, each wavelength gives a separate fringe pattern. The minimum resolvable wavelength difference is determined by the ability to resolve close fringes. The ratio of the wavelength λ to the least resolvable wavelength difference $\Delta\lambda$ is known as the chromatic resolving power $\mathcal{R}$. At nearly normal incidence it is given by Eq. (5), where n is the refractive index between the two mirrors separated a distance d. For a wave-

$$\mathcal{R} = \frac{\lambda}{(\Delta\lambda)_{min}} = \mathcal{F}\frac{2nd}{\lambda} \tag{5}$$

length of 500 nm, $nd = 10$ mm, and $R = 90\%$, the resolving power is well over 10^6. SEE REFLECTION OF ELECTROMAGNETIC RADIATION; RESOLVING POWER.

When Fabry-Perot interferometers are used with lasers, they are generally used in the central spot scanning mode. The interferometer is illuminated with a collimated laser beam and all the light transmitted through the Fabry-Perot is focused onto a detector, whose output is dis-

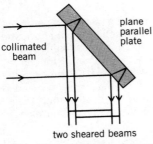

Fig. 8. Lateral shear interferometer.

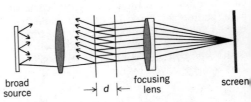

Fig. 9. Fabry-Perot interferometer.

played on an oscilloscope. Often one of the mirrors is on a piezoelectric mirror mount. As the voltage to the piezoelectric crystal is varied, the mirror separation is varied. The light output as a function of mirror separation gives the spectral frequency content of the laser source.

Holographic interferometry. A wave recorded in a hologram is effectively stored for future reconstruction and use. Holographic interferometry is concerned with the formation and interpretation of the fringe pattern which appears when a wave, generated at some earlier time and stored in a hologram, is later reconstructed and caused to interfere with a comparison wave. It is the storage or time-delay aspect which gives the holographic method a unique advantage over conventional optical interferometry.

A hologram can be made of an arbitrarily shaped, rough scattering surface, and after suitable processing, if the hologram is illuminated with the same reference wavefront used in recording the hologram, the hologram will produce the original object wavefront. If the hologram is placed back into its original position, a person looking through the hologram will see both the original object and the image of the object stored in the hologram. If the object is now slightly deformed, interference fringes will be produced which tell how much the surface is deformed. Between adjacent fringes the optical path between the source and viewer has changed by one wavelength. While the actual shape of the object is not determined, the change in the shape of the object is measured to within a small fraction of a wavelength, even though the object's surface is rough compared to the wavelength of light.

Double-exposure. Double-exposure holographic interferometry (**Fig. 10**) is similar to real-time holographic interferometry described above, except now two exposures are made before processing: one exposure with the object in the undeformed state and a second exposure after deformation. When the hologram reconstruction is viewed, interference fringes will be seen which show how much the object was deformed between exposures.

The advantage of the double-exposure technique over the real-time technique is that there is no critical replacement of the hologram after processing. The disadvantage is that continuous comparison of surface displacement relative to an initial state cannot be made, but rather only the difference between two states is determined.

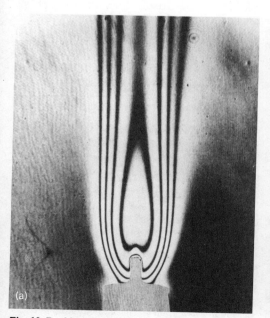

Fig. 10. Double-exposure holographic interferograms. (a) Interferogram of candle flame. (b) Interferogram of debanded region of honeycomb construction panel. (*From C. M. Vest, Holographic interferometry, John Wiley and Sons, 1978*)

Time-average. In time-average holographic interferometry (**Fig. 11**) a time-average hologram of a vibrating surface is recorded. If the maximum amplitude of the vibration is limited to some tens of light wavelengths, illumination of the hologram yields an image of the surface on which is superimposed several interference fringes which are contour lines of equal displacement of the surface. Time-average holography enables the vibrational amplitudes of diffusely reflecting surfaces to be measured with interferometric precision. SEE HOLOGRAPHY.

Speckle interferometry. A random intensity distribution, called a speckle pattern, is generated when light from a highly coherent source, such as a laser, is scattered by a rough surface. The use of speckle patterns in the study of object displacements, vibration, and distortion is becoming of more importance in the nondestructive testing of mechanical components. For example, time-averaged speckle photographs can be used to analyze the vibrations of an object in its plane. In physical terms the speckles in the image are drawn out into a line as the surface vibrates, instead of being double as in the double-exposure technique. The diffraction pattern of this smeared-out speckle-pattern recording is related to the relative time spent by the speckle at each point of its trajectory (**Fig. 12**).

Speckle interferometry can be used to perform astronomical measurements similar to those performed by the Michelson stellar interferometer. Stellar speckle interferometry is a technique for obtaining diffraction-limited resolution of stellar objects despite the presence of the turbulent atmosphere that limits the resolution of ground-based telescopes to approximately 1 second of arc. For example, the diffraction limit of the 200-in.-diameter (5-m) Palomar Mountain telescope is approximately 0.02 second of arc, 1/50 the resolution limit set by the atmosphere. SEE OPTICAL TELESCOPE.

The first step of the process is to take a large number, perhaps 100, of short exposures of the object, where each photo is taken for a different realization of the atmosphere. Next the optical diffraction pattern, that is, the squared modulus of the Fourier transform of all the short-exposure photographs, is added. By taking a further Fourier transform of each ensemble average diffraction pattern, the ensemble average of the spatial autocorrelation of the diffraction-limited images of each object is obtained. SEE SPECKLE.

Phase-shifting interferometry. Electronic phase-measurement techniques can be used in interferometers such as the Twyman-Green, where the phase distribution across the interferogram is being measured. Phase-shifting interferometry is often used for these measurements since it provides for rapid precise measurement of the phase distribution. In phase-shifting interferometry, the phase of the reference beam in the interferometer is made to vary in a known manner. This can be achieved, for example, by mounting the reference mirror on a piezoelectric transducer. By varying the voltage on the transducer, the reference mirror is moved a known amount to change the phase of the reference beam a known amount. A solid-state detector array is used to

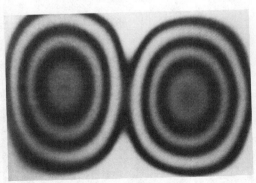

Fig. 11. Photograph of time-average holographic interferogram. (*From C. M. Vest, Holographic Interferometry, John Wiley and Sons, 1978*)

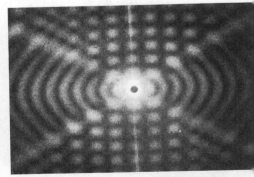

Fig. 12. Diffraction patterns from time-averaged speckle interferogram of a surface vibrating in its own plane with a figure-eight motion. (*From J. C. Dainty, Laser Speckle and Related Phenomena, Springer-Verlag, 1975*)

detect the intensity distribution across the interference pattern. This intensity distribution is read into computer memory three or more times, and between each intensity measurement the phase of the reference beam is changed a known amount. From these three or more intensity measurements, the phase across the interference pattern can be precisely determined to within a fraction of a degree.

Bibliography. J. C. Dainty (ed.), *Laser Speckle and Related Phenomena*, 1975; R. K. Erf (ed.), *Speckle Metrology*, 1978; E. Hecht and A. Zajac, *Optics*, 2d ed., 1987; G. Hernandez, *Fabry-Perot Interferometers*, 1984; R. Jones and C. M. Wykes, *Holographic and Speckle Interferometry*, 1983; D. Malacara (ed.), *Optical Shop Testing*, 1978; W. H. Steel, *Interferometry*, 2d ed., 1984; C. Vest, *Holographic Interferometry*, 1979; J. C. Wyant, Interferometric optical metrology: Basic principles and new systems, *Laser Focus*, 18(5):65–71, 1982.

INTERFERENCE FILTER
Bruce H. Billings

An optical filter in which the wavelengths that are not transmitted are removed by interference phenomena rather than by absorption or scattering. In addition to being able to duplicate most of the spectral characteristics of absorption color filters, these devices can be made to transmit a very narrow band of wavelengths. They can thus be used as monochromators to examine a radiation source at the wavelength of a single spectrum line. For example, the solar disk can be observed in light of the hydrogen line Hα and thus the distribution of excited hydrogen over the disk can be determined. *See* Color filter; Interference of waves.

Most narrow-band interference filters are based on the Fabry-Perot interferometer. The Fabry-Perot interference filter differs from the interferometer only in the thickness of the space between the partially reflecting layers. In the interferometer, this space can be several centimeters. In the filter, it is normally a few thousand angstroms. In the simplest filter, a glass plate is coated with a layer of silver which is covered by a layer of dielectric followed in turn by another evaporated layer of semitransparent silver. *See* Interferometry.

Basic properties. At all wavelengths at which the dielectric layer has an optical thickness of an integral number of half waves, the filter will have a passband. The number of half waves corresponding to a given passband is called the order of the passband. The transmission T of the filter can be represented by Eq. (1), where r is the reflectivity of the silver film and t the transmission of the film; δ is defined in Eq. (2), where d is the thickness of the dielectric layer, n

$$T = \frac{t^2}{(1-r)^2 + 4r \sin^2(\delta/2)} \quad (1) \qquad \delta = \frac{4\pi d}{\lambda}(n^2 - \sin^2\theta)^{1/2} + 2y \quad (2)$$

its refractive index, λ the wavelength, y the phase shift experienced by the light at the metal-dielectric boundary, and θ the angle of incidence.

By inspection of Eq. (1), it is apparent that maxima occur when $\delta/2 = m\pi$, where m is an integer.

Some of the quantities which are of interest to the user of these filters are (1) the peak transmission, (2) the transmission between peaks, (3) the bandwidth, and (4) the angular field of view, that is, the angle through which the filter must be tilted to shift the wavelength of peak transmission a distance equal to the bandwidth.

Each of these quantities can be determined theoretically from Eq. (1). A typical filter has a peak transmission of 40% at its peak wavelength of 546.1 nanometers, a transmission between peaks of 0.2%, a bandwidth of 10 nm, and an angular field of view of 20°. These numbers represent nearly the best that can be done with the simple metal-dielectric filter.

Multilayer types. An increase in reflectivity can result in narrower bandwidths. There are techniques by which high reflectivities can be achieved which are lossless, that is, which have no absorption. This results in higher peak transmission and lower off-peak transmission. The first is the multilayer filter. In this device, the metal layers are replaced by a series of dielectric layers. The boundary between two dielectric layers of refractive indices n_1 and n_2 has a reflecting power of perhaps 4% in the case of glass and air, or less for two dielectrics whose indices are close together. The value of the reflectivity r is given by the standard Fresnel reflection law, Eq. (3). SEE REFLECTION OF ELECTROMAGNETIC RADIATION.

$$r = \left(\frac{n_1 - n_2}{n_1 + n_2}\right)^2 \qquad (3)$$

By making several layers of alternate high- and low-index dielectric, it is possible to reinforce the reflectivity of a single boundary and build it up by multiple reflection to any desired value. It is necessary only that the layers be of such thickness that the reflections from successive boundaries are in phase. When each layer is optically a quarter wavelength in thickness, this reinforcement takes place. A complete filter is sketched in **Fig. 1**. It might consist of seven alternate layers of high- and low-index dielectric of a thickness of a quarter wavelength apiece, fol-

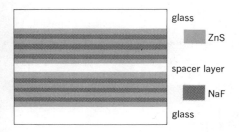

Fig. 1. Schematic diagram of seven-layer solid Fabry-Perot filter. (*After D. E. Gray, ed., American Institute of Physics Handbook, McGraw-Hill, 1957*)

lowed by the dielectric spacer which is an integral number of half waves and which is followed by seven more quarter-wavelength layers. The characteristics of such a filter are shown in **Fig. 2**. For a seven-layer reflection filter, the reflectivity can be built up to 95%. One would expect improvement over the metal filter, and in fact, the peak transmission of such a filter is as high as 80% and the bandwidth as low as 0.5 nm.

Frustrated reflection. A second technique is the use of frustrated total internal reflection for the partially reflecting layers of the filter. Light is totally internally reflected when incident on the hypotenuse of a right angle prism, as in **Fig. 3**. If another prism is brought up to the first, as in **Fig. 4**, a part of the light is transmitted through the combination. The fraction transmitted depends on the separation between the prisms. The reflectivity can be made as high as desired. In a filter, the prism hypotenuse is coated with a low-index layer of a thickness chosen to give the proper value of the reflectivity. This is covered with a high-index layer whose thickness determines the wavelength of the passband, as in the normal Fabry-Perot filter. This is followed by a second low-index layer and a second prism. The first frustrated reflection is at the boundary between the first prism and the low-index layer. The second reflection is at the boundary of the high-index layer and the second low-index layer. The thickness of the low-index layer can be adjusted to give any value of reflectivity. The filter is lossless unless there is absorption within the layers. The resulting filter can be made to have a bandwidth of less than 0.6 nm.

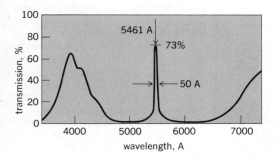

Fig. 2. Transmission of filter shown in Fig. 1 as a function of wavelength. (*After D. E. Gray, ed., American Institute of Physics Handbook, McGraw-Hill, 1957*)

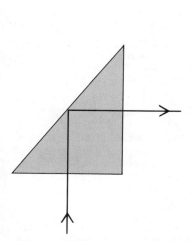

Fig. 3. Total internal reflection in a right-angle prism.

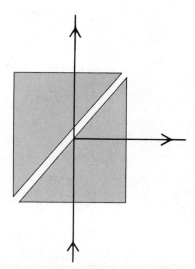

Fig. 4. Frustrated total internal reflection.

Bibliography. M. Born and E. Wolf, *Principles of Optics*, 6th ed., 1980; D. E. Gray (ed.), *American Institute of Physics Handbook*, 3d ed., 1972, reprint 1982; R. Kingslake, *Applied Optics and Optical Engineering*, vol. 1: *Light: Its Generation and Modification*, 1965; Optical Society of America, *Handbook of Optics*, 1978.

DIFFRACTION
Francis A. Jenkins AND William W. Watson

The bending of light, or other waves, into the region of the geometrical shadow of an obstacle. More exactly, diffraction refers to any redistribution in space of the intensity of waves that results from the presence of an object that causes variations of either the amplitude or phase of the waves. Most diffraction gratings cause a periodic modulation of the phase across the wavefront rather than a modulation of the amplitude. Although diffraction is an effect exhibited by all types of wave motion, this article will deal only with electromagnetic waves, especially those of visible light. Some important differences that occur with microwaves will also be mentioned. SEE ELECTROMAGNETIC WAVE.

Diffraction is a phenomenon of all electromagnetic radiation, including radio waves; microwaves; infrared, visible, and ultraviolet light; and x-rays. The effects for light are important in connection with the resolving power of optical instruments.

There are two main classes of diffraction, which are known as Fraunhofer diffraction and Fresnel diffraction. The former concerns beams of parallel light, and is distinguished by the simplicity of the mathematical treatment required and also by its practical importance. The latter class includes the effects in divergent light, and is the simplest to observe experimentally. A complete explanation of Fresnel diffraction has challenged the most able physicists, although a satisfactory approximate account of its main features was given by A. Fresnel in 1814. At that time, it played an important part in establishing the wave theory of light.

To illustrate the difference between the methods of observation of the two types of diffraction, **Fig. 1** shows the experimental arrangements required to observe them for a circular hole in a screen s. The light originates at a very small source O, which can conveniently be a pinhole illuminated by sunlight. In Fraunhofer diffraction, the source lies at the principal focus of a lens L_1, which renders the light parallel as it falls on the aperture. A second lens L_2 focuses parallel

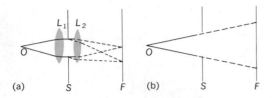

Fig. 1. Observation of the two prinicipal types of diffraction, in the case of a circular aperture. (a) Fraunhofer and (b) Fresnel diffraction.

diffracted beams on the observing screen F, situated in the principal focal plane of L_2. In Fresnel diffraction, no lenses intervene. The diffraction effects occur chiefly near the borders of the geometrical shadow, indicated by the broken lines. An alternative way of distinguishing the two classes, therefore, is to say that Fraunhofer diffraction concerns the effects near the focal point of a lens or mirror, while Fresnel diffraction concerns those effects near the edges of shadow. Photographs of some diffraction patterns of each class are shown in **Fig. 2**. All of these may be demonstrated especially well by using the light beam from a neon-helium laser. SEE LASER.

FRAUNHOFER DIFFRACTION

This class of diffraction is characterized by a linear variation of the phases of the Huygens secondary waves with distance across the wavefront, as they arrive at a given point on the observing screen. At the instant that the incident plane wave occupies the plane of the diffracting screen, it may be regarded as sending out, from each element of its surface, a multitude of secondary waves, the joint effect of which is to be evaluated in the focal plane of the lens L_2. The analysis of these secondary waves involves taking account of both their amplitudes and their

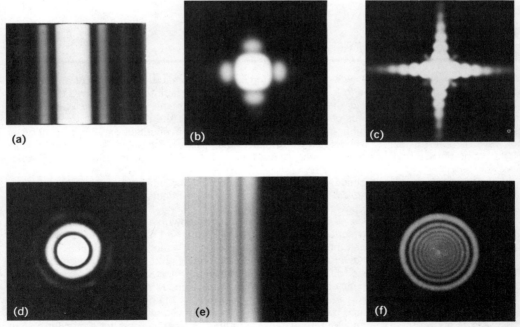

Fig. 2. Diffraction patterns, photographed with visible light. (a) Fraunhofer pattern, for a slit; (b) Fraunhofer pattern, square aperture, short exposure; (c) Fraunhofer pattern, square aperture, long exposure (courtesy of F. S. Harris); (d) Fraunhofer pattern, circular aperture (courtesy of R. W. Ditchburn); (e) Fresnel pattern, straight edge; (f) Fresnel pattern, circular aperture.

WAVE OPTICS **155**

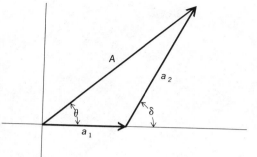

Fig. 3. Graphical addition of two amplitudes.

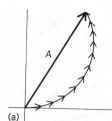
(a)

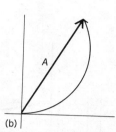
(b)

Fig. 4. Vibration curves. (a) Addition of many equal amplitudes differing in phase by equal amounts. (b) Equivalent curve, when amplitudes and phase differences become infinitesimal.

phases. The simplest way to do this is to use a graphical method, the method of the so-called vibration curve, which can readily be extended to cases of Fresnel diffraction. *See* Huygens' principle.

Vibration curve. The basis of the graphical method is the representation of the amplitude and phase of a wave arriving at any point by a vector, the length of which gives the magnitude of the amplitude, and the slope of which gives the value of the phase. In **Fig. 3** are shown two vectors of amplitudes a_1 and a_2, pertaining to two waves having a phase difference δ of 60°. That is, the waves differ in phase by one-sixth of a complete vibration. The resultant amplitude A and phase θ (relative to the phase of the first wave) are then found from the vector sum of a_1 and a_2, as indicated. A mathematical proof shows that this proposition is rigorously correct and that it may be extended to cover the addition of any number of waves.

The vibration curve results from the addition of a large (really infinite) number of infinitesimal vectors, each representing the contribution of the Huygens secondary waves from an element of surface of the wavefront. If these elements are assumed to be of equal area, the magnitudes of the amplitudes to be added will all be equal. They will, however, generally differ in phase, so that if the elements were small but finite each would be drawn at a small angle with the preceding one, as shown in **Fig. 4**a. The resultant of all elements would be the vector A. When the individual vectors represent the contributions from infinitesimal surface elements (as they must for the Huygens wavelets), the diagram becomes a smooth curve, the vibration curve, shown in Fig. 4b. The intensity on the screen is then proportional to the square of this resultant amplitude. In this way, the distribution of the intensity of light in any Fraunhofer diffraction pattern may be determined.

The vibration curve for Fraunhofer diffraction by screens having slits with parallel, straight edges is a circle. Consider, for example, the case of a slit of width b illustrated in **Fig. 5**. The edges of the slit extend perpendicular to the plane of the figure, and the slit is illuminated by plane waves of light coming from the left. If s is the distance from O of a surface element ds (ds actually being a strip extending perpendicular to the figure), the extra distance that the wavelet from ds must travel in reaching a point on the screen lying at the angle θ from the center is $s \sin \theta$. Since this extra distance determines the phase difference, the latter varies linearly with s. This condition necessitates that the vibration curve be a circle.

The intensity distribution for Fraunhofer diffraction by a slit as a function of the angle θ may be simply calculated as follows. The extra distances traveled by the wavelets from the upper and lower edges of the slit, as compared with those from the center, are $+(b/2) \sin \theta$ and $-(b/2) \sin \theta$. The corresponding phase differences are $2\pi/\lambda$ times these quantities, λ being the wavelength of the light. Using the symbol β for $(\pi b \sin \theta)/\lambda$, it is seen that the end points of the effective part of the vibration curve must differ in slope by $\pm \beta$ from the slope at its center, where it is taken as zero. **Figure 6** shows the form of the vibration curve for $\beta = \pi/4$, that is, for $\sin \theta = \frac{1}{4}\lambda/b$. The resultant is $A = 2r \sin \beta$, where r is the radius of the arc. The amplitude A_0 that would be obtained if all the secondary waves were in phase at the center of the diffraction pattern,

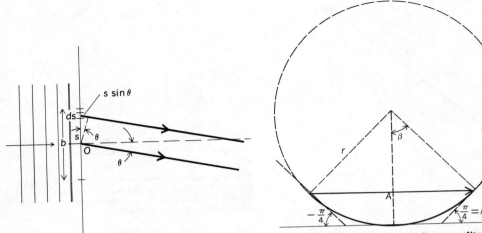

Fig. 5. Analysis of Fraunhofer diffraction by a slit.

Fig. 6. Vibration curve and resultant amplitude for a particular point in Fraunhofer diffraction by a slit.

where $\theta = 0$, is the length of the arc. Thus, Eq. (1) holds. The intensity at any angle is given by Eq. (2), where I_0 is the intensity at the center of the pattern. **Figure 7** shows a graph of this

$$\frac{A}{A_0} = \frac{\text{chord}}{\text{arc}} = \frac{2r \sin \beta}{2r\beta} = \frac{\sin \beta}{\beta} \quad (1) \qquad\qquad I = I_0 \frac{\sin^2 \beta}{\beta^2} \quad (2)$$

function. The central maximum is twice as wide as the subsidiary ones, and is about 21 times as intense as the strongest of these. A photograph of this pattern is shown in Fig. 2a.

The dimensions of the pattern are important, since they determine the angular spread of the light behind the slit. The first zeros occur at values $\beta = (\pi b \sin \theta)/\lambda = \pm \pi$. In most cases the angle θ is extremely small, so that Eq. (3) holds. For a slit 1 mm wide, for example, and green

$$\sin \theta \approx \theta = \pm \frac{\lambda}{b} \quad (3)$$

light of wavelength 5×10^{-5} cm, Eq. (3) gives the angle as only 0.0005 radian, or 1.72 minutes of arc. The slit would have to be much narrower than this, or the wavelength much longer, for the approximation to cease to be valid.

The main features of Fraunhofer diffraction patterns of other shapes can be understood with the aid of the vibration curve. Thus for a rectangular or square aperture, the wavefront may

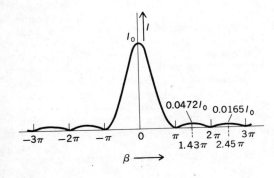

Fig. 7. Intensity distribution curve for Fraunhofer diffraction by a slit.

be subdivided into elements parallel to either of two adjacent sides, giving an intensity distribution which follows the curve of Fig. 7 in the directions parallel to the two sides. Photographs of such patterns appear in Fig. 2*b* and *c*. In Fig. 2*c* it will be seen that there are also faint subsidiary maxima lying off the two principal directions. These have intensities proportional to the products of the intensities of the side maxima in the slit pattern. The fact that these subsidiary intensities are extremely low compared with that of the central maximum has an important application to the apodization of lenses, to be discussed later.

Diffraction grating. An idealized diffraction grating consists of a large number of similar slits, equally space. Equal segments of the vibration curve are therefore effective, as shown in **Fig. 8***a*. The resultants *a* of each segment are then to be added to give *A*, the amplitude due to the whole grating, as shown in Fig. 8*b*. The phase difference between successive elements is here assumed to be very small. As it is increased, by going to a larger angle θ, the resultant *A* first goes to zero at an angle corresponding to λ/W, where *W* is the total width of the grating. After going through numerous low-intensity maxima, *A* again rises to a high value when the phase difference between the successive vectors for the individual slits approaches a whole vibration. These small vectors *a* are then all lined up again, as they were at the center of the pattern ($\theta = 0$). The resulting strong maximum represents the "first-order spectrum," since its position depends on the wavelength. A similar condition occurs when the phase difference becomes two, three, or more whole vibrations, giving the higher-order spectra. By means of this diagram, it is possible not only to predict the intensities of successive orders for an ideal grating, but also to find the sharpness of the maxima which represent the spectrum lines. SEE DIFFRACTION GRATING.

Determination of resolving power. Fraunhofer diffraction by a circular aperture determines the resolving power of instruments such as telescopes, cameras, and microscopes, in which the width of the light beam is usually limited by the rim of one of the lenses. The method of the vibration curve may be extended to find the angular width of the central diffraction maximum for this case. **Figure 9** compares the treatments of square and circular apertures by showing, above, the elements of equal phase difference into which the wavefront may be divided, and, below, the corresponding vibration curves. For the square aperture shown in Fig. 9*a*, the areas of the surface elements are equal, and the curve forms a complete circle at the first zero of intensity. In Fig. 9*b* these areas, and hence the lengths of the successive vectors, are not equal, but increase

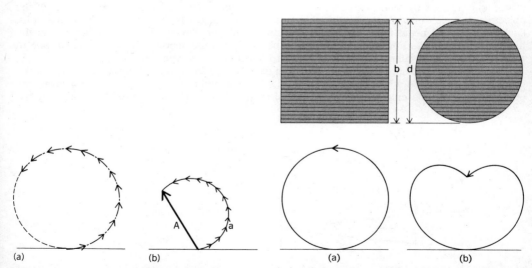

Fig. 8. Graphical analysis of diffraction grating. (**a**) Vibration curve for grating of 12 slits. (**b**) Resultant amplitude *A* formed by adding amplitudes *a* from individual slits. Each *a* represents the chord of one of the short arcs in part *a*.

Fig. 9. Condition of vibration curves at first minimum for (**a**) a square aperture and (**b**) a circular aperture.

as the center of the curve is approached, and then decrease again. The result is that the curve must show somewhat greater phase differences at its extremes in order to form a closed figure. An exact construction of the curve or, better, a mathematical calculation shows that the extreme phase differences required are $\pm 1.220\pi$, yielding Eq. (4) for the angle θ at the first zero of inten-

$$\sin \theta \approx \theta = \pm \frac{1.220\lambda}{d} \tag{4}$$

sity. Here d is the diameter of the circular aperture. When this result is compared with that of Eq. (3), it is apparent that, relative to the pattern for a rectangle of side $b = d$, this pattern is spread out by 22%. Obviously it now has circular symmetry and consists of a diffuse central disk, called the Airy disk, surrounded by faint rings (Fig. 2d). The angular radius of the disk, given by Eq. (4), may be extremely small for an actual optical instrument, but it sets the ultimate limit to the sharpness of the image, that is, to the resolving power. SEE RESOLVING POWER.

Other applications. Among the applications of Fraunhofer diffraction. other than the calculation of resolving power, are its use in (1) the theory of certain types of interferometer in which the interfering light beams are brought together by diffraction, (2) the theory of microscopic imaging, and (3) the design of apodizing screens for lenses.

Michelson's stellar interferometer. This instrument was devised by A. A. Michelson to overcome the limitation expressed by Eq. (4). In front of a telescope are placed two fixed mirrors, shown at M_2 and M_3 in **Fig. 10**, and two others, M_1 and M_4, movable so as to vary their separation D. The light of a star is directed by these mirrors into a telescope as shown, and the diffraction patterns, of size corresponding to the aperture of the mirrors, are superimposed in the focal plane P of the lens. When D is small the resulting pattern is crossed by interference fringes. If the light source has a finite width, increasing D causes these fringes to become indistinct and eventually to disappear. For a circular disk source, such as a star, the angle subtended by the disk must be $1.22\lambda/D$ for the disappearance of the fringes. The resolving power of the telescope has thus effectively been increased in the ratio D/d. In this way angular diameters as small as 0.02 second of arc have been measured, corresponding to $D = 7$ m or 24 ft. SEE INTERFERENCE OF WAVES; INTERFEROMETRY.

Microscopic imaging. E. Abbe's theory of the microscope evaluates the resolving power of this instrucment by considering not only the diffraction caused by the limited aperture of the objective, but also that caused by the object itself. If this object is illuminated by coherent light, such as a parallel beam from a point source, its Fraunhofer diffraction pattern is formed in the rear focal plane of the microscope objective. Abbe took as an object a diffraction grating, and in this case the pattern consists of a series of sharp maxima representing the various orders $m = 0$, $\pm 1, \pm 2, \ldots$. If the light of all these orders is collected by the objective, a perfect image of the grating can be formed where the light is reunited in the image plane. In practice, however, the objective can include only a limited number of them, as is indicated in **Fig. 11**. Here only the

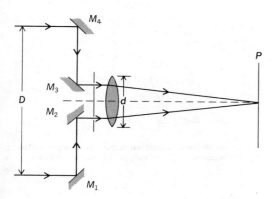

Fig. 10. Michelson stellar interferometer, which is used for measuring the diameters of stars.

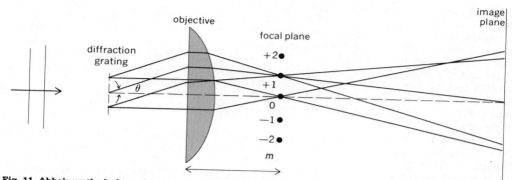

Fig. 11. Abbe's method of treating the resolving power of a microscope.

orders +2 to −2 are shown entering the objective. The higher orders, involving greater values of θ, would miss the lens.

The final image must be produced by interference in the image plane of the Huygens secondary waves coming from the various orders. To obtain a periodic variation of intensity in that plane, at least two orders must be accepted by the objective. The angle θ_1 at which the first order occurs is given by $\sin \theta_1 = \lambda/d$, where d is the spacing of the lines in the grating. The limit of resolution of the microscope, that is, the smallest value of d that will produce an indication of separated lines in the image, may thus be found from the angular aperture of the objective. In order for it to accept only the orders 0, ±1, the lens aperture 2α must equal at least $2\theta_1$, giving $d = \lambda/\sin \alpha$. This resolving limit may be decreased by illuminating the grating from one side, so that the zero-order light falls at one edge of the lens, and that of one of the first orders at the other edge. Then $\theta_1 = 2\alpha$, and the limit of resolution is approximately $0.5\lambda/\sin \alpha$. SEE OPTICAL MICROSCOPE.

Apodization. This is the name given to a procedure by which the effect of subsidiary maxima (such as those shown in Fig. 7) may be partially suppressed. Such a suppression is desirable when one wishes to observe the image of a very faint object adjacent to a strong one. If the fainter object has, for example, only 1/1000 of the intensity of the stronger one, the two images will have to be far enough apart so that the principal maximum for the fainter one is at least comparable in intensity to the secondary maxima for the stronger object at that point. In the pattern of a rectangular aperture it is not until the tenth secondary maximum that the intensity of these falls below 1/1000 of the intensity of the principal maximum. Here it has been assumed, however, that the fainter image lies along one of the two principal directions of diffraction of the rectangular aperture, perpendicular to two of its adjacent sides. At 45° to these directions the subsidiary maxima are much fainter (Fig. 2c), and even the second one has an intensity of only 1/3700, the square of the value of the second subsidiary maximum indicated in Fig. 7.

The simplest apodizing screen is a square aperture placed over a lens, the diagonal of the square being equal to the diameter of the lens. If the lens is the objective of an astronomical telescope, for example, the presence of a fainter companion in a double-star system can often be detected by turning the square aperture until the image of the companion star lies along its diagonal. Apodizing screens may be of various shapes, depending on the purpose to be achieved. It has been found that a screen of graded density, which shades the lens from complete opacity at the rim to complete transparency at a small distance inward, is effective in suppressing the circular diffraction rings surrounding the Airy disk. In all types of apodization there is some sacrifice of true resolving power, so that it would not be used if the two images to be resolved were of equal intensity.

FRESNEL DIFFRACTION

The diffraction effects obtained when the source of light or the observing screen are at a finite distance from the diffracting aperture or obstacle come under the classification of Fresnel diffraction. This type of diffraction requires for its observation only a point source, a diffracting screen of some sort, and an observing screen. The latter is often advantageously replaced by a magnifier or a low-power microscope. The observed diffraction patterns generally differ according to the radius of curvature of the wave and the distance of the point of observation behind the screen. If the diffracting screen has circular symmetry, such as that of an opaque disk or a round hole, a point source of light must be used. If it has straight, parallel edges, it is desirable from the standpoint of brightness to use an illuminated slit parallel to these edges. In the latter case, it is possible to regard the wave emanating from the slit as a cylindrical one. For the purpose of deriving the vibration curve, the appropriate way of dividing the wavefront into infinitesimal elements is to use annular rings in the first case, and strips parallel to the axis of the cylinder in the second case.

Figure 12 illustrates the way in which the radii of the rings, or the distances to the edges of the strips, must be chosen in order that the phase difference may increase by an equal amount from one element to the next. Figure 12a shows a section of the wavefront diverging from the source S, and the paths to the screen of two secondary wavelets. The shortest possible path is b, while r is that for another wavelet originating at a distance s above the "pole" O. Since all points on the wavefront are at the same distance a from S, the path difference between the two routes from S to P is $r - b$. When this is evaluated, to terms of the first order in s/a and s/b, the phase difference is given by Eq. (5). The phase difference across an elementary zone of radius s and

$$\delta = \frac{2\pi}{\lambda}(r - b) \approx \frac{\pi(a + b)}{ab\lambda}s^2 = Cs^2 \tag{5}$$

width ds then becomes $d\delta = 2Csds$, so that for equal increments of δ the increment of s must be proportional to $1/s$. The annular zones and the strips drawn in this way on the spherical or cylindrical wave, respectively, looking toward the pole from the direction of P, are shown in Fig. 12b and c.

For the annular zonal elements, the areas $2\pi sds$ are all equal, and hence the amplitude elements of the vibration curve should have the same magnitude. Actually, they must be regarded as falling off slowly, due to the influence of the "obliquity factor" of Huygens' principle. The resulting vibration curve is nearly, but not quite, circular, and is illustrated in **Fig. 13a**. It spirals in toward the center C, at a rate that has been considerably exaggerated in the figure. The intensity at any point P on the axis of a circular screen centered on O can now be determined as the square of the resultant amplitude A for the appropriate part or parts of this curve. The curve shown in Fig. 13a is for a circular aperture which exposes five Fresnel zones, each one represented by a half-turn of the spiral. The resultant amplitude is almost twice as great (and the intensity four times as great) as it would be if the whole wave were exposed, in which case the vector would terminate at C. The diffraction by other circular screens may be determined in this same manner.

Zone plate. This is a special screen designed to block off the light from every other half-period zone, and represents an interesting application of Fresnel diffraction. The Fresnel half-period zones are drawn, with radii proportional to the square roots of whole numbers, and alternate ones are blackened. The drawing is then photographed on a reduced scale. When light from

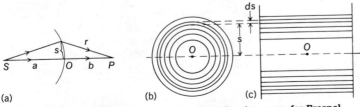

Fig. 12. Division of wavefront for constructing vibration curve for Fresnel diffraction. (a) Section of wave diverging from S and paths to screen of two secondary wavelets. (b) Annular zones. (c) Strips.

WAVE OPTICS

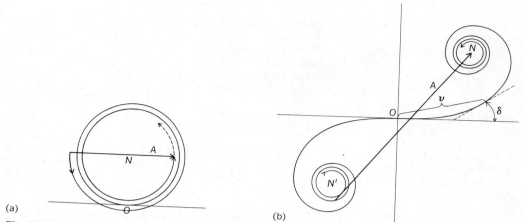

Fig. 13. Vibration curves for Fresnel diffraction patterns (*a*) Circular division of wavefront. (*b*) Strip division (Cornu's spiral).

a point source is sent through the negative, an intense point image is produced, much like that formed by a lens. The zone plate has the effect of removing alternate half-turns of the spiral, the resultants of the others all adding in the same phase. By putting $\delta = \pi$ and $a = \infty$ in Eq. (5), it is found that the "focal length" b of a zone plate is s_1^2/λ, where s_1 is the radius of the first zone.

Cornu's spiral. This is the vibration curve for a cylindrical wavefront, and is illustrated in Fig. 13*b*. The areas of the elementary zones, and hence the magnitudes of the component vectors of the vibration curve, decrease rapidly, being proportional to ds, and hence to $1/s$ (Fig. 12*c*). The definition of Cornu's spiral requires that its slope δ at any point be proportional to the square of the corresponding distance s measured up the wavefront [Fig. 12*a* and Eq. (5)]. The length of the spiral from the origin is proportional to s, but it is usually drawn in terms of the dimensionless variable v, defined by $v = s\sqrt{2C/\pi}$ in the notation of Eq. (5). The coordinates of any point on the curve may then be found from tables of Fresnel's integrals.

As an example of the application of Cornu's spiral, consider the diffraction of an opaque straight edge, such as a razor edge, illuminated by light from a narrow slit parallel to the edge. At some point outside the edge of the shadow, say one which exposes three half-period strips beyond the pole, the resultant amplitude will be that labeled A in Fig. 13*b*. The intensity will be greater than that given by the amplitude NN', which represents the amplitude for the whole (unobstructed) wave. On going further away from the edge of the shadow, the tail of the vector A will move along the spiral inward toward N', and the intensity will pass through maxima and minima. At the edge of the geometrical shadow the amplitude is ON, and the intensity is just one-fourth of that due to the unobstructed wave. Further into the shadow the intensity approaches zero regularly, without fluctuations, as the tail of the vector moves up toward N. A photograph of the straight-edge pattern is shown in Fig. 2*e*.

Babinet's principle. This states that the diffraction patterns produced by complementary screens are identical. Two screens are said to be complementary when the opaque parts of one correspond to the transparent parts of the other, and vice versa. Babinet's principle is not very useful in dealing with Fresnel diffraction, except that it may furnish a short-cut method in obtaining the pattern for a particular screen from that of its complement. The principle has an important application for Fraunhofer diffraction, in parts of the field where there is zero intensity without any screen. Under this condition the amplitudes produced by the complementary screens must be equal and opposite, since the sum of the effects of their exposed parts gives no light. The intensities, being proportional to the squares of the amplitudes, must therefore be equal. In Fraunhofer diffraction the pattern due to a disk is the same as that due to a circular hole of the same size.

DIFFRACTION OF MICROWAVES

The diffraction of microwaves, which have wave-lengths in the range of millimeters to centimeters, has been intensively studied since World War II because of its importance in radar work. Many of the characteristics of optical diffraction can be strikingly demonstrated by the use of microwaves. Microwave diffraction shows certain features, however, that are not in agreement with the Huygens-Fresnel theory, because the approximations made in that theory are no longer valid. Most of these approximations, for example, that made in deriving Eq. (5), depend for their validity on the assumption that the wavelength is small compared to the dimensions of the apparatus. Furthermore, it is not legitimate to postulate that the wave has a constant amplitude across an opening, and zero intensity behind the opaque parts, except for the very minute waves of light.

As an example of the failure of classical diffraction theory when applied to microwaves, the results of the diffraction by a circular hole in a metal screen may be mentioned. The observed patterns begin to show deviations from the Fresnel theory, and even from the more rigorous Kirchhoff theory, when the point of observation is within a few wavelengths' distance from the plane of the aperture. Even in this plane itself there are detection effects that could not have been predicted from the earlier theories, which treat light as a scalar, rather than a vector, wave motion. An exact vector theory of diffraction, developed by A. Sommerfeld, has been applied only in a few simple cases, but the measurements at microwave frequencies agree with it wherever it has been tested. SEE LIGHT.

Bibliography. C. J. Ball, *Introduction to the Theory of Diffraction*, 1971; M. Born and E. Wolf, *Principles of Optics*, 6th ed., 1980; F. A. Jenkins, *Fundamentals of Optics*, 4th ed., 1976; J. Meyer-Arendt, *Introduction to Classical and Modern Optics*, 2d ed., 1984; A. Sommerfeld, *Lectures on Theoretical Physics*, vol. 4: *Optics*, 1954.

DIFFRACTION GRATING
GEORGE R. HARRISON

An optical device consisting of an assembly of narrow slits or grooves, which by diffracting light produces a large number of beams which can interfere in such a way as to produce spectra. Since the angles at which constructive interference patterns are produced by a grating depend on the lengths of the waves being diffracted, the waves of various lengths in a beam of light striking the grating will be separated into a number of spectra, produced in various orders of interference on either side of an undeviated central image. By controlling the shape and size of the diffracting grooves when producing a grating and by illuminating the grating at suitable angles, a beam of light can be thrown into a single spectrum whose purity and brightness may exceed that produced by a prism. Gratings can now be made with much larger apertures than prisms, and in such form that they waste less light and give higher intrinsic dispersion and resolving power. A single grating can be used over a much broader range of spectrum than can any single prism, and its dispersion will vary less rapidly with wavelength. Gratings are used in large spectrographs and for highly precise spectroscopic work, as well as in monochromators and analytical spectrographs. SEE DIFFRACTION; INTERFERENCE OF WAVES; OPTICAL PRISM.

Transmission gratings consist of a large number of narrow transparent and opaque slits alternating side by side in regular order and with uniform separation, through which a beam of light will appear as a series of spectra in various orders of interference. Such gratings are conveniently used in small spectroscopes and spectrometers, but only for visible light, since they are usually not transparent to ultraviolet or infrared radiation. They are commonly made by contact molding from a master grating.

Reflection gratings, either plane or concave, are used in most spectrographs. Such a grating may consist of an original ruling or of a metal-coated replica from an original. Large grating replicas practically indistinguishable in performance or permanence from an original can now be made.

Production of gratings. Gratings are engraved by highly precise ruling engines, which use a diamond tool to press into a highly polished mirror surface a series of many thousands of fine shallow burnished grooves. Gratings for the range 150–1000 nanometers are commonly ruled with 5000–30,000 grooves per inch or 200–1200 grooves per millimeter (the usual value is near 15,000 per inch or 600 per millimeter), on a thin layer of aluminum deposited on glass by evaporation in vacuum. Gratings for the infrared region are often ruled on gold, silver, copper, lead, or tin mirrors, with coarser groove spacings.

If a grating is to give resolution approaching the theoretical limit, its grooves must be ruled straight, parallel, and equally spaced to within a few tenths of the shortest incident wavelength. The proper overall spacing of grooves must also be maintained if changes in focal properties are not to result. Scattered light and false images may arise from local spacing error and groove shape variations of only a few hundredths of the diffracted wavelength.

Among the false lines produced by imperfect gratings are Rowland ghosts, which arise from periodic errors in groove position; Lyman ghosts, which come from a combination of periodicities in ruling; and satellites, caused by sets of irregularly placed grooves, which may seriously reduce resolution. Target pattern arises from unequal contribution of light from all parts of a grating, and is especially prevalent in concave gratings, in which the shape of the grooves may change as the cutting angle of the diamond changes.

Gratings of 2, 4, or 6 in. (5, 10, or 15 cm) ruled width are commonly used in commercial spectrographs, with projection distances of 20–180 in. (50–450 cm). In large research instruments, gratings of 6 to 10 in. (15 to 25 cm) ruled width are used with projection distances of 10–50 ft (3–15 m) or more. The largest modern gratings, used in their highest orders, show resolving power $\lambda/\delta\lambda$ in excess of 900,000 in the green region of the spectrum, and in excess of 1.5×10^6 at shorter wavelengths. Here λ is the mean wavelength of two closely spaced, just resolvable spectral lines, and $\delta\lambda$ is their wavelength difference. Such gratings give resolution equal to that of most interferometers, and in addition provide greater photographic speed, are easier to adjust, follow more simple laws of wavelength distribution, and permit a wider range of wavelengths to be photographed at one time without crossed dispersion. *See* INTERFEROMETRY; RESOLVING POWER.

Properties of gratings. A grating illuminated at angle α (measured from the normal) will direct wavelength λ toward angle β in accordance with the formula $m\lambda = d(\sin \alpha \pm \sin \beta)$, where m is an integral order of interference, d is the grating constant, or distance between consecutive grooves, and the + and − signs refer to orders on opposite sides of the normal. The linear dispersion produced by a grating on a photographic plate depends on its intrinsic angular dispersion multiplied by the distance P from grating to plate. The intrinsic angular dispersion is given by the formula shown below. Theoretically the resolving power of a grating is mN, where N

$$\frac{d\beta}{d\lambda} = \frac{1}{\lambda}\left(\frac{\sin \alpha}{\cos \beta} + \tan \beta\right)$$

is the number of grooves in the grating. Resolving power is not directly dependent on the number of grooves, since for gratings of a given size m and N are inversely related. It is basically dependent on the number of wavelengths of optical retardation the grating introduces between the extreme rays leaving it. Another useful concept is the resolving limit $d\sigma$, the smallest wave number difference the grating can resolve, which remains essentially constant for a given angle of illumination of a given grating at all wavelengths, except as errors in groove spacing become more important for shorter wavelengths.

The manner in which incident light will be distributed among the various orders of interference depends upon the shape and orientation of the groove sides and on the relation of wavelength to groove separation. When $d \lesssim \lambda$, diffraction effects predominate in controlling the intensity distribution among orders, but when $d > \lambda$, optical reflection from the sides of the grooves is more strongly involved. It is possible to "blaze" a grating by ruling its grooves so that their sides reflect a large fraction of the incoming light of suitably short wavelengths in one general direction. Controlled groove shape is especially important in the gratings known as echelettes and echelles, in which as much as 80% of the incoming light may be sent into one particular order for a given wavelength. Many ordinary gratings are blazed.

Grating spectroscopes. These consist usually of a slit, a lens or mirror to collimate the light sent through the slit into a parallel beam, a transmission or reflection grating to disperse the light, a lens or mirror to focus the light into spectrum lines (which are monochromatic images of

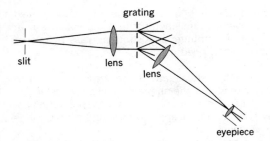

Fig. 1. Transmission grating spectroscope.

the slit in the light of each wavelength passing through it), and an eyepiece for viewing the spectrum (**Fig. 1**). If a camera is substituted for the telescope, the instrument becomes a grating spectrograph. If a photoelectric cell, a thermocouple, or other radiation-detecting device is used instead of a camera or telescope, the device becomes a grating spectrometer.

Echelette grating. This has coarse groove spacing and is designed for the infrared region, the grooves being so shaped and of such size that most of the radiation is concentrated by reflection into a small angular coverage. Radiation of any given wavelength is thus concentrated largely into one order by shaping the point of the ruling diamond to give grooves with comparatively flat sides and by choosing a groove separation which minimizes diffraction effects.

Echelle grating. This is designed for use in high orders and at angles of illumination greater than 45° to obtain high dispersion and resolving power by the use of high orders of interference. Echelles have properties lying midway between those of plane gratings of the ordinary type and interferometers of the reflection echelon type, orders of interference ranging from 100 to 1000 being used. Overlapping orders are separated by using crossed dispersion.

Concave grating. This is a widely used form of reflection grating with which a spectrograph can be formed that has no auxiliary optical parts except a slit and a camera. Being ruled on a concave mirror, this type of grating can both collimate and focus the light that falls upon it. It is made by spacing straight grooves equally along the chord (rather than the arc) of a spherical or paraboloidal mirror surface. Light which passes through a slit and falls on such a grating is dispersed by it into spectra which are in focus on the Rowland circle, a circle drawn tangent to the face of the grating at its midpoint, having a diameter equal to the radius of curvature of the grating surface.

A great advantage of the concave grating is that it provides a dispersing and focusing system free of refracting material, so that it can be used with ultraviolet, visible, or infrared radiation interchangeably so long as its grooves diffract radiation and its surface has adequate reflecting power. A disadvantage is its astigmatism at high angles of incidence or reflection, which can, however, be diminished with various optical devices. A plane reflection grating used with two concave mirrors avoids this difficulty.

Grating mountings. The slit, grating, and camera of a concave grating spectrograph can be placed anywhere on the Rowland circle so that any desired wavelength range can be photographed in the desired order (**Fig. 2**).

The various possible combinations of fixed and moving parts give rise to a number of different grating mountings. In the Rowland mounting, camera and grating are connected by a bar forming a diameter of the Rowland circle, the two running on tracks placed at right angles with the slit fixed at their junction. A spectrum of limited extent having uniform dispersion is then produced at the camera, and camera and grating can be moved on the tracks to shift wavelength coverage. In the Paschen-Runge mounting, slit and grating are fixed, and photographic plates can be clamped to a fixed track almost anywhere on the Rowland circle. In the Eagle mounting, most suited to long, narrow housing, the grating can be rotated and moved toward or away from the slit. The slit is placed close to the camera, which is arranged to rotate so that it can be kept on the Rowland circle.

All these mountings suffer from astigmatism arising from using the grating off-axis. Although this does not markedly reduce the sharpness of the spectrum lines, it may result in a great

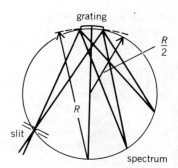

Fig. 2. Rowland circle.

loss of light intensity when the grating is used at high angles, and it makes difficult the sharp focusing of step filters, sector disks, and interferometer patterns that are placed at the slit.

Astigmatism is greatly reduced in the Wadsworth mounting, in which the slit is placed at the principal focus of a concave mirror, so that the light falling on the grating is in a parallel beam. The grating can be illuminated at any desired angle up to about 40°, and light is taken off along the grating normal, the spectrum being focused on a photographic plate at half the usual distance. The usual dispersion of the grating is halved, but the speed of the spectrograph is increased fourfold, and at high angles the speed is increased much more because of reduction of astigmatism.

Most modern commercial grating spectrographs, because of the need for portability, are based either on the Eagle mounting, with which a rather limited portion of the spectrum can be photographed at one time, or the Wadsworth mounting, which can give greater spectral coverage without resetting but is bulkier and cannot be used at such high values of $m\lambda$. In order to obtain more complete spectrum coverage in a single exposure, an echelle with crossed dispersion or a grating with some other device for separating the orders may be used.

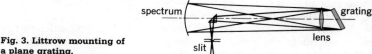

Fig. 3. Littrow mounting of a plane grating.

Plane reflection gratings are ordinarily used in the Littrow mounting (**Fig. 3**), in which a single lens serves for collimating and focusing, or in the Ebert mounting in which there are two concave mirrors.

Bibliography. R. Chang, *Basic Principles of Spectroscopy*, 1978; M. C. Hutley, *Diffraction Gratings*, 1982; F. A. Jenkins, *Fundamentals of Optics*, 4th ed., 1976; D. L. Pavia et al., *Introduction to Spectroscopy*, 1979; D. A. Ramsay, *Spectroscopy*, 1976.

POLARIZATION OF WAVES
Bruce H. Billings

Polarization is the phenomenon which is exhibited when a transverse wave is polarized. The term polarization is also used to describe the process of polarizing a wave.

In an unpolarized wave, the vibrations in a plane perpendicular to the ray appear to be oriented in all directions with equal probability. In a polarized wave the displacement direction of the vibrations is completely predictable. For certain disturbances, such as the transverse acoustic wave produced when a steel bar is struck, the polarization is complete. Electromagnetic radiation is normally unpolarized if it is generated by atomic processes. Thus ultraviolet, visible, and infrared radiations produced by heated bodies or electrical discharges are generally unpolarized. Radiation generated by vacuum-tube oscillators or transistor oscillators is always polarized. The probability waves (matter waves) associated with atomic or nuclear particles are generally unpolarized. SEE ELECTROMAGNETIC RADIATION.

Some of the different types of polarization, as well as the technique of producing polarization in an unpolarized wave, are described in another article. The electric vector can lie in a plane or it can follow a path whose projection at right angles to the direction of propagation is a circle or an ellipse. The same types of polarization can be produced in any transverse wave. SEE POLARIZED LIGHT.

Electromagnetic radiation is difficult to polarize in certain spectral regions, and few techniques exist for analysis. This is true in the ultraviolet below 190 nanometers. No dichroic polarizers have been found for this region, and transparent birefringent materials from which Nicol or Wollaston polarizing prisms could be made do not seem to exist. Polarization by reflection is possible, but very little work has been done with this technique. In the infrared region from the end of the visible spectrum to approximately 2 micrometers, sheet polarizers exist. To around 4 μm, polarizing prisms can be made. From 4 to 80 μm, reflection from a single plate or transmission through a pile of transparent plates is the common procedure. All these techniques produce linear polarization. Elliptical or circular polarization is more difficult to achieve.

X-ray photons, electrons, neutrons, and other particles can be polarized most easily by scattering.

POLARIZED LIGHT
Bruce H. Billings

Light which has its electric vector oriented in a predictable fashion with respect to the propagation direction. In unpolarized light, the vector is oriented in a random, unpredictable fashion. Even in short time intervals, it appears to be oriented in all directions with equal probability. Most light sources seem to be partially polarized so that some fraction of the light is polarized and the remainder unpolarized. It is actually more difficult to produce a completely unpolarized beam of light than one which is completely polarized.

The polarization of light differs from its other properties in that human sense organs are essentially unable to detect the presence of polarization. The Polaroid Corp. with its polarizing sunglasses and camera filters has made millions of people conscious of phenomena associated with polarization. Light from a rainbow is completely linearly polarized; that is, the electric vector lies in a plane. The possessor of polarizing sunglasses discovers that with such glasses, the light from a section of the rainbow is extinguished.

According to all available theoretical and experimental evidence, it is the electric vector rather than the magnetic vector of a light wave that is responsible for all the effects of polarization and other observed phenomena associated with light. Therefore, the electric vector of a light wave, for all practical purposes, can be identified as the light vector. SEE CRYSTAL OPTICS; ELECTROMAGNETIC RADIATION; LIGHT; POLARIZATION OF WAVES.

One of the simplest ways of producing linearly polarized light is by reflection from a dielectric surface. At a particular angle of incidence, the reflectivity for light whose electric vector is in the plane of incidence becomes zero. The reflected light is thus linearly polarized at right angles to the plane of incidence. This fact was discovered by E. Malus in 1808. Brewster's law shows that at the polarizing angle the refracted ray makes an angle of 90° with the reflected ray. By combining this relationship with Snell's law of refraction, it is found that Eq. (1) holds, where i is

$$\tan i = n \qquad (1)$$

the angle of incidence and n is the refractive index. This provides a simple way of measuring refractive indices. SEE REFRACTION OF WAVES.

Law of Malus. If linearly polarized light is incident on a dielectric surface at Brewster's angle (the polarizing angle), then the reflectivity of the surface will depend on the angle between the incident electric vector and the plane of incidence. When the vector is in the plane of incidence, the reflectivity will be at a maximum. To compute the complete relationship, the incident light vector **A** is broken into components, one vibrating in the plane of incidence and one at right angles to the plane, as in Eqs. (2) and (3), where θ is the angle between the light vector and a

$$A_{\parallel} = A \sin \theta \qquad (2) \qquad\qquad A_{\perp} = A \cos \theta \qquad (3)$$

plane perpendicular to the plane of incidence. Since the component in the plane of incidence is not reflected, the reflected ray can be written as Eq. (4), where r is the reflectivity at Brewster's angle. The intensity is given by Eq. (5). This is the mathematical statement of the law of Malus.

$$B = Ar \cos \theta \qquad (4) \qquad\qquad I = B^2 = A^2 r^2 \cos^2 \theta \qquad (5)$$

Linear polarizing devices. The angle θ can be considered as the angle between the transmitting axes of a pair of linear polarizers. When the polarizers are parallel, they are transparent. When they are crossed, the combination is opaque. The first polarizers were glass plates inclined so that the incident light was at Brewster's angle. Such polarizers are quite inefficient since only a small percentage of the incident light is reflected as polarized light. More efficient polarizers can be constructed.

Dichroic crystals. Certain natural materials absorb linearly polarized light of one vibration direction much more strongly than light vibrating at right angles. Such materials are termed dichroic. For a description of them SEE DICHROISM.

Tourmaline is one of the best-known dichroic crystals, and tourmaline plates were used as polarizers for many years. A pair was usually mounted in what were known as tourmaline tongs.

Birefringent crystals. Other natural materials exist in which the velocity of light depends on the vibration direction. These materials are called birefringent. The simplest of these structures are crystals in which there is one direction of propagation for which the light velocity is independent of its state of polarization. These are termed uniaxial crystals, and the propagation direction mentioned is called the optic axis. For all other propagation directions, the electric vector can be split into two components, one lying in a plane containing the optic axis and the other at right angles. The light velocity or refractive index for these two waves is different. SEE BIREFRINGENCE.

One of the best-known of these birefringent crystals is transparent calcite (Iceland spar), and a series of polarizers have been made from this substance. W. Nicol (1829) invented the Nicol prism, which is made of two pieces of calcite cemented together as in **Fig. 1**. The cement is Canada balsam, in which the wave velocity is intermediate between the velocity in calcite for the fast and the slow ray. The angle at which the light strikes the boundary is such that for one ray the angle of incidence is greater than the critical angle for total reflection. Thus the rhomb is transparent for only one polarization direction.

Canada balsam is not completely transparent in the ultraviolet at wavelengths shorter than 400 nanometers. Furthermore, large pieces of calcite material are exceedingly rare. A series of polarizers has been made using quartz, which is transparent in the ultraviolet and which is more commonly available in large pieces. Because of the small difference between the refractive indices of quartz and Canada balsam, a Nicol prism of quartz would be tremendously long for a given linear aperture.

A different type of polarizer, made of quartz, was invented by W. H. Wollaston and is shown in **Fig. 2**. Here the vibration directions are different in the two pieces so that the two rays are

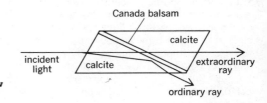

Fig. 1. Nicol prism. The ray for which Snell's law holds is called the ordinary ray.

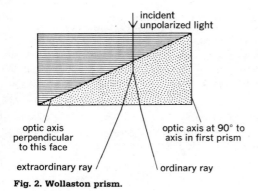
Fig. 2. Wollaston prism.

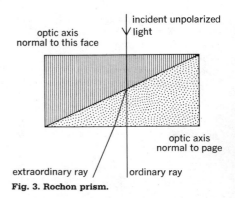
Fig. 3. Rochon prism.

deviated as they pass through the material. The incoming light beam is thus separated into two oppositely linearly polarized beams which have an angular separation between them. By using appropriate optical stops (obstacles that restrict light rays) in the system, it is possible to select either beam.

In the Wollaston prism, both beams are deviated; since the quartz produces dispersion, each beam is spread into a spectrum. This is not the case of a prism which was invented by A. Rochon. Here the two pieces are arranged as in **Fig. 3**. One beam proceeds undeviated through the device and is thus achromatic. SEE OPTICAL PRISM.

Sheet polarizers. A third mechanism for obtaining polarized light is the Polaroid sheet polarizer invented by E. H. Land. Sheet polarizers fall into three types. The first is a microcrystalline polarizer in which small crystals of a dichroic material are oriented parallel to each other in a plastic medium. Typical microcrystals, such as needle-shaped quinine iodosulfate, are embedded in a viscous plastic and are oriented by extruding the material through a slit.

The second type depends for its dichroism on a property of an iodine-in-water solution. The iodine appears to form a linear high polymer. If the iodine is put on a transparent oriented sheet of material such as polyvinyl alcohol (PVA), the iodine chains apparently line themselves parallel to the PVA molecules and the resulting dyed sheet is strongly dichroic. A third type of sheet polarizer depends for its dichroism directly on the molecules of the plastic itself. This plastic consists of oriented polyvinylene. Because these polarizers are commercially available and can be obtained in large sheets, many experiments involving polarized light have been performed which would have been quite difficult with the reflection polarizers or the birefringent crystal polarizer.

Characteristics. There are several characteristics of linear polarizers which are of interest to the experimenter. First is the transmission for light polarized parallel and perpendicular to the axis of the polarizer; second is the angular field; and third is the linear aperture. A typical sheet polarizer has a transmittance of 48% for light parallel to the axis and 2×20^{-4}% for light perpendicular to the axis at a wavelength of 550 nm. The angular field is 60°, and sheets can be many feet in diameter. The transmittance perpendicular to the axis varies over the angular field.

The Nicol prism has transmittance similar to that of the Polaroid sheeting but a much reduced linear and angular aperture.

Polarization by scattering. When an unpolarized light beam is scattered by molecules or small particles, the light observed at right angles to the original beam is polarized. The light vector in the original beam can be considered as driving the equivalent oscillators (nuclei and electrons) in the molecules. There is no longitudinal component in the original light beam. Accordingly, the scattered light observed at right angles to the beam can only be polarized with the electric vector at right angles to the propagation direction of the original beam. In most situations, the scattered light is only partially polarized because of multiple scattering. The best-known example of polarization by scattering is the light of the north sky. The percentage polarization can be quite high in clean country air. A technique was invented for using measurements of sky polarization to determine the position of the sun when it is below the horizon. SEE SCATTERING OF ELECTROMAGNETIC RADIATION.

Types of polarized light. Polarized light is classified according to the orientation of the electric vector. In linearly polarized light, the electric vector remains in a plane containing the propagation direction. For monochromatic light, the amplitude of the vector changes sinusoidally with time. In circularly polarized light, the tip of the electric vector describes a circular helix about the propagation direction. The amplitude of the vector is constant. The frequency of rotation is equal to the frequency of the light. In elliptically polarized light, the vector also rotates about the propagation direction, but the amplitude of the vector changes so that the projection of the vector on a lane at right angles to the propagation direction describes an ellipse.

These different types of polarized light can all be broken down into two linear components at right angles to each other. These are defined by Eqs. (6) and (7), where A_x and A_y are the

$$E_x = A_x \sin(\omega t + \varphi_x) \qquad (6) \qquad\qquad E_y = A_y \sin(\omega t + \varphi_y) \qquad (7)$$

amplitudes, φ_x and φ_y are the phases, ω is 2π times the frequency, and t is the time. For linearly polarized light Eqs. (8) hold. For circularly polarized light Eqs. (9) hold. For elliptically polarized

$$\varphi_x = \varphi_y \quad A_x \neq A_y \qquad (8) \qquad\qquad \varphi_x = \varphi_y \pm \frac{\pi}{2} \quad A_x = A_y \qquad (9)$$

light Eqs. (10) hold. In the last case, it is always possible to find a set of orthogonal axes inclined at an angle α to x and y along which the components will be E'_x and E'_y, such that Eqs. (11) hold.

$$\varphi_x \neq \varphi_y \quad A_x \neq A_y \qquad (10) \qquad\qquad \varphi'_x = \varphi'_y \pm \frac{\pi}{2} \quad A'_x \neq A'_y \qquad (11)$$

In this new system, the x' and y' amplitudes will be the major and minor axes a and b of the ellipse described by the light vector and α will be the angle of orientation of the ellipse axes with respect to the original coordinate system. The relationships between the different quantities can be written as in Eqs. (12) and (13). The terms are defined by Eqs. (14)–(17). These same types

$$\tan 2\alpha = \tan 2\gamma \cos \varphi \qquad (12) \qquad\qquad \sin 2\beta = \sin 2\gamma \sin \varphi \qquad (13)$$

$$\tan \gamma = A_y/A_x \qquad (14) \qquad\qquad \varphi = \varphi_x - \varphi_y \qquad (15)$$

$$\tan \beta = \pm b/a \qquad (16) \qquad\qquad A_x^2 + A_y^2 = a^2 + b^2 \qquad (17)$$

of polarized light can also be broken down into right and left circular components or into two orthogonal elliptical components. These different vector bases are useful in different physical situations.

Production of polarized light. Linear polarizers have already been discussed. Circularly and elliptically polarized light are normally produced by combining a linear polarizer with a wave plate. A Fresnel rhomb can be used to produce circularly polarized light.

Wave plate. A plate of material which is linearly birefringent is called a wave plate or retardation sheet. Wave plates have a pair of orthogonal axes which are designated fast and slow. Polarized light with its electric vector parallel to the fast axis travels faster than light polarized parallel to the slow axis. The thickness of the material can be chosen so that for light traversing the plate, there is a definite phase shift between the fast component and the slow component. A plate with a 90° phase shift is termed a quarter-wave plate. The retardation in waves is given by Eq. (18), where $n_s - n_f$ is the birefringence; n_s is the slow index at wavelength λ; n_f is the fast index; and d is the plate thickness.

$$\delta = \frac{(n_s - n_f)d}{\lambda} \qquad (18)$$

Wave plates can be made by preparing X-cut sections of quartz, calcite, or other birefringent crystals. For retardations of less than a few waves, it is easiest to use sheets of oriented plastics or of split mica. A quarter-wave plate for the visible or infrared is easy to fabricate from mica. The plastic wrappers from many American cigarette packages seem to have almost exactly a half-wave retardation from green light. Since mica is not transparent in the ultraviolet, a small retardation in this region is most easily achieved by crossing two quartz plates which differ by the requisite thickness.

Linearly polarized light incident normally on a quarter-wave plate and oriented at 45° to the fast axis can be split into two equal components parallel to the fast and slow axes. These can be represented, before passing through the plate, by Eqs. (19) and (20), where x and y are parallel to

$$E_x = A_x \sin(\omega t + \varphi_x) \quad (19) \qquad E_y = A_x \sin(\omega t + \varphi_x) \quad (20)$$

the wave-plate axes. After passing through the plate, the two components can be written as Eqs. (21) and (22), where E_x is now advanced one quarter-wave with respect to E_y.

$$E_x = A_x \sin\left(\omega t + \varphi_x + \frac{\pi}{2}\right) \quad (21) \qquad E_y = A_x \sin(\omega t + \varphi_x) \quad (22)$$

It is possible to visualize the behavior of the light by studying the sketches in **Fig. 4**, which show the projection on a plane $z=0$ at various times. It is apparent that the light vector is of constant amplitude, and that the projection on a plane normal to the propagation direction is a circle. If the linearly polarized light is oriented at $-45°$ to the fast axis, the light vector will revolve in the opposite direction. Thus it is possible with a quarter-wave plate and a linear polarizer to make either right or left circularly polarized light. If the linearly polarized light is at an angle other than 45° to the fastest axis, the light vector will revolve in the opposite direction. Thus it is possible with a quarter-wave plate and a linear polarizer to make either right or left circularly polarized light. If the linearly polarized light is at an angle of 45° to the wave-plate axes. This polarization is independent of the orientation of the wave-plate axis. For elliptically polarized light, the behavior of the quarter-wave plate is much more complicated. However, as was mentioned earlier, the elliptically polarized light can be considered as composed of two linear components parallel to the major and minor axes of the ellipse and with a quarter-wave phase difference between them. If the quarter-wave plate is oriented parallel to the axes of the ellipse, the two transmitted components will either have zero phase difference or a 180° phase difference and will be linearly polarized. At other angles, the transmitted light will still be elliptically polarized, but

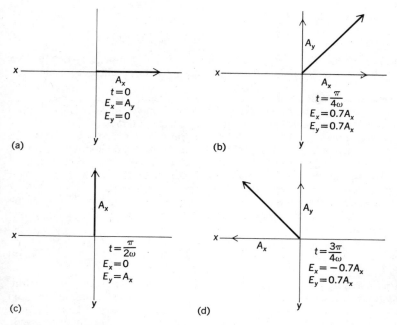

Fig. 4. Projection of light vector of constant amplitude on plane $z=0$ for circularly polarized light. (a) $t = 0$. (b) $t = \pi/4\omega$. (c) $t = \pi/2\omega$. (d) $t = 3\pi/4\omega$.

with different major and minor axes. Similar treatment for a half-wave plate shows that linearly polarized light oriented at an angle of θ to the fast axis is transmitted as linearly polarized light oriented at an angle $-\theta$ to the fast axis.

Wave plates all possess a different retardation at each wavelength. This appears immediately from Eq. (18). It is conceivable that a substance could have dispersion of birefringence, such as to make the retardation of a plate independent of wave-length. However, no material having such a characteristic has as yet been found.

Fresnel rhomb. A quarter-wave retardation can be provided achromatically by the Fresnel rhomb. This device depends on the phase shift which occurs at total internal reflection. When linearly polarized light is totally internally reflected, it experiences a phase shift which depends on the angle of reflection, the refractive index of the material, and the orientation of the plane of polarization. Light polarized in the plane of incidence experiences a phase shift which is different from that of light polarized at right angles to the plane of incidence. Light polarized at an intermediate angle can be split into two components, parallel and at right angles to the plane of incidence, and the two components mathematically combined after reflection.

The phase shifts can be written as Eqs. (23) and (24), where $\varphi_\parallel$ is the phase shift parallel

$$\tan \frac{\varphi_\parallel}{2} = \frac{\sqrt{n^2 \sin^2 i - 1}}{\cos i} \qquad (23) \qquad \tan \frac{\varphi_\perp}{2} = \frac{\sqrt{n^2 \sin^2 i - 1}}{n \cos i} \qquad (24)$$

to the plane of incidence; $\varphi_\perp$ is the phase shift at right angles to the plane of incidence; i is the angle of incidence on the totally reflecting internal surface; and n is the refractive index. The difference $\varphi_\parallel - \varphi_\perp$ reaches a value of about $\pi/4$ at an angle of 52° for $n = 1.50$. Two such reflections give a retardation of $\pi/2$. The Fresnel rhomb shown in **Fig. 5** is cut so that the incident light is

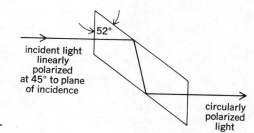

Fig. 5. Fresnel rhomb.

reflected twice at 52°. Accordingly, light polarized at 45° to the principal plane will be split into two equal components which will be shifted a quarter-wave with respect to each other, and the transmitted light will be circularly polarized. Nearly achromatic wave plates can be made by using a series of wave plates in series with their axes oriented at different specific angles with respect to a coordinate system.

Analyzing devices. Polarized light is one of the most useful tools for studying the characteristics of materials. The absorption constant and refractive index of a metal can be calculated by measuring the effect of the metal on polarized light reflected from its surface. SEE REFLECTION OF ELECTROMAGNETIC RADIATION.

The analysis of polarized light can be performed with a variety of different devices. If the light is linearly polarized, it can be extinguished by a linear polarizer and the direction of polarization of the light determined directly from the orientation of the polarizer. If the light is elliptically polarized, it can be analyzed with the combination of a quarter-wave plate and a linear polarizer. Any such combination of polarizer and analyzer is called a polariscope. As explained previously, a quarter-wave plate oriented parallel to one of the axes of the ellipse will transform elliptically polarized light to linearly polarized light. Accordingly, the quarter-wave plate is rotated until the beam is extinguished by the linear polarizer. At this point, the orientation of the quarter-wave plate gives the orientation of the ellipse and the orientation of the polarizer gives the ratio of the major to minor axis. Knowledge of the origin of the elliptically polarized light usually gives the orientation of the components which produced it, and from these various items, the phase shifts and attenuations produced by the experiment can be deduced.

One of the best-known tools for working with polarized light is the Babinet compensator. This device is normally made of two quartz prisms put together in a rhomb. One prism is cut with the optic axis in the plane of incidence of the prism, and the other with the optic axis perpendicular to the plane of incidence. The retardation is a function of distance along the rhomb; it will be zero at the center, varying to positive and negative values in opposite directions along the rhomb. It can be used to cancel the known or unknown retardation of any wave plate.

Retardation theory. It is difficult to see intuitively the effect of a series of retardation plates or even of a single plate of general retardation δ on light which is normally incident on the plate and which is polarized in a general fashion. This problem is most easily solved algebraically. The single-wave plate is assumed to be oriented normal to the direction of propagation of the light, which is taken to be the z direction of a set of cartesian coordinates. Its fast axis is at an angle α to the x axis. The incident light can be represented by Eqs. (25) and (26). A first step is to break the light up into components $E_{x'}$ and $E_{y'}$ parallel to the axes of the plate. It is possible to write Eqs. (27) and (28). These components can also be written as Eqs. (29) and (30). After passing through the plate, the components become Eqs. (31) and (32).

$$E_x = A_x \sin(\omega t + \varphi_x) \qquad (25)$$

$$E_y = A_y \sin(\omega t + \varphi_y) \qquad (26)$$

$$E_{x'} = E_x \cos \alpha - E_y \sin \alpha \qquad (27)$$

$$E_{y'} = E_x \sin \alpha + E_y \cos \alpha \qquad (28)$$

$$E_{x'} = A_{x'} \sin(\omega t + \varphi_{x'}) \qquad (29)$$

$$E_{y'} = A_{y'} \sin(\omega t + \varphi_{y'}) \qquad (30)$$

$$E_{x''} = A_{x'} \sin(\omega t + \varphi_{x'} + \delta) \qquad (31)$$

$$E_{y''} = A_{y'} \sin(\omega t + \varphi_{y'}) \qquad (32)$$

In general, it is of interest to compare the output with the input. The transmitted light is thus broken down into components along the original axes. This results in Eqs. (33) and (34).

$$E_{x'''} = E_{x''} \cos \alpha + E_{y''} \sin \alpha \qquad (33)$$

$$E_{y'''} = -E_{x''} \sin \alpha + E_{y''} \cos \alpha \qquad (34)$$

With the set of equations (25)–(34), it is possible to compute the effect of a wave plate on any form of polarized light.

Jones calculus. Equations (33) and (34) still become overwhelmingly complicated in any system involving several optical elements. Various methods have been developed to simplify the problem and to make possible some generalizations about systems of elements. One of the most straightforward, proposed by R. C. Jones, involves reducing Eqs. (33) and (34) to matrix form. The Jones calculus for optical systems involves the polarized electric components of the light vector and is distinguished from other methods in that it takes cognizance of the absolute phase of the light wave.

The Jones calculus writes the light vector in the complex form as in Eqs. (35) and (36).

$$E_x = A_x e^{i(\omega t + \varphi_x)} \quad E_y = A_y e^{i(\varphi t + \varphi_y)} \qquad (35)$$

$$E = \begin{vmatrix} A_x e^{i\varphi_x} \\ A_y e^{i\varphi_y} \end{vmatrix} e^{i\omega t} \qquad (36)$$

Matrix operators are developed for different optical elements. From Eqs. (25)–(34), the operator for a wave plate can be derived directly.

The Jones calculus is ordinarily used in a normalized form which simplifies the matrices to a considerable extent. In this form, the terms involving the actual amplitude and absolute phase of the vectors and operators are factored out of the expressions. The intensity of the light beam is reduced to unity in the normalized vector so that Eq. (37) holds. Under this arrangement, the matrices for various types of operations can be written as Eq. (38). This is the operator for a wave

$$A_x^2 + A_y^2 = 1 \qquad (37)$$

$$G(\delta) = \begin{vmatrix} e^{i(\delta/2)} & 0 \\ 0 & e^{-i(\delta/2)} \end{vmatrix} \qquad (38)$$

plate of retardation δ and with axes along x and y. Equation (39) gives the operator for a rotator which rotates linearly polarized light through an angle α. Equation (40) gives the operator for a

$$S(\alpha) = \begin{vmatrix} \cos\alpha & -\sin\alpha \\ \sin\alpha & \cos\alpha \end{vmatrix} \quad (39) \qquad\qquad P_h = \begin{vmatrix} 1 & 0 \\ 0 & 0 \end{vmatrix} \quad (40)$$

perfect linear polarizer parallel to the x axis. A wave plate at an angle α can be represented by Eq. (41). A series of optical elements can be represented by the product of a series of matrices.

$$G(\delta,\alpha) = S(\alpha)G(\delta)S(-\alpha) \quad (41)$$

This simplifies enormously the task of computing the effect of many elements. It is also possible with the Jones calculus to derive a series of general theorems concerning combinations of optical elements. Jones has described three of these, all of which apply only for monochromatic light.

    1. An optical system consisting of any number of retardation plates and rotators is optically equivalent to a system containing only two elements, a retardation plate and a rotator.

    2. An optical system containing any number of partial polarizers and rotators is optically equivalent to a system containing only two elements—one a partial polarizer and the other a rotator.

    3. An optical system containing any number of retardation plates, partial polarizers, and rotators is optically equivalent to a system containing four elements—two retardation plates, one partial polarizer, and one rotator.

As an example of the power of the calculus, a rather specific theorem can be proved. A rotator of any given angle α can be formed by a sequence of three retardation plates, a quarter-wave plate, a retardation plate at 45° to the quarter-wave plate, and a second quarter-wave plate crossed with the first, as in Eq. (42), where β is the angle between the axis of the first quarter-

$$S(\alpha) = S(\beta)G\left(-\frac{\pi}{2}\right)S(-\beta)S\left(\beta+\frac{\pi}{4}\right) \cdot G(\delta)S\left(-\beta-\frac{\pi}{4}\right)S(\beta)G\left(\frac{\pi}{2}\right)S(-\beta) \quad (42)$$

wave plate and the x axis, and δ is the retardation of the plate in the middle of the sandwich.

    The first simplification arises from the fact that the axis rotations can be done in any order. This reduces Eq. (43) to Eq. (44). Now Eqs. (45) and (46) hold. When the multiplication is carried through, Eq. (47) is obtained.

$$S\left(\beta+\frac{\pi}{4}\right) = S(\beta)S\left(\frac{\pi}{4}\right) = S\left(\frac{\pi}{4}\right)S(\beta) \quad (43)$$

$$S(\alpha) = S(\beta)G\left(-\frac{\pi}{2}\right)S\left(\frac{\pi}{4}\right) \cdot G(\delta)S\left(-\frac{\pi}{4}\right)G\left(\frac{\pi}{2}\right) \quad (44)$$

$$S\left(-\frac{\pi}{4}\right)G\left(\frac{\pi}{2}\right) = \frac{1}{\sqrt{2}}\begin{vmatrix} 1 & -i \\ -i & 1 \end{vmatrix} \quad (45)$$

$$G\left(-\frac{\pi}{2}\right)S\left(\frac{\pi}{4}\right) = \frac{1}{\sqrt{2}}\begin{vmatrix} 1 & i \\ i & 1 \end{vmatrix} \quad (46)$$

$$S(\alpha) = S(\beta)\frac{1}{2}\begin{vmatrix} e^{i\delta/2}+e^{-i\delta/2} & -ie^{i\delta/2}+ie^{-i\delta/2} \\ ie^{i\delta/2}-ie^{-i\delta/2} & e^{i\delta/2}+e^{-i\delta/2} \end{vmatrix} S(-\beta) = S(\beta)S\left(\frac{\delta}{2}\right)S(-\beta) = S\left(\frac{\delta}{2}\right) \quad (47)$$

    The rotation angle is therefore equal to one-half the phase angle of the retardation. This combination is a true rotator in that the rotation is independent of the azimuth angle of the incident polarized light.

    A variable rotator can be made by using a Soleil compensator for the central element. This consists of two quartz wedges joined to form a plane parallel quartz plate.

Mueller matrices. In the Jones calculus, the intensity of the light passing through the system must be obtained by calculation from the components of the light vector. A second calculus is frequently used in which the light vector is split into four components. This also uses matric operators which are termed Mueller matrices. In this calculus, the intensity I of the light is one component of the vector and thus is automatically calculated. The other components of the vector are given by Eqs. (48)–(50). The matrix of a perfect polarizer parallel to the x axis can be written as Eq. (51).

$$M = A_x^2 - A_y^2 \qquad (48) \qquad\qquad C = 2A_xA_y \cos(\varphi_x - \varphi_y) \qquad (49)$$

$$S = 2A_xA_y \sin(\varphi_x - \varphi_y) \qquad (50) \qquad\qquad P = \frac{1}{2}\begin{vmatrix} 1 & 1 & 0 & 0 \\ 1 & 1 & 0 & 0 \\ 0 & 0 & 0 & 0 \\ 0 & 0 & 0 & 0 \end{vmatrix} \qquad (51)$$

The calculus can treat unpolarized light directly. Such a light vector is given by Eq. (52). The vector for light polarized parallel to the x axis is written as expression (53). In the same

$$\begin{vmatrix} I \\ M \\ C \\ S \end{vmatrix} = \begin{vmatrix} 1 \\ 0 \\ 0 \\ 0 \end{vmatrix} \qquad (52) \qquad\qquad\qquad \begin{vmatrix} 1 \\ 1 \\ 0 \\ 0 \end{vmatrix} \qquad (53)$$

manner as in the Jones calculus, matrices can be derived for retardation plates, rotators, and partial polarizers. This calculus can also be used to derive various general theorems about various optical systems. SEE FARADAY EFFECT; INTERFERENCE OF WAVES; OPTICAL ACTIVITY; ROTATORY DISPERSION.

Bibliography. M. Born and E. Wolf, *Principles of Optics*, 6th ed., 1980; E. Hecht and A. Zajac, *Optics*, 2d ed., 1987; F. A. Jenkins, *Fundamentals of Optics*, 4th ed., 1976; G. P. Konnen, *Polarized Light in Nature*, 1985; W. Swindell (ed.), *Polarized Light*, 1975.

5
INTERACTION OF LIGHT WITH MATTER

Physical optics	**177**
Reflection of electromagnetic radiation	**177**
Albedo	**180**
Ellipsometry	**182**
Reflection and transmission coefficients	**184**
Absorption of electromagnetic radiation	**185**
Shadow	**194**
Opaque medium	**195**
Transparent medium	**195**
Translucent medium	**195**
Color filter	**195**
Refraction of waves	**196**
Dispersion (radiation)	**200**
Optical materials	**200**
Optical activity	**206**
Rotatory dispersion	**209**
Cotton effect	**210**
Crystal optics	**213**
Birefringence	**220**
Pleochroism	**220**
Dichroism	**220**

Trichroism 221
Scattering of electromagnetic radiation 222
Opalescence 225
Tyndall effect 226
Raman effect 226
Quasielastic light scattering 232
Nonlinear optics 236
Meteorological optics 241

PHYSICAL OPTICS
RICHARD C. LORD

The study of the interaction of electromagnetic waves in the optical range with material systems. The optical range of wavelengths may be taken as the range from about 1 nanometer (4×10^{-8} in.) to about 1 millimeter (0.04 in.). More narrowly, physical optics deals with the relationship between the atomic structure of a system and the manner in which the system affects light sent into it. The chief founder of this branch of science was Michael Faraday, who in 1845 provided the first clue to the electromagnetic nature of light by showing the optical properties of glass could be altered by a magnetic field. SEE FARADAY EFFECT.

The explanation of the absorption, reflection, scattering, polarization, and dispersion of light by a material medium in terms of the properties of the atoms and molecules making up the medium is the objective of physical optics. In the course of seeking this objective, physicists have found that optical investigations are powerful methods of determining the structures of atoms and molecules and of larger systems composed thereof. SEE ABSORPTION OF ELECTROMAGNETIC RADIATION; CRYSTAL OPTICS; DIFFRACTION; DISPERSION (RADIATION); ELECTROMAGNETIC RADIATION; ELECTROOPTICS; INTERFERENCE OF WAVES; LASER; LIGHT; MAGNETOOPTICS; POLARIZED LIGHT; REFLECTION OF ELECTROMAGNETIC RADIATION; REFRACTION OF WAVES; SCATTERING OF ELECTROMAGNETIC RADIATION.

Bibliography. M. Born and E. Wolf, *Principles of Optics*, 6th ed., 1980; F. A. Jenkins, *Fundamentals of Optics*, 4th ed., 1976; R. S. Longhurst, *Geometrical and Physical Optics*, 2d ed., 1974.

REFLECTION OF ELECTROMAGNETIC RADIATION
HERWIG KOGELNIK

The returning or throwing back of electromagnetic radiation such as light, ultraviolet rays, radio waves, or microwaves by a surface upon which the radiation is incident. In general, a reflecting surface is the boundary between two materials of different electromagnetic properties, such as the boundary between air and glass, air and water, or air and metal. Devices designed to reflect radiation are called reflectors or mirrors.

Reflection angle. The simplest reflection laws are those that govern plane waves of radiation. The law of reflection concerns the incident and reflected rays (as in the case of a beam from a flashlight striking a mirror) or, more precisely, the wave normals of the incident and reflected rays and the normal to the reflecting surface all lie in one plane, called the plane of incidence, and that the reflection angle θ_{refl} equals the angle of incidence θ_{inc} as in Eq. (1) [see **Fig. 1**].

$$\theta_{refl} = \theta_{inc} \qquad (1)$$

The angles θ_{inc} and θ_{refl} are measured between the surface normal and the incident and reflected rays, respectively. The surface (in the above example, that of the mirror) is assumed to be smooth, with surface irregularities small compared to the wavelength of the radiation. This results in so-called specular reflection. In contrast, when the surface is rough, the reflection is diffuse. An example of this is the diffuse scattering of light from a screen or from a white wall where light is returned through a whole range of different angles.

Reflectivity. The reflectivity of a surface is a measure of the amount of reflected radiation. It is defined as the ratio of the intensities of the reflected and incident radiation. The reflectivity depends on the angle of incidence, the polarization of the radiation, and the electromagnetic properties of the materials forming the boundary surface. These properties usually change with the wavelength of the radiation. Reflecting materials are divided into two groups: transparent materials and opaque conducting materials. Radiation penetrating a transparent material propagates essentially unattenuated, while radiation penetrating a conducting material is heavily attenuated. Transparent materials are also called dielectrics. In the wavelength range of visible light, typical dielectrics are glass, quartz, and water. Conducting materials are usually metals such as gold, silver, or aluminum, which are good reflectors at almost all wavelengths. SEE ABSORPTION OF ELECTROMAGNETIC RADIATION.

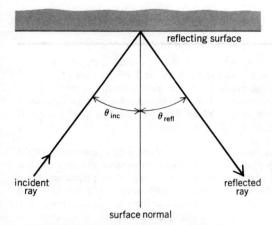

Fig. 1. Reflection of electromagnetic radiation from a smooth surface.

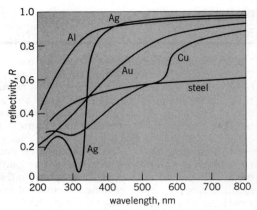

Fig. 2. Reflectivity of some common metals for normal incidence as a function of wavelength. (*After F. A. Jenkins, Fundamentals of Optics, 4th ed., McGraw-Hill, 1976*)

Reflection from metals. The reflectivity of polished metal surfaces is usually quite high. Silver and aluminum, for example, reflect more than 90% of visible light. The reflectivity of some common metals as a function of wavelength is given in **Fig. 2** for normal incidence ($\theta_{inc} = 0$). The reflectivities vary considerably with wavelength, generally falling off toward the shorter wavelengths, with silver exhibiting a reflection "window" at 320 nanometers. The reflectivity values depend somewhat on the way the metal surface was prepared; for example, whether or not it was polished or was produced by evaporation. The presence of an oxidation layer is also a factor influencing (and usually decreasing) the reflectivity. In ordinary mirrors the reflecting surface is the interface between metal and glass, which is thus protected from oxidation, dirt, and other forms of deterioration. When it is not permissible to use this protection for technical reasons, one uses "front-surface" mirrors, which are usually coated with evaporated aluminum. The aluminum has a high reflectivity and deteriorates relatively little in the atmosphere. Front-surface mirrors are used in scientific applications such as interferometry and in large reflecting telescopes.

Reflection from dielectrics. The material property that determines the amount of radiation reflected from an interface between two dielectric media is the phase velocity v of the electromagnetic radiation in the two materials. In optics one uses as a measure for this velocity the refractive index n of the material, which is defined by Eq. (2) as the ratio of the velocity of

$$n = c/v \tag{2}$$

light c in vacuum and the phase velocity in the material. For visible light, for example, the refractive index of air is about $n = 1$, the index of water is about $n = 1.33$, and the index of glass is about $n = 1.5$. *See* Refraction of Waves.

For normal incidence ($\theta_{inc} = 0$) the reflectivity R of the interface is given by Eq. (3), in

$$R = \left(\frac{v_1 - v_2}{v_1 + v_2}\right)^2 = \left(\frac{n_2 - n_1}{n_2 + n_1}\right)^2 \tag{3}$$

which the material constants are labeled 1 and 2 as shown in **Fig. 3**, where the radiation is incident in material 1. The reflectivity of an air-water interface is about 2% ($R = 0.02$) and that of an air-glass interface about 4% ($R = 0.04$); the other 98% or 96% are transmitted through the water or glass, respectively.

A ray incident upon the interface at an oblique, nonnormal angle θ_1 is deviated as it penetrates material 2 as shown in Fig. 3. This is called refraction, and the refracted angle θ_2 follows from Snell's law of refraction, Eq. (4). Again, there is partial reflection, with the reflectivity de-

$$n_1 \sin \theta_1 = n_2 \sin \theta_2 \tag{4}$$

pending on the angle of incidence θ_1 and the polarization of the radiation. When the electric field is polarized parallel to the plane of incidence, the reflectivity $R_\parallel$ is given by Eq. (5), and when the electric field is polarized perpendicular to the plane of incidence, the reflectivity $R_\perp$ is given by Eq. (6). These formulas are known as the "Fresnel Formulas." For unpolarized radiation, in which

$$R_\parallel = \frac{\tan^2(\theta_1 - \theta_1)}{\tan^2(\theta_1 + \theta_2)} \tag{5}$$

$$R_\perp = \frac{\sin^2(\theta_1 - \theta_2)}{\sin^2(\theta_1 + \theta_2)} \tag{6}$$

the electric field varies rapidly in a random, unpredictable manner, the reflectivity $\bar{R}$ is the average of $R_\parallel$ and $R_\perp$. **Figure 4** shows $R_\parallel$, $R_\perp$, and $\bar{R}$ for an air-glass interface as a function of angle. The reflectivity approaches 100% at grazing incidence ($\theta_1 \approx 90°$) for both polarizations. SEE REFLECTION AND TRANSMISSION COEFFICIENTS.

At an angle of about $\theta_1 = \theta_B = 56°$ the reflectivity $R_\parallel$ assumes zero value. This angle is called Brewster's angle, which is, in general, obtained from formula (7). At this angle, which is

$$\tan \theta_B = n_2/n_1 \tag{7}$$

also called the polarizing angle, only radiation polarized perpendicular to the plane of incidence is reflected.

Total internal reflection. Total reflection, that is, reflection of 100% of the incident radiation, occurs at the interface of two dielectrics when the radiation is incident in the denser medium, that is, when $n_1 > n_2$ and when the angle of incidence θ_1 is larger than the critical angle θ_0 given by Eq. (8). Total internal reflection can be observed, for example, by a submerged diver

$$\sin \theta_0 = n_2/n_1 \tag{8}$$

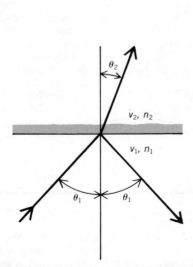

Fig. 3. Reflection and refraction of electromagnetic radiation at an interface between two dielectric media.

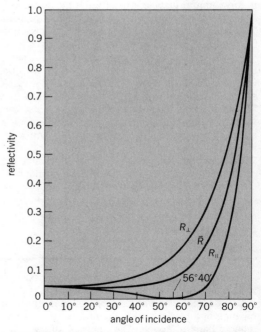

Fig. 4. Reflectivity as a function of angle of incidence for an air-glass interface. (*After M. Born and E. Wolf, Principles of Optics, 5th ed., Pergamon Press, 1975*)

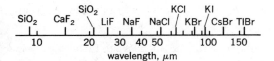

Fig. 5. Residual rays for various crystals.

looking up at the water-air interface for which the critical angle is about $\theta_0 = 49°$. For a glass-air interface the critical angle is approximately $\theta_0 = 42°$.

Selective reflection from crystals. The discussion of reflection from dielectrics to this point has been concerned with reflectivity of nonabsorbing media far removed from absorption bands. These bands are located in the spectral regions where the frequency of the radiation corresponds to a resonance frequency of the atoms, molecules, or crystal lattice of the medium. Since this radiation is strongly absorbed, it is also strongly reflected. The metallic sheen of dye crystals, which have very strong absorption bands in the visible spectrum, is caused by selective reflection. Crystalline solids such as rocksalt or quartz, the lattices of which are built up of atoms bearing net electric charges, show strong selective reflection in the infrared region at wavelengths near those of the strong absorption bands associated with lattice vibrations in the crystal. By reflecting an infrared beam several times from such a material, highly monochromatic radiation can be obtained at the specific wavelengths. These monochromatic beams are referred to as residual rays or reststrahlen. **Figure 5** shows residual rays for some crystals.

Antireflection coatings. In order to reduce undesired reflections from surfaces of optical components such as photographic lenses, one can coat the surface with a thin film. To minimize reflection, the refractive index n_f of this film should be the geometric mean of the indices n_1 and n_2 of the incident and refracting media, Eq. (9), and the film thickness d_f should be equal to a quarter of the wavelength of light in the film material, that is, it should satisfy Eq. (10), where λ

$$n_f^2 = n_1 n_2 \qquad (9) \qquad\qquad n_f d_f = \lambda/4 \qquad (10)$$

is the vacuum wavelength of the light for which the reflectivity of the surface is minimized. Such a film is called a quarter-wave layer. Materials such as magnesium fluoride with an index of $n = 1.38$ are used for antireflection coatings have been made which reduce the reflectivity to less than 0.1% on certain materials; on glass surfaces less than 1/2% has been achieved.

High-reflectivity multilayers. A series of dielectric quarter-wave layers of alternating high and low refractive index can be used to produce reflectors of high reflectivity at a specified wavelength. This is called a quarter-wave stack. The high reflectivity is due to an interference effect. Stacks with about 20 layers are used to make reflectors with reflectives higher than 99%. Mirrors such as these are used for the construction of laser resonators. Multilayer coatings are also used to enhance the reflectivity of metals such as aluminum. About four quarter-wave layers can enhance the metal reflectivity to values better than 99%. SEE ALBEDO; GEOMETRICAL OPTICS; INTERFERENCE FILTER; MIRROR OPTICS.

Bibliography. *American Institute of Physics Handbook*, 3d ed., 1972; M. Born and E. Wolf, *Principles of Optics*, 6th ed., 1980; F. A. Jenkins, *Fundamentals of Optics*, 4th ed., 1976; J. Strong, *Concepts of Classical Optics*, 1958.

ALBEDO
JOSEPH VEVERKA

A term used to describe the reflecting properties of surfaces. White surfaces have albedos close to 1; black surfaces have albedos close to 0.

Several types of albedos are in common use. The Bond albedo (A_B) determines the energy balance of a planet and is defined as the fraction of the total incident solar energy that the planet reflects back to space. The "normal albedo" of a surface, more properly called the normal reflectance (r_n), is a measure of the relative brightness of the surface when viewed and illuminated vertically. Such measurements are referred to as a perfectly white Lambert surface — a surface which absorbs no light, and scatters the incident energy isotropically — usually approximated by magnesium oxide (MgO), magnesium carbonate (MgCO$_3$), or some other bright powder. SEE PHOTOMETRY.

Bond albedos for solar system objects range from 0.76 for cloud-shrouded Venus to values as low as 0.01–0.02 for some asteroids and satellites (**Table 1**). The value for Earth is 0.33. The Bond albedo is defined over all wavelengths, and its value therefore depends on the spectrum of the incident radiation. For objects in the outer solar system, the values of A_B in Table 1 are estimates derived indirectly, since for these bodies it is impossible to measure the scattered radiation in all directions from Earth.

Normal reflectances of some common materials are listed in **Table 2**. The normal reflectances of many materials are strongly wavelength-dependent, a fact which is commonly used in planetary science to infer the composition of surfaces remotely. While the Bond albedo cannot exceed unity, the normal reflectance of a surface can if the material is more backscattering at opposition than the reference surface.

In the case of solar system objects, a third type of albedo, the geometric albedo (p), is commonly defined. It is the ratio of incident sunlight reflected in the backscattering direction (zero phase angle or opposition) by the object, to that which would be reflected by a circular disk of the same size but covered with a perfectly white Lambert surface. Objects like the Moon, covered with dark, highly textured surfaces, show uniformly bright disks at opposition (no limb darkening); for these, $p = r_n$. At the other extreme, a planet covered with a Lambert-like visible surface (frost or bright cloud) will be limb-darkened as the cosine of the incidence angle at opposition, and $p = \frac{2}{3} r_n$. Table 1 lists the visual geometric albedo p_v, which is the geometric albedo at a wavelength of 550 nanometers.

For solar system objects, the ratio A_B/p is called the phase integral and is denoted by q. Here p is the value of the geometric albedo averaged over the incident spectrum, and does not generally equal p_v given in Table 1. Values range from near 1.5 for cloud-covered objects (1.3 for Venus) to 0.2 for the very dark and rugged satellites of Mars: Phobos and Deimos. The value for the Moon is about 0.6.

It is possible to define other types of albedos based on the geometry and nature (diffuse or collimated) of the incident beam and scattered beams. For such bidirectional albedos or reflectances, the angles of incidence and scattering as well as the angle between these two directions (phase angle) must be specified. Finally, in determining the energy balance of surfaces, one is

Table 1. Bond albedos and visual geometric albedos for solar system objects

Object	Bond albedo, A_B	Visual geometric albedo, p_v
Mercury	0.1	0.12
Venus	0.76	0.59
Earth	0.33	0.37
Mars	0.25	0.15
Jupiter	0.34	0.45
Saturn	0.34	0.46
Uranus	0.4	0.48
Neptune	0.4	0.50
Pluto	?	0.4
Moon	0.12	0.11
Io (J1)	0.56	0.63
Europa (J2)	0.58	0.68
Ganymede (J3)	0.38	0.43
Callisto (J4)	0.13	0.17
Titan (S6)	0.30	0.16
Ceres	0.03	0.05
Vesta	0.12	0.23

Table 2. Normal reflectances of materials*

Material	Albedo
Lampblack	0.02
Charcoal	0.04
Carbonaceous meteorites	0.05
Volcanic cinders	0.06
Basalt	0.10
Iron meteorites	0.18
Chondritic meteorites	0.29
Granite	0.35
Olivine	0.40
Quartz	0.54
Pumice	0.57
Snow	0.70
Sulfur	0.85
Magnesium oxide	1.00

*Powders; for wavelengths near 0.5 micrometer.

often interested in the analog for a flat surface of the bond albedo, that is, the fraction of the incident energy that the surface reflects. For a plane parallel beam incident at an angle i, this quantity, $A(i)$, is generally close to the value A_B, but can be significantly larger than A_B for some surfaces, for large values of i.

ELLIPSOMETRY
David Beaglehole

A technique for determining the properties of a material from the characteristics of light reflected from its surface. The materials studied include semiconductors, liquids, and metals.

Principles. Electromagnetic waves passing through a medium cause the electrons attached to the atoms of the medium to oscillate at the frequency of the wave. This slows the wave so that its velocity v in the medium is different from its velocity c in empty space. The refractive index of the medium n is a measure of this change in velocity where $n = c/v$. Alternatively the relative permittivity or dielectric constant $\epsilon_r = n^2$ is often used. When the wave is of high frequency, the electrons can also be excited to higher energy states, absorbing the radiation. Thus a quantity measuring the absorption is also needed to characterize the medium, and this can be done by making the refractive index and the relative permittivity complex numbers; the refractive index becomes $\mathbf{n} = n + ik$, where n measures the change in velocity as before, and k the absorption of the wave. See Absorption of electromagnetic radiation; Light; Refraction of waves.

When the electromagnetic wave is reflected from the surface of a material, the amplitude of the reflected wave depends upon the properties of the material, the angle of incidence, and the polarization of the wave. An s wave has its electric vector parallel to the surface, while a p wave has its electric vector in the plane of incidence (the plane formed by the incident and reflected waves). The reflection amplitudes are written as r_s and r_p. These are the ratios of the reflected- to incident-wave amplitudes and include a phase shift in the wave.

That the state of polarization affects the reflection properties can be seen readily in **Fig. 1**. The incident s wave causes the electrons in the medium to oscillate at right angles to the reflection direction for all angles of incidence. These oscillating electrons thus generate the reflected s wave with its electric field transverse to the direction of propagation. An incident p wave, however, will cause the medium electrons to oscillate partly along the reflection direction. At a particular angle of incidence called the Brewster angle θ_B, where $\tan \theta_B = n_2/n_1$, the oscillation is completely along the reflection direction. Since oscillations along the propagation direction do not generate radiating transverse electromagnetic waves, the p reflection amplitude r_p is less than r_s, falling to zero at the Brewster angle (the case shown in Fig. 1). See Reflection of electromagnetic radiation.

The properties of the medium can be determined by measuring the intensities of the s and p reflected waves, that is, by measuring r_s^2 and r_p^2. Ellipsometry is an alternative technique which measures $r = r_p/r_s$. This can be done by shining light which is linearly polarized at 45° to both the s and p directions so that the s and p incident amplitudes are equal and in phase. The electric vector of the reflected light in general traces an ellipse (**Fig. 2**), where the orientation and the

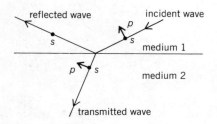

Fig. 1. Incident, transmitted and reflected electromagnetic waves at the surface of a material. The angle of incidence shown is the Brewster angle at which the p polarized wave in medium 2 points along the reflection direction. The s wave is polarized perpendicular to the plane of the page.

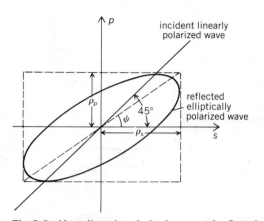
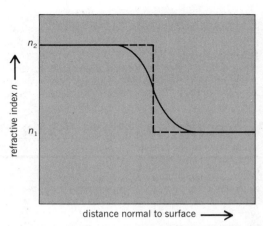

Fig. 2. Incident, linearly polarized wave and reflected elliptically polarized wave in the ellipsometry technique. The paths traced by the tip of the electric vector in a plane perpendicular to the propagation direction are shown.

Fig. 3. Refractive index variation near the surface of a material. The broken line indicates an abrupt step in refractive index, while the solid curve indicates the actual gradual variation with a transition region usually a few molecular diameters in thickness.

lengths of the major and minor axes depend on the magnitudes, ρ_p and ρ_s, of r_p and r_s, and on the relative phase difference Δ between the two waves. Thus r can be expressed in alternative forms by Eq. (1), where ρ and ψ are given by Eq. (2) and r' and r'' are the real and imaginary parts of r.

$$r = \rho e^{i\Delta} = \tan \psi e^{i\Delta} = r' + r'' \qquad (1) \qquad \rho = \tan \psi = \frac{\rho_p}{\rho_s} \qquad (2)$$

The major and minor axes of the ellipse are determined by ρ and Δ. Traditional ellipsometric techniques use a polarizer, an analyzer, and a compensator to determine ψ and Δ, while modern methods based upon periodic modulation of the incident wave can give r' and r'' directly. *See* POLARIZED LIGHT.

Applications. The two chief applications of ellipsometry are the study of surface properties and the area of spectroscopic ellipsometry.

Surface properties. When the media on each side of the surface are nonabsorbing, and the dielectric constant undergoes an abrupt step at the surface, then both r_p and r_s are real; that is, the s and p reflected waves are exactly in phase or exactly out of phase. Then r is a real number that varies from 1 at normal incidence, passes through 0 at the Brewster angle θ_B where $r_p = 0$, and is equal to -1 at grazing incidence. As a result, linearly polarized incident light gives linearly polarized reflected light. In practice, however, no interface is abrupt, and there is always a transition region in the variation from medium 1 to medium 2 (**Fig. 3**). Then both r_p and r_s are altered by an amount proportional to d/λ, where d is the transition region thickness and λ is the wavelength of the light. At the Brewster angle, $r' = 0$ and the p wave has a small part which is 90° out of phase with the s wave. This quantity $r(\theta_B) = r''(\theta_B)$ is called the coefficient of ellipticity and is usually denoted by $\bar{\rho}$. It has been used to study surfaces of glass, interfaces between two liquids and interfaces between a liquid and its vapor. Near the triple point the latter have been found to be only a few molecular diameters thick. Another use has been to study layers of organic molecules floating on water.

Spectroscopic ellipsometry. When the medium reflecting the wave is absorbing, r is complex even for an abrupt boundary, depending upon both the real and imaginary points of **n**. A common use for ellipsometry is to study the electronic excitations of materials by deducing the absorptive part of **n** (or ϵ_r) from measurements of r as a function of incident wavelength. This is called spectroscopic ellipsometry. The technique has also been used to study how the properties of materials vary as films are formed (for instance, by vacuum evaporation), and to monitor the formation of oxides or other chemical reactions at the surface of a reactive material.

Bibliography. F. Abeles (ed.), Proceedings of the international conference on ellipsometry and other optical methods for the analysis of surfaces and thin films, Paris, 1983, *Journal de Physique*, Supplement C-10, 1983; P. Drude, *Theory of Optics*, 1959; F. A. Jenkins, *Fundamentals of Optics*, 4th ed., 1976; B. O. Seraphin (ed.), *Optical Properties of Solids: New Developments*, 1976.

REFLECTION AND TRANSMISSION COEFFICIENTS
William R. Smythe

When an electromagnetic wave passes from a medium of permeability μ_1 and dielectric constant ϵ_1 to one with values μ_2 and ϵ_2, part of the wave is reflected at the boundary and part transmitted. The ratios of the amplitudes in the reflected wave and the transmitted wave to that in the incident wave are called the reflection and transmission coefficients, respectively. For oblique incidence, the reflection and refraction formulas of optics are most convenient, but for normal incidence of plane waves on plane boundaries, such as occur with transmission lines, waveguides, and some free waves, the concept of wave impedance and characteristic impedance is useful.

For a z-directed wave with electric intensity $\mathbf{E}$ in the x direction and magnetic intensity $\mathbf{H}$ in the y direction, the total phasor fields on the incident side are given by Eqs. (1) and (2), where

$$\check{E}_x = E_0 e^{-jkz} + \check{E}'_0 e^{jkz} \quad (1) \qquad \check{H}_y = (\eta)^{-1}(E_0 e^{-jkz} - E'_0 e^{jkz}) \quad (2)$$

primes are used for reflected quantities and η is the wave impedance. The sign difference in Eqs. (1) and (2) is due to the fact that Poynting's vector, $\tfrac{1}{2}\mathbf{E} \times \mathbf{H}$, is positive for the incident and negative for the reflected wave. For the transmitted wave, Eqs. (3) hold. Since the tangential

$$\check{E}''_x = E''_0 e^{-jkz} \qquad \check{H}''_y = (\eta'')^{-1} E''_0 e^{-jkz} \quad (3)$$

components of $\mathbf{E}$ and $\mathbf{H}$ are continuous across the boundary at $z = 0$, $\check{E}_x = \check{E}''_x$ and $\check{H}_y = \check{H}''_y$, so that Eqs. (4) hold. The ratios for the reflected and transmitted fields obtained by solving these

$$E_0 + E'_0 = E''_0 \qquad \eta''(E_0 - E'_0) = \eta E''_0 \quad (4)$$

equations are given by Eqs. (5), which are the reflection and transmission coefficients, respectively. *See* Electromagnetic radiation; Poynting's vector.

$$\frac{E'_0}{E_0} = \frac{\eta'' - \eta}{\eta'' + \eta} \qquad \frac{E''_0}{E_0} = \frac{2\eta''}{\eta'' + \eta} \quad (5)$$

Coefficients for optics. Equations (5) hold for normal incidence in optics, if the velocities v and v'' are written for η and η''. For a plane wave whose electric vector is normal to the plane of incidence and whose direction makes an acute angle θ with the normal to the interface, the reflection and transmission coefficients are given by Eqs. (6), where $v'' \sin \theta''$. When the electric

$$\frac{E'_0}{E_0} = -\frac{\sin(\theta - \theta'')}{\sin(\theta + \theta'')} \qquad \frac{E''_0}{E_0} = \frac{2 \sin \theta'' \cos \theta}{\sin(\theta + \theta'')} \quad (6)$$

vector lies in the plane of incidence, the coefficients are given by Eqs. (7). The ratio v/v'' is the index of refraction. *See* Refraction of waves.

$$\frac{E'_0}{E_0} = \frac{\tan(\theta - \theta'')}{\tan(\theta + \theta'')} \qquad \frac{E''_0}{E_0} = \frac{2 \sin \theta'' \cos \theta}{\sin(\theta + \theta'') \sin(\theta - \theta'')} \quad (7)$$

Waveguides. In waveguides, as in free space, the characteristic impedance is defined as the ratio of the transverse electric field E_t to the transverse magnetic field H_t. For waveguides, this radio depends on the frequency and the dimensions of the waveguide, as well as on the permeabilities and dielectric constants. For a transverse interface, the boundary conditions used for Eqs. (4) on the tangential fields still hold. Thus, Eqs. (5) for the reflection and transmission coefficients are valid if η and η'' are replaced by the characteristic impedances on the incident and emergent sides, respectively.

Transmission lines and networks. Let $\check{Z}$ and $\check{Z}''$ be the characteristic impedances on the incident and emergent sides of a discontinuity in a transmission line or on the two sides of a junction between two networks. Then the relations between the potentials and currents of the incident, reflected, and transmitted waves are given by Eqs. (8a)–(8c), respectively. At the discon-

$$\check{V} = \check{Z}\check{I} \quad (8a) \qquad \check{V}' = -\check{Z}\check{I}' \quad (8b) \qquad \check{V}'' = \check{Z}''\check{I}'' \quad (8c)$$

tinuity, potential and current must be continuous so that Eqs. (9) hold. Solution for the ratios gives

$$\check{V} + \check{V}' = \check{V}'' \qquad \check{I} + \check{I}' = \check{I}'' \quad (9)$$

Eqs. (10), which are the reflection and transmission coefficients.

$$\frac{\check{V}'}{\check{V}} = \frac{\check{Z}'' - \check{Z}}{\check{Z}'' + \check{Z}} \qquad \frac{\check{V}''}{\check{V}} = \frac{2\check{Z}''}{\check{Z}'' + \check{Z}} \quad (10)$$

Coefficients for acoustics. Equations (10) hold in acoustics, provided acoustic impedance is substituted for electrical impedance. Acoustic impedance is defined as the product of the density of a medium by the speed of sound in it. There are two types of waves in solid mediums, longitudinal waves and shear waves, and thus there are two impedances.

Bibliography. D. Dearholt and W. McSpadden, *Electromagnetic Wave Propagation*, 1973; J. D. Kraus and K. R. Carver, *Electromagnetics*, 3d ed., 1984; P. Lorrain and D. R. Corson, *Electromagnetic Fields and Waves*, 2d ed., 1970; S. Ramo and J. R. Whinnery, *Fields and Waves in Communication Electronics*, 2d ed., 1984; W. R. Smythe, *Static and Dynamic Electricity*, 3d ed., 1968.

ABSORPTION OF ELECTROMAGNETIC RADIATION
WILLIAM WEST

The process whereby the intensity of a beam of electromagnetic radiation is attenuated in passing through a material medium by conversion of the energy of the radiation to an equivalent amount of energy which appears within the medium; the radiant energy is converted into heat or some other form of molecular energy. A perfectly transparent medium permits the passage of a beam of radiation without any change in intensity other than that caused by the spread or convergence of the beam, and the total radiant energy emergent from such a medium equals that which entered it, whereas the emergent energy from an absorbing medium is less than that which enters, and, in the case of highly opaque media, is reduced practically to zero.

No known medium is opaque to all wavelengths of the electromagnetic spectrum, which extends from radio waves, whose wavelengths are measured in kilometers, through the infrared, visible, and ultraviolet spectral regions, to x-rays and gamma rays, of wavelengths down to 10^{-13} m. Similarly, no material medium is transparent to the whole electromagnetic spectrum. A medium which absorbs a relatively wide range of wavelengths is said to exhibit general absorption, while a medium which absorbs only restricted wavelength regions of no great range exhibits selective absorption for those particular spectral regions. For example, the substance pitch shows general absorption for the visible region of the spectrum, but is relatively transparent to infrared radiation of long wavelength. Ordinary window glass is transparent to visible light, but shows general absorption for ultraviolet radiation of wavelengths below about 310 nanometers, while colored glasses show selective absorption for specific regions of the visible spectrum. The color of objects which are not self-luminous and which are seen by light reflected or transmitted by the object is usually the result of selective absorption of portions of the visible spectrum. Many colorless substances, such as benzene and similar hydrocarbons, selectively absorb within the ultraviolet region of the spectrum, as well as in the infrared. *See* COLOR; ELECTROMAGNETIC RADIATION.

Laws of absorption. The capacity of a medium to absorb radiation depends on a number of factors, mainly the electronic and nuclear constitution of the atoms and molecules of the medium, the wavelength of the radiation, the thickness of the absorbing layer, and the variables which determine the state of the medium, of which the most important are the temperature and the concentration of the absorbing agent. In special cases, absorption may be influenced by electric or magnetic fields. The state of polarization of the radiation influences the absorption of media containing certain oriented structures, such as crystals of other than cubic symmetry.

Lambert's law. Lambert's law, also called Bouguer's law or the Lambert-Bouguer law, expressed the effect of the thickness of the absorbing medium on the absorption. If a homogeneous medium is thought of as being constituted of layers of uniform thickness set normally to the beam, each layer absorbs the same fraction of radiation incident on it. If I is the intensity to which a monochromatic parallel beam is attenuated after traversing a thickness d of the medium, and I_0 is the intensity of the beam at the surface of incidence (corrected for loss by reflection from this surface), the variation of intensity throughout the medium is expressed by Eq. (1), in which α is a constant for the medium called the absorption coefficient. This exponential relation can be expressed in an equivalent logarithmic form as in Eq. (2), where $k = \alpha/2.303$ is called the extinc-

$$I = I_0 e^{-\alpha d} \qquad (1) \qquad \qquad \log_{10}(I_0/I) = (\alpha/2.303)d = kd \qquad (2)$$

tion coefficient for radiation of the wavelength considered. The quantity $\log_{10}(I_0/I)$ is often called the optical density, or the absorbance of the medium.

Equation (2) shows that as monochromatic radiation penetrates the medium, the logarithm of the intensity decreases in direct proportion to the thickness of the layer traversed. If experimental values for the intensity of the light emerging from layers of the medium of different thicknesses are available (corrected for reflection losses at all reflecting surfaces), the value of the extinction coefficient can be readily computed from the slope of the straight line representing the logarithms of the emergent intensities as functions of the thickness of the layer.

Equations (1) and (2) show that the absorption and extinction coefficients have the dimensions of reciprocal length. The extinction coefficient is equal to the reciprocal of the thickness of the absorbing layer required to reduce the intensity to one-tenth of its incident value. Similarly, the absorption coefficient is the reciprocal of the thickness required to reduce the intensity to $1/e$ of the incident value, where e is the base of the natural logarithms, 2.718.

Beer's law. This law refers to the effect of the concentration of the absorbing medium, that is, the mass of absorbing material per unit of volume, on the absorption. This relation is of prime importance in describing the absorption of solutions of an absorbing solute, since the solute's concentration may be varied over wide limits, or the absorption of gases, the concentration of which depends on the pressure. According to Beer's law, each individual molecule of the absorbing material absorbs the same fraction of the radiation incident upon it, no matter whether the molecules are closely packed in a concentrated solution or highly dispersed in a dilute solution. The relation between the intensity of a parallel monochromatic beam which emerges from a plane parallel layer of absorbing solution of constant thickness and the concentration of the solution is an exponential one, of the same form as the relation between intensity and thickness expressed by Lambert's law. The effects of thickness d and concentration c on absorption of monochromatic radiation can therefore be combined in a single mathematical expression, given in Eq. (3), in which k' is a constant for a given absorbing substance (at constant wavelength and temperature), independent of the actual concentration of solute in the solution. In logarithms, the relation becomes Eq. (4). The values of the constants k' and ϵ in Eqs. (3) and (4) depend on the units of

$$I = I_0 e^{-k'cd} \qquad (3) \qquad \qquad \log_{10}(I_0/I) = (k'/2.303)cd = \epsilon cd \qquad (4)$$

concentration. If the concentration of the solute is expressed in moles per liter, the constant ϵ is called the molar extinction coefficient. Some authors employ the symbol a_M, which is called the molar absorbance index, instead of ϵ.

If Beer's law is adhered to, the molar extinction coefficient does not depend on the concentration of the absorbing solute, but usually changes with the wavelength of the radiation, with the temperature of the solution, and with the solvent.

The dimensions of the molar extinction coefficient are reciprocal concentration multiplied by reciprocal length, the usual units being liters/(mole)(cm). If Beer's law is true for a particular solution, the plot of $\log(I_0/I)$ against the concentrations for solutions of different concentrations, measured in cells of constant thickness, will yield a straight line, the slope of which is equal to the molar extinction coefficient.

While no true exceptions to Lambert's law are known, exceptions to Beer's law are not uncommon. Such exceptions arise whenever the molecular state of the absorbing solute depends on the concentration. For example, in solutions of weak electrolytes, whose ions and undissociated molecules absorb radiation differently, the changing ratio between ions and undissociated molecules brought about by changes in the total concentration prevents solutions of the electrolyte from obeying Beer's law. Aqueous solutions of dyes frequently deviate from the law because of dimerization and more complicated aggregate formation as the concentration of dye is increased.

Absorption measurement. The measurement of the absorption of homogeneous media is usually accomplished by absolute or comparative measurements of the intensities of the incident and transmitted beams, with corrections for any loss of radiant energy caused by processes other than absorption. The most important of these losses is by reflection at the various surfaces of the absorbing layer and of vessels which may contain the medium, if the medium is liquid or gaseous. Such losses are usually automatically compensated for by the method of measurement employed. Losses by reflection not compensated for in this manner may be computed from Fresnel's laws of reflection. *See Reflection of electromagnetic radiation.*

Scattering. Absorption of electromagnetic radiation should be distinguished from the phenomenon of scattering, which occurs during the passage of radiation through inhomogeneous media. Radiant energy which traverses media constituted of small regions of refractive index different from that of the rest of the medium is diverted laterally from the direction of the incident beam. The diverted radiation gives rise to the hazy or opalescent appearance characteristic of such media, exemplified by smoke, mist, and opal. If the centers of inhomogeneity are sufficiently dilute, the intensity of a parallel beam is diminished in its passage through the medium because of the sidewise scattering, according to a law of the same form as the Lambert-Bouguer law for absorption, given in Eq. (5), where I is the intensity of the primary beam of initial intensity I_0,

$$I = I_0 e^{-\tau d} \tag{5}$$

after it has traversed a distance d through the scattering medium. The coefficient τ, called the turbidity of the medium, plays the same part in weakening the primary beam by scattering as does the absorption coefficient in true absorption. However, in true scattering, no loss of total radiant energy takes place, energy lost in the direction of the primary beam appearing in the radiation scattered in other directions. In some inhomogeneous media, both absorption and scattering occur together. *See Scattering of electromagnetic radiation.*

Physical nature. Absorption of radiation by matter always involves the loss of energy by the radiation and a corresponding gain in energy by the atoms or molecules of the medium.

The energy of an assembly of gaseous atoms consists partly of kinetic energy of the translational motion which determines the temperature of the gas (thermal energy), and partly of internal energy, associated with the binding of the extranuclear electrons to the nucleus, and with the binding of the particles within the nucleus itself. Molecules, composed of more than one atom, have, in addition, energy associated with periodic rotations of the molecule as a whole and with oscillations of the atoms within the molecule with respect to one another.

The energy absorbed from radiation appears as increased internal energy, or in increased vibrational and rotational energy of the atoms and molecules of the absorbing medium. As a general rule, translational energy is not directly increased by absorption of radiation, although it may be indirectly increased by degradation of electronic energy or by conversion of rotational or vibrational energy to that of translation by intermolecular collisions.

Quantum theory. In order to construct an adequate theoretical description of the energy relations between matter and radiation, it has been necessary to amplify the wave theory of radiation by the quantum theory, according to which the energy in radiation occurs in natural units called quanta. The value of the energy in these units, expressed in ergs or calories, for example, is the same for all radiation of the same wavelength, but differs for radiation of different wavelengths. The energy E in a quantum of radiation of frequency ν (where the frequency is equal to the velocity of the radiation in a given medium divided by its wavelength in the same medium) is directly proportional to the frequency, or inversely proportional to the wavelength, according to the relation given in Eq. (6), where h is a universal constant known as Planck's constant. The

$$E = h\nu \tag{6}$$

value of h is 6.63×10^{-34} joule-second, and if ν is expressed in s^{-1}, E is given in joules per quantum.

The most energetic type of change that can occur in an atom involves the nucleus, and increase of nuclear energy by absorption therefore requires quanta of very high energy, that is, of high frequency or low wavelength. Such rays are the γ-rays, whose wavelength varies downward from 0.01 nm. Next in energy are the electrons nearest to the nucleus and therefore the most tightly bound. These electrons can be excited to states of higher energy by absorption of x-rays, whose range in wavelength is from about 0.01 to 1 nm. Less energy is required to excite the more loosely bound valence electrons. Such excitation can be accomplished by the absorption of quanta of visible radiation (wavelength 700 nm for red light to 400 nm for blue) or of ultraviolet radiation, of wavelength down to about 100 nm. Absorption of ultraviolet radiation of shorter wavelengths, down to those on the border of the x-ray region, excites electrons bound to the nucleus with intermediate strength.

The absorption of relatively low-energy quanta of wavelength from about 1 to 10 micrometers suffices to excite vibrating atoms in molecules to higher vibrational states, while changes in rotational energy, which are of still smaller magnitude, may be excited by absorption of radiation of still longer wavelength, from the short-wavelength radio region of about 1 cm to long-wavelength infrared radiation, some hundredths of a centimeter in wavelength.

Gases. The absorption of gases composed of atoms is usually very selective. For example, monatomic sodium vapor absorbs very strongly over two narrow wavelength regions in the yellow part of the visible spectrum (the so-called D lines), and no further absorption by monatomic sodium vapor occurs until similar narrow lines appear in the near-ultraviolet. The valence electron of the sodium atom can exist only in one of a series of energy states separated by relatively large energy intervals between the permitted values, and the sharp-line absorption spectrum results from transitions of the valence electron from the lowest energy which it may possess in the atom to various excited levels. Line absorption spectra are characteristic of monatomic gases in general.

The visible and ultraviolet absorption of vapors composed of diatomic or polyatomic molecules is much more complicated than that of atoms. As for atoms, the absorbed energy is utilized mainly in raising one of the more loosely bound electrons to a state of higher energy, but the electronic excitation of a molecule is almost always accompanied by simultaneous excitation of many modes of vibration of the atoms within the molecule and of rotation of the molecule as a whole. As a result, the absorption, which for an atom is concentrated in a very sharp absorption line, becomes spread over a considerable spectral region, often in the form of bands. Each band corresponds to excitation of a specific mode of vibration accompanying the electronic change, and each band may be composed of a number of very fine lines close together in wavelength, each of which corresponds to a specific rotational change of the molecule accompanying the electronic and vibrational changes. Band spectra are as characteristic of the absorption of molecules in the gaseous state, and frequently in the liquid state, as line spectra are of gaseous atoms.

Liquids. Liquids usually absorb radiation in the same general spectral region as the corresponding vapors. For example, liquid water, like water vapor, absorbs infrared radiation strongly (vibrational transitions), is largely transparent to visible and near-ultraviolet radiation, and begins to absorb strongly in the far-ultraviolet. A universal difference between liquids and gases is the disturbance in the energy states of the molecules in a liquid caused by the great number of intermolecular collisions; this has the effect of broadening the very fine lines observed in the absorption spectra of vapors, so that sharp-line structure disappears in the absorption bands of liquids.

Solids. Substances which can exist in solid, liquid, and vapor states without undergoing a temperature rise to very high values usually absorb in the same general spectral regions for all three states of aggregation, with differences in detail because of the intermolecular forces present in the liquid and solid. Crystalline solids, such as rock salt or silver chloride, absorb infrared radiation of long wavelength, which excites vibrations of the electrically charged ions of which these salts are composed; such solids are transparent to infrared radiations of shorter wavelengths. In colorless solids, the valence electrons are two tightly bound to the nuclei to be excited by visible radiation, but all solids absorb in the near- or far-ultraviolet region. SEE CRYSTAL OPTICS.

The use of solids as components of optical instruments is restricted by the spectral regions to which they are transparent. Crown glass, while showing excellent transparency for visible light and for ultraviolet radiation immediately adjoining the visible region, becomes opaque to radiation of wavelength about 300 nm and shorter, and is also opaque to infrared radiation longer than about 2000 nm in wavelength. Quartz is transparent down to wavelengths about 180 nm in the ultraviolet, and to about 4 μm in the infrared. The most generally useful material for prisms and windows for the near-infrared region is rock salt, which is highly transparent out to about 15 μm. For a detailed discussion of the properties of optical glass SEE OPTICAL MATERIALS.

Fluorescence. The energy acquired by matter by absorption of visible or ultraviolet radiation, although primarily used to excite electrons to higher energy states, usually ultimately appears as increased kinetic energy of the molecules, that is, as heat. It may, however, under special circumstances, be reemitted as electromagnetic radiation. Fluorescence is the reemission, as radiant energy, of absorbed radiant energy, normally at wavelengths the same as or longer than those absorbed. The reemission, as ordinarily observed, ceases immediately when the exciting radiation is shut off. Refined measurements show that the fluorescent reemission persists, in different cases, for periods of the order of 10^{-9} to 10^{-5} s. The simplest case of fluorescence is the resonance fluorescence of monatomic gases at low pressure, such as sodium or mercury vapors, in which the reemitted radiation is of the same wavelength as that absorbed. In this case, fluorescence is the converse of absorption: Absorption involves the excitation of an electron from its lowest energy state to a higher energy state by radiation, while fluorescence is produced by the return of the excited electron to the lower state, with the emission of the energy difference between the two states as radiation. The fluorescent radiation of molecular gases and of nearly all liquids, solids, and solutions contains a large component of wavelengths longer than those of the absorbed radiation, a relationship known as Stokes' law of fluorescence. In these cases, not all of the absorbed energy is reradiated, a portion remaining as heat in the absorbing material. The fluorescence of iodine vapor is easily seen on projecting an intense beam of visible light through an evacuated bulb containing a few crystals of iodine, but the most familiar examples are provided by certain organic compounds in solution—for instance, quinine sulfate, which absorbs ultraviolet radiation and reemits blue, or fluorescein, which absorbs blue-green light and fluoresces with an intense, bright-green color.

Phosphorescence. The radiant reemission of absorbed radiant energy at wavelengths longer than those absorbed, for a readily observable interval after withdrawal of the exciting radiation, is called phosphorescence. The interval of persistence, determined by means of a phosphoroscope, usually varies from about 0.001 s to several seconds, but some phosphors may be induced to phosphorescence by heat days or months after the exciting absorption. An important and useful class of phosphors is the impurity phosphors, solids such as the sulfides of zinc or calcium which are activated to the phosphorescent state by incorporating minute amounts of foreign material (called activators), such as salts of manganese or silver. So-called fluorescent lamps contain a coating of impurity phosphor on their inner wall which, after absorbing ultraviolet radiation produced by passage of an electrical discharge through mercury vapor in the lamp, reemits visible light. The receiving screen of a television tube contains a similar coating, excited not by radiant energy but by the impact of a stream of electrons on the surface.

Luminescence. Phosphorescence and fluorescence are special cases of luminescence, which is defined as light emission that cannot be attributed merely to the temperature of the emitting body. Luminescence may be excited by heat (thermoluminescence), by electricity (electroluminescence), by chemical reaction (chemiluminescence), or by friction (triboluminescence), as well as by radiation.

Absorption and emission coefficients. The absorption and emission processes of atoms were examined from the quantum point of view by Albert Einstein in 1916, with some important results that have been realized practically in the invention of the maser and the laser. Consider an assembly of atoms undergoing absorption transitions of frequency ν s^{-1} from the ground state 1 to an excited state 2 and emission transitions in the reverse direction, the atoms and radiation being at equilibrium at temperature T. The equilibrium between the excited and unexcited atoms is determined by the Boltzmann relation $N_2/N_1 = \exp(-h\nu/kT)$, where N_1 and N_2 are the equilibrium numbers of atoms in states 1 and 2, respectively, and the radiational equilibrium is determined by equality in the rate of absorption and emission of quanta. The number of quanta absorbed per second is $B_{12}N_1\rho(\nu)$, where $\rho(\nu)$ is the density of radiation of frequency ν (proportional to the intensity), and B_{12} is a proportionality constant called the Einstein coefficient for absorption. Atoms in state 2 will emit radiation spontaneously (fluorescence), after a certain mean life, at a rate of $A_{21}N_2$ per second, where A_{21} is the Einstein coefficient for spontaneous emission from state 2 to state 1. To achieve consistency between the density of radiation of frequency ν at equilibrium calculated from these considerations and the value calculated from Planck's radiation law, which is experimentally true, it is necessary to introduce, in addition to the spontaneous emission, an emission of intensity proportional to the radiation density of frequency ν in which the atoms are immersed. The radiational equilibrium is then determined by Eq. (7), where B_{21} is the Einstein coefficient of stimulated emission. The Einstein radiation coeffi-

$$B_{12}N_1\rho(\nu) = A_{21}N_2 + B_{21}N_2\rho(\nu) \tag{7}$$

cients are found to be related by Eqs. (8a) and (8b).

$$B_{12} = B_{21} \tag{8a} \qquad A_{21} = (8\pi h\nu^3/c^3) \cdot B_{21} \tag{8b}$$

In the past when one considered radiation intensities available from terrestrial sources, stimulated emission was very feeble compared with the spontaneous process. Stimulated emission is, however, the fundamental emission process in the laser, a device in which a high concentration of excited molecules is produced by intense illumination from a "pumping" source, in an optical system in which excitation and emission are augmented by back-and-forth reflection until stimulated emission swamps the spontaneous process. *See* Laser; Optical pumping.

There are also important relations between the absorption characteristics of atoms and their mean lifetime τ in the excited state. Since A_{21} is the number of times per second that a given atom will emit a quantum spontaneously, the mean lifetime before emission in the excited state is $\tau = 1/A_{21}$. It can also be shown that A_{21} and τ are related, as shown in Eq. (9), to the f

$$A_{21} = 1/\tau = \frac{(8\pi^2\nu^2 e^2)}{mc^3} \cdot f = 7.42 \times 10^{-22} f\nu^2 \qquad (\nu \text{ in s}^{-1}) \tag{9}$$

number or oscillator strength for the transition that occurs in the dispersion equations shown as Eqs. (13) to (17). The value of f can be calculated from the absorption integrated over the band according to Eq. (18).

Dispersion. A transparent material does not abstract energy from radiation which it transmits, but it always decreases the velocity of propagation of such radiation. In a vacuum, the velocity of radiation is the same for all wavelengths, but in a material medium, the velocity of propagation varies considerably with wavelength. The refractive index μ of a medium is the ratio of the velocity of light in vacuum to that in the medium, and the effect of the medium on the velocity of radiation which it transmits is expressed by the variation of refractive index with the wavelength λ of the radiation, $d\mu/d\lambda$. This variation is called the dispersion of the medium. For radiation of wavelengths far removed from those of absorption bands of the medium, the refractive index increases regularly with decreasing wavelength or increasing frequency; the dispersion is then said to be normal.

In regions of normal dispersion, the variation of refractive index with wavelength can be expressed with considerable accuracy by Eq. (10), known as Cauchy's equation, in which A, B,

$$\mu = A + \frac{B}{\lambda^2} + \frac{C}{\lambda^4} \tag{10}$$

and C are constants with positive values. As an approximation, C may be neglected in comparison with A and B, and the dispersion, $d\mu/d\lambda$, is then given by Eq. (11). Thus, in regions of normal

$$\frac{d\mu}{d\lambda} = \frac{-2B}{\lambda^3} \tag{11}$$

dispersion, the dispersion is approximately inversely proportional to the cube of the wavelength.

Dispersion by a prism. The refraction, or bending, of a ray of light which enters a material medium obliquely from vacuum or air (the refractive index of which for visible light is nearly unity) is the result of the diminished rate of advance of the wavefronts in the medium. Since, if the dispersion is normal, the refractive index of the medium is greater for violet than for red light, the wavefront of the violet light is retarded more than that of the red light. Hence, white light entering obliquely into the medium is converted within the medium to a continuously colored band, of which the red is least deviated from the direction of the incident beam, the violet most, with orange, yellow, green, and blue occupying intermediate positions. On emergence of the beam into air again, the colors remain separated. The action of the prism in resolving white light into its constituent colors is called color dispersion. SEE OPTICAL PRISM; REFRACTION OF WAVES.

The angular dispersion of a prism is the ratio, $d\theta/d\lambda$, of the difference in angular deviation $d\theta$ of two rays of slightly different wavelength which pass through the prism to the difference in wavelength $d\lambda$ when the prism is set for minimum deviation.

The angular dispersion of the prism given in Eq. (12) is the product of two factors, the

$$\frac{d\theta}{d\lambda} = \frac{d\theta}{d\mu} \cdot \frac{d\mu}{d\lambda} \qquad (12)$$

variation, $d\theta/d\mu$, the deviation θ with refractive index μ, and the variation of refractive index with wavelength, the dispersion of the material of which the prism is made. The latter depends solely on this material, while $d\theta/d\mu$ depends on the angle of incidence and the refracting angle of the prism. The greater the dispersion of the material of the prism, the greater is the angular separation between rays of two given wavelengths as they leave the prism. For example, the dispersion of quartz for visible light is lower than that of glass; hence the length of the spectrum from red to violet formed by a quartz prism is less than that formed by a glass prism of equal size and shape. Also, since the dispersion of colorless materials such as glass or quartz is greater for blue and violet light than for red, the red end of the spectrum formed by prisms is much more contracted than the blue.

The colors of the rainbow result from dispersion of sunlight which enters raindrops and is refracted and dispersed in passing through them to the rear surface, at which the dispersed rays are reflected and reenter the air on the side of the drop on which the light was incident. SEE METEOROLOGICAL OPTICS.

Anomalous dispersion. The regular increase of refractive index with decreasing wavelength expressed by Cauchy's equation breaks down as the wavelengths approach those of strong absorption bands. As the absorption band is approached from the long-wavelength side, the refractive index becomes very large, then decreases within the band to assume abnormally small values on the short-wavelength side, values below those for radiation on the long-wavelength side. A hollow prism containing an alcoholic solution of the dye fuchsin, which absorbs green light strongly, forms a spectrum in which the violet rays are less deviated than the red, on account of the abnormally low refractive index of the medium for violet light. The dispersion of media for radiation of wavelengths near those of strong absorption bands is said to be anomalous, in the sense that the refractive index decreases with decreasing wavelength instead of showing the normal increase. The theory of dispersion shows, however, that both the normal and anomalous variation of refractive index with wavelength can be satisfactorily described as aspects of a unified phenomenon, so that there is nothing fundamentally anomalous about dispersion in the vicinity of an absorption band. SEE DISPERSION (RADIATION).

Normal and anomalous dispersion of quartz are illustrated in **Fig. 1**. Throughout the near-infrared, visible, and near-ultraviolet spectral regions (between P and R on the curve), the dispersion is normal and adheres closely to Cauchy's equation, but it becomes anomalous to the right of R. From S to T, Cauchy's equation is again valid.

Relation to absorption. Figure 1 shows there is an intimate connection between dispersion and absorption; the refractive index rises to high values as the absorption band is approached from the long-wavelength side and falls to low values on the short-wavelength side of the band. In fact, the theory of dispersion shows that the complete dispersion curve as a function of wavelength is governed by the absorption bands of the medium. In classical electromagnetic theory, electric charges are regarded as oscillating, each with its appropriate natural frequency ν_0, about positions of equilibrium within atoms or molecules. Placed in a radiation field of frequency ν per second, the oscillator in the atom is set into forced vibration, with the same frequency as that of the radiation. When ν is much lower or higher than ν_0, the amplitude of the forced vibration is small, but the amplitude becomes large when the frequency of the radiation equals the natural frequency of the oscillator. In much the same way, a tuning fork is set into vibration by sound

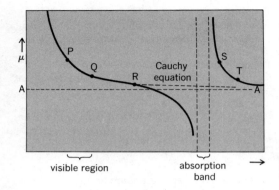

Fig 1. Curve showing anomalous dispersion of quartz. A is limiting value of μ as λ approaches infinity. (*After F. A. Jenkins*, **Fundamentals of Optics**, *4th ed., McGraw-Hill, 1976*)

waves corresponding to the same note emitted by another fork vibrating at the same frequency. To account for the absorption of energy by the medium from the radiation, it is necessary to postulate that in the motion of the atomic oscillator some frictional force, proportional to the velocity of the oscillator, must be overcome. For small amplitudes of forced oscillation, when the frequency of the radiation is very different from the natural period of the oscillator, the frictional force and the absorption of energy are negligible. Near resonance between the radiation and the oscillator, the amplitude becomes large, with a correspondingly large absorption of energy to overcome the frictional resistance. Radiation of frequencies near the natural frequency therefore corresponds to an absorption band.

To show that the velocity of the radiation within the medium is changed, it is necessary to consider the phase of the forced vibration, which the theory shows to depend on the frequency of the radiation. The oscillator itself becomes a source of secondary radiation waves within the medium which combine to form sets of waves moving parallel to the original waves. Interference between the secondary and primary waves takes place, and because the phase of the secondary waves, which is the same as that of the atomic oscillators, is not the same as that of the primary waves, the wave motion resulting from the interference between the two sets of waves is different in phase from that of the primary waves incident on the medium. But the velocity of propagation of the waves is the rate of advance of equal phase; hence the phase change effected by the medium, which is different for each frequency of radiation, is equivalent to a change in the velocity of the radiation within the medium. When the frequency of the radiation slightly exceeds the natural frequency of the oscillator, the radiation and the oscillator become 180° out of phase, which corresponds to an increase in the velocity of the radiation and accounts for the observed fall in refractive index on the short-wavelength side of the absorption band.

The theory leads to Eqs. (13) through (17) for the refractive index of a material medium as a function of the frequency of the radiation. In the equations the frequency is expressed as angular frequency, $\omega = 2\pi\nu \text{ s}^{-1} = 2\pi c/\lambda$, where c is the velocity of light. When the angular frequency ω of the radiation is not very near the characteristic frequency of the electronic oscillator, the refractive index of a homogeneous medium containing N molecules per cubic centimeter is given by Eq. (13a), where e and m are the charge and mass of the electron, and f is the number of oscilla-

$$\mu^2 = 1 + \frac{4\pi Ne^2}{m} \cdot \frac{f}{\omega_0^2 - \omega^2} \quad (13a) \qquad \mu^2 = 1 + 4\pi Ne^2 \sum_i \frac{f_i/m_i}{\omega_i^2 - \omega^2} \quad (13b)$$

tors per molecule of characteristic frequency ω_0. The f value is sometimes called the oscillator strength. If the molecule contains oscillators of different frequencies and mass (for example, electronic oscillators of frequency corresponding to ultraviolet radiation and ionic oscillators corresponding to infrared radiation), the frequency term becomes a summation, as in Eq. (13b), where ω_i is the characteristic frequency of the ith type of oscillator, and f_i and m_i are the corresponding f value and mass. In terms of wavelengths, this relation can be written as Eq. (14), where A_i is a

$$\mu^2 = 1 + \sum_i \frac{A_i \lambda^2}{\lambda^2 - \lambda_i^2} \quad (14)$$

constant for the medium, λ is the wavelength of the radiation, and $\lambda_i = c/\nu_i$ is the wavelength corresponding to the characteristic frequency ν_i per second (Sellmeier's equation).

If the medium is a gas, for which the refractive index is only slightly greater than unity, the dispersion formula can be written as Eq. (15).

$$\mu = 1 + 2\pi Ne^2 \sum_i \frac{f_i/m_i}{\omega_i^2 - \omega^2} \tag{15}$$

So long as the absorption remains negligible, these equations correctly describe the increase in refractive index as the frequency of the radiation begins to approach the absorption band determined by ω_i or λ_i. They fail when absorption becomes appreciable, since they predict infinitely large values of the refractive index when ω equals ω_i, whereas the refractive index remains finite throughout an absorption band.

The absorption of radiant energy of frequency very close to the characteristic frequency of the medium is formally regarded as the overcoming of a frictional force when the molecular oscillators are set into vibration, related by a proportionality constant g to the velocity of the oscillating particle; g is a damping coefficient for the oscillation. If the refractive index is determined by a single electronic oscillator, the dispersion equation for a gas at radiational frequencies within the absorption band becomes Eq. (16). At the same time an absorption constant κ enters the equa-

$$\mu = 1 + \frac{2\pi Ne^2}{m} \frac{f(\omega_0^2 - \omega^2)}{(\omega_0^2 - \omega^2)^2 + \omega^2 g^2} \tag{16}$$

tions, related to the absorption coefficient α of Eq. (1) by the expression $\kappa = \alpha c/2\omega\mu$. Equation (17) shows the relationship. For a monatomic vapor at low pressure, Nf is about 10^{17} per cubic

$$\kappa = \frac{2\pi Ne^2}{m} \frac{f\omega g}{(\omega_0^2 - \omega^2)^2 + \omega^2 g^2} \tag{17}$$

centimeter, ω_0 is about 3×10^{15} per second, and g is about 10^{11} per second. These data show that, when the frequency of the radiation is not very near ω_0, ωg is very small in comparison with the denominator and the absorption is practically zero. As ω approaches ω_0, κ increases rapidly to a maximum at a radiational frequency very near ω_0 and then falls at frequencies greater than ω_0. When the absorption is relatively weak, the absorption maximum is directly proportional to the oscillator strength f. In terms of the molar extinction coefficient ϵ of Eq. (4), it can be shown that this direct relation holds, as seen in Eq. (18). The integration in Eq. (18) is carried out over

$$f = 4.319 \times 10^{-9} \int \epsilon d\bar{\nu} \tag{18}$$

the whole absorption spectrum. The integral can be evaluated from the area under the curve of ϵ plotted as a function of wave number $\bar{\nu}$ cm^{-1} = $\nu(\text{s}^{-1})/c = 1/\lambda$.

The width of the absorption band for an atom is determined by the value of the damping coefficient g; the greater the damping, the greater is the spectral region over which absorption extends.

The general behavior of the refractive index through the absorption band is illustrated by the dotted portions of **Fig. 2**. The presence of the damping term $\omega^2 g^2$ in the denominator of Eq.

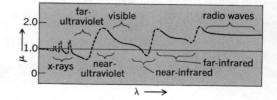

Fig. 2. Complete dispersion curve through the electromagnetic spectrum for a substance. (After F. A. Jenkins, Fundamentals of Optics, 4th ed., McGraw-Hill, 1976)

(17) prevents the refractive index from becoming infinite when $\omega = \omega_0$. Its value increases to a maximum for a radiation frequency less than ω_0, then falls with increasing frequency in the center of the band (anomalous dispersion) and increases from a relatively low value on the high-frequency side of the band.

Figure 2 shows schematically how the dispersion curve is determined by the absorption bands throughout the whole electromagnetic spectrum. The dotted portions of the curve correspond to absorption bands, each associated with a distinct type of electrical oscillator. The oscillators excited by x-rays are tightly bound inner electrons; those excited by ultraviolet radiation are more loosely bound outer electrons which control the dispersion in the near-ultraviolet and visible regions, whereas those excited by the longer wavelengths are atoms or groups of atoms.

It will be observed in Fig. 2 that in regions of anomalous dispersion the refractive index of a substance may assume a value less than unity; the velocity of light in the medium is then greater than in vacuum. The velocity involved here is that with which the phase of the electromagnetic wave of a single frequency ω advances, for example, the velocity with which the crest of the wave advances through the medium. The theory of wave motion, however, shows that a signal propagated by electromagnetic radiation is carried by a group of waves of slightly different frequency, moving with a group velocity which, in a material medium, is always less than the velocity of light in vacuum. The existence of a refractive index less than unity in a material medium is therefore not in contradiction with the theory of relativity.

In quantum theory, absorption is associated not with the steady oscillation of a charge in an orbit but with transitions from one quantized state to another. The treatment of dispersion according to quantum theory is essentially similar to that outlined, with the difference that the natural frequencies ν_0 are now identified with the frequencies of radiation which the atom can absorb in undergoing quantum transitions. These transition frequencies are regarded as virtual classical oscillators, which react to radiation precisely as do the oscillators of classical electromagnetic theory.

Selective reflection. Nonmetallic substances which show very strong selective absorption also strongly reflect radiation of wavelengths near the absorption bands, although the maximum of reflection is not, in general, at the same wavelength as the maximum absorption. The infrared rays selectively reflected by ionic crystals are frequently referred to as reststrahlen, or residual rays. For additional information on selective reflection SEE REFLECTION OF ELECTROMAGNETIC RADIATION.

Bibliography. M. Born and E. Wolf, *Principles of Optics*, 6th ed., 1980; R. W. Ditchburn, *Light*, 3d ed., 1977; F. A. Jenkins, *Fundamentals of Optics*, 4th ed., 1976; Optical Society of America, *Handbook of Optics*, 1978; A. Sommerfeld, *Lectures of Theoretical Physics*, vol. 4, 1954; J. A. Stratton, *Electromagnetic Theory*, 1941; J. Strong, *Concepts of Classical Optics*, 1958.

SHADOW
Francis A. Jenkins and William W. Watson

A region of darkness caused by the presence of an opaque object interposed between such a region and a source of light. A shadow can be totally dark only in that part called the umbra, in which all parts of the source are screened off. With a point source, the entire shadow consists of an umbra, since there can be no region in which only part of the source is eclipsed. If the source has an appreciable extent, however, there exists a transition surrounding the umbra, called the penumbra, which is illuminated by only part of the source. Depending on what fraction of the source is exposed, the illumination in the penumbra varies from zero at the edge of the full shadow to the maximum where the entire source is exposed. The edge of the umbra is not perfectly sharp, even with an ideal point source, because of the wave character of light. SEE DIFFRACTION.

For example, the shadow of the Moon shown in the **illustration** has an umbra which barely touches the Earth during a total eclipse of the Sun. The eclipse is only partial for an observer situated in the penumbra. Astronauts circling the Earth or the Moon observe this shadow pattern during each orbit.

The term shadow is also used with other types of radiation, such as sound or x-rays. In the case of sound, pronounced shadows are formed only for high frequencies. The high frequency of

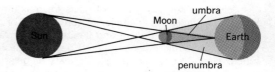

Shadow of the Moon at the time of an eclipse of the Sun. Relative sizes are not to scale.

visible light waves makes possible shadows that are nearly black. X-ray photographs are shadowgrams in which the bones and tissues appear by virtue of their different opacities or degrees of absorption. The bending of light rays by reflection or refraction may also produce shadow patterns. An important practical example is the schlieren method of observing flow patterns in wind tunnels.

OPAQUE MEDIUM
M. G. Mellon

One which is impervious to rays of light, that is, not transparent to the human eye. By extension, a medium may be described as opaque if it does not transmit infrared waves or other regions of the electromagnetic spectrum, such as the x-ray, ultraviolet, and microwave regions. The property of zero transmittance does not necessarily imply total reflectance; that is, opacity can result both from reflection and from absorption of incident rays. SEE ABSORPTION OF ELECTROMAGNETIC RADIATION; REFLECTION OF ELECTROMAGNETIC RADIATION.

TRANSPARENT MEDIUM
M. G. Mellon

Ordinarily, a medium which has the property of transmitting rays of light in such a way that the human eye may see through the medium distinctly. It is pervious to light, that is, to the visible region of the electromagnetic spectrum. By extension, a medium may be described as transparent to other regions of the spectrum, such as x-rays and microwaves. Just as a blue filter passes blue rays, an ultraviolet filter may be considered as passing ultraviolet rays, or being transparent to them. SEE ABSORPTION OF ELECTROMAGNETIC RADIATION; TRANSLUCENT MEDIUM.

TRANSLUCENT MEDIUM
M. G. Mellon

A medium which transmits rays of light so diffused that objects cannot be seen distinctly; that is, the medium is only partially transparent. Familiar examples are various forms of glass which admit considerable light but impede vision. Inasmuch as the term translucent seems to imply seeing, usage of the term is ordinarily limited to the visible region of the spectrum. SEE TRANSPARENT MEDIUM.

COLOR FILTER
William L. Wolfe

An optical element that partially absorbs incident radiation, often called an absorption filter. The absorption is selective with respect to wavelength, or color, limiting the colors that are transmitted by limiting those that are absorbed. Color filters absorb all the colors not transmitted. They are used in photography, optical instruments, and illuminating devices to control the amount and spectral composition of the light.

Color filters are made of glass for maximum permanence, of liquid solutions in cells with transparent faces for flexibility of control, and of dyed gelatin or plastic (usually cellulose acetate) for economy, convenience, and flexibility. The plastic filters are often of satisfactory permanence, but they are sometimes cemented between glass plates for greater toughness and scratch resistance. They do not have as good quality as gelatin filters.

Color filters are sometimes classified according to their type of spectral absorption: short-wavelength pass, long-wavelength pass or band-pass; diffuse or sharp-cutting; monochromatic or conversion. The short-wavelength pass transmits all wavelengths up to the specified one and then absorbs. The long-wavelength pass is the opposite. Every filter is a band-pass filter when considered generally. Even an ordinary piece of glass does not transmit in the ultraviolet or infrared parts of the spectrum. Color filters, however, are usually discussed in terms of the portion of the visible part of the spectrum. Sharp and diffuse denote the sharpness of the edges of the filter band pass. Monochromatic filters are very narrow band pass filters. Conversion filters alter the spectral response or distribution of one selective detector or source to that of another, for example, from that of a light bulb to that of the Sun (see **illus**.).

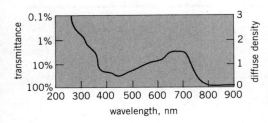

Transmission of a conversion filter, used to convert the spectral distribution of a light bulb (color temperature of 2360 K or 3790°F) to that of the Sun (color temperature of 5500 K or 9440°F). (*After Eastman Kodak Co., Kodak Filters for Scientific and Technical Uses, 1972*)

Neutral-density filters transmit a constant fraction of the light across the spectrum considered. They are made either by including an appropriate absorbent like carbon that is spectrally nonselective, or by a thin, partially transparent reflective layer, often of aluminum.

The transmittance of a filter is the ratio of the transmitted flux to the incident flux, expressed as either a ratio or a percentage. The density of a filter is the logarithm to base 10 of the reciprocal of the transmittance. For example, a filter with a 1% transmittance has a density of 2.

Other, closely related filters are the Christiansen, Lyot or birefringent, residual-ray, and interference filters. They use the phenomena of refraction, polarization, reflection, and interference. SEE ABSORPTION OF ELECTROMAGNETIC RADIATION; COLOR; INTERFERENCE FILTER; REFLECTION OF ELECTROMAGNETIC RADIATION.

REFRACTION OF WAVES
JOHN W. STEWART

The change of direction of propagation of any wave phenomenon which occurs when the wave velocity changes. The term is most frequently applied to visible light, but it also applies to all other electromagnetic waves, as well as to sound and water waves.

The physical basis for refraction can be readily understood with the aid of **Fig. 1**. Consider a succession of equally spaced wavefronts approaching a boundary surface obliquely. The direction of propagation is in ordinary cases perpendicular to the wavefronts. In the case shown, the velocity of propagation is less in medium 2 than in medium 1, so that the waves are slowed down as they enter the second medium. Thus the direction of travel is bent toward the perpendicular to the boundary surface (that is, $\theta_2 < \theta_1$). If the waves enter a medium in which the velocity of propagation is faster than in their original medium, they are refracted away from the normal.

Snell's law. The simple mathematical relation governing refraction is known as Snell's law. If waves traveling through a medium at speed v_1 are incident on a boundary surface at angle

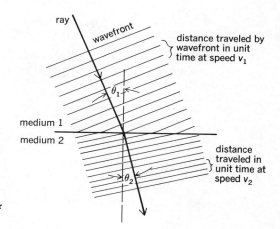

Fig. 1. Physical basis for Snell's law.

θ_1 (with respect to the normal), and after refraction enter the second medium at angle θ_2 (with the normal) while traveling at speed v_2, then Eq. (1) holds. The index of refraction n of a medium is

$$\frac{v_1}{v_2} = \frac{\sin \theta_1}{\sin \theta_2} \tag{1}$$

defined as the ratio of the speed of waves in vacuum c to their speed in the medium. Thus $c = n_1 v_1 = n_2 v_2$, and therefore Eq. (2) holds. The refracted ray, the normal to the surface, and the incident ray always lie in the same plane.

$$n_1 \sin \theta_1 = n_2 \sin \theta_2 \tag{2}$$

The relative index of refraction of medium 2 with respect to that of medium 1 may be defined as $n = n_1/n_2$. Snell's law then becomes Eq. (3). For sound and other elastic waves requiring a medium in which to propagate, only this last form has meaning. Equation (3) is frequently

$$\sin \theta_1 = n \sin \theta_2 \tag{3}$$

used for light when one medium is air, whose index of refraction is very nearly unity.

When the wave travels from a region of low velocity (high index) to one of high velocity (low index), refraction occurs only if $(n_1/n_2) \sin \theta_1 \leqq 1$. If θ_1 is too large for this relation to hold, then $\sin \theta_2 > 1$, which is meaningless. In this case the waves are totally reflected from the surface back into the first medium. The largest value that θ_1 can have without total internal reflection taking place is known as the critical angle θ_c. Thus $\sin \theta_c = n_2/n_1$. When the angle of incidence $\theta_1 < \theta_c$, refraction occurs, as in **Fig. 2a**. When $\theta_1 = \theta_c$, the emergent ray just grazes the surface (Fig. 2b). Total internal reflection (Fig. 2c) represents the only practical case for which 100% of the

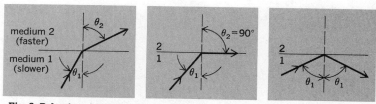

Fig. 2. Behavior of ray traveling from medium of high refractive index to medium of low refractive index. (a) When $\theta_1 < \theta_{c'}$, ray is refracted. (b) When $\theta_1 = \theta_{c'}$, ray grazes surface. (c) When $\theta_1 > \theta_{c'}$, ray is reflected.

incident energy is reflected and none is absorbed. When it is desired to change the direction of a beam of light without loss of energy, totally reflecting prisms are often used, as in prism binoculars.

If waves travel through a medium having a continuously varying index of refraction, the rays follow smooth curves with no abrupt changes of direction. Suppose (**Fig. 3**) that $n = n(y)$, and that the incident ray lies in the xy plane. If θ is the angle between the direction of the ray and the y axis, then Snell's law can be written in the differential form given by Eq. (4). In a particular case Eq. (4) can be integrated to give the path of the ray.

$$\frac{d\theta}{dn} = -\frac{1}{n}\tan\theta \tag{4}$$

Visible light. Many interesting cases of refraction occur for visible light. The refraction of light by a prism in air affords a particularly simple and useful example. SEE LIGHT.

As a ray passes through the prism of **Fig. 4**, its total deflection or deviation $D = \theta_1 + \theta_4 - A$, where A is the vertex angle of the prism. Also, by Snell's law, Eq. (5) holds. It is found

$$n = \frac{\sin\theta_1}{\sin\theta_2} = \frac{\sin\theta_4}{\sin\theta_3} \tag{5}$$

that the deviation is a minimum when the ray passes through the prism symmetrically (that is, when $\theta_1 = \theta_4$). For minimum deviation, Eq. (6) holds. For a given prism, the dispersion, or lateral

$$n = \frac{\sin\tfrac{1}{2}(A+D)}{\sin\tfrac{1}{2}A} \tag{6}$$

spread of the spectrum formed is maximum for that wavelength of light which passes through the prism at minimum deviation. SEE OPTICAL PRISM.

For most optical materials, the dispersion $dn/d\lambda$ (λ is the wavelength) is negative; therefore red light is bent less than blue light. Typical values of n for optical materials range from 1.5 for ordinary crown glass, 1.7 or 1.8 for dense flint glass, and up to 2.42 for diamond. For water, n is 1.33. Some special substances have even higher values. Many substances show anisotropy in the refraction of light, with different indices of refraction in different directions. SEE OPTICAL MATERIALS.

For a lens (**Fig. 5**), refraction occurs at both surfaces. If the lens is thin and the rays all make small angles with the axis of the system, application of Snell's law to the two spherical surfaces yields the well-known lens formula, Eq. (7), where s is the object distance from the lens,

$$\frac{1}{s} + \frac{1}{s'} = \frac{1}{f} \tag{7}$$

s' the image distance, and f the focal length of the lens. Magnifying instruments such as binoc-

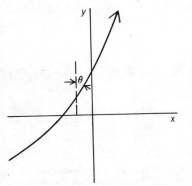

Fig. 3. Path of light in medium having continuously varying index of refraction $n = n(y)$.

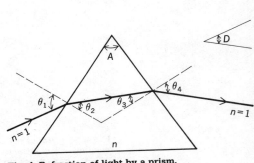

Fig. 4. Refraction of light by a prism.

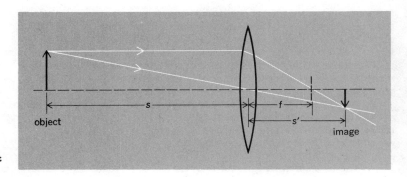

Fig. 5. Refraction of light by a lens.

ulars, telescopes, microscopes, and projectors make use of refraction by lenses or prisms in their operation. *See* Lens.

Double refraction. Some anisotropic single crystals such as those of calcite and quartz are birefringent, or doubly refracting. If one looks through such a crystal at a dot on a piece of paper, he sees two images. As the crystal is rotated in the plane of the paper, one image remains stationary while the other appears to rotate about it. *See* Birefringence.

Two separate rays propagate through the crystal; they are called the ordinary ray and the extraordinary ray. These rays are linearly polarized at right angles to each other. The ordinary ray obeys Snell's law; the extraordinary ray in general does not. The extraordinary ray does not propagate perpendicularly to its wavefronts. The separation between the two rays depends upon the direction in which the light travels through the crystal relative to that of the optic axis of the crystal. Light traveling parallel to the optic axis is only singly refracted.

Birefringent crystals are either uniaxial or biaxial, depending upon whether they have one optic axis or two. They are said to be positive or negative, depending upon whether the velocity of propagation (within the crystal) of the extraordinary wave is greater or less than that of the ordinary wave. Calcite is a uniaxial negative crystal, quartz a uniaxial positive crystal. The most commonly used biaxial crystal is mica. Doubly refracting crystals are frequently employed as polarizers, such as the nicol prism. *See* Crystal optics; Polarized light.

Refractometry. The measurement of indices of refraction, called refractometry, can be made in several ways. A very accurate technique is to determine in a prism spectrometer the minimum deviation D for a prism made from the material in question. The value of n is then calculated from Eq. (6). Hollow prisms can be used in this manner to determine the values of indices of refraction of various liquids. Alternatively, the critical angle for total internal reflection may be measured. Another method is to observe visually the apparent thickness of a slab of material by looking straight through it, and to compare the apparent thickness with the real thickness as measured with a micrometer. Then Eq. (8) holds.

$$n = \frac{\text{real thickness}}{\text{apparent thickness}} \qquad (8)$$

Interferometric methods are particularly convenient for gases. In the Jamin refractometer, for example, a simple count of fringes as the gas is slowly admitted to an initially evacuated tube in the optical path yields n. These techniques can also be used for solids, particularly when the material is available in the form of thin films. *See* Interferometry.

Refractometry is an important tool in analytical chemistry. For example, information about the composition of an unknown solution can frequently be obtained by measurement of its index of refraction.

Atmospheric refraction. Gases have indices of refraction only slightly greater than unity. In general, $n - 1$ is proportional to the density of the gas, or to the ratio of pressure to absolute temperature. The index of refraction of the Earth's atmosphere increases continuously from 1.000000 at the edge of space to 1.000293 (yellow light) at 32°F (0°C) and 760 mmHg (101.325

kilopascals) pressure. Thus celestial bodies as seen in the sky are actually nearer to the horizon than they appear to be. The effect decreases from a maximum of about 35 minutes of arc for an object on the horizon to zero at the zenith, where the light enters the atmosphere at perpendicular incidence. Thus the Sun (and all other bodies) appears to rise 2 or more minutes earlier (depending upon latitude) and to set 2 or more minutes later than would be the case without refraction. This must be taken into account when the altitude of a celestial body is observed for navigational purposes.

Other manifestations of atmospheric refraction are the mirages and "looming" of distant objects which occur over oceans or deserts, where the vertical density gradient of the air is quite uniform over a large area. The twinkling of stars is caused by the rapid small fluctuations in density along the light path in the atmosphere. Rainbows are produced by the multiple reflections, refraction, and dispersion of sunlight by spherical raindrops. SEE METEOROLOGICAL OPTICS.

Other electromagnetic waves. Although refraction is most frequently encountered for the visible portion of the spectrum, it is of importance for other electromagnetic radiation. For very long-wavelength radiation, the index of refraction of many materials is equal to the square root of the dielectric constant k. In general, $dn/d\lambda$ is negative except in the regions of so-called anomalous dispersion near absorption bands. On the short-wavelength side of an absorption band, n can be less than 1.00. Since it is the phase velocity of the wave rather than the group velocity which is involved in the definition of the index of refraction, this does not represent a violation of the principle of relativity, that is, that energy cannot be propagated at a velocity faster than the velocity of light in vacuum. At very high frequencies, the index of refraction of all materials is also slightly less than unity. SEE ABSORPTION OF ELECTROMAGNETIC RADIATION.

Refraction plays a role in the propagation beyond the line of sight of radio waves in the Earth's atmosphere.

The interaction of electromagnetic radiation with more or less opaque substances is often described in terms of a complex index of refraction. The real part of this quantity has the usual meaning for the small amount of light which penetrates into the material before it is absorbed. The imaginary part is a measure of the absorption.

Bibliography. M. Born and E. Wolf, *Principles of Optics*, 6th ed., 1980; D. Halliday and R. Resnick, *Physics*, 3rd ed., 1981; F. A. Jenkins, *Fundamentals of Optics*, 4th ed., 1976.

DISPERSION (RADIATION)
WILLIAM WEST

The separation of a complex of electromagnetic or sound waves into its various frequency components. For example, a beam of white light can be separated into its monochromatic components by virtue of the different velocities of rays of different wavelength of the beam as it passes through a prism. The dispersion of a material, such as glass or water, at a given wavelength in the electromagnetic spectrum is defined as the rate of change of refractive index with wavelength at the wavelength in question. For an extended discussion of the dispersion of light SEE ABSORPTION OF ELECTROMAGNETIC RADIATION.

OPTICAL MATERIALS
WILLIAM L. WOLFE

Generally, all substances used to reflect, refract, filter, polarize, modulate, detect, or disperse infrared, visible, or ultraviolet radiation or light; more specifically, the transparent materials usually used for windows and lenses. These materials are often some formulation of glass, a plastic in either bulk or thin-film form, or crystals either single or in aggregates. The most important optical properties are the degree and spectral region of transparency, the value of the refractive index over the same spectral region, and the uniformity of the sample. Desirable materials are hard, strong, and insensitive to temperature variations. SEE ABSORPTION OF ELECTROMAGNETIC RADIATION; COLOR

FILTER; INFRARED RADIATION; INTERFERENCE FILTER; LENS; OPTICAL DETECTORS; OPTICAL MODULATORS; POLARIZED LIGHT; REFLECTION OF ELECTROMAGNETIC RADIATION; REFRACTION OF WAVES; ULTRAVIOLET RADIATION.

Transmission loss is caused by reflection at the surface, absorption in the bulk, and scattering in both places. Surface reflection is given by Fresnel's equation, which for normal incidence is $[(n-1)/(n+1)]^2$ for each surface, where n is the refractive index. The reflection varies from about 4% per surface for a material such as glass ($n = 1.5$) to 36% per surface for germanium ($n = 4$). The change with wavelength is quite gradual between regions of strong absorption, being about $5 \times 10^{-6}/°F$ ($10^{-5}/°C$) for many materials. In general, denser materials have higher refractive indices. Absorption is the process by which incident radiation is converted to some other form of energy in the material. Absorption may be due to increased motion of electrons in the solid or the vibration of atoms or ions in the lattice or partially ordered network. The absorption by electrons is a smooth function, increasing gradually in proportion to the square of the wavelength. Absorption by vibration is characterized by relatively narrow bands at distinct wavelengths. Scattering of radiation occurs when a local disturbance, such as a fleck of dirt, impurity ion, grain boundary, or imperfection of the surface, is encountered by the incoming wave. If the wavelength is large compared to the disturbance, then the transmission varies as λ^{-4}; if small, a more complicated oscillating dependence is observed. SEE SCATTERING OF ELECTROMAGNETIC RADIATION.

Optical glass. The optical material most widely used is optical glass, which is available in a wide range of refractive indices and the change of refractive index with wavelength, called dispersion. Optical glass differs from ordinary glass in its freedom from imperfections. It must be free from unmelted particles or "stones," from bubbles, and from chemical inhomogeneity, which gives rise to regions of variable refractivity known as cords or striae.

The refractive index is sometimes followed by a subscript indicating the wavelength of the light in vacuum; n_D indicated the refractive index for the mean D line of sodium, the radiation that gives the bright yellow color to most flames. It has a wavelength of 589.3 nanometers. In optical glass catalogs, it is customary to give the refractive indices for a group of spectral lines chosen by E. Abbe for the convenience with which they could be obtained for spectrometric work. Two more wavelengths, one in the near-ultraviolet and one in the near-infrared, have been added, chiefly to accommodate calculations for photographic systems sensitive in these areas. The source, designation, and wavelength in vacuum of these lines are given in **Table 1**, together with some other lines which are being used increasingly because of their greater convenience.

Optical glass catalogs also give the mean dispersion, commonly designated by ($n_C - n_F$) and other partial dispersions, as well as the dispersion ratios ($n_D - n_C$)/($n_F - n_C$), and so forth, in which the symbols C, D, and F refer to the refractive indices for the spectral lines as designated in Table 1. Another number given under various names is the nu value, $\nu = (n_D - 1)/(n_F - n_C)$. This expression occurs in the calculation of the color correction of lenses. SEE CHROMATIC ABERRATION.

Glass types. The early optical glasses were crowns and flints. The crown glasses were essentially of the same type as window glass, with low index and dispersion; the flint glasses contained lead oxide and had higher index and dispersion. These glasses, approximated by the long line of **Fig. 1** that starts in the lower left and runs to the upper right corner (the flint line), did not have sufficient range in optical properties to enable desirable corrections to be made in optical systems. A great step forward was made by O. Schott in the introduction of the borosilicate, fluoride, and barium crown and flint types, and a second advance was made by G. W. Morey

Table 1. Designation, source, and wavelength of spectral lines used in spectrometric measurements*

Source	Hg	Hg	Hg	H	H	He	Hg
Designation	—	h	g	G'	F	—	e
Wavelength, nm	365.0	404.7	435.8	434.1	486.1	492.2	546.1
Source	He	Na (mean)		H	He	K (mean)	Hg
Designation	d	D		C	b	A	—
Wavelength, nm	587.6	589.3		656.3	706.5	768.2	1014.0

*From G. W. Morey, *Properties of Glass*, 2d ed., Reinhold, 1954, supplemented, 1968.

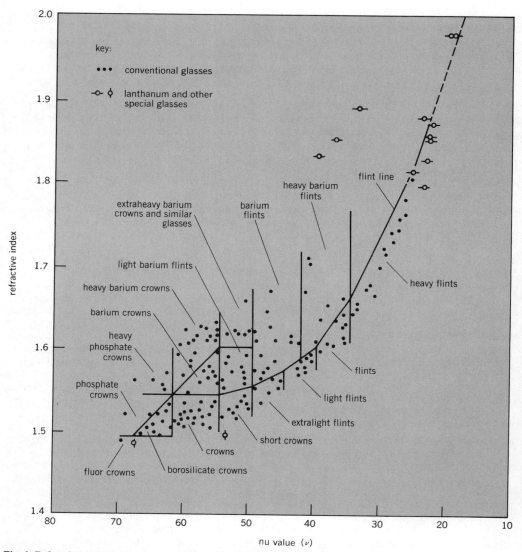

Fig. 1. Refractive index and nu values of some liquids, crystals, and conventional, lanthanum, and other special glasses. Glasses are grouped into types. (*After G. W. Morey, The Properties of Glass, 2d ed., Reinhold, 1954*)

in the rare-earth glasses, now usually called lanthanum crowns and flints, also indicated by Fig. 1. They had different refractive indices and dispersions, so they could be used in conjunction with lenses of other materials to correct for the color aberrations of optical systems.

Effect of absorption. When the absorption of light in glass is uniformly distributed throughout the visible spectrum and the amount of absorption is small, the glass appears colorless when viewed in white light; when the amount of uniform absorption increases, the glass takes on a grayish hue. If the absorption is significantly greater for light of any particular color, the transmitted light will appear of a complementary color; since the absorption is never monochromatic, the color will depend on the thickness of the sample. Comparisons among materials are best made by means of curves showing the change of absorption with wavelength for samples of standard thickness.

Silica glass, in thicknesses such as are used in cameras and optical instruments, transmits all the radiation to which ordinary photographic plates are sensitive, that is, down to a wavelength of 220 nm, but wavelengths of 193 nm and less are almost completely absorbed. The transmission of the usual colorless glass is limited chiefly by absorptive bands. The limit of transmission in ultraviolet light is determined largely by the content of ferric oxide (Fe_2O_3), which shows strong absorption in the near-ultraviolet. Special glasses should contain, and have been made to contain, as little as 5 parts per million of iron oxide. The limit of transmission in the very-near-infrared is determined largely by the content of ferrous oxide (FeO), which shows strong absorption at about 1 μm. The best optical glasses have a transparency of over 99% throughout the visible spectrum (380–780 nm), a result achieved only by using the greatest care in excluding impurities, especially iron. Ordinary window glass removes the 313-nm line of mercury vapor and shorter wavelengths.

Colored glass. The coloring agent or colorant in glass is usually produced by: substances dissolved in the glass which absorb characteristic frequencies; particles of submicroscopic dimensions, such as gold or copper in ruby glass; or particles of microscopic or larger dimensions, either themselves colored or colorless. The coloring agents which act by virtue of characteristic absorp-

Table 2. NATO optical glasses

Six-digit code	Schott name	n_d	V_d[a]	P_{Fd}[b]	Stain[c]	Bubble[d]	Density[e]	Internal transmittance[f]
511604	K-7	1.51112	60.41	0.6950	0	0–1	2.53	0.986
517642	BK-7	1.51680	64.17	0.6928	0	0	2.51	0.972
522595	K-5	1.52249	59.48	0.6956	0	2	2.63	0.98
523602	K-50	1.52257	60.18	0.6956	0	0	2.62	0.986
527511	KzF-6	1.52682	51.13	0.6969	2	1	2.54	0.88
529517	KzF-2	1.52944	51.68	0.6970	2	0–1	2.55	0.91
548458	LLF-1	1.54814	45.75	0.7020	0	0–1	2.94	0.985
573575	BaK-1	1.57250	57.55	0.6966	1	1	3.19	0.97
581409	LF-5	1.58144	40.85	0.7040	0	0–1	3.22	0.988
613443	KzFS-N4	1.61340	44.30	0.6997	2	0–1	3.20	0.65
613574	SK-19	1.61342	57.37	0.6967	1	0–1	3.56	0.91
613586	SK-4	1.61272	58.63	0.6966	2	0–1	3.57	0.92
618551	SSK-4	1.61765	55.14	0.6982	1	0–1	3.63	0.94
620364	F-2	1.62004	36.37	0.7062	0	0	3.61	0.90
620603	SK-16	1.62041	60.33	0.6953	4	0–1	3.58	0.88
648339	SF-2	1.64769	33.85	0.7076	0	0–1	3.86	0.84
650392	BaSF-10	1.65016	39.15	0.7057	2–3	2	3.91	0.80
652585	LaK-N7	1.65160	58.52	0.6961	2	1	3.84	0.845
653397	KzFS-5	1.65332	39.71	0.7013	4–5	0–1	3.44	0.14
670471	BaF-N10	1.67003	47.11	0.7017	3	0–1	3.76	0.29
691547	LaK-N9	1.69100	54.71	0.6959	1	1	3.55	0.745
699301	SF-15	1.69895	30.07	0.7102	0	1	4.06	0.01
702410	BaSF-52	1.70181	41.01	0.7030	5	1	3.96	0.43
717480	LaF-N3	1.71700	47.98	0.7007	5	1	4.34	0.62
720504	LaK-10	1.72000	50.41	0.6974	2	1	3.81	0.755
744448	LaF-N2	1.74400	44.77	0.7023	4–5	1	4.46	0.47
755276	SF-4	1.75520	27.58	0.7114	2	1	4.78	0.31
805254	SF-6	1.80518	25.43	0.7129	3–4	0–1	5.18	0.03

[a] $V_d = (n_d - 1)/(n_F - n_C)$.
[b] $P_{Fd} = (n_F - n_d)/(n_F - n_C)$.
[c] Stain scale ranges from class 0 (practically no stain even after 100 h exposure to a standard acid solution) to class 5 (stain occurs in less than 12 min).
[d] Bubble designation ranges from class 0 (total cross section of all apparent inclusions, such as bubbles and stria, does not exceed 0.029 mm^2 in a volume of 100 cm^3) to class 5 (this value is 2 mm^2).
[e] Relative to water.
[f] For a wavelength of 350 nm and a thickness of 5 mm (0.197 in.).

tion spectra are all elements belonging to the transition rows of the periodic system, and especially to the first of these rows, to which belong titanium, vanadium, chromium, manganese, iron, cobalt, nickel, and copper, the commonest and most effective colorants of glass.

Available glasses and limitations. Most companies that sell glass for use in lenses for optical instruments also produce glass charts, like that shown in Fig. 1, of their products. Designers often search for very special substances for extra color correction of lenses, but these materials are usually scarce and expensive. Agreement has been reached by many users to utilize only the NATO (North Atlantic Treaty Organization) glasses listed in **Table 2** for all but the most unusual requirements. NATO glasses manufactured with a set of glass formulations for which almost all optical designs will be done, for reasons of economy.

Most glasses are uniformly transparent over the entire visible spectrum, suffering only reflection losses, the limits of transmission being determined by the content of ferric oxide and ferrous oxide. However, even for completely pure glass, the vibrations between silicon and oxygen atoms in the silicate structure limit transmission to about 2.5 micrometers, and freeing electrons from the lattice determines the short-wavelength limit. Research on other matrices such as germanium oxide and calcium aluminate has produced somewhat softer substances that transmit to about 5 μm. Monoxide compositions of combinations of silicon, germanium, selenium, arsenic, tellurium, and sulfur are softer still and melt at low temperatures (about 900°F or 500°C), but transmit to about 14 μm.

Crystalline materials. Because of the limitations of glasses, single-crystal or polycrystalline aggregates are used for special applications. The most useful of these are of cubic symmetry, so that the refractive index and other physical properties are isotropic. The properties of some of the more useful of these materials are listed in **Table 3**. The short-wavelength edge of the transmission region λ_{min} is determined by the width of the energy gap; the long-wavelength edge is determined by the onset of molecular vibrations or rotations. The long-wavelength limit λ_{max} is proportional to the square-root of the mass of the constituent molecules and inversely proportional to the square root of the binding energy. Many single-crystal materials are available in cylinders which are about 12 in. (30 cm) in diameter by 12 in. (30 cm) long. Some polycrystalline samples are available in even larger diameters (3 ft or 1 m) in thinner sections.

Table 3. Optical materials for the ultraviolet and infrared and selected properties

Material		λ_{min}, μm	λ_{max}, μm	T_m, K (°F)*	H_K, kg/mm²†
Name	Symbol				
Quartz	SiO_2	0.12	5	1740 (2670)	740–460
Sapphire	Al_2O_3	0.14	6	2300 (3680)	1370
Rutile	MgO	0.45	6	3070 (5070)	690
Diamond	C	0.25	80+		
Salt	NaCl	0.21	2	1075 (1475)	15
Potassium iodide	KI	0.25	45	1000 (1340)	
Cesium bromide	CsBr	0.25	80	900 (1160)	20
KRS-5	TlBr-TlI	0.6	40	690 (780)	
Magnesium fluoride	MgF_2	0.11	8		
Calcium fluoride	CaF_2	0.13	12	1630 (2470)	160
Cadmium telluride	CdTe	0.9	16	1310 (1900)	
Zinc sulfide	ZnS	2	20	1293 (1868)	
Calcium aluminate (glass)	$CaAl_2O_3$	0.4	6		
Lithium fluoride	LiF	0.12	9	1140 (1590)	100
Calcite	$CaCO_3$	0.2	6	1170 (1650)	
Arsenic trisulfide glass	As_2S_3	0.6	13	480 (400)	100
Silicon	Si	1.2	15	1690 (2580)	1150
Germanium	Ge	1.8	20	1210 (1720)	
Barium fluoride	BaF_2	0.25	15	1550 (2330)	82
Potassium chloride	KCl	0.21	30	1060 (1450)	

*T_m = melting temperature.   †H_K = Knoop hardness.

Plastics. The most significant properties of plastics (compared to glass) are their softness and molecular complexity. In the visible spectrum they can be used for relatively cheap lenses but must be protected from abrasion. They can be used in the infrared at relatively short and very long wavelengths. The resonance of the molecular bond between carbon and hydrogen prevents the use of most plastics beyond about 3 μm in all but the thinnest membranes. Polyethylene and polystyrene are particularly good choices for thin window coverings through the infrared.

Materials for infrared and ultraviolet. The lack of a good transparent glass for these spectral regions caused the long and continuing search for appropriate substitutes, which included single crystals, polycrystalline compacts, and plastics. These are listed in Table 3 along with some of their more important properties. The nu value or Abbe number is meaningless for materials that are used for infrared or ultraviolet work, but a generalized coefficient can be used $(n - 1)/(\Delta n/\Delta \lambda)$, where $\Delta n/\Delta \lambda$ is the slope of the refractive index versus wavelength curve in the useful region well between the absorption bands. Some representative materials are shown in **Fig. 2** for the main infrared regions, and the data of Fig. 1 are contained within the shaded area of Fig. 2. In Fig. 2, $\nu_{1.0-2.7} = [n(1.85\ \mu m) - 1]/[n(1.0\ \mu m) - n(2.7\ \mu m)]$, and $\nu_{3.0-5.5}$ and ν_{8-15} are similarly defined.

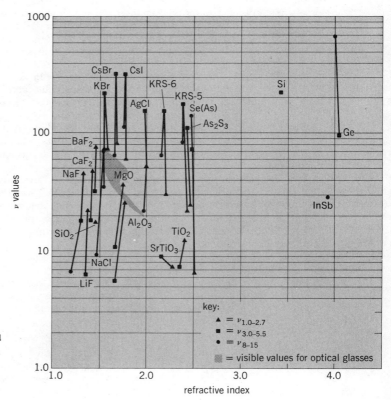

Fig. 2. Reciprocal dispersion of some infrared optical materials. (*After W. Wolfe, ed., Handbook of Military Infrared Technology, Government Printing Office, 1966*)

For the ultraviolet region of the spectrum, the main refractors are lithium fluoride (LiF), calcium fluoride (CaF_2), and silicon dioxide (SiO_2); the shortest wavelength attainable is with lithium fluoride and is about 0.12 μm. The region short of this is known as the vacuum ultraviolet, where the atmosphere and all known materials are opaque. In the very far infrared (which usually means beyond about 100 μm) most of the alkali halides "open up" and transmit reasonably well.

Many plastics are useful if not thicker than several thousandths of an inch. These are used primarily as windows in the form of membranes to maintain a prescribed interior atmosphere.

Cautions. The required quality of an optical material depends, of course, on its use. The properties that should be considered in general include the transmission, refractive index, melting or softening temperature, strength, hardness, possibility of good surface finish, stain, scratch, internal scatter, homogeneity, and strength. The influence of each of these properties can be complex, and the degree of importance varies with the application.

Bibliography. D. D. Gray (ed.), *American Institute of Physics Handbook*, 3d ed., 1972; *The Infrared Handbook*, Office of Naval Research, 1978; R. Kingslake (ed.), *Applied Optics and Optical Engineering*, vol. 1, 1965; Optical Society of America, *Handbook of Optics*, 1978.

OPTICAL ACTIVITY
VINCENT MADISON

The effect of asymmetric compounds on polarized light. To exhibit this effect, a molecule must be nonsuperimposable on its mirror image, that is, must be related to its mirror image as the right hand is to the left hand. An optically active compound and its mirror image are called enantiomers or optical isomers. Enantiomers differ only in their geometric arrangements; they have identical chemical and physical properties. The right-handed and left-handed forms of a molecule can be distinguished only by their optical activity or by their interactions with other asymmetric molecules. Optical activity can be used to probe other aspects of molecular geometry, as well as to identify which enantiomer is present and its purity.

As an example of optical isomers, consider tartaric acid (**Fig. 1**), which was one of the first synthetic molecules to be separated into its enantiomers. In this case the asymmetry of each isomer is magnified when trillions of molecules form a crystal; two types of asymmetric crystals are formed.

The physical basis of optical activity is the differential interaction of asymmetric substances with left versus right circularly polarized light. If solids and substances in strong magnetic fields are excluded, optical activity is an intrinsic property of the molecular structure and is one of the best methods of obtaining structural information from a sample in which the molecules are randomly oriented. The relationship between optical activity and molecular structure results from the interaction of polarized light with electrons in the molecule. Thus the molecular groups that contribute most directly to optical activity are those that have mobile electrons which can interact with light. Such groups are called chromophores, since their absorption of light is responsible for the color of objects. For example, the chlorophyll chromophore makes plants green. SEE FARADAY EFFECT; MAGNETOOPTICS; POLARIZED LIGHT.

Methods of measurement. Optical activity is measured by two methods, optical rotation and circular dichroism.

Optical rotation. This method depends on the different velocities of left and right circularly polarized light beams in the sample. The velocities are not measured directly, but both beams are passed through the sample simultaneously. This is equivalent to using plane-polarized light.

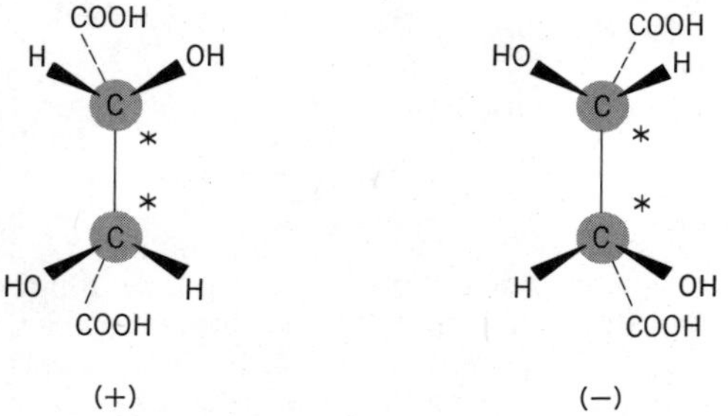

Fig. 1. Enantiomers of tartaric acid.

The differing velocities of the left and right circularly polarized components yield a rotation of the plane of polarization. A polarimeter for observing optical rotation consists of a light source, a fixed polarizer, a sample compartment, and a rotatable polarizer. A cell containing solvent is placed between the polarizers, and one of them is adjusted to be perpendicular to the other, excluding the passage of light. The solvent in the cell is then replaced by a solution of the sample, and the polarizer is rotated to again exclude passage of light. The optical rotation a is the number of degrees the polarizer was rotated. A positive or negative sign indicates the direction of rotation. Enantiomers have rotations of equal magnitude, but opposite signs. The optical rotation depends on the substance, solvent, concentration, cell path length, wavelength of the light, and temperature. Standardized specific rotations $[\alpha]$ are reported as defined in Eq. (1), where T is the temper-

$$[\alpha]_\lambda^T = \frac{a}{cl} \qquad (1)$$

ature (°C), λ the wavelength (often the orange sodium D line), l the cell path length in decimeters, and c the concentration in grams per milliliter. Alternatively, M_ϕ is defined by normalizing to the rotation for a 1-molar solution, Eq. (2), where M_ϕ is the molar rotation and MW the molecular weight. For polymers, the mean residue rotation, m_ϕ, may be defined by the right side of Eq. (2)

$$M_\phi = [\alpha]_\lambda^T \text{MW}/100 \qquad (2)$$

by using the mean residue (monomer unit) weight for MW. The variation of optical rotation with wavelength is known as optical rotatory dispersion (ORD).

Circular dichroism. Circular dichroism (CD) is the difference in absorption of left and right circularly polarized light. Since this difference is about a millionth of the absorption of either polarization, special techniques are needed to determine it accurately. Circular dichroism spectrometers consist of a light source, a monochromator to select a single wavelength, a modulator to produce circularly polarized light, a sample compartment, a phototube to detect transmitted light, and associated electronic components. The modulator rapidly switches (typically 50,000 times per second) between left and right circular polarization of the light beam. The absorption of an optically inactive sample is independent of polarization, so that the light intensity at the phototube is constant; thus a constant direct current is generated. The absorption of an optically active sample depends on the polarization, so that the light intensity at the phototube varies at the frequency of the modulator; thus an alternating current is generated. The circular dichroism is proportional to the amplitude of the alternating current. The proportionality constant is determined through calibration by using a compound of known circular dichroism.

Circular dichroism is reported as a difference in absorption, Eq. (3), or as an ellipticity (a measure of the elliptical polarization of the emergent beam), Eq. (4), for a 1-molar solution, where

$$\Delta\epsilon = \epsilon_L - \epsilon_R = (A_L - A_R)/(c'l') \qquad (3) \qquad M_\theta = 3300\Delta\epsilon \qquad (4)$$

ϵ is the extinction coefficient, A is the absorbance [log (I_0/I)], subscripts L and R indicate left or right circular polarization, c' is the concentration in moles per liter, l' is the path length in centimeters, I_0 and I are the light intensities in the absence and presence of the sample, respectively, and M_θ is the molar ellipticity. Either $\Delta\epsilon$ or ellipticity, m_θ, may be expressed per residue by making c' the concentration of residues (monomer units). As in the case of optical rotation, enantiomers have circular dichroism spectra of equal magnitude but opposite signs.

Variation with wavelength. Optical rotation and circular dichroism are two manifestations of the same interactions between polarized light and molecules. They are related by a mathematical transformation. An important difference between the two measurements is the way in which they vary with wavelength. Optical rotation extends to wavelengths far from any absorption of light. Thus colorless substances still have significant optical rotation at the sodium line. However, all groups which absorb light (chromophores) contribute at all wavelengths, and it can be difficult to extract the contribution of a single group. On the other hand, circular dichroism is confined to the narrow absorption band of each chromophore. Thus it is easier to determine the contribution of individual chromophores, information vital to structural analysis. SEE COTTON EFFECT; DICHROISM; ROTATORY DISPERSION.

Correlation with molecular structure. In synthesizing enantiomers, chemists focus on an asymmetric center, that is, a locus which imparts asymmetry to the whole molecule. A common asymmetric center is a tetrahedral carbon atom with four different groups attached, such as the carbons marked with asterisks in tartaric acid (Fig. 1). However, in correlating optical activity with molecular structure, the focus is on the three-dimensional arrangement of the chromophores which interact most strongly with light.

As examples, consider the nucleoside adenosine and its dimer (**Fig. 2**). The most mobile electrons are in the aromatic ring system, the chromophore (Fig. 2a). The electrons in the sugar ribose are more tightly bound and interact less strongly with visible and ultraviolet light. However, all the asymmetric centers are in the ribose part of the molecule. For adenosine, light interacting with the aromatic chromophore is only weakly influenced by the asymmetric centers in ribose, so that small circular dichroism bands are observed (Fig. 2c).

In the covalently linked dimer of adenosine, the observed circular dichroism bands are about 10 times larger than those of the monomer. In the 240- to 300-nm region of the spectra (Fig. 2c), two bands are observed for the dimer, but only one for the monomer. This indicates strong interaction of the two aromatic chromophores, and hence their close proximity in the dimer. Analysis of the circular dichroism spectra expected for various arrangements of the two chromophores, as well as other types of experimental data, indicates that the aromatic rings are stacked (Fig. 2b). The asymmetric centers in ribose cause the formation of the stacked arrangement shown rather than its mirror image.

Stacking of aromatic rings, as exemplified by the adenosine dimer, is a common feature of nucleic acid polymers (deoxyribonucleic acid and ribonucleic acid) isolated from biological sources. Slight differences in the stacking geometry gives each of these polymers a characteristic circular dichroism spectrum. Alterations in the stacking arrangement caused by some pharmacologically active agents can be detected through alterations in the circular dichroism spectra. These structural changes may in turn be related to the pharmacological action.

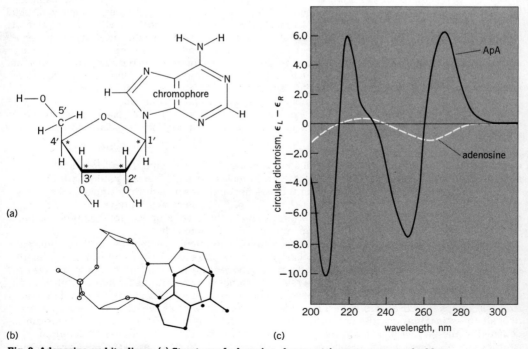

Fig. 2. Adenosine and its dimer. (a) Structure of adenosine. Asymmetric centers are marked by an asterisk. (b) Stacked arrangement of adenosine dimer (ApA). The 3' carbon of one adenosine is linked to the 5' carbon of the other by a phosphate group. (c) Circular dichroism spectra of ApA and adenosine at neutral pH in aqueous solution at room temperature.

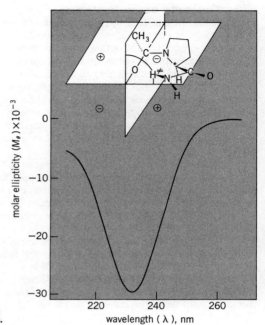

Fig. 3. Folded arrangement of *L*-proline derivative (*N*-acetyl-*L*-proline-amide) and its circular dichroism spectrum in *p*-dioxane solution at room temperature.

A derivative of the amino acid proline (**Fig. 3**) can be used to illustrate another way in which optical activity depends on molecular structure. In this molecule only the OCN group (amide chromophore) which is in the horizontal plane of the drawing and the hydrogen which is marked $H^{\neq}$ need be considered. By forming the N-H bond, $H^{\neq}$ acquires a charge of about $+\frac{1}{3}$ electron. It has been predicted that such a positive charge will perturb the motion of the electrons in the amide chromophore in a manner which will produce a negative circular dichroism band when the charge is above the plane of the amide group and to the right of the oxygen. Only for the arrangement shown is the magnitude of the circular dichroism band expected to be as large as observed. Furthermore, it has been shown that there will be no circular dichroism if $H^{\neq}$ is in either of the two planes shown, and that for $H^{\neq}$ in adjacent quadrants the sign of the circular dichroism band alternates (Fig. 3). For this compound, reflection through the horizontal plane will generate the enantiomer. This would place $H^{\neq}$ in the lower right quadrant and generate a positive circular dichroism band with magnitude equal to that of Fig. 3.

Bibliography. L. D. Barron, *Molecular Light Scattering and Optical Activity*, 1983; V. A. Bloomfield, D. M. Crothers, and I. Tinoco, Jr., *Physical Chemistry of Nucleic Acids*, 1974; S. F. Mason, *Molecular Optical Activity and the Chiral Discriminations*, 1982; J. A. Schellman, Symmetry rules for optical rotation, *Accounts Chem. Res.*, 1:144–155, 1968; I. Tinoco, Jr., and C. R. Cantor, Application of optical rotatory dispersion and circular dichroism to the study of biopolymers, *Meth. Biochem. Anal.*, 18:81–203, 1970; R. W. Woody, Optical rotatory properties of biopolymers, *J. Polym. Sci. Macromol. Rev.*, 12:181–321, 1977.

ROTATORY DISPERSION
Bruce H. Billings

A term used to describe the change in rotation as a function of wavelength experienced by linearly polarized light as it passes through an optically active substance. *See* Optical activity; Polarized light.

Optically active materials. Substances that are optically active can be grouped into two classes. In the first the substances are crystalline and the optical activity depends on the arrangement of nonoptically active molecular units. When these crystals are dissolved or melted, the resulting liquid is not optically active. In the second class the optical activity is a characteristic of the molecular units themselves. Such materials are optically active as liquids or solids. A typical substance in the first category is quartz. This crystal is optically active and rotates the plane of polarization by an amount which depends on the direction in which the light is propagated with respect to the optic axis. Along the axis the rotation is 29.73°/mm (755°/in.) for light of wavelength 508.6 nanometers. At other angles the rotation is less and is obscured by the crystal's linear birefringence. Molten quartz, or fused quartz, is isotropic. Turpentine is a typical material of the second class. It gives rotation of $-37°$ in a 10-cm (3.94-in.) length for the sodium D lines. SEE CRYSTAL OPTICS.

Reasons for variation. In all materials the rotation varies with wavelength. The variation is caused by two quite different phenomena. The first accounts in most cases for the majority of the variation in rotation and should not strictly be termed rotatory dispersion. It depends on the fact that optical activity is actually circular birefringence. In other words, a substance which is optically active transmits right circularly polarized light with a different velocity from left circularly polarized light.

Any type of polarized light can be broken down into right and left components. Let these components be R and L. The lengths of the rotating light vectors will then be $R/\sqrt{2}$ and $L/\sqrt{2}$. At $t=0$, the R vector may be at an angle ψ_r with the x axis and the L vector at an angle ψ_l. Since the vectors are rotating at the same velocity, they will coincide at an angle β which bisects the difference as in Eq. (1). If $R=L$, the sum of these two waves will be linearly polarized light vibrating at an angle α to the axes given by Eq. (2).

$$\beta = \frac{\psi_r + \psi_l}{2} \quad (1) \qquad \alpha = \frac{\psi_r - \psi_l}{2} \quad (2)$$

If, in passing through a material, one of the circularly polarized beams is propagated at a different velocity, the relative phase between the beams will change in accordance with Eq. (3), where d is the thickness of the material, λ is the wavelength, and n_r and n_l are the indices of refraction for right and left circularly polarized light. The polarized light incident at an angle α has, according to this equation, been rotated an angle given by Eq. (4).

$$\psi'_r - \psi'_l = \frac{2\pi d}{\lambda}(n_r - n_l) + \psi_r - \psi_l \quad (3) \qquad \gamma = \frac{\pi d}{\lambda}(n_r - n_l) \quad (4)$$

This shows that the rotation would depend on wavelength, even in a material in which n_r and n_l were constant and which thus had no circular dipersion. It is for this reason that the term rotatory dispersion is perhaps ill-defined in much of the literature.

In addition to this pseudodispersion which depends on the material thickness, there is a true rotatory dispersion which depends on the variation with wavelength of n_r and n_l.

From Eq. (4) it is possible to compute the circular birefringence for various materials. This quantity is of the order of magnitude of 10^{-8} for many solutions and 10^{-5} for crystals. It is 10^{-1} for linear birefringent crystals.

Bibliography. L. D. Barron, *Molecular Light Scattering and Optical Activity*, 1983; P. Crabble, *ORD and CD in Chemistry and Biochemistry*, 1972; T. M. Lowry, *Optical Rotatory Power*, 1935, reprint 1964; S. F. Mason, *Molecular Optical Activity and the Chiral Discriminations*, 1982; G. Snatzke (ed.), *Optical Rotatory Dispersion and Circular Dichroism in Organic Chemistry*, 1976.

COTTON EFFECT
ALBERT MOSCOWITZ

The characteristic wavelength dependence of the optical rotatory dispersion curve or the circular dichroism curve or both in the vicinity of an absorption band.

When an initially plane-polarized light wave traverses an optically active medium, two prin-

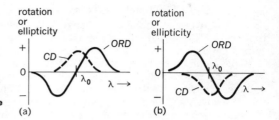

Fig. 1. Behavior of the ORD and CD curves in the vicinity of an absorption band at wavelength λ_0 (idealized). (a) Positive Cotton effect. (b) Negative Cotton effect.

cipal effects are manifested: a change from planar to elliptic polarization, and a rotation of the major axis of the ellipse through an angle relative to the initial direction of polarization. Both effects are wavelength dependent. The first effect is known as circular dichroism, and a plot of its wavelength (or frequency) dependence is referred to as a circular dichroism (CD) curve. The second effect is called optical rotation and, when plotted as a function of wavelength, is known as an optical rotatory dispersion (ORD) curve. In the vicinity of absorption bands, both curves take on characteristic shapes, and this behavior is known as the Cotton effect, which may be either positive or negative (**Fig. 1**). There is a Cotton effect associated with each absorption process, and hence a partial CD curve or partial ORD curve is associated with each particular absorption band or process. SEE POLARIZED LIGHT; ROTATORY DISPERSION.

Measurements. Experimental results are commonly reported in either of two sets of units, termed specific and molar (or molecular). The specific rotation [α] is the rotation in degrees produced by a 1-decimeter path length of material containing 1 g/ml of optically active substance, and the specific ellipticity θ is the ellipticity in degrees for the same path length and same concentration. Molar rotation [φ] (sometimes [M]) and molar ellipticity [θ] are defined by Eqs. (1) and (2). For comparisons among different compounds, the molar quantities are more useful, since they allow direct comparison on a mole-for-mole basis.

$$[\varphi] = [\alpha]M/100 \qquad (1) \qquad\qquad [\theta] = \theta M/100 \qquad (2)$$

The ratio of the area under the associated partial CD curve to the wavelength of the CD maximum is a measure of the rotatory intensity of the absorption process. Moreover, for bands appearing in roughly the same spectral region and having roughly the same half-width (**Fig. 2**), the peak-to-trough rotation of the partial ORD curve is roughly proportional to the wavelength-weighted area under the corresponding partial CD curve. In other words, relative rotatory intensities can be gaged from either the pertinent partial ORD curves or pertinent partial CD curves. A convenient quantitative measure of the rotatory intensity of an absorption process is the rotational strength. The rotational strength R_i of the ith transition, whose partial molar CD curve is $[\theta_i(\lambda)]$, is given by relation (3).

$$R_i \approx 6.96 \times 10^{-43} \int_0^\infty \frac{[\theta_i(\lambda)]}{\lambda} d\lambda \qquad (3)$$

Molecular structure. The rotational strengths actually observed in practice vary over quite a few orders of magnitude, from $\sim 10^{-38}$ down to 10^{-42} cgs and less; this variation in

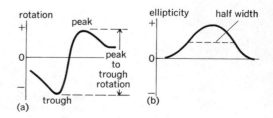

Fig. 2. Curves used to determine relative rotatory intensities. (a) Partial ORD curve. (b) Partial CD curve.

magnitude is amenable to stereochemical interpretation. In this connection it is useful to classify optically active chromophores, which are necessarily dissymmetric, in terms of two limiting types: the inherently dissymmetric chromophore, and the inherently symmetric but dissymmetrically perturbed chromophore. SEE OPTICAL ACTIVITY.

A symmetric chromophore is one whose inherent geometry has sufficiently high symmetry so that the isolated chromophoric group is superimposable on its mirror image, for example, the carbonyl group >C=O. The transitions of such a chromophore can become optically active, that is, exhibit a Cotton effect, only when placed in a dissymmetric molecular environment. Thus, in symmetrical formaldehyde, $H_2C=O$, the carbonyl transitions are optically inactive; in ketosteroids, where the extrachromophoric portion of the molecule is dissymmetrically disposed relative to the symmetry planes of the >C=O group, the transitions of the carbonyl group exhibit Cotton effects. In such instances the signed magnitude of the rotational strength will depend both upon the chemical nature of the extrachromophoric perturbing atoms and their geometry relative to that of the inherently symmetric chromophore. In a sense, the chromophore functions as a molecular probe for searching out the chemical dissymmetries in the extrachromophoric portion of the molecule.

The type of optical activity just described is associated with the presence of an asymmetric carbon (or other) atom in a molecule. The asymmetric atom serves notice to the effect that, if an inherently symmetric chromophore is present in the molecule, it is almost assuredly in a dissymmetric environment, and hence it may be anticipated that its erstwhile optically inactive transitions will exhibit Cotton effects. Moreover, the signed magnitude of the associated rotational strengths may be interpreted in terms of the stereochemistry of the extrachromophoric environment, as compared with that of the chromophore. But an asymmetric atom is not essential for the appearance of optical activity. The inherent geometry of the chromophore may be of sufficiently low symmetry so tht the isolated chromophore itself is chiral, that is, not superimposable on its mirror image, for example, in hexahelicene.

In such instances the transitions of the chromophore can manifest optical activity even in the absence of a dissymmetric environment. In addition, it is very often true that the magnitudes of the rotational strengths associated with inherently dissymmetric chromophores will be one or more orders of magnitude greater ($\sim 10^{-38}$ cgs, as opposed to $< 10^{-39}$ cgs) than those associated with inherently symmetric chromophores. Hence, in the spectral regions of the transitions of the inherently dissymmetric chromophore, it will be the sense of handedness of the inherently dissymmetric chromophore itself that will determine the sign of the rotational strength, rather than perturbations due to any dissymmetric environment in which the inherently dissymmetric chromophore may be situated.

The sense of handedness of an inherently dissymmetric chromophore may be of considerable significance in determining the absolute configuration or conformations of the entire molecule containing that chromophore. Accordingly, the absolute configuration or conformation can often be found by focusing attention solely on the handedness of the chromophore itself. For example, in the chiral molecule shown in **Fig. 3** there is a one-to-one correspondence between the sense of helicity of the nonplanar diene chromophore present and the absolute configuration at the asymmetric carbon atoms. Hence there exists a one-to-one correspondence between the handedness of the diene and the absolute configuration of the molecule. Since it is known that a right-handed diene helix (**Fig. 4**) associates a positive rotational strength with the lowest diene singlet

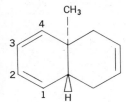

Fig. 3. Structural formula of (+)-*trans*-9-methyl-1,4,9,10-tetrahydronaphthalene.

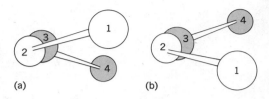

Fig. 4. Schematic representation of the twisted diene chromophore showing the two possible handednesses; the numbering is as indicated in Fig. 3. (*a*) Right-handed. (*b*) Left-handed.

transition in the vicinity of 260 nanometers, by examination of the pertinent experimental Cotton effect (positive), the absolute configuration of the molecule is concluded to be as shown.

Other examples of inherently dissymmetric chromophores are provided by the helical secondary structures of proteins and polypeptides. Here the inherent dissymmetry of the chromophoric system arises through a coupling of the inherently symmetric monomers, which are held in a comparatively fixed dissymmetric disposition relative to each other through internal hydrogen bonding. The sense of helicity is then related to the signs of the rotational strengths of the coupled chromophoric system. The destruction of the hydrogen bonding destroys the ordered dissymmetric secondary structure, and there is a concomitant decrease in the magnitude of the observed rotational strengths.

Bibliography. L. D. Barron, *Molecular Light Scattering and Optical Activity*, 1983; S. F. Mason, *Molecular Optical Activity and the Chiral Discriminations*, 1982; *Proc. Roy. Soc. London*, ser. A, 297:1–172, 1976; L. Velluz et al., *Optical Circular Dichroism: Principles, Measurements, and Applications*, 1969.

CRYSTAL OPTICS
Bruce H. Billings

The study of the propagation of light, and associated phenomena, in crystalline solids. The propagation of light in crystals can actually be so complicated that not all the different phenomena are yet completely understood, and not all theoretically predicted phenomena have been demonstrated experimentally.

For a simple cubic crystal the atomic arrangement is such that in each direction through the crystal the crystal presents the same optical appearance. The atoms in anisotropic crystals are closer together in some planes through the material than in others. In anisotropic crystals the optical characteristics are different in different directions. In classical physics the progress of an electromagnetic wave through a material involves the periodic displacement of electrons. In anisotropic substances the forces resisting these displacements depend on the displacement direction. Thus the velocity of a light wave is different in different directions and for different states of polarization. The absorption of the wave may also be different in different directions. *See* DICHROISM; TRICHROISM.

In an isotropic medium the light from a point source spreads out in a spherical shell. The light from a point source embedded in an anisotropic crystal spreads out in two wave surfaces, one of which travels at a faster rate than the other. The polarization of the light varies from point to point over each wave surface, and in any particular direction from the source the polarization of the two surfaces is opposite. The characteristics of these surfaces is opposite. The characteristics of these surfaces can be determined experimentally by making measurements on a given crystal.

For a transparent crystal the theoretical optical behavior is well enough understood so that only a few measurements need to be made in order to predict the behavior of a light beam passing through the crystal in any direction. It is important to remember that the velocity through a crystal is not a function of position in the crystal but only of the direction through the lattice. For information closely related to the ensuing discussion *see* POLARIZED LIGHT. *See also* REFRACTION OF WAVES.

Index ellipsoid. In the most general case of a transparent anisotropic medium, the dielectric constant is different along each of three orthogonal axes. This means that when the light

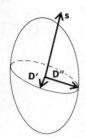

Fig. 1. Index ellipsoid, showing construction of directions of vibrations of D vectors belonging to a wave normal s. (*After M. Born and E. Wolf, Principles of Optics, 6th ed., Pergamon Press, 1980*)

vector is oriented along each direction, the velocity of light is different. One method for calculating the behavior of a transparent anisotropic material is through the use of the index ellipsoid, also called the reciprocal ellipsoid, optical indicatrix, or ellipsoid of wave normals. This is the surface obtained by plotting the value of the refractive index in each principal direction for a linearly polarized light vector lying in that direction (**Fig. 1**). The different indices of refraction, or wave velocities associated with a given propagation direction, are then given by sections through the origin of the coordinates in which the index ellipsoid is drawn. These sections are ellipses, and the major and minor axes of the ellipse represent the fast and slow axes for light proceeding along the normal to the plane of the ellipse. The length of the axes represents the refractive indices for the fast and slow wave, respectively. The most asymmetric type of ellipsoid has three unequal axes. It is a general rule in crystallography that no property of a crystal will have less symmetry than the class in which the crystal belongs. In other words, if a property of the crystal had lower symmetry, the crystal would belong in a different class. Accordingly, there are many crystals which, for example, have four- or sixfold rotation symmetry about an axis, and for these the index ellipsoid cannot have three unequal axes but is an ellipsoid of revolution. In such a crystal, light will be propagated along this axis as though the crystal were isotropic, and the velocity of propagation will be independent of the state of polarization. The section of the index ellipsoid at right angles to this direction is a circle. Such crystals are called uniaxial and the mathematics of their optical behavior is relatively straightforward.

Ray ellipsoid. The normal to a plane wavefront moves with the phase velocity. The Huygens wavelet, which is the light moving out from a point disturbance, will propagate with a ray velocity. Just as the index ellipsoid can be used to compute the phase or wave velocity, so can a ray ellipsoid be used to calculate the ray velocity. The length of the axes of this ellipsoid is given by the velocity of the linearly polarized ray whose electric vector lies in the axis direction.

The ray ellipsoid in the general case for anisotropic crystal is given by Eq. (1), where α, β,

$$\alpha^2 x^2 + \beta^2 y^2 + \gamma^2 z^2 = 1 \qquad (1)$$

and γ are the three principal indices of refraction and where the velocity of light in a vacuum is taken to be unity. From this ellipsoid the ray velocity surfaces or Huygens wavelets can be calculated as just described. These surfaces are of the fourth degree and are given by Eq. (2). In the uniaxial case $\alpha = \beta$ and Eq. (2) factors into Eqs. (3). These are the equations of a sphere and an

$$(x^2 + y^2 + z^2)\left(\frac{x^2}{\alpha^2} + \frac{y^2}{\beta^2} + \frac{z^2}{\gamma^2}\right) - \frac{1}{\alpha^2}\left(\frac{1}{\beta^2} + \frac{1}{\gamma^2}\right)x^2$$
$$- \frac{1}{\beta^2}\left(\frac{1}{\gamma^2} + \frac{1}{\alpha^2}\right)y^2 - \frac{1}{\gamma^2}\left(\frac{1}{\alpha^2} + \frac{1}{\beta^2}\right)z^2 + \frac{1}{\alpha^2\beta^2\gamma^2} = 0 \qquad (2)$$

$$x^2 + y^2 + z^2 = \frac{1}{\alpha^2} \qquad\qquad \gamma^2(x^2 + y^2) + \alpha^2 z^2 = 1 \qquad (3)$$

ellipsoid. The z axis of the ellipsoid is the optic axis of the crystal.

The refraction of a light ray on passing through the surface of an anisotropic uniaxial crystal can be calculated with Huygens wavelets in the same manner as in an isotropic material. For the ellipsoidal wavelet this results in an optical behavior which is completely different from that normally associated with refraction. The ray associated with this behavior is termed the extraordinary ray. At a crystal surface where the optic axis is inclined at an angle, a ray of unpolarized light incident normally on the surface is split into two beams: the ordinary ray, which proceeds through the surface without deviation; and the extraordinary ray, which is deviated by an angle determined by a line drawn from the center of one of the Huygens ellipsoidal wavelets to the point at which the ellipsoid is tangent to a line parallel to the surface. The construction is shown in **Fig. 2**. The two beams are oppositely linearly polarized. When the incident beam is inclined at an

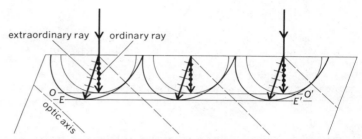

Fig. 2. Huygens construction for a plane wave incident normally on transparent calcite. If one proceeds to find the common tangents to the secondary wavelets shown, the two plane waves labeled OO' and EE' are obtained. (After F. A. Jenkins, *Fundamentals of Optics*, 4th ed., McGraw-Hill, 1976)

angle ϕ to the normal, the ordinary ray is deviated by an amount determined by Snell's law of refraction, the extraordinary ray by an amount which can be determined in a manner similar to that of the normal incidence case already described. The plane wavefront in the crystal is first found by constructing Huygens wavelets as shown in **Fig. 3**. The line from the center of the wavelet to the point of tangency gives the ray direction and velocity.

The relationship between the normal to the wavefront and the ray direction can be calculated algebraically in a relatively straightforward fashion. The extraordinary wave surface, or Huygens wavelet, is given by Eq. (3), which can be rewritten as Eq. (4), where ϵ is the extraordinary

$$\epsilon^2(x^2 + y^2) + \omega^2 z^2 = 1 \tag{4}$$

index of refraction and ω the ordinary index. A line from the center of this ellipsoid to a point (x_1, y_1, z_1) on the surface gives the velocity of a ray in this direction. The wave normal correspond-

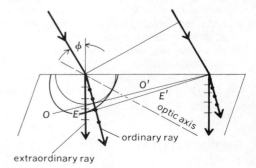

Fig. 3. Huygens construction when the optic axis lies in the plane of incidence. (After F. A. Jenkins, *Fundamentals of Optics*, 4th ed., McGraw-Hill, 1976)

ing to this ray is found by dropping a perpendicular from the center of the ellipsoid to the plane tangent at the point (x_1, y_1, z_1). For simplicity consider the point in the plane $y = 0$. The tangent at the point $x_1 z_1$ is given by Eq. (5). The slope of the normal to this line is given by Eq. (6). The tangent of the angle between the optic axis and the wave normal is the reciprocal of this number, as shown in Eq. (7), where ψ is the angle between the ray and the optic axis. The difference

$$\epsilon^2 x x_1 + \omega^2 z z_1 = 1 \quad (5) \qquad \frac{z}{x} = \frac{\omega^2 z_1}{\epsilon^2 x_1} \quad (6) \qquad \tan \varphi = \frac{\epsilon^2 x_1}{\omega^2 z_1} = \frac{\epsilon^2}{\omega^2} \tan \psi \quad (7)$$

between these two angles, π, can be calculated from Eq. (8). This quantity is a maximum when Eq. (9) holds.

$$\tan \tau = \frac{\tan \varphi - \tan \psi}{1 + \tan \varphi \tan \psi} \quad (8) \qquad \tan \varphi = \pm \frac{\epsilon}{\omega} \quad (9)$$

One of the first doubly refracting crystals to be discovered was a transparent variety of calcite called Iceland spar. This uniaxial crystal cleaves into slabs in which the optic axis makes an angle of 45° with one pair of surfaces. An object in contact with or a few inches from such a slab will thus appear to be doubled. If the slab is rotated about a normal to the surface, one image rotates about the other. For the sodium D lines at 589 mμ the indices for calcite are given by Eqs. (10). From these, the maximum angle τ_{max} between the wave normal and the ray direction is

$$\epsilon = 1.486 \qquad \omega = 1.659 \quad (10)$$

computed to be 6°16′. The wave normal at this value makes an angle of 41° 52′ with the optic axis. This is about equal to the angle which the axis makes with the surface in a cleaved slab. Accordingly, the natural rhomb gives nearly the extreme departure of the ray direction from the wave normal.

Interference in polarized light. One of the most interesting properties of plates of crystals is their appearance in convergent light between pairs of linear, circular, or elliptical polarizers. An examination of crystals in this fashion offers a means of rapid identification. It can be done with extremely small crystals by the use of a microscope in which the illuminating and viewing optical systems are equipped with polarizers. Such a polarizing microscope is a common tool for the mineralogist and the organic chemist.

In convergent light the retardation through a birefringent plate is different for each direction. The slow and fast axes are also inclined at a different angle for each direction. Between crossed linear polarizers the plate will appear opaque at those angles for which the retardation is an integral number of waves. The locus of such points will be a series of curves which represent the characteristic interference pattern of the material and the angle at which the plate is cut. In order to calculate the interference pattern, it is necessary to know the two indices associated with different angles of plane wave propagation through the plate. This can be computed from the index ellipsoid. SEE INTERFERENCE OF WAVES.

Ordinary and extraordinary indices. For the uniaxial crystal there is one linear polarization direction for which the wave velocity is always the same. The wave propagated in this direction is called the ordinary wave. This can be seen directly by inspection of the index ellipsoid. Since it is an ellipsoid of revolution, each plane passing through the center will produce an ellipse which has one axis equal to the axis of the ellipsoid. The direction of polarization will always be at right angles to the plane of incidence and the refractive index will be constant. This constant index is called the ordinary index. The extraordinary index will be given by the other axis of the ellipse and will depend on the propagation direction. When the direction is along the axis of the ellipsoid, the extraordinary index will equal the ordinary index.

From the equation of the ellipsoid, Eq. (11), one can derive the expression for the ellipse and in turn Eq. (12) for the extraordinary index n_e as a function of propagation direction. Since

$$\frac{x^2}{\omega^2} + \frac{y^2}{\omega^2} + \frac{z^2}{\epsilon^2} = 1 \quad (11) \qquad n_e = \frac{\omega \epsilon}{(\epsilon^2 \cos^2 r + \omega^2 \sin^2 r)^{1/2}} \quad (12)$$

the ellipsoid is symmetrical it is necessary to define this direction only with respect to the ellipsoid axis. Here r is the angle in the material between the normal to the wavefront and the axis of the

INTERACTION OF LIGHT WITH MATTER

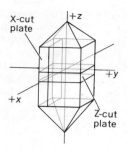

Fig. 4. X-cut and Z-cut plates in $NH_4H_2PO_4$ crystal.

ellipsoid, n_e is the extraordinary index associated with this direction, ω is the ordinary index, and ϵ is the maximum value of the extraordinary index (usually referred to simply as the extraordinary index).

A slab cut normal to the optic axis is termed a Z-cut or C-cut plate (**Fig. 4**). When such a plate is placed between crossed linear polarizers, such as Nicol prisms, it gives a pattern in monochromatic light shown in **Fig. 5**. The explanation of this pattern is obtained from the equations given for the indices of refraction. To a first approximation the retardation for light of wavelength λ passing through the plate at an angle r can be written as Eq. (13), where d is the

$$\Gamma = \frac{(n_e - \omega)d}{\lambda \cos r} \qquad (13)$$

thickness of the plate. The axis of the equivalent retardation plate at an angle r will be in a plane containing the optic axis of the plate and the direction of light propagation. Wherever Γ is a whole number of waves, the light leaving the plate will have the same polarization as the incident light. The ordinary index is constant. The extraordinary index is a function of r alone, as seen from Eq. (12). Accordingly, the locus of direction of constant whole wave retardation will be a series of cones. If a uniform white light background is observed through such a plate, a series of rings of constant whole wave retardation will be seen. Since the retardation is a function of wavelength, the rings will appear colored. The innermost ring will have the least amount of color. The outermost ring will begin to overlap and disappear as the blue end of the spectrum for one ring covers

Fig. 5. Interference figure from fluorspar that has been cut perpendicular to the optic axis and placed between crossed Nicol prisms. (*After M. Born and E. Wolf, Principles of Optics, 6th ed., Pergamon Press, 1980*)

the red end of its neighbor. The crystal plate has no effect along the axes of the polarizers since here the light is polarized along the axis of the equivalent retardation plate. The family of circles is thus bisected by a dark cross. When the crystal is mounted between like circular polarizers, the dark cross is not present and the center of the system of rings is clear. During World War II crystal plates were used in this fashion as gunsights. The rings appear at infinity even when the plate is close to the eye. Lateral movement of the plate causes no angular motion of the ring system. The crystal plate could be rigidly fastened to a gun mount and adjusted so as to show at all times the direction in space in which the gun was pointing.

Uniaxial crystal plates in which the axis lies in the plane of the plate are termed X-cut or Y-cut. When such plates are observed between crossed polarizers, a pattern of hyperbolas is observed.

When crystal plates are combined in series, the patterns become much more complex. In addition to the fringes resulting from the individual plates, a system of so-called moiré fringes appears.

Negative and positive crystals. In calcite the extraordinary wave travels faster than the ordinary wave. Calcite and other materials in which this occurs are termed negative crystals. In positive crystals the extraordinary wave travels slower than the ordinary wave. In the identification of uniaxial minerals one of the steps is the determination of sign. This is most easily demonstrated in a section cut perpendicular to the optic axis. Between crossed linear polarizers the pattern in convergent light, as already mentioned, is a series of concentric circles which are bisected by a dark cross. When a quarter-wave plate is inserted between one polarizer and the crystal, with its axis at 45° to the polarizing axis, the dark rings are displaced outward or inward in alternate quadrants. If the rings are displaced outward along the slow axis of the quarter-wave plate, the crystal is positive. If the rings are displaced inward along the axis, then the crystal is negative.

If a Z-cut plate of a positive uniaxial crystal is put in series with a similar plate of a negative crystal, it is possible to cancel the birefringence so that the combination appears isotropic.

Biaxial crystals. In crystals of low symmetry the index ellipsoid has three unequal axes. These crystals are termed biaxial and have two directions along which the wave velocity is independent of the polarization direction. These correspond to the two sections of the ellipsoid which are circular. These sections are inclined at equal angles to the major and intermediate axes. In convergent light between polarizers, these crystals show a pattern which is quite different from that which appears with uniaxial crystals. A plate cut normal to the major axis of the index ellipsoid has a pattern of a series of lemniscates and ovals. The directions corresponding to the optic axes appear as black spots between crossed circular polarizers. The interference pattern between crossed linear polarizers is shown in **Fig. 6**.

Angle between optic axes. One of the quantities used to describe a biaxial crystal or to identify a biaxial mineral is the angle between the optic axes. This can be calculated directly from the equation for the index ellipsoid, Eq. (14), where α, β, and γ are the three indices of refraction

$$\frac{x^2}{\alpha^2} + \frac{y^2}{\beta^2} + \frac{z^2}{\gamma^2} = 1 \tag{14}$$

of the material. The circular sections of the ellipsoid must have the intermediate index as a radius. If the relative sizes of the indices are so related that $\alpha > \beta > \gamma$, the circular sections will have β as a radius and the normal to these sections will lie in the xz plane. The section of the ellipsoid cut by this plane will be the ellipse of Eq. (15). The radius of length β will intersect this ellipse at a point $x_1 z_1$ where Eq. (16) holds. The solution of these two equations gives for the points $x_1 z_1$

$$\frac{x^2}{\alpha^2} + \frac{z^2}{\gamma^2} = 1 \tag{15} \qquad\qquad x_1^2 + z_1^2 = \beta^2 \tag{16}$$

Eq. (17). These points and the origin define the lines in the xz plane which are also in the planes

$$x_1 = \pm\sqrt{\frac{\alpha^2(\gamma^2 - \beta^2)}{\gamma^2 - \alpha^2}} \qquad\qquad z_1 = \pm\sqrt{\frac{\gamma^2(\alpha^2 - \beta^2)}{\alpha^2 - \gamma^2}} \tag{17}$$

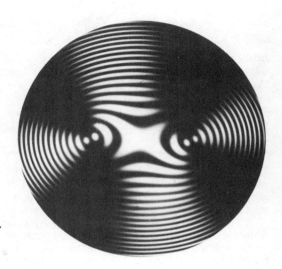

Fig. 6. Interference figure from Brazil topaz. (*After M. Born and E. Wolf, Principles of Optics, 6th ed., Pergamon Press, 1980*)

of the circular sections of the ellipsoid. Perpendiculars to these lines will define the optic axes. The angle Ω between the axes and the z direction will be given by Eq. (18).

$$\tan^2 \Omega = \frac{z_1^2}{x_1^2} = \frac{(1/\alpha^2) - (1/\beta^2)}{(1/\beta^2) - (1/\gamma^2)} \tag{18}$$

The polarization direction for light passing through a biaxial crystal is computed in the same way as for a uniaxial crystal. The section of the index normal to the propagation direction is ordinarily an ellipse. The directions of the axes of the ellipse represent the vibration direction and the lengths of the half axes represent the indices of refraction. One polarization direction will lie in a plane which bisects the angle made by the plane containing the propagation direction and one optic axis and the plane containing the propagation direction and the other optic axis. The other polarization direction will be at right angles to the first. When the two optic axes coincide, this situation reduces to that which was demonstrated earlier for the case of uniaxial crystals.

Conical refraction. In biaxial crystals there occurs a set of phenomena which have long been a classical example of the theoretical prediction of a physical characteristic before its experimental discovery. These are the phenomena of internal and external conical refraction. They were predicted theoretically in 1832 by William Hamilton and experimentally demonstrated in 1833 by H. Lloyd.

As shown earlier, the Huygens wavelets in a biaxial crystal consist of two surfaces. One of these has its major axis at right angles to the major axis of the other. The two thus intersect. In making the geometrical construction to determine the ray direction, two directions are found where the two wavefronts coincide and the points of tangency on the two ellipsoids are multiple. In fact, the plane which represents the wavefront is found to be tangent to a circle which lies partly on one surface and partly on the other. The directions in which the wavefronts coincide are the optic axes. A ray incident on the surface of a biaxial crystal in such a direction that the wavefront propagates along an axis will split into a family of rays which lie on a cone. If the crystal is a plane parallel slab, these rays will be refracted at the second surface and transmitted as a hollow cylinder parallel to the original ray direction. Similarly, if a ray is incident on the surface of a biaxial crystal plate at such an angle that it passes along the axes of equal ray velocity, it will leave the plate as a family of rays lying on the surface of a cone. The first of these phenomena is internal conical refraction; the second is external conical refraction. Equation (19) gives the half angle ψ of the external cone.

$$\tan \psi = \frac{\beta}{\sqrt{\alpha\gamma}} \sqrt{(\beta - \alpha)(\gamma - \beta)} \qquad (19)$$

Bibliography. M. Born and E. Wolf, *Principles of Optics*, 6th ed., 1980; F. A. Jenkins, *Fundamentals of Optics*, 4th ed., 1976; E. E. Wahlstrom, *Optical Crystallography*, 5th ed., 1979; E. A. Wood, *Crystals and Light*, 1977.

BIREFRINGENCE
Bruce H. Billings

The splitting which a wavefront experiences when a wave disturbance is propagated in an anisotropic material; also called double refraction. In anisotropic substances the velocity of a wave is a function of displacement direction. Although the term birefringence could apply to transverse elastic waves, it is usually applied only to electromagnetic waves.

In birefringent materials either the separation between neighboring atomic structural units is different in different directions, or the bonds tying such units together have different characteristics in different directions. Many crystalline materials, such as calcite, quartz, and topaz, are birefringent. Diamonds on the other hand, are isotropic and have no special effect on polarized light of different orientations. Plastics composed of long-chain molecules become anisotropic when stretched or compressed. Solutions of long-chain molecules become birefringent when they flow. This first phenomenon is called photoelasticity; the second, streaming birefringence. *See* Photoelasticity.

For each propagation direction with linearly polarized electromagnetic waves, there are two principal displacement directions for which the velocity is different. These polarization directions are at right angles. The difference between the two indices of refraction is also called the birefringence. When the plane of polarization of a light beam does not coincide with one of the two principal displacement directions, the light vector will be split into components parallel to each direction. At the surface of such materials the angle of refraction is different for light polarized parallel to the two principal directions.

For additional information on birefringence and birefringent materials *see* Crystal optics; Polarized light; Refraction of waves.

PLEOCHROISM
Bruce H. Billings

In some colored transparent crystals, the effect wherein the color is quite different in different directions through the crystals. In such a crystal the absorption of light is different for different polarization directions. Tourmaline offers one of the best-known examples of this phenomenon. In colored transparent tourmaline the effect may be so strong that one polarized component of a light beam is wholly absorbed, and the crystal can be used as a polarizer. For a fuller discussion of the effect *see* Dichroism; Polarized light; Trichroism.

DICHROISM
Bruce H. Billings

In certain anisotropic materials, the property of having different absorption coefficients for light polarized in different directions.

There are few natural materials which exhibit strong dichroism. One of the first to be discovered was tourmaline. Light transmitted by thin plates of dark forms of tourmaline is almost completely polarized. *See* Polarized light.

In isotropic optical materials the optical density is defined as in Eq. (1), where I_0 is the

$$d = \log \frac{I_0}{I} \quad (1)$$

intensity of the incident light, and I that of the transmitted light. In anisotropic materials that are dichroic, the value of d can vary as a function of the vibration direction of the electric vector of the light wave. Just as the index ellipsoid is used to define the birefringence of a material, a density surface can be used to define the dichroism. *See Crystal Optics.*

Compared to the literature on birefringence and optical activity, there has been relatively little material on dichroism. This is partly because of the difficulty in making measurements. The Kramers-Kronig relationship shows that any material whose refractive index is different from unity and varies as a function of wavelength will absorb radiation at some wavelength. From the Kramers-Kronig relationship it is apparent that all optically anisotropic materials should be dichroic. From the values of the refractive index at different wavelengths, the spectral positions and intensity of the absorption can be calculated. In a linear birefringent material the refractive index depends on the polarizing direction or electric vector of the radiation. For each direction of propagation there are two perpendicular vibration directions with different refractive indices. It can be inferred that, at some wavelength, the absorption for light vibrating in these two directions will be different. For transparent materials this wavelength is in the ultraviolet and the absorption is thus difficult to measure. The absorption difference is also frequently so small that it cannot be detected. In other words, the absorption is apparently the same in each direction. Furthermore, the dichroic band may coincide with a region of isotropic absorption.

If the absorption in a dichroic material is different for different linear states of polarization, the material is termed linear dichroic. If it is different for right and left circularly polarized light, it is termed circular dichroic. Similarly, there can be elliptically dichroic crystals. In biaxial crystals there are three different refractive indices corresponding to an electric vector lying along each of the three orthogonal axes of the index ellipsoid. Such a crystal has dichroism which is different for light traveling along each of these three principal axes.

Most dichroic materials exist in the form of relatively thin sheets. Here one is dealing with a section of the dichroic surface. Associated with the sheet of material will be a direction of maximum absorption and one of minimum absorption. The density equation can be rewritten as Eqs. (2), where the densities are for light vibrating parallel or perpendicular to the axis of the section of dichroic surface. *See Trichroism.*

$$d_\| = \log \frac{I_{0\|}}{I_\|} \qquad d_\perp = \log \frac{I_{0\perp}}{I_\perp} \quad (2)$$

TRICHROISM
Bruce H. Billings

When certain optically anisotropic transparent crystals are subjected to white light, a cube of the material is found to transmit a different color through each of the three pairs of parallel faces. Such crystals are sometimes termed trichroic, and the phenomenon is called trichroism. This expression is used only rarely today since the colors in a particular crystal can appear quite different if the cube is cut with a different orientation with respect to the crystal axes. Accordingly, the term is frequently replaced by the more general term pleochroism. Even this term is being replaced by the phrase linear dichroism or circular dichroism to correspond with linear birefringence or circular birefringence. *See Birefringence; Dichroism; Pleochroism.*

Cordierite is a typical trichroic crystal. In light with a vibration direction parallel to the X axis of the index ellipsoid the crystal appears yellow. With the vibration direction parallel to the Y axis the crystal is dark violet. In the Z direction the crystal is clear.

The phenomena of trichroism can be explained crudely as follows. Classically, one can consider an electron in a biaxial crystal as having three different force constants associated with a displacement directed along each of the principal axes. Linear polarized light traveling along the X axis with its electric vector parallel to the Y axis will displace the electron against the Y force constant and will experience a certain absorption and retardation. It will be unaffected by the force constants in the X and Z directions. Similarly, polarized light traveling in the Y direction will experience absorption and retardation. Unpolarized light will also be absorbed in a different fashion depending on the direction of propagation. In this case light traveling in the X direction can be considered as composed of an equal mixture of light polarized parallel to the Y axis and the Z axis. The absorption will be intermediate between the two polarization directions. SEE CRYSTAL OPTICS; POLARIZED LIGHT.

Bibliography. N. H. Hartshorne and A. Stuart, *Practical Optical Crystallography*, 2d ed., 1970; E. E. Wahlstrom, *Optical Crystallography*, 5th ed., 1979; E. A. Wood, *Crystals and Light: An Introduction to Optical Crystallography*, 1977.

SCATTERING OF ELECTROMAGNETIC RADIATION
J. F. SCOTT

The process in which energy is removed from a beam of electromagnetic radiation and reemitted with a change in direction, phase, or wavelength. All electromagnetic radiation is subject to scattering by the medium (gas, liquid, or solid) through which it passes. In the short-wavelength, high-energy regime in which electromagnetic radiation is most easily discussed by means of a particle description, these processes are termed photon scattering. At slightly longer wavelengths, the scattering of x-rays provides the most effective means of determining the structure of crystalline solids. In the visible wavelength region, scattering of light produces the blue sky, red sunsets, and white clouds. At longer wavelengths, scattering of radio waves determines their characteristics as they pass through the atmosphere. SEE LIGHT; METEOROLOGICAL OPTICS.

It has been known since the work of J. Maxwell in the nineteenth century that accelerating electric charges radiate energy and, conversely, that electromagnetic radiation consists of fields which accelerate charged particles. Light in the visible, infrared, or ultraviolet region interacts primarily with the electrons in gases, liquids, and solids—not the nuclei. The scattering process in these wavelength regions consists of acceleration of the electrons by the incident beam, followed by reradiation from the accelerating charges. SEE ELECTROMAGNETIC RADIATION.

Scattering processes may be divided according to the time between the absorption of energy from the incident beam and the subsequent reradiation. True "scattering" refers only to those processes which are essentially instantaneous. Mechanisms in which there is a measurable delay between absorption and reemission are usually termed luminescence. If the delay is longer than a microsecond or so, the process may be called fluorescence; and mechanisms involving very long delays (seconds) are usually termed phosphorescence. SEE ABSORPTION OF ELECTROMAGNETIC RADIATION.

Inelastic scattering. Instantaneous scattering processes may be further categorized according to the wavelength shifts involved. Some scattering is "elastic"; there is no wavelength change, only a phase shift. In 1928 C. V. Raman discovered the process in which light was inelastically scattered and its energy was shifted by an amount equal to the vibrational energy of a molecule or crystal. Such scattering is usually called the Raman effect. This term has been used in a more general way, however, often to describe inelastic scattering of light by spin waves in magnetic crystals, by plasma waves in semiconductors, or by such exotic excitations as "rotons," the elementary quanta of superfluid helium. SEE RAMAN EFFECT.

Brillouin and Rayleigh scattering. In liquids or gases two distinct processes generate inelastic scattering with small wavelength shifts. The first is Brillouin scattering from pressure waves. When a sound wave propagates through a medium, it produces alternate regions of high compression (high density) and low compression (or rarefaction). A picture of the density distribution in such a medium is shown in **Fig. 1**. The separation of the high-density regions is equal to one wavelength λ for the sound wave propagating. Brillouin scattering of light to higher (or lower) frequencies occurs because the medium is moving toward (or away from) the light source.

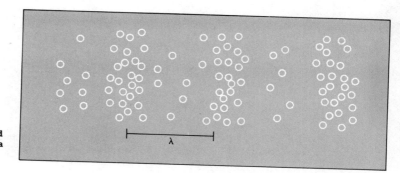

Fig. 1. Pressure-induced density fluctuations in a fluid.

This is an optical Doppler effect, in which the frequency of sound from a moving object is shifted up in frequency as the object moves toward the observer.

The second kind of inelastic scattering studied in fluids is due to entropy and temperature fluctuations. In contrast to the pressure fluctuations due to sound waves, these entropy fluctuations do not generate scattering at sharp, well-defined wavelength shifts from the exciting wavelength; rather, they produce a broadening in the scattered radiation centered about the exciting wavelength. This is because entropy fluctuations in a normal fluid are not propagating and do not, therefore, have a characteristic frequency; unlike sound they do not move like waves through a liquid—instead, they diffuse. Under rather special circumstances, however, these entropy or temperature fluctuations can propagate; this has been observed in super-fluid helium and in crystalline sodium fluoride at low temperatures, and is called second sound.

Scattering from entropy or temperature fluctuations is called Rayleigh scattering. In solids this process is obscured by scattering from defects and impurities. Under the assumption that the scattering in fluids is from particles much smaller than the wavelength of the exciting light, Lord Rayleigh derived in 1871 an equation, shown below, for such scattering; here $I(\theta)$ is the intensity

$$r^2 I(\theta)/I_0 = \pi\, d\, \lambda^{-4} v^2 (1 + \cos^2 \theta)(n-1)^2$$

of light scattered from an incident beam of wavelength λ and intensity I_0 at a distance r; d is the number of scattering particles; v is the volume of the disturbing particle; and n is the index of refraction of the fluid. The $\cos \theta$ term is present for unpolarized incident light, where θ is the scattering angle. Measurement of the ratio in Rayleigh's equation allows the determination of either Avogadro's number N or the molecular weight M of the fluid, if the other is known. The dependence of scattering intensity upon the inverse fourth power of the wavelength given in Rayleigh's equation is responsible for the fact that daytime sky looks blue and sunsets red: blue light is scattered out of the sunlight by the air molecules more strongly than red; at sunset, more red light passes directly to the eyes without being scattered.

Large particles. Rayleigh's derivation of his scattering equation relies on the assumption of small, independent particles. Under some circumstances of interest, both of these assumptions fail. Colloidal suspensions provide systems in which the scattering particles are comparable to or larger than the exciting wavelengths. Such scattering is called the Tyndall effect and results in a nearly wavelength-independent (that is, white) scattering spectrum. The Tyndall effect is the reason clouds are white (the water droplets become larger than the wavelengths of visible light). SEE TYNDALL EFFECT.

The breakdown of Rayleigh's second assumption—that of independent particles—occurs in all liquids. There is strong correlation between the motion of neighboring particles. This leads to fixed phase relations and destructive interference for most of the scattered light. The remaining scattering arises from fluctuations in particle density discussed above and was first analyzed theoretically by A. Einstein in 1910 and by M. Smoluchowski in 1908.

Rayleigh's basic theory has been extended by several authors. Rayleigh in 1911 and R. Gans in 1925 derived scattering formulas appropriate for spheres of finite size, and in 1947 P.

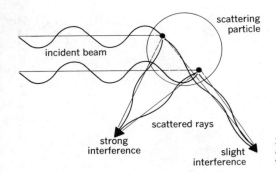

Fig. 2. The role of interference in determining the angular dependence of light scattering from particles that are comparable in size to the wavelength.

Debye extended the theory to include random coil polymers. The combined results of these three workers, generally called the Rayleigh-Gans-Debye theory, is valid for any size of particle, provided the refractive index n is near unity; whereas the Rayleigh theory is valid for any index, provided the particles are very small. The index n of any medium would be unity if there were no scattering.

A more complete theory than Rayleigh-Gans-Debye was actually developed earlier by G. Mie in 1908; however, Mie's theory generally requires numerical solution. Mie's theory is valid even for particles larger than the wavelength of light used. For such particles large phase shifts occur for the scattered light. Mie scattering exhibits several maxima and minima as a function of scattering angle; the positions of these maxima depend upon particle size, as indicated in **Fig. 2.** These secondary maxima are essentially higher-order Tyndall scattering.

Critical opalescence. The most striking example of light scattering is that of critical opalescence, discovered by T. Andrews in 1869. As the critical temperature of a fluid is approached along the critical isochore, fluctuations in the density of the medium become larger than a wavelength and stay together for longer times. The independent particle approximation becomes very bad, and the intensity of light scattered in all directions becomes very large. The fluid, which might have been highly transparent at a temperature 1°C (1.8°F) higher, suddenly scatters practically all of the light incident upon it. The regions of liquidlike density in the fluid have molecules moving in phase, or "coherently," within them. Such coherent regions in fluids near their critical temperatures behave very much like the large spherical particles diagrammed in Fig. 2, thus establishing the relationship between Mie scattering of colloids and critical opalescence. The theory of scattering from correlated particles was developed primarily by L. S. Ornstein and F. Zernike in 1914–1926. *See* COHERENCE; OPALESCENCE.

Nonlinear scattering. At the power densities available with pulsed lasers, a variety of other scattering mechanisms can become important. (For nonlaser excitation these processes exist but are generally too weak to be detected.) Such processes include second-harmonic generation, in which the energy from two quanta (photons) in the incident beam produces one doubly energetic photon. A practical example is the production of green light from the red light of a ruby laser. Third-harmonic generation (and similar higher-order processes) also occurs as an intense scattering mechanism at sufficiently high power densities. Whereas second-harmonic generation can occur only in certain crystals, for reasons of symmetry, third-harmonic generation occurs in any medium. Other more exotic nonlinear scattering mechanisms include sum generation and difference generation, in which two beams of light are combined to produce scattering having frequencies which are the sum or difference of the frequencies in the two incident beams, and rather weak processes such as the hyper-Raman effect, in which the scattered light has a frequency which is twice that of the incident beam plus (or minus) some vibrational energy. All these effects are called nonlinear because their intensities vary according to the square or higher power of the incident intensity. These nonlinear processes may be very important in astrophysics (for example, in stellar interiors), in addition to laboratory laser experiments. *See* NONLINEAR OPTICS.

Bibliography. W. Hayes and R. Loudon, *Scattering of Light by Crystals*, 1978; B. J. Berne and R. Pecora, *Dynamic Light Scattering*, 1976.

OPALECENCE
BENJAMIN CHU

The milky iridescent appearance of a dense transparent medium when the system (or medium) is illuminated by polychromatic radiation in the visible range, such as sunlight. Slight changes in the rainbowlike color of the system can occur, depending on the scattering angle, that is, the angle between the directions of incident radiation and of observation. All dense transparent mediums have local density fluctuations due to the thermal motions of molecules, or concentration fluctuations due to the presence of a second component, such as colloidal suspensions or macromolecules in solution. Local fluctuations in density (or concentration) are accompanied by local fluctuations in the refractive index. Since the system is optically inhomogeneous, some of the light is scattered to the side. Normally, the amount of light scattered is very small, perhaps of the order of magnitude of 10^{-4} or less of its incident radiation. Whenever the amplitude of fluctuation becomes large, a significant portion of the incident light may be scattered. The transmitted light is then visibly weakened, and the medium look turbid.

Opalescence is a general term which applies to the optical phenomenon of intense scattering in the visible range of the electromagnetic radiation by a system with strong local optical inhomogeneities. The iridescence, or rainbowlike display of interference of colors, arises because the intensity of scattered light is approximately proportional to the reciprocal fourth power of the wavelength of incident light (Rayleigh's law). SEE SCATTERING OF ELECTROMAGNETIC RADIATION.

Critical opalescence. A classical view of the critical point in gas-liquid phase transitions is that it is the state at which the densities of the coexisting gas-liquid phases are equal, and also that it is represented by a characteristic critical temperature above which gas cannot be liquified, no matter how great the applied pressure is. The corresponding pressure required to liquefy the gas at the critical temperature is the critical pressure. For a one-component system the compressibility becomes very large in the neighborhood of the critical point and infinite at the critical point itself. Thus the energy required in the compression of a gas to a given amplitude of fluctuations becomes smaller the closer one approaches the critical point. There the thermal motions of molecules can produce strong density fluctuations, resulting in a very impressive scattering, the so-called critical opalescence. Apart from the critical points of gas-liquid transitions, several other types of second-order phase transitions at which the second derivatives of the free energy are discontinuous, such as critical mixing (consolute) points of binary liquid mixtures, exhibit critical opalescent behavior.

One can study the nature of phase transitions by observing the size and shape of local fluctuations and their time-dependent changes. By using statistical models, one can then relate fluctuations to thermodynamic and transport properties of the system, such as isothermal compressibility and thermal conductivity. In the critical region any such properties become very difficult to measure by conventional means. The changes at the critical region in thermodynamic and transport properties are very dramatic; for example, both isothermal compressibility and heat capacity diverge at the critical point. While the scattered intensity is related to the amplitude of local fluctuations in the refractive index, the angular dependence of scattered light reveals the extent of such fluctuations. As one approaches the critical point, the scattered intensity increases because the isothermal compressibility becomes larger; the system consequently looks very turbid. Further analysis shows that the scattered light is concentrated more and more in the forward direction. This indicates that the extensions of fluctuations approach the wavelength of the incident radiation, which is a few thousand angstroms for light in the visible region. These large fluctuations result from long-range molecular interactions in the system. A suitable approach is to consider the incident electromagnetic radiation as a measuring scale. Thus, for smaller fluctuations or fluctuations in metal alloys near the consolute point, one can use x-rays with a wavelength of 0.15 nanometer or less instead of visible light with a wavelength of several thousand angstroms, to investigate fluctuation sizes ranging from tens to hundreds of angstroms, even when the system is barely opalescent to the naked eye or does not transmit visible light.

Time dependency. The density (or concentration) fluctuations that are produced by thermal motions of molecules are time-dependent, so that light is quasi-elastically or inelastically scattered. The spectral distribution of the scattered light characterizes the time dependence of such fluctuations and can be resolved by means of interferometric and optical beat frequency techniques, using laser as a light source. The divergence in heat capacity and probably also in thermal conductivity indicates a slowdown in the relaxation times of thermal fluctuations and an increasing difficulty in reaching thermal equilibrium. Experiments and theory are both difficult. Care should therefore be exercised when studying the phenomenon of critical opalescence. SEE ABSORPTION OF ELECTROMAGNETIC RADIATION; LASER.

TYNDALL EFFECT
QUENTIN VAN WINKLE

Visible scattering of light along the path of a beam of light as it passes through a system containing discontinuities. The luminous path of the beam of light is called a Tyndall cone. An example is shown in the **illustration**. In colloidal systems the brilliance of the Tyndall cone is directly

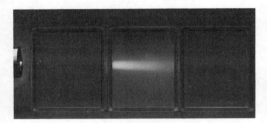

Luminous light path known as the Tyndall cone or Tyndall effect. (*Courtesy of H. Steeves and R. G. Babcock*)

dependent on the magnitude of the difference in refractive index between the particle and the medium. In aqueous gold sols, where the difference in refractive index is high, strong Tyndall cones are observed.

For systems of particles with diameters less than one-twentieth the wavelength of light, the light scattered from a polychromatic beam is predominantly blue in color and is polarized to a degree which depends on the angle between the observer and the incident beam. The blue color of tobacco smoke is an example of Tyndall blue. As particles are increased in size, the blue color of scattered light disappears and the scattered radiation appears white. If this scattered light is received through a nicol prism which is oriented to extinguish the vertically polarized scattered light, the blue color appears again in increased brilliance. This is called residual blue, and its intensity varies as the inverse eighth power of the wavelength. SEE SCATTERING OF ELECTROMAGNETIC RADIATION.

RAMAN EFFECT
RICHARD C. LORD

A phenomenon observed in the scattering of light as it passes through a material medium, whereby the light suffers a change in frequency and a random alteration in phase. Raman scattering differs in both these respects from Rayleigh and Tyndall scattering, in which the scattered light has the same frequency as the unscattered and bears a definite phase relation to it. The intensity of normal Raman scattering is roughly one-thousandth that of Rayleigh scattering in liquids and smaller still in gases. For an extended discussion of Rayleigh scattering SEE SCATTERING OF ELECTROMAGNETIC RADIATION. SEE ALSO TYNDALL EFFECT.

Discovery. Because of its low intensity, the Raman effect was not discovered until 1928, although the scattering of light by transparent solids, liquids, and gases had been investigated for many years before. Prompted by A. H. Compton's observation of frequency changes in x-rays scattered by electrons (Compton effect), the Indian physicists C. V. Raman and K. S. Krishnan examined sunlight scattered by a number of liquids. With the help of complementary filters, they found that there were frequencies in the scattered light that were lower than the frequencies in the filtered sunlight. They then showed, by using light of a single frequency from a mercury arc, that the new frequencies in the scattered radiation were characteristic of the scattering medium. Within a few months of Raman and Krishnan's first announcement of their discovery, the Soviet physicists G. Landsberg and L. Mandelstam communicated their independent discovery of the existence of the effect in crystals. In Soviet literature the phenomenon is referred to as combination scattering, and not Raman effect.

The development of the laser has led to a resurgence of interest in the Raman effect and to the discovery of a number of related phenomena. A beam of laser radiation is intense, polarized, and coherent; it can be made monochromatic, small in diameter, and highly collimated. The laser is therefore nearly ideal for the production of the Raman effect, and other kinds of sources are seldom employed. Many different wavelengths in the visible spectrum and adjacent regions are available. The argon-ion and krypton-ion lasers are most commonly used, since they have high continuous-wave power (1 to 10 W), but tunable dye lasers are also often employed in excitation of resonance Raman scattering. SEE LASER.

Raman spectroscopy. Raman scattering is analyzed by spectroscopic means. The collection of new frequencies in the spectrum of monochromatic radiation scattered by a substance is characteristic of the substance and is called its Raman spectrum. Although the Raman effect can be made to occur in the scattering of radiation by atoms, it is of greatest interest in the spectroscopy of molecules and crystals.

Because of the laser beam's small diameter and high collimation, it can easily be used to excite the Raman effect. A typical optical arrangement is shown in **Fig. 1**. Monochromatic radiation from the laser impinges on the sample S in an appropriate transparent cell. It may be desirable to condense or expand the laser beam by means of a lens system L_1 and to remove unwanted radiation from the beam by a narrow-band optical filter F. A concave mirror M_1 can return unscattered radiation for a second passage through the sample.

Raman scattering is approximately uniform in all directions and is usually studied at right angles, (Fig. 1). In this way the intense radiation of the laser beam interferes least with the observation of the weak scattered light. This light is collected by a lens system L_2 and focused on the slit of a scanning monochromator, which analyzes it spectroscopically. As the spectrum is scanned, the dispersed radiation from the monochromator is detected by a photomultiplier PM, further amplified and processed electronically at A, and then recorded by a strip-chart recorder CR. The recorder is driven in synchronism with the monochromator by a suitable mechanism D. The concave mirror M_2 may be used to augment the amount of scattered radiation by collecting light scattered at $-90°$ and returning it to the $+90°$ direction. The polarization characteristics of

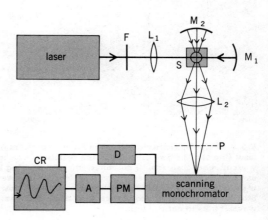

Fig. 1. Laser-Raman spectroscopic system.

the scattered radiation are frequently of interest, especially since the laser radiation itself is linearly polarized. An analyzing device for evaluating the degree of polarization of the scattered radiation may be inserted at Point P.

The appearance of a photoelectrically recorded Raman spectrum of liquid carbon tetrachloride as excited by the red line of the helium-neon laser at 632.8 nm (6328 A, power incident on the sample of about 50 mW) is shown in **Fig. 2**. Intensity of the scattered light on an arbitrary scale is plotted vertically against the wave number in cm^{-1} measured with respect to the wave number of the exciting line taken as zero. For convenience, it is usual to express the data of Raman spectroscopy in cm^{-1} rather than frequency units (s^{-1}). (Frequency v in $s^{-1} = c\tilde{v}$ in cm^{-1}, where c is the velocity of light in vacuum in cm/s.)

A spectrum of the most intense line in Fig. 2 is shown at higher resolution (smaller spectral slit width $\Delta \tilde{v}$) in **Fig. 3**. The line is seen to consist of several closely spaced components. These result from the presence of the two isotopes of chlorine, ^{35}Cl and ^{37}Cl, which produce five isotopic species $C^{35}Cl_n{}^{37}Cl_{4-n}$, $n = 0, 1, 2, 3, 4$. The line due to the least abundant species, $n = 0$, is not visible, but the other four are readily identified.

Theory. The mechanism of the Raman effect can be envisaged either by the corpuscular picture of light or from the point of view of the wave theory. Both pictures merge in the basic quantum theory of radiation. The corpuscular model of light scattering envisages light quanta or photons as particles which have linear and angular momenta. On passing through a material medium, these particles collide with atoms or molecules. If the collision is elastic, the photons bounce off the molecules with unchanged energy E and momentum, and hence with unchanged frequency v. Such a process gives rise to Rayleigh scattering. If the collision is inelastic, the photons may gain energy from, or lose it to, the molecules. A change ΔE in the photon energy by Planck's relationship, $E = hv$, must produce a change in the frequency $\Delta v = \Delta E/h$. Such inelastic collisions are rare compared to the elastic ones, and the Raman effect is correspondingly much weaker than Rayleigh scattering. SEE LIGHT.

In the wave picture of the effect, the electromagnetic waves which constitute the incoming monochromatic radiation sweep through the material medium. Since the atoms and molecules

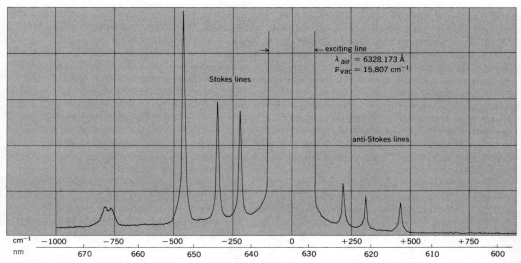

Fig. 2. Photoelectric recording of the Raman spectrum of carbon tetrachloride excited by He-Ne laser line at 632.8 nm. Intensity of the radiation is recorded vertically against the horizontal wave-number scale (cm^{-1}) measured from the exciting line as zero. The Rayleigh-scattered exciting line is three orders of magnitude more intense than the Raman lines, and its maximum is therefore far off-scale. The Stokes lines appear at lower frequencies, and the less intense anti-Stokes lines at higher frequencies, than those of the exciting line. The lower scale shows wavelengths in nanometers.

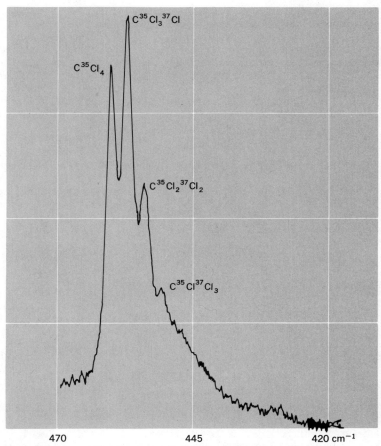

Fig. 3. The Stokes line at -460 cm^{-1} in the Raman spectrum of carbon tetrachloride. This spectrum, taken at about 10 times better resolution than that of Fig. 2, shows lines due to four or five isotopic species resulting from the 3:1 ratio of the chlorine isotopes ^{35}Cl and ^{37}Cl. The instrumental wave-number scale requires a calibration correction to increase the wave numbers by 1.4 cm^{-1}.

composing the medium are made up of negatively charged electrons and positively charged nuclei, the electric field of the light waves sets the electrons to oscillating, chiefly with the frequency of the incoming radiation. The oscillating electrons re-create the alternating electric field of the incoming light, thus passing the light wave along through the medium. This process is analogous to the elastic collisions which are given by the corpuscular picture.

The ability of the electrons and nuclei in a molecule to be displaced by an electric field is called the molecular polarizability α. It is not a simple property of the molecule, but depends in a complicated way on the frequency of the electric field, on the orientation of the molecule, and on the internal motions of the nuclei and electrons. Thus the molecular polarizability α varies periodically with molecular rotation and vibration, and thereby the effect of a light wave on the electrons and nuclei of a molecule can be changed.

When a monochromatic light wave sweeps through a transparent medium containing rotating and vibrating molecules, most of the wave is recreated unchanged by the oscillating electrons, but because of the periodic changes produced in α by rotation and vibration, new frequencies are added to the light wave. The appearance of these new frequencies, whose values are determined by the rotational and vibrational energies of the molecules, is analogous to the result of the inelastic collisions of the corpuscular model. For the wave picture of the Raman effect, the quantity α is the basic quantity. The intensity of the Raman effect depends on the magnitude of the changes produced in α by molecular rotation and vibration, and the number and values of new frequencies (usually expressed as frequency shifts $\Delta\nu$ from the original monochromatic frequency) depend on the variation of α with the frequencies of rotation and vibration.

The temperature of the scattering molecules is an additional factor which affects the intensity of Raman frequencies higher than the exciting frequency (the anti-Stokes lines of Fig. 2). The anti-Stokes lines, having higher frequencies, correspond to photons which have higher energy than that of the exciting light, and this energy must come from the molecules. If the molecules do not have any available vibrational or rotational energy, that is, if they are at the absolute zero of temperature, there is no possibility of inelastic collisions in which energy is transferred from a molecule to a photon. So, anti-Stokes lines vanish at absolute zero. At nonzero temperatures the intensity ratio of an anti-Stokes line to a Stokes line is approximated by the ratio of the number of molecules which can give up the corresponding energy to the number which can accept it from the light wave.

Special forms. The development of lasers resulted in the discovery of a number of kinds of Raman scattering.

Resonance Raman effect. When the exciting radiation falls within the frequency range of a molecule's absorption band in the visible or ultraviolet spectrum, the radiation may be scattered by two different processes, resonance fluorescence or the resonance Raman effect. Both these processes give much more intense scattering than the normal nonresonant Raman effect. Resonance fluorescence differs from the resonance Raman effect in that the absolute frequencies of the fluorescent spectrum do not shift when the exciting radiation's frequency is changed, so long as the latter does not move outside the absorption band. The absolute frequencies of the resonance Raman effect, on the contrary, shift by exactly the amount of any shift in the exciting frequency, just as do those of the normal Raman effect. Thus the main characteristic of the resonance as compared to the normal Raman effect is its intensity, which may be greater by two or three orders of magnitude.

The resonance Raman effect was anticipated by G. Placzek in 1934 in his pioneering development of the polarizability theory of Raman scattering. It was actually observed before the discovery of lasers, but tunable lasers are the most effective sources for the study of its various aspects. A typical resonance Raman spectrum is shown in **Fig. 4**, in which oxyhemoglobin is excited by the 568.2-nm (5682 A) wavelength of singly ionized krypton. The top spectrum $I_\parallel$ is taken with the polarizer P of Fig. 1 set to pass the components parallel to the direction of laser polarization; the bottom spectrum $I_\perp$ is taken with P set to pass perpendicular components. Lines in which $I_\perp$ is much greater than $I_\parallel$ are said to have inverse polarization and are seen only in the resonance Raman effect; these include the lines at 1305, 1342, and 1589 cm^{-1} and numerous others.

Hyper-Raman effect. The nature of this effect is most easily described in terms of the corpuscular picture of the Raman effect. With an intense laser source, the number of monochromatic photons impinging on the molecules of a medium per unit volume and unit time may be extremely large. If so, the probability that two photons will collide simultaneously with the same molecule is very much larger than in normal scattering, and there is considerable chance that the two photons will unite and be scattered as a single photon of approximately twice the frequency. The rules governing the scattering in such three-photon processes (two incoming and one outgoing photons) are quite different from those for normal (two-photon) Rayleigh and Raman scattering. For example, in molecules that are centrosymmetric, the collision must be inelastic, that is, the molecule must absorb or give up an amount of energy ΔE during the process. The frequency of the scattered photon will therefore not be exactly twice the frequency of the incident photons but will differ from it by $\Delta\bar{\nu} = \Delta E/hc$. Such scattered radiation is called the hyper-Raman effect. Even in molecules that are not centrosymmetric, the likelihood of elastic collisions is much smaller than in the normal case, so that the intensity of hyper-Rayleigh scattering may be substantially weaker than the hyper-Raman scattering.

As implied above, the selection rules for the vibrational and rotational transitions in the

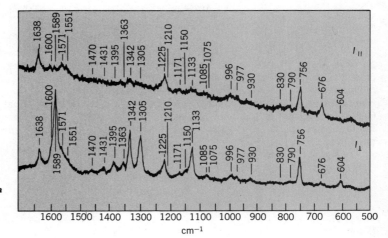

Fig. 4. Resonance Raman spectrum of oxyhemoglobin. (After T. G. Spiro and T. C. Strekas, *Resonance Raman spectra of hemoglobin and cytochrome c*, Proc. Nat. Acad. Sci. USA, 69:2622–2626, 1972)

hyper-Raman effect are different from those of the normal Raman effect. Thus certain transitions are observable in the hyper-Raman effect that are normally forbidden. This is one virtue of the hyper-Raman effect; the other is that it is observed in a spectral region whose frequency is far removed from that of the incoming radiation (and is, in fact, twice that of the latter). The effect is therefore observable without interference from the normal Rayleigh line.

Stimulated Raman effect. The mechanism of the stimulated Raman effect depends on the coherent pumping of the molecules of the sample into an excited vibrational state by the powerful electric field of the laser beam. In view of the large discrepancy of one or two orders of magnitude between the frequency of the vibration and the frequency of the laser, this can be accomplished only if the field of the light wave has a very high value (the threshold power) and if the mismatch in frequency is compensated by the generation of coherent radiation with a frequency equal to that of the laser minus the vibrational frequency. The coherent radiation so produced is called stimulated Raman scattering. It was first observed by R. Woodbury and A. Ng in 1962. They found the effect in liquid nitrobenzene, which they were using as an electrooptical shutter within a laser system. *See* OPTICAL PUMPING.

In addition to its high intensity and its coherence, there are other new features of stimulated Raman scattering. Since the pumping power of the incident laser beam must exceed a certain threshold for the scattering to take place, when the laser power is used up in exciting one vibrational mode, there is insufficient power available to excite other modes. Therefore the stimulated Raman effect usually contains only one frequency, though in rare cases the power may be divided between two vibrational modes of roughly the same threshold. However, the power in the scattered radiation may itself produce further stimulated Raman emission by a repetition of the intitial process. This results in a new frequency, which is the laser frequency minus exactly twice the frequency of vibrational mode that is being scattered. This fact shows that the mechanism does not involve a double jump in the vibrational levels; such a double jump would give a frequency shift that is not exactly twice that of the vibrational fundamental because of vibrational anharmonicity.

Another striking and unusual effect in stimulated Raman scattering is the excitation of intense anti-Stokes radiation. This radiation may be even stronger than the Stokes radiation in certain circumstances. Moreover, it can be observed at such low temperatures that the initial populations of the excited vibrational levels needed for normal anti-Stokes Raman scattering are zero. It arises from the above-mentioned pumping of molecules from the ground vibrational state into upper excited states by the intitial laser power. These excited molecules can then be pumped by further radiation back into the ground state, with a simultaneous stimulated emission of coherent radiation at a frequency that equals the laser plus the molecular vibrational frequency.

The development of tunable lasers has led to a special technique for stimulated Raman scattering called coherent anti-Stokes Raman spectroscopy (CARS). In this technique, two lasers are used, one of fixed and the other of tunable frequency. The two beams enter the sample at angles differing only by some appropriate small amount (approximately 2°) and simultaneously impinge on the sample molecules. Whenever the frequency difference between the two lasers coincides with the frequency of a Raman-active vibration of the molecules, emission of coherent radiation (both Stokes and anti-Stokes) is stimulated. Thus the total Raman spectrum can be scanned in stimulated emission by varying the frequency of the tunable laser. An advantage of CARS, in addition to the high intensity of the scattering, is that its elevated frequency avoids interference from sample fluorescence, which always has frequencies below that of the exciting radiation.

Applications. Raman spectroscopy is of considerable value in determining molecular structure and in chemical analysis. Molecular rotational and vibrational frequencies can be determined directly, and from these frequencies it is sometimes possible to evaluate the molecular geometry, or at least to find the molecular symmetry.

Even when a precise determination of structure is not possible, much can often be said about the arrangement of atoms in a molecule from empirical information about the characteristic Raman frequencies of groups of atoms. This kind of information is closely similar to that provided by infrared spectroscopy; in fact, Raman and infrared spectra often provide complementary data about molecular structure. The complex structures of biologically important molecules, for example, are the subjects of current spectroscopic research. Both normal and resonance Raman spectroscopy are valuable techniques in molecular biology (see Fig. 4). Raman spectra also provide information for solid-state physicists, particularly with respect to lattice dynamics but also concerning the electronic structures of solids.

Bibliography. N. Bloembergen, *Non-Linear Optics*, 1965; A. J. Clark and R. E. Hester (eds.), *Advances in Infrared and Raman Spectroscopy*, vols. 1–11, 1975–1984; G. Herzberg, *Infrared and Raman Spectra of Polyatomic Molecules*, vol. 2, 2d ed., 1945; D. A. Long, *Raman Spectroscopy*, 1977; M. C. Tobin, *Laser Raman Spectroscopy*, 1971, reprint 1981; E. B. Wilson, J. C. Decius, and P. C. Cross, *Molecular Vibrations*, 1955, reprint 1980.

QUASIELASTIC LIGHT SCATTERING
ROBERT PECORA

Small frequency shifts or broadening from the frequency of the incident radiation in the light scattered from a liquid, gas, or solid. The term quasielastic arises since the frequency changes are usually so small that, without instrumentation specifically designed for their detection, they would not be observed and the scattering process would appear to occur with no frequency changes at all, that is, elastically. The technique is used by chemists, biologists, and physicists to study the dynamics of molecules in fluids, mainly liquids and liquid solutions.

Several distinct experimental techniques are grouped under the heading of quasielastic light scattering (QLS). Intensity fluctuation spectroscopy (IFS) is the technique most often used to study such systems as macromolecules in solution and critical phenomena where the molecular motions to be studied are rather slow. This technique, also called photon correlation spectroscopy and, less frequently, optical mixing spectroscopy, is used to measure the dynamical constants of processes with relaxation time scales slower than about 10^{-6} s. For faster processes, dynamical constants are obtained by utilizing techniques known as filter methods, which obtain direct measurements of the frequency changes of the scattered light by utilizing a monochromator or filter much as in Raman spectroscopy. SEE RAMAN EFFECT; SCATTERING OF ELECTROMAGNETIC RADIATION.

Static light scattering. If light is scattered by a collection of scatterers, the scattered intensity at a point far from the scattering volume is the result of interference between the wavelets scattered from each of the scatterers and, consequently, will depend on the relative positions and orientations of the scatterers, the scattering angle θ, and the wavelength λ of the light used. The structure of scatterers in solution whose size is comparable to $(4\pi\lambda)\sin\theta/2\ (\equiv q)$ where q is the length of the scattering vector, may be studied by this technique, variously called static light scattering, integrated intensity light scattering, or in the older literature simply light scattering. It was, in fact, developed in the 1940s and 1950s to measure equilibrium properties of polymers both in solution and in bulk. Molecular weights, radii of gyration, solution virial coefficients, molecular optical anisotropies, and sizes and structure of heterogeneities in bulk polymers are routinely obtained from this type of experiment. Static light scattering is a relatively mature field, although continued improvements in instrumentation (mainly the use of lasers and associated techniques) are steadily increasing its reliability and range of application.

Both static and quasielastic light scattering experiments may be performed with the use of polarizers to select the polarizations of both the incident and the scattered beams. The plane containing the incident and scattered beams is called the scattering plane. If an experiment is performed with polarizers selecting both the incident and final polarizations perpendicular to the scattering plane, the scattering is called polarized scattering. If the incident polarization is perpendicular to the scattering plane and the scattered polarization lies in that plane, the scattering is called depolarized scattering. Usually the intensity associated with the polarized scattering is much larger than that associated with the depolarized scattering. The depolarized scattering from relatively small objects is zero unless the scatterer is optically nonspherical.

Intensity fluctuation spectroscopy. The average intensity of light scattered from a system at a given scattering angle depends, as stated above, on the relative positions and orientations of the scatterers. However, molecules are constantly in motion due to thermal forces, and are constantly translating, rotating and, for some molecules, undergoing internal rearrangements. Because of these thermal fluctuations, the scattered light intensity will also fluctuate. The intensity will fluctuate on the same time scale as the molecular motion since they are proportional to each other.

Figure 1 shows a schematic diagram of a typical intensity fluctuation apparatus. Light from a laser source traverses a polarizer to ensure a given polarization. It is then focused on a small volume of the sample cell. Light from the scattering volume at scattering angle θ is passed through an analyzer to select the polarization of the scattered light, and then through pinholes and lenses to the photomultiplier (PM) tube. The output of the photomultiplier is amplified, discriminated, sent to a photon counter, and then to a hard-wired computer called an autocorrelator, which computes the time autocorrelation function of the photocounts. The autocorrelator output is then sent to a computer for further data analysis.

The scattered light intensity as a function of time will resemble a noise signal. In order to facilitate interpretation of experimental data in terms of molecular motions, the time correlation function of the scattered intensity is usually computed by the autocorrelator. The autocorrelation function obtained in one of these experiments is often a single exponential decay, $C(t) = \exp(-t/\tau_r)$, where τ_r is the relaxation time.

The upper limit on decay times τ_r that can be measured by intensity fluctuation spectroscopy is about a microsecond, although with special variations of the technique somewhat faster

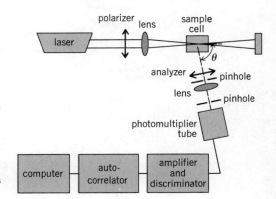

Fig. 1. Schematic diagram of an intensity fluctuation spectroscopy apparatus.

decay times may be measured. For times faster than this, filter experiments are usually performed by using a Fabry-Perot interferometer.

Fabry-Perot interferometry. Light scattered from scatterers which are moving exhibits Doppler shifts or broadening due to the motion. Thus, an initially monochromatic beam of light from a laser will be frequency-broadened by scattering from a liquid, gas, or solid, and the broadening will be a measure of the speed of the motion. For a dilute gas the spectrum will usually be a gaussian. For a liquid, however, the most common experiment of this type yields a single lorentzian line with its maximum at the laser frequency $I(\omega) = A/\pi[(1/\tau_r)/(\omega^2 + 1/\tau_r^2)]$. **Figure 2** shows a schematic of a typical Fabry-Perot interferometry apparatus. The Fabry-Perot interferometer acts as the monochromator and is placed between the scattering sample and the photomultiplier. Fabry-Perot interferometry measures the (average) scattered intensity as a function of frequency change from the laser frequency. This intensity is the frequency Fourier transform of the time correlation function of the scattered electric field. Intensity fluctuation spectroscopy experiments utilizing an autocorrelator measure the time correlation function of the intensity (which equals the square of the scattered electric field). For scattered fields with gaussian amplitude distributions the results of these two types of experiment are easily related. Sometime intensity fluctuation spectroscopy experiments are performed in what is sometimes called a heterodyne mode. In this case, some unscattered laser light is mixed with the scattered light on the surface of the photodetector. Intensity fluctuation spectroscopy experiments in the heterodyne mode measure the frequency Fourier transform of the time correlation function of the scattered electric field. SEE INTERFEROMETRY.

Translational diffusion coefficients. The most widespread application of quasielastic light scattering is the measurement of translational diffusion coefficients of macromolecules and particles in solution. For particles in solution whose characteristic dimension R is small compared to $q^{-1} = (4\pi/\lambda \sin \theta/2)^{-1}$, that is, $qR < 1$, it may be shown that the time correlation function measured in a polarized intensity fluctuation spectroscopy experiment is a single exponential with relaxation time $1/\tau_r = 2q^2D$, where D is the particle translational diffusion coefficient. For rigid, spherical particles of any size an intensity fluctuation spectroscopy experiment also provides a measure of the translational diffusion coefficient.

Translational diffusion coefficients of spherical particles in dilute solution may be used to obtain the particle radius R through use of the Stokes-Einstein relation [Eq. (1)], where k_B is

$$D = \frac{k_B T}{6\pi\eta R} \tag{1}$$

Boltzmann's constant, T the absolute temperature, and η the solvent viscosity. If the particles are

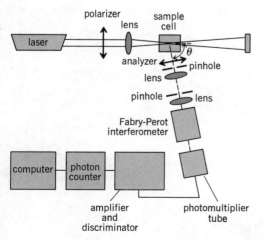

Fig. 2. Typical filter apparatus.

shaped like ellipsoids of revolution or long rods, relations known, respectively, as the Perrin and Broersma equations may be used to relate the translational diffusion coefficient to particle dimensions. For flexible macromolecules in solution and also for irregularly shaped rigid particles, the Stokes-Einstein relation is often used to define a hydrodynamic radius (R_H).

This technique is routinely used to study such systems as flexible coil macromolecules, proteins, micelles, vesicles, viruses, and latexes. Size changes such as occur, for instance, in protein denaturation may be followed by intensity fluctuation spectroscopy studies of translational diffusion. In addition, the concentration and, in some cases, the ionic strength dependence of D are monitored to yield information on particle interactions and solution structure.

Intensity fluctuation spectroscopy experiments are also used to obtain mutual diffusion coefficients of mixtures of small molecules (for example, benzene–carbon disulfide mixtures) and are also used to measure the behavior of the mutual diffusion coefficient near the critical (consolute) point of a binary liquid mixture. Experiments of this type have proved to be very important in formulating theories of phase transitions.

Rotational diffusion coefficients. Rotational diffusion coefficients are most easily measured by depolarized quasi-elastic light scattering. The instantaneous depolarized intensity for a nonspherical scatterer depends upon the orientation of the scatterer. Rotation of the scatterer will then modulate the depolarized intensity. In a similar way, the frequency distribution of the depolarized scattered light will be broadened by the rotational motion of the molecules. Thus, for example, for dilute solutions of diffusing cylindrically symmetric scatterers, a depolarized intensity fluctuation spectroscopy experiment will give an exponential intensity time correlation function with the decay constant containing a term dependent on the scatterer rotational diffusion coefficient D_R [Eq. (2)]. A depolarized filter experiment on a similar system will give a single lorentzian

$$1/\tau_r = 2(q^2 D + 6 D_R) \quad (2)$$

with $1/\tau_r$ equal to one-half that given in Eq. (2). For small molecules (for example, benzene) and relatively small macromolecules (for example, proteins with molecular weight less than 30,000) in solution, filter experiments are used to determine rotational diffusion coefficients. In these cases, the contribution of the translational diffusion to τ_r is negligible. For larger, more slowly rotating macromolecules, depolarized intensity fluctuation spectroscopy experiments are used to determine D_R.

Quasielastic light scattering is the major method of studying the rotation of small molecules in solution. Studies of the concentration dependence, viscosity dependence, and anisotropy of the molecular rotational diffusion times have been performed on a wide variety of molecules in liquids as well as liquid crystals.

Rotational diffusion coefficients of very large (≥ 100 nm) nonspherical particles may also be measured from polarized intensity fluctuation spectroscopy experiments at high values of q.

Other applications. There are many variations on quasi-elastic light scattering experiments. For instance, polarized filter experiments on liquids also give a doublet symmetrically placed about the laser frequency. Known as the Brillouin doublet, it is separated from the incident laser frequency by $\pm C_s q$, where C_s is the hypersonic sound velocity in the scattering medium. Measurement of the doublet spacing then yields sound velocities. This technique is being extensively utilized in the study of bulk polymer systems as well as of simple liquids.

In a variation of the intensity fluctuation spectroscopy technique, a static electric field is imposed upon the sample. If the sample contains charged particles, the molecule will acquire a drift velocity proportional to the electric field strength $v = \mu E$, where μ is known as the electrophoretic mobility. Light scattered from this system will experience a Doppler shift proportional to v. Thus, in addition to particle diffusion coefficients, quasielectric light scattering can be used to measure electrophoretic mobilites.

Quasielectric light scattering may also be used to study fluid flow and motile systems. Intensity fluctuation spectroscopy, for instance, is a widely used technique to study the motility of microorganisms (such as sperm cells); it is also used to study blood flow.

Bibliography. L. P. Bayvel and A. R. Jones, *Electromagnetic Scattering and its Applications*, 1981; B. J. Berne and R. Pecora, *Dynamic Light Scattering*, 1976; B. Chu, *Laser Light Scattering*, 1974.

NONLINEAR OPTICS
John F. Reintjes

A field of study concerned with the interaction of electromagnetic radiation and matter in which the matter responds in a nonlinear manner to the incident radiation fields. The nonlinear response can result in intensity-dependent variation of the propagation characteristics of the radiation fields or in the creation of radiation fields that propagate at new frequencies or in new directions. Nonlinear effects can take place in solids, liquids, gases, and plasmas, and may involve one or more electromagnetic fields as well as internal excitations of the medium. The wavelength range of interest generally coincides with the spectrum covered by lasers, extending from the far infrared to the vacuum ultraviolet, but some nonlinear interactions have been observed at wavelengths extending from the microwave to the x-ray ranges. Historically, nonlinear optics precedes the laser, but most of the work done in the field has made use of the high powers available from lasers. SEE LASER.

Nonlinear materials. Nonlinear effects of various types are observed at sufficiently high light intensities in all materials. It is convenient to characterize the response of the medium mathematically by expanding it in a power series in the electric and magnetic fields of the incident optical waves. The linear terms in such an expansion give rise to the linear index of refraction, linear absorption, and the magnetic permeability of the medium, while the higher-order terms give rise to nonlinear effects. Direct calculations of some nonlinear susceptibilities, the coefficients in the expansion that describe the strength of the nonlinear interactions, have been made in some materials, notably atomic and molecular gases and some crystals, using various theories. More generally, however, the nonlinear coefficients are determined by measurement. SEE ABSORPTION OF ELECTROMAGNETIC RADIATION; MAGNETIC PERMEABILITY; REFRACTION OF WAVES.

In general, nonlinear effects associated with the electric field of the incident radiation dominate over magnetic interactions. The even-order dipole susceptibilities are zero except in media which lack a center of symmetry, such as certain classes of crystals, certain symmetric media to which external forces have been applied, or at boundaries between certain dissimilar materials. Odd-order terms can be nonzero in all materials regardless of symmetry. Generally the magnitudes of the nonlinear susceptibilities decrease rapidly as the order of the interaction increases. Second- and third-order effects have been the most extensively studied of the nonlinear interactions, although effects up to order 30 have been observed in a single process. In some situations multiple low-order interactions occur, resulting in a very high effective order for the overall nonlinear process. For example, ionization through absorption of effectively 100 photons has been observed. In other situations, such as dielectric breakdown, or saturation of absorption, effects of different order cannot be separated, and all orders must be included in the response.

Second-order effects. Second-order effects involve a polarization with the dependence $P_{nl}^{(2)} = dE^2$, where E is the electric field of the optical waves and d is a nonlinear susceptibility. The second-order polarization has components that oscillate at sum and difference combinations of the incident frequencies, and also a component that does not oscillate. The oscillating components produce a propagating polarization wave in the medium with a propagation vector that is the appropriate sum or difference of the propagation vectors of the incident waves. The nonlinear polarization wave serves as a source for an optical wave at the corresponding frequency in a process that is termed three-wave parametric mixing.

Phase matching. The strongest interaction occurs when the phase velocity of the polarization wave is the same as that of a freely propagating wave of the same frequency. The process is then said to be phase-matched. Dispersion in the refractive indices, which occurs in all materials, usually prevents phase matching from occurring unless special steps are taken. Phase matching in crystals can be achieved by using noncollinear beams, materials with periodic structures, anomalous dispersion near an absorption edge, compensation using free carriers in a magnetic field, or by using the birefringence possessed by some crystals. In the birefringence technique, the one used most commonly for second-order interactions, one or two of the interacting waves propagates as an extraordinary wave in the crystal. The phase-matching conditions are achieved by choosing the proper temperature and propagation direction. For a given material these conditions depend on the wavelengths and direction of polarization of the individual waves.

The conditions for phase-matched three-wave parametric mixing can be summarized by Eqs. (1) and (2), where the ν's are the frequencies of the waves and the **k**'s are the propagation constants.

$$\nu_3 = \nu_1 \pm \nu_2 \qquad (1) \qquad\qquad \mathbf{k}_3 = \mathbf{k}_1 \pm \mathbf{k}_2 \qquad (2)$$

When phase matching is accomplished, the power in the generated wave is many orders of magnitude greater than that generated in non-phase-matched interactions. SEE CRYSTAL OPTICS.

Frequency mixing. In three-wave sum- and difference-frequency mixing, two incident waves at ν_1 and ν_2 are converted to a third wave at ν_3 according to Eq. (1). The simplest interaction of this type is second-harmonic generation in which $\nu_3 = 2\nu_1$. For this interaction the phase-matching condition reduces to $n(\nu_3) = n_1(\nu_1)/2 + n_2(\nu_1)/2$, where $n(\nu_3)$ is the refractive index at the harmonic wavelength and $n_1(\nu_1)$ and $n_2(\nu_1)$ are the refractive indices of the incident waves at the fundamental frequency. By using radiation from pulsed lasers, conversion efficiencies of over 90% have been achieved in second-harmonic generation. With continuous-wave lasers, it is more efficient if the nonlinear crystal is placed in the laser cavity. Conversion efficiencies of over 30% of the internal laser power have been obtained in this way for continuous-wave lasers.

Second-harmonic generation and second-order frequency mixing have been demonstrated at wavelengths ranging from the infrared to the ultraviolet, generally coinciding with the transparency range of the nonlinear crystals. Sum-frequency mixing has been used at wavelengths as short as 185 nm. Difference-frequency mixing can be used to generate both visible and infrared radiation. Radiation has been generated in this way to wavelengths of about 2 mm. If one or more of the incident wavelengths is tunable, the generated wavelength will also be tunable, providing a source of radiation that is useful for high-resolution spectroscopy. Radiation can be converted from longer-wavelength ranges, such as the infrared, to shorter-wavelength regions, such as the visible, where more sensitive and more convenient detectors are available.

Parametric generation. Parametric generation is the reverse process of sum frequency mixing. In parametric generation a single input wave at ν_3 is converted to two lower-frequency waves according to the relation $\nu_3 = \nu_1 + \nu_2$. The individual values of ν_1 and ν_2 are determined by the simultaneous satisfaction of the phase-matching condition in Eq. (2). Generally this condition can be satisfied for only one pair of frequencies at a time for a given propagation direction. By changing the phase-matching conditions, the individual longer wavelengths can be tuned. Parametric generation can be used to amplify waves at lower frequencies than the pump wave or, in an oscillator, as a source of tunable radiation in the visible or infrared over a wavelength range similar to that covered in difference-frequency mixing.

Optical rectification. The component of the second-order polarization that does not oscillate produces an electrical voltage. The effect is called optical rectification and is a direct analog of the rectification that occurs in electrical circuits at much lower frequencies. It has been used in conjunction with ultrashort mode-locked laser pulses to produce electrical pulses that, at a duration of several picoseconds, are among the shortest electrical pulses that have been generated. SEE OPTICAL PULSES.

Third-order interactions. Third-order interactions give rise to several types of nonlinear effects. Four-wave parametric mixing involves interactions which generate waves at sum- and difference-frequency combinations of the form $\nu_4 = \nu \pm \nu_2 \pm \nu_3$ with the corresponding phase-matching condition $\mathbf{k}_4 = \mathbf{k}_1 \pm \mathbf{k}_2 \pm \mathbf{k}_3$. Phase matching in liquids is usually accomplished by using noncollinear pump beams. Phase matching in gases can also be accomplished by using anomalous dispersion near absorption resonances or by using mixtures of gases, one of which exhibits anomalous dispersion. The use of gases, which are usually transparent to longer and shorter wavelengths than are the solids used for second-order mixing, allows the range of wavelengths covered by parametric mixing interactions to be extended considerably. Four-wave mixing processes of the type $\nu_4 = \nu_1 - \nu_2 - \nu_3$ have been used to generate far-infrared radiation out to wavelengths of the order of 25 micrometers. Sum-frequency mixing and third-harmonic generation have been used to generate radiation extending into the vacuum ultraviolet. Radiation down to 35.5 nm has been obtained with still higher-order processes.

The nonlinear susceptibility is greatly increased, sometimes by four to eight orders of magnitude, through resonant enhancement that occurs when the input frequencies or their multiples or sum or difference combinations coincide with appropriate energy levels in the nonlinear medium. Two-photon resonances are of particular importance since they do not involve strong absorption of the incident or generated waves. Resonant enhancement in gases has allowed tunable dye lasers to be used for nonlinear interactions, providing a source of tunable radiation in the vacuum ultraviolet and in the far infrared. Such radiation is useful in spectroscopic studies of atoms and molecules.

Just as with three-wave mixing, four-wave sum-frequency generation can be used to convert infrared radiation to the visible where it is more easily detected. These interactions have been used for infrared spectroscopic studies and for infrared image conversion.

Self-action and related effects. Nonlinear polarization components at the same frequencies as those in the incident waves can result in effects that change the index of refraction or the absorption coefficient, quantities that are constants in linear optical theory.

Multiphoton absorption and ionization. Materials that are transparent at low optical intensities can have their absorption increase at high intensities. This effect involves simultaneous absorption of multiple numbers of photons from one or more incident waves. When the process involves transitions to discrete upper levels in gases or to conduction bands in solids that obey certain quantum-mechanical selection rules, it is usually termed multiphoton absorption. When transitions to the continuum of gases are involved, the process is termed multiphoton ionization.

Saturable absorption. In materials which have a strong linear absorption at the incident frequency, the absorption can decrease at high intensities. This effect, termed saturable absorption, was observed long before lasers were available. Saturable absorbers are useful in operating Q-switched and mode-locked lasers.

Self-focusing and self-defocusing. Intensity-dependent changes in refractive index can affect the propagation characteristics of a laser beam. For many materials, for example, many solids, Kerr active liquids, and some gases, the index of refraction increases with increasing optical intensity. If the laser beam has a profile that is more intense in the center than at the edges, the profile of the refractive index corresponds to that of a positive lens, causing the beam to focus. This effect, termed self-focusing, can cause an initially uniform laser beam to break up into many smaller spots with diameters of the order of a few micrometers and intensity levels that are high enough to damage many solids. This is a primary mechanism that limits the maximum intensities that can be obtained from some high-power pulsed solid-state lasers. SEE KERR EFFECT.

In other materials the refractive index decreases as the optical intensity increases. The resulting nonlinear lens causes the beam to defocus, an effect termed self-defocusing. When encountered in media that are weakly absorbing, the effect is termed thermal blooming. It is prominent, for example, in the propagation of high-power infrared laser beams through the atmosphere.

Broadening of spectrum. When pulsed laser fields are involved, the nonlinear refractive index can lead to a broadening of the laser spectrum. In some situations the broadened spectrum can extend from the ultraviolet to the infrared and has been used with picosecond-duration pulses for time-resolved spectroscopic studies. It can also be combined with dispersive delay lines to shorten the pulse duration. Some of the shortest-duration optical pulses, of the order of 8 femtoseconds, have been generated in this manner.

Degenerate four-wave mixing. The same interactions that give rise to the self-action effects can also cause nonlinear interactions between waves that are at the same frequency but are otherwise distinguishable, for example, by their direction of polarization or propagation. Termed degenerate four-wave mixing, this interaction gives rise to a number of effects such as amplification of a weak probe wave and generation of a fourth wave that propagates in a direction opposite to that of a probe wave. In one of the most interesting aspects of this interaction, the backward wave has the same distribution of phase variations as the probe wave but with their sense reversed. This effect, termed phase conjugation or time reversal, can allow the correction of phase distortions of light beams caused by propagation through inhomogeneous media, by aberrated optics, or by mode dispersion in optical waveguides. The reconstruction of an image that is viewed through a distorting piece of glass is illustrated in **Fig. 1**. Phase conjugation has important applications in the area of image transmission through optical fibers and in various forms of optical processing and information transfer and can be used for real-time holography. The four-wave mixing phase-conjugation interaction has been used inside dye laser ring cavities to generate pulses with subpicosecond duration. SEE HOLOGRAPHY; IMAGE PROCESSING; OPTICAL FIBERS; OPTICAL PHASE CONJUGATION.

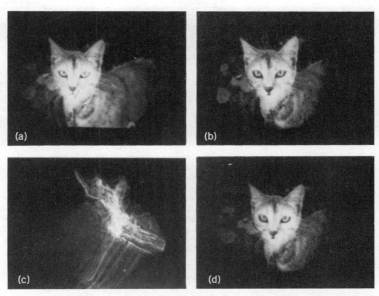

Fig. 1. Use of nonlinear optical phase conjugation for correction of distortions on an image. (a) Image obtained with a normal mirror and no distortion. (b) Image obtained with an optical conjugator and no distortion. (c) Image obtained with a normal mirror when the object is viewed through a distorting piece of glass. (d) Image obtained with an optical conjugator when the object is viewed through a distorting piece of glass. (*From J. Feinberg, Self-pumped, continuous-wave phase conjugator using internal reflection, Opt. Lett., 7:486–488, 1982*)

Control by low-frequency fields. Low-frequency electric, magnetic, or acoustic fields can be used to control the polarization or propagation of an optical wave. These effects, termed electro-, magneto-, or acoustooptic effects, are useful in the modulation and deflection of light waves and are used in information-handling systems. SEE ACOUSTOOPTICS; ELECTROOPTICS; MAGNETOOPTICS; OPTICAL INFORMATION SYSTEMS; OPTICAL MODULATORS.

Nonlinear spectroscopy. The variation of the nonlinear susceptibility near the resonances that correspond to sum- and difference-frequency combinations of the input frequencies forms the basis for various types of nonlinear spectroscopy which allow study of energy levels that are not normally accessible with linear optical spectroscopy.

Nonlinear spectroscopy can be performed with many of the interactions discussed earlier. Multiphoton absorption spectroscopy can be performed by using two strong laser beams, or a strong laser beam and a weak broadband light source. If two counterpropagating laser beams are used, spectroscopic studies can be made of energy levels in gases with spectral resolutions much smaller than the Doppler limit. Nonlinear optical spectroscopy has been used to identify many new energy levels with principal quantum numbers as high as 150 in several elements.

Many types of four-wave mixing interactions can also be used in nonlinear spectroscopy. The most widespread of these processes is a four-wave interaction of the form $v_4 = 2v_1 - v_2$, and utilizes Raman resonances which coincide with the energy difference $v_1 - v_2$. Termed coherent anti-Stokes Raman spectroscopy (CARS), it offers the advantage of greatly increased signal levels over linear Raman spectroscopy for the study of certain classes of materials. CARS has been used in areas such as the study of combustion dynamics, molecular temperatures, the viewing of selected types of chemical species in biological materials on a microscopic spatial scale, and measurement of nonliner optical susceptibilities. An example of the use of CARS for distinguishing different molecular species on a microscopic spatial scale is illustrated in **Fig. 2**. The spectroscopy

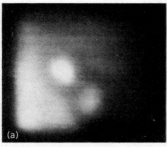 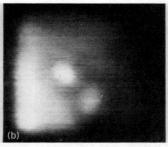 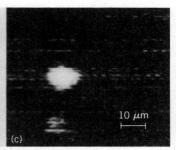

Fig. 2. Use of coherent anti-Stokes-Raman scattering (CARS) for molecular selective microscopy. (a) Image of a sample containing both deuterated and nondeuterated liposomes (artificial membranes) as seen when the lasers are tuned to the Raman resonance in the deuterated material. Both deuterated and nondeuterated liposomes are visible. (b) Image of the same sample when the lasers are detuned from the Raman resonance. Both deuterated and nondeuterated liposomes are again visible. (c) Image obtained when the nonresonant image in b is subtracted from the resonant image in a. Only the deuterated material is visible. The approximate diameter of the liposomes is 5 micrometers. (From M. D. Duncan, Molecular discrimination and contrast enhancement using a scanning coherent anti-Stokes Raman microscope, Opt. Comm., 50:307–312, 1984)

of energy levels near the generated frequency in the vacuum ultraviolet has been studied by using four-wave mixing interactions of the type $\nu_4 = 2\nu_1 + \nu_2$. Other types of nonlinear spectroscopy involve polarization changes or higher-order processes. SEE RAMAN EFFECT.

Time measurements. Various nonlinear interactions, such as second harmonic generation, multiphoton absorption followed by fluorescence, or coherent Raman scattering, have provided a means for measurement of pulse duration and excited-state lifetimes in the picosecond and subpicosecond time regimes.

Inelastic scattering. Light can scatter inelastically from fundamental excitations in the medium, resulting in the production of radiation at new frequencies that differ from the incident frequencies by an amount that is equal to the frequency of the fundamental excitation. The difference in photon energy between the incident and scattered fields is taken up or given up by the medium. Inelastic scattering can take place from any of the possible excitations in the medium, and include: Brillouin scattering from acoustic vibrations; various forms of Raman scattering involving internal vibrations or rotational modes of molecules, optical phonons, electronic states in atoms or molecules, polaritons, plasmons, spin waves, Landau levels in semiconductors, and transitions that involve spin flips in semiconductors; Rayleigh scattering involving entropy fluctuations; and scattering from density variations or concentration fluctuations in gases. Inelastic scattering may occur in liquids, solids, gases, or plasmas and may be influenced by the application of external fields or other externally applied forces.

At power levels that are available from pulsed lasers, the scattered radiation can become stimulated, resulting in exponential growth of the scattered power. In stimulated scattering the incident light can be almost completely scattered. Stimulated scattering has been observed for most of the interactions that were mentioned above. The stimulated scattered light has many of the properties of the incident laser light, including collimation and narrow line width. By using different materials, stimulated scattering can provide laserlike sources in spectral regions that are not directly covered by lasers. In some interactions the use of external electric or magnetic fields can allow the production of tunable stimulated scattered radiation. Under certain conditions the scattered field is phase-conjugate to the incident field and can be used to correct phase aberrations. SEE SCATTERING OF ELECTROMAGNETIC RADIATION.

Coherent effects. Another class of effects involves a coherent interaction between the optical field and an atom in which the phase of the atomic wave functions is preserved during the interaction. These interactions involve the transfer of a significant fraction of the atomic population to an excited state. As a result, they cannot be described with the simple perturbation expansion used for the other nonlinear optical effects. Rather they require that the response be described by using all powers of the incident fields. These effects are generally observed only for short light pulses, of the order of several nanoseconds or less. In one interaction, termed self-induced transparency, a pulse of light of the proper shape, magnitude, and duration can propagate unattenuated in a medium which is otherwise absorbing.

Other coherent effects involve changes of the propagation speed of a light pulse or production of a coherent pulse of light at a characteristic time after two pulses of light spaced by a time τ have entered the medium. The generated pulse in this last effect is termed a photon echo. Still other coherent interactions involve oscillations of the atomic polarization, giving rise to effects known as optical nutation and free induction decay. Two-photon coherent effects are also possible. SEE NONLINEAR OPTICAL DEVICES.

Bibliography. D. H. Auston, in S. L. Shapiro (ed.), *Ultra Short Light Pulses*, 1976; D. C. Hanna, M. A. Yuratich, and D. Cotter, *Nonlinear Optics of Free Atoms and Molecules*, 1979; H. Rabin and C. L. Tang (eds.), *Nonlinear Optics, Quantum Electronics*, vol. 1, 1975; J. Reintjes, *Nonlinear Optical Parametric Processes in Liquids and Gases*, 1984.

METEOROLOGICAL OPTICS
ROBERT GREENLER

The study of optical phenomena occurring in the atmosphere. Many light effects can be seen by looking skyward, and all of them, resulting from the interaction of light with the atmosphere, lie in the province of atmospheric optics or meteorological optics. The subject also includes the effect of light waves too long or too short to be detected by the human eye—light-type radiation in the infrared or ultraviolet regions of the spectrum. Light interacts with the different components of the atmosphere by a variety of physical processes, the most important being scattering, reflection, refraction, diffraction, absorption, and emission. Some other processes involving photochemistry and ionization are not considered in this article.

Scattering. An observer of light in the sky, while looking in a direction away from the Sun, will see evidence that some process has changed the path of the sunlight to direct it to the observer's eye. For most of the light in the sky this process is that of scattering, by which some of the incident light is sent off in all directions. Scattering by dust in the air makes visible the beam of light coming into a room through a window. Scattering by dust particles or liquid droplets in the atmosphere can add to the brightness of the sky, but the sky would appear light even if the atmosphere were free of all such particles. On a submicroscopic scale, air is not a continuous, uniform fluid but is composed of molecules, which are smaller than the wavelength of visible light by a factor of about a thousand. Such small particles scatter light but do not scatter all wavelengths with equal efficiency. They scatter the short light waves (blue) more strongly than the long waves (red), with the result that the clear sky appears blue.

The Sun's disk, seen high in the sky on a clear day, appears white. As it moves closer to the horizon, its color changes to yellow, and to orange, and perhaps to red. When the low sun is observed, the light rays that reach the eyes travel a much longer path through the Earth's atmosphere than when the Sun is overhead. Throughout this path, the shorter (blue) wavelengths are selectively scattered in all directions, with the result that the unscattered light that reaches the observer from the Sun is depleted in the blue end of the spectrum and appears orange or red. Thus the blue sky and the red setting Sun are both consequences of the same scattering process.

Larger particles in the atmosphere scatter light more strongly than the gas molecules. If they are of sizes less than the wavelength of visible light, they will selectively scatter the shorter, visible wavelength and contribute to red sunsets. Smoke from large forest fires or fine ejected material from volcanic eruptions can enhance the colors of sunsets thousands of miles away. In fact, following major volcanic explosions, sunsets may be enhanced over the entire Earth as fine ejected material mixes with the atmosphere and circles the Earth with a fallout time of up to a few years.

White clouds are composed of either small transparent water droplets or small transparent ice crystals, but cloud particles are small on a scale different from the scatterers of the blue sky. The droplets are typically 50 times larger than the wavelength of visible light, and so would be considered large particles in light scattering. Such large particles scatter light of all visible wavelengths equally well so that the scattered light would have the same spectral distribution as the incident sunlight, which is known as white light. Even if there is some wavelength dependence of the scattered light from one particle, multiple scattering from many particles in a cloud of the appropriate thickness will result in the scattered light appearing white. If the atmosphere of the Earth were made thicker (that is, if air were added), the sky would appear whiter. This effect can be seen even on a clear day: the sky near the horizon is whiter than the sky overhead. There is more air along a sighting path near the horizon than along a vertical path through the atmospheric layer.

The blue color of distant mountains, hazes, or smogs demonstrates the wavelength dependence of light scattering from small particles. Because red or infrared radiation is scattered less by some hazes, red filters or infrared-sensitive film or detectors are sometimes used in aerial photography or satellite image recording for better penetration of haze layers. SEE SCATTERING OF ELECTROMAGNETIC RADIATION.

Reflection. When light encounters the smooth surface of an ice crystal, some of the light is reflected from the surface and some is transmitted. Sometimes many small ice crystals grow in the atmosphere and slowly fall through the air. Their presence is seen as clouds or as a haze or (ice) fog. The very complicated, many branched ice crystals known as snowflakes usually occur at ground level in temperate regions; simple ice-crystal forms can be found at higher elevation in the sky.

As hexagonal flat-plate crystals fall through the air, they tend to orient with their wide, flat surfaces nearly horizontal. Reflection of sunlight from the nearly horizontal surfaces of such crystals can give rise to a vertical column of light that appears above or below the Sun (or both), when the Sun is near the horizon. These sun pillars are similar in origin to the streak of sunlight (called a glitter path) reflected from the slightly rough surface of a lake or ocean.

Sun pillars can also be produced by hexagonal-column crystals that have the shape of a wooden pencil (before sharpening). Such crystals tend to fall with their long axes horizontal, and light reflection from the long side faces produce sun pillars. Reflection from the vertically oriented faces of falling crystals (the side faces of falling plates or the end faces of falling pencils) can result in the parhelic circle, a band of light parallel to the horizon, passing through the Sun and extending all the way around the sky. Many other effects can result from a combination of reflection and refraction in falling ice crystals or raindrops.

A particular form of a sun pillar, called the subsun, can be seen when looking down from an airplane flying over a layer of ice-crystal clouds (see **illus**.). When the Sun is high in the sky and the crystals are well oriented, a somewhat elongated, bright spot can be seen. It is just a reflected image of the Sun in the nearly horizontal ice-crystal surfaces. SEE REFLECTION OF ELECTROMAGNETIC RADIATION.

Refraction. When a ray of light passes from one transparent medium to another, it is refracted at the surface, that is, its direction changes abruptly. Light from the Sun is thus refracted as it enters an ice crystal and again as it leaves. As a result of these two refractions, light deviates from its original direction. For a hexagonal ice crystal the deviation can bring sunlight to an observer's eye from different directions, producing a variety of spots, arcs, or streaks of light intensity in the sky. Sun dogs and the 22° halo are two of the more commonly observed effects that result from light refracted by falling ice crystals. Light that enters an ice crystal and is internally reflected from one or more crystal faces before emerging can produce a wide variety of halo effects that are observable with the naked eye.

In the case of water, where surface tension acts as an elastic skin, squeezing a small falling drop into a spherical shape, light rays that enter the transparent sphere are reflected internally before they emerge to produce a rainbow.

When a light ray passes through air whose density is not uniform but changes gradually with position, the ray gradually changes its direction, traveling in a smooth curve. The normal variation in atmospheric pressure (and hence of air density) with height causes stars near the horizon to appear elevated above their true position. The curved paths of the light rays enable observation of the Sun, apparently sitting just above the horizon, when it actually is located geometrically just below the horizon. The density of air also changes as a result of the air temperature.

INTERACTION OF LIGHT WITH MATTER 243

The subsun, resulting from sunlight reflected off the nearly horizontal faces of falling flat-plate ice crystals. (*From R. Greenler, Rainbows, Halos, and Glories, Cambridge University Press, 1980*)

Temperature variation of air near the Earth's surface can produce a number of optical distortions that are referred to as mirages. Small-scale variations and temporal fluctuations of air density resulting from air turbulence or temperature variations result in the twinkling of stars or the shimmering of distant scenes. *See* REFRACTION OF WAVES.

Diffraction. Diffraction depends on both the wavelength of the light and the size of the particles. If the Moon is viewed through a thin cloud containing water droplets or small ice crystals, it seems to be surrounded by colored rings. The rings are a diffraction effect, with the long (red) wavelengths giving rise to a larger set of rings than the shorter (blue) wavelengths. The resulting display is called the corona. A similar display, called the glory, can be seen by looking in exactly the opposite direction from the Sun or Moon; an observer flying in the sunlight over a cloud layer can see the colored rings of glory around the shadow of the airplane on the clouds below. *See* DIFFRACTION.

Emission and absorption. Small particles in the air, and the gas molecules that constitute the air, can absorb and emit light. In a heavily polluted atmosphere the color of the sky may be affected by the selective color absorption of smoke particles, but in general there is not much absorption or emission of visible-light in the atmosphere. The absorption and emission of infrared radiation, however, are very important processes in establishing the temperature of the Earth's atmosphere and the Earth's surface. *See* ABSORPTION OF ELECTROMAGNETIC RADIATION.

Bibliography. C. F. Bohren and D. R. Huffman, *Absorption and Scattering of Light by Small Particles*, 1983; R. Greenler, *Rainbows, Halos, and Glories*, 1980; E. J. McCartney, *Optics of the Atmosphere*, 1976; M. Minnaert, *Light and Colour in the Open Air*, 1954; R. A. R. Tricker, *Introduction to Meteorological Optics*, 1970.

6

INTERACTION OF LIGHT WITH ENERGY FIELDS

Electrooptics	246
Kerr effect	247
Magnetooptics	248
Faraday effect	250
Acoustooptics	252
Optical modulators	253
Photoelasticity	257

ELECTROOPTICS

MICHEL A. DUGUAY AND JANIS A. VALDMANIS

The branch of physics that deals with the influence of an electric field on the optical properties of matter, especially in crystalline form. These properties include transmission, emission, and absorption of light.

An electric field applied to a transparent crystal can change its refractive indexes and, therefore, alter the state of polarization of light propagating through it. When the refractive-index changes are directly proportional to the applied field, the phenomenon is termed the Pockels effect. When they are proportional to the square of the applied field, it is called the Kerr effect. SEE KERR EFFECT; POLARIZED LIGHT; REFRACTION OF WAVES.

Pockels cell. The Pockels effect is used in a light modulator called the Pockels cell. This device (**Fig. 1**) consists of a crystal C placed between two polarizers P_1 and P_2 whose transmission axes are crossed. A crystal often used for this application is potassium dihydrogen phosphate (KH_2PO_4). Ring electrodes bonded to two crystal faces allow an electric field to be applied parallel to the axis OZ, along which a light beam (for example, a laser beam) is made to propagate. The crystal has been cut and oriented in such a way that, in the absence of an electric field, the polarization of light propagating along OZ does not change, and therefore no light is transmitted past P_2. However, when an electronic driver V applies an electric field of the proper magnitude (2000 V/cm or 5000 V/in. is a typical value) across the crystal, the crystal becomes birefringent, and light that has been vertically polarized by P_1 undergoes a 90° change in polarization in the crystal, and is now well transmitted past P_2. Pockels cells of this type can be switched on and off in well under 1 nanosecond. SEE BIREFRINGENCE; CRYSTAL OPTICS.

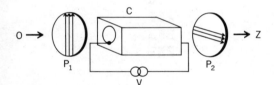

Fig. 1. Typical Pockels cell light modulator. The arrows (*OZ*) represent a light beam.

Electrooptic sampling. The linearity and high-speed response of the Pockels effect within an electrooptic crystal make possible a unique optical technique for measuring the amplitude of repetitive high-frequency (greater than 1 GHz) electric signals that cannot be measured by conventional means. The technique is known as electrooptic sampling and employs a special traveling-wave Pockels cell between crossed polarizers, P_1 and P_2 (**Fig. 2**). The electrooptic crystal (commonly lithium tantalate, $LiTaO_3$) is placed between electrodes that are part of a high-speed transmission line along which the unknown electric signal travels. Ultrashort optical pulses (on the order of 1 picosecond) from a mode-locked laser, synchronized with the electric signals and directed through the crystal, are used to repetitively sample a small portion of the electrical wave-

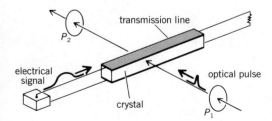

Fig. 2. An electrooptic sampling arrangement using a transmission-line Pockels cell. Examples of optical and electrical waveforms are shown.

form. After passing through the second polarizer P_2, the optical pulses have an intensity proportional to the amplitude of that portion of the electrical waveform experienced as the pulse passed through the crystal. By varying the relative delay between the arrival of the optical pulse and electric signal in the crystal and graphing the transmitted intensity as a function of delay, an accurate replica of the electrical waveform is obtained. The availability of subpicosecond optical pulses makes it possible to measure electric signals containing frequencies up to 1 THz with similar temporal resolution. Electrooptic sampling is an important technique for the analysis of ultrafast electric signals such as those generated by high-speed transistors and optical detectors. SEE LASER; OPTICAL DETECTORS; OPTICAL MODULATORS; OPTICAL PULSES.

Bibliography. J. A. Valdmanis, G. A. Mourou, and C. W. Gabel, Subpicosecond electrical sampling, *IEEE J. Quant. Electr.*, QE-19:664–667, April 1983; A. Yariv, *Optical Electronics*, 3d ed., 1985.

KERR EFFECT
MICHEL A. DUGUAY

Electrically induced birefringence that is proportional to the square of the electric field. When a substance (especially a liquid or a gas) is placed in an electric field, its molecules may become partly oriented. This renders the substance anisotropic and gives it birefringence, that is, the ability to refract light differently in two directions. This effect, which was discovered in 1875 by John Kerr, is called the electrooptical Kerr effect, or simply the Kerr effect. SEE BIREFRINGENCE.

When a liquid is placed in an electric field, it behaves optically like a uniaxial crystal with the optical axis parallel to the electric lines of force. The Kerr effect is usually observed by passing light between two capacitor plates inserted in a glass cell containing the liquid. Such a device is known as a Kerr cell. There are two principal indices of refraction, n_o and n_e (known as the ordinary and extraordinary indices), and the substance is called a positively or negatively birefringent substance, depending on whether $n_e - n_o$ is positive or negative.

Light passing through the medium normal to the electric lines of force (that is, parallel to the capacitor plates) is split into two linearly polarized waves traveling with the velocities c/n_o and c/n_e, respectively, where c is the velocity of light, and with the electric vector vibrating perpendicular and parallel to the lines of force.

The difference in propagation velocity causes a phase difference δ between the two waves, which, for monochromatic light of wavelength λ_0, is $\delta = (n_e - n_o)x/\lambda_0$, where x is the length of the light path in the medium.

Kerr constant. Kerr found empirically that $(n_e - n_o) = \lambda_0 B E^2$, where E is the electric field strength and B a constant characteristic of the material, called the Kerr constant. Havelock's law states that $B\lambda n/(n - 1)^2 = k$, where n is the refractive index of the substance in the absence of the field and k is a constant characteristic of the substance but independent of the wavelength λ. Roughly speaking, the Kerr constant is inversely proportional to the absolute temperature. The phase difference δ is determined experimentally by standard optical techniques. If the wavelength λ_0 is expressed in meters, and the field strength E in volts/m, the Kerr constant for carbon disulfide, which has been determined most accurately, is $B = 3.58 \times 10^{-14}$. Values of B range from -25×10^{-14} for paraldehyde to $+384 \times 10^{-14}$ for nitrobenzol.

The theory of the Kerr effect is based on the fact that individual molecules are not electrically isotropic but have permanent or induced electric dipoles. The electric field tends to orient these dipoles, while the normal agitation tends to destroy the orientation. The balance that is struck depends on the size of the dipole moment, the magnitude of the electric field, and the temperature. This theory accounts well for the observed properties of the Kerr effect.

In certain crystals there may be an electrically induced birefringence that is proportional to the first power of the electric field. This is called the Pockels effect. In these crystals the Pockels effect usually overshadows the Kerr effect, which is nonetheless present. In crystals of cubic symmetry and in isotropic solids (such as glass) only the Kerr effect is present. In these substances the electrically induced birefringence (Kerr effect) must be carefully distinguished from that due to mechanical strains induced by the same field. SEE ELECTROOPTICS.

Kerr shutter. An optical Kerr shutter or Kerr cell consists of a cell containing a liquid (for example, nitrobenzene) placed between crossed polarizers. As such, its construction very much resembles that of a Pockels cell. With a Kerr cell (see **illus**.), an electric field is applied by means of an electronic driver v and capacitorlike electrodes in contact with the liquid; the field is perpendicular to the axis of light propagation and at 45° to the axis of either polarizer. In the absence of a field, the optical path through the crossed polarizers is opaque. When a field is applied, the liquid becomes birefringent, opening a path through the crossed polarizers. In commercial Kerr cell shutters, the electric field (a typical value is 10 kV/cm) is turned on and off in a matter of several nanoseconds (1 ns = 10^{-9} s). For laser-beam modulation the Pockels cell is preferred because it requires smaller voltage pulses. Kerr cell shutters have the advantage over the Pockels cell of a wider acceptance angle for the incoming light. SEE OPTICAL MODULATORS.

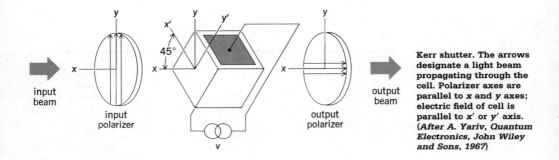

Kerr shutter. The arrows designate a light beam propagating through the cell. Polarizer axes are parallel to x and y axes; electric field of cell is parallel to x' or y' axis. (*After A. Yariv, Quantum Electronics, John Wiley and Sons, 1967*)

A so-called ac Kerr effect or optical Kerr effect has also been observed and put to use in connection with lasers. When a powerful plane-polarized laser beam propagates through a liquid, it induces a birefringence through a mechanism that is very similar to that of the ordinary, or dc, Kerr effect. In this case, it is the ac electric field of the laser beam (oscillating at a frequency of several hundred terahertz) which lines up the molecules. By using laser pulses with durations of only a few picoseconds (1 ps = 10^{-12} s), extremely fast optical Kerr shutters have been built in the laboratory. SEE LASER.
Bibliography. A. Yariv, *Optical Electronics*, 3d ed., 1985.

MAGNETOOPTICS
G. H. DIEKE AND WILLIAM W. WATSON

That branch of physics which deals with the influence of a magnetic field on optical phenomena. Considering the fact that light is electromagnetic radiation, an interaction between light and a magnetic field would seem quite plausible. It is, however, not the direct interaction of the magnetic field and light that produces the known magnetooptic effects, but the influence of the magnetic field upon matter which is in the process of emitting or absorbing light.

Zeeman effect. This produces a splitting of spectrum lines when the emitting light source is placed in a magnetic field. The inverse Zeeman effect refers to a similar splitting of absorption lines when the absorbing substance is in a magnetic field. The Zeeman effect for spectrum lines originating from closely spaced levels is called the Paschen-Back effect. The explanation of most other magnetooptical phenomena is based on the Zeeman effect, which therefore may be regarded as the basic magnetooptic effect.

Faraday effect. This is the rotation of the plane of polarization of light when light traverses certain substances in a magnetic field. SEE FARADAY EFFECT.

Voigt effect. An anisotropic substance placed in a magnetic field becomes birefringent (doubly refracting), and its optical properties are similar to those of a uniaxial crystal. The Faraday effect is the result of this birefringence when observations are made parallel to the magnetic field. The analogous observations perpendicular to the magnetic lines of force are more difficult and were not successfully carried out until 1898 because of the smallness of the effect. The transverse magnetooptic birefringence is called the Voigt effect after its discoverer, W. Voigt.

The Voigt effect (also called magnetic double refraction) can easily be calculated for substances having a normal Zeeman effect. For more complicated Zeeman effects the results can also be theoretically predicted but are less simple quantitatively, though not essentially different from those in the simpler cases.

The Voigt effect depends on the fact that the indices n_s and n_p for light polarized perpendicular or parallel to the magnetic lines of force respectively are different from one another in a magnetic field where the absorption line shows a Zeeman effect. The value of n_p is independent of the magnetic field because the central component does not change while $n_s = \frac{1}{2}(n^+ + n^-)$. The appropriately labeled curve in the **illustration** gives $n_s - n_p$, on which the observed effects of the transverse magnetic double refraction depend.

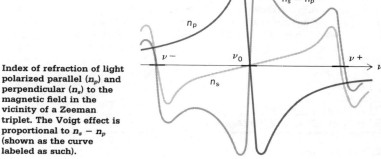

Index of refraction of light polarized parallel (n_p) and perpendicular (n_s) to the magnetic field in the vicinity of a Zeeman triplet. The Voigt effect is proportional to $n_s - n_p$ (shown as the curve labeled as such).

The theoretical formulas which represent the double refraction, though readily derived, are quite complicated. When the wavelength is considerably removed from the Zeeman triplet, the phase difference is as given by Eq. (1), where ν_0 is absorption line frequency, ν is frequency of

$$\delta = \frac{2\pi x}{c}(n_p - n_s) = \frac{e^4 f x}{32\pi^2 c^3 n_0 (\nu - \nu_0)^3} NH^2 \qquad (1)$$

transmitted light, e is charge of electron, x is path length, c is velocity of light, n_0 is index of refraction without field, N is number of absorbing atoms per unit volume, H is magnetic field strength, and f is so-called oscillator strength, measure of strength of absorption.

Contrary to the situation existing in the Faraday effect, first-order effects of magnetic double refraction are canceled out because of the presence of two perpendicular, symmetrically placed Zeeman components of combined strength equal to that of the parallel component. Because of this the Voigt effect can be observed only in the vicinity of sharply defined absorption lines, that is, in gases and in certain crystals having sharp lines, such as rare-earth salts. In the case of the rare-earth salts a linear Voigt effect is also possible at extremely low temperatures.

Since the formula for phase difference contains the oscillator strength f, the Voigt effect may be used to measure this important quantity.

Cotton-Mouton effect. This effect is concerned with the double refraction of light in a liquid when the liquid is placed in a transverse magnetic field. It is analogous to the electrooptical Kerr effect and is observed in liquids with complicated molecular structure. If the molecule has a magnetic moment, the field tries to orient the molecule, but the thermal motion tends to oppose this action. There is thus a degree of orientation, which is dependent on the temperature. If the molecule itself is optically anisotropic, the liquid also will be anisotropic and will exhibit double refraction. SEE KERR EFFECT.

The Cotton-Mouton effect is observed chiefly in nitrobenzene and aromatic organic liquids. Aliphatic compounds have a considerably smaller effect.

The phase difference of the Cotton-Mouton effect is expressed by Eq. (2), where x is the

$$\delta = C_m x H^2 \tag{2}$$

path length and C_m is called the Cotton-Mouton constant. For nitrobenzene at a temperature of 61.3°F (16.3°C) and a wavelength λ of 578 nanometers, $C_m = 2.53 \times 10^{-2}/(\text{m} \cdot \text{T}^2) = 2.53 \times 10^{-12}/(\text{cm} \cdot \text{gauss}^2)$. With large magnets ($H = 4.65$ teslas $= 46{,}500$ gauss) under the most favorable circumstances, rotations of the plane of polarization up to 27° have been observed.

The dispersion of the Cotton-Mouton effect is given by Havelock's law as in the Kerr effect.

Magnetooptic Kerr effect. This deals with the changes that are produced in the optical properties of a reflecting surface of a ferromagnetic substance when the substance is magnetized. In a typical case this will result in elliptically polarized light appearing in reflection, when the ordinary rules of metallic reflection would given only plane-polarized light. The component produced by the magnetic field when the magnetization is close to saturation is only of the order 10^{-3} of that normally present. The explanation must be sought in the fact that the conduction electrons made to vibrate by the incident light will have a curved path in the magnetic field.

Majorana effect. This deals with optical anisotropy of colloidal solutions. The effect probably is caused by the orientation of the particles in the magnetic field.

Magnetooptic effects have played an increasingly important role in microwave spectroscopy, where transitions between the Zeeman components of a single level can be observed directly.

FARADAY EFFECT
G. H. DIEKE AND WILLIAM W. WATSON

Rotation of the plane of polarization of a beam of linearly polarized light when the light passes through matter in the direction of the lines of force of an applied magnetic field. Discovered by M. Faraday in 1846, the effect is often called magnetic rotation. The magnitude α of the rotation depends on the strength of the magnetic field H, the nature of the transmitting substance, the frequency v of the light, the temperature, and other parameters. In general, $\alpha = VxH$, where x is the length of the light path in the magnetized substance and V the so-called Verdet constant. The constant V is a property of the transmitting substance, its temperature, and the frequency of the light.

The Faraday effect is particularly simple in substances having sharp absorption lines, that is, in gases and in certain crystals, particularly at low temperatures. Here the effect can be fully explained from the fundamental properties of the atoms and molecules involved. In other substances the situation may be more complex, but the same principles furnish the explanation.

Rotation of the plane of polarization occurs when there is a difference between the indices of refraction n^+ for right-handed polarized light and n^- for left-handed polarized light. Most substances do not show such a difference without a magnetic field, except optically active substances such as crystalline quartz or a sugar solution. It should be noted that the index of refraction in the vicinity of an absorption line changes with the frequency (**Fig. 1**a). SEE POLARIZED LIGHT.

When the light travels parallel to the lines of force in a magnetic field, an absorption line splits up into two components which are circularly polarized in opposite directions; that is the normal Zeeman effect. This means that, for one line, only right-handed circularly polarized light is absorbed, and for the other one, only left-handed light. The indices of refraction n^- and n^+ bear to their respective absorption frequencies the same relation as indicated in Fig. 1a; that is, they are identical in shape but displaced by the frequency difference between the two Zeeman components. It is evident that $n^+ - n^-$ is different from zero (Fig. 1b), and the magnetic rotation is proportional to this difference. The magnitude of the rotation is largest in the immediate vicinity of the absorption line and falls off rapidly as the frequency of the light increases or decreases.

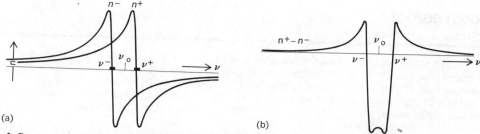

Fig. 1. Curves used in explaining the Faraday effect. (a) Index of refraction for left-handed circularly polarized light (n^-) and right-handed light (n^+) in the vicinity of an absorption line split into a doublet (ν^-, ν^+) in a magnetic field. (b) Difference between two curves, $n^+ - n^-$. Magnetic rotation is proportional to this difference.

The Faraday effect may be complicated by the fact that a particular absorption line splits into more than two components or that there are several original absorption lines in a particular region of the spectrum.

The case represented in Fig. 1 is independent of the temperature, and the rotation is symmetric on both sides of an absorption line. This case is called, not quite correctly, the diamagnetic Faraday effect. It occurs when the substance is diamagnetic, which means the splitting of the absorption line is due to the splitting of the upper level only (**Fig. 2a**), and the lower level of the line is not split. The same situation prevails in general when the intensity of the two Zeeman components is equal. This holds for all substances except paramagnetic salts at very low temperatures (Fig. 2b), in which case the ν^- component is absent.

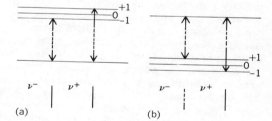

Fig. 2. Two cases of Faraday effect. (a) Diamagnetic case, which is temperature-independent, and (b) paramagnetic case, which is temperature-dependent.

In the latter case at high temperatures there are an equal number of ions in the +1 and −1 states, and the two Zeeman components have equal intensities, as in the previously discussed situation. When the temperature is lowered, however, the ions concentrate more and more in the lower level (−1), and therefore absorption by ions in the upper level (+1) disappears. At very low temperatures only the high-frequency component in the Zeeman pattern is left. In this case the n^- refraction coefficient is not affected by the presence of the absorption line and is a constant. The difference $n^+ - n^-$ will therefore have the shape of the n^+ curve in Fig. 1a. Here the rotation is not symmetric with respect to the absorption line. If it is right-handed on one side of the line, it will be left-handed on the other. As the temperature is raised, the other line comes in with increasing strength until the two are nearly equal. In the transition region the Faraday effect depends strongly on the temperature. This case has been called the paramagnetic Faraday effect.

It is possible to modulate laser light by use of the Faraday effect with a cylinder of flint glass wrapped with an exciting coil. But since this coil must produce a high magnetic field of about 19,000 gauss, this method of modulation has been little used. For a discussion of other phenomena related to the Faraday effect SEE MAGNETOOPTICS.

ACOUSTOOPTICS

Michael J. Brienza

That field of scientific investigation dealing with the interaction of acoustical and optical energy. The interaction usually occurs in a medium capable of supporting the propagation of both forms of energy, that is, one which is reasonably transparent to both. The chief acoustooptic effect is the change of the optical index of refraction of the material under the influence of strain caused by the acoustic wave. As the pressure on a material changes, so does its density and thus its index of refraction. *See* Light.

Although most acoustooptic effects were described and observed before 1935, their practical application awaited their successful union with lasers.

Magnitude of effect. In general, a small change Δn in the index of refraction n is proportional to $n^2 \Delta \rho$, where $\Delta \rho$ is the change in the density ρ. The relationship between the initial stress or pressure and the resulting changes in the index of refraction defines the photoelastic coefficient p. It can be shown that Δn equals the acoustic strain multiplied by $-n^3 p/2$. The ability of a low-intensity acoustic wave to interact with a light beam passing through it is proportional to the acoustic power and to the quantity $M = n^6 p^2 / v^3 \rho$, where v is the acoustic propagation velocity in the material. M is often referred to as a figure of merit since it depends on the properties of the medium alone. The properties which have the greatest influence on the effectiveness of acoustooptical materials are n and v. Liquids, in which v is relatively low, make good acoustooptical materials. However, due to the rapid increase in acoustic attenuation with frequency, liquids typically are useful only up to about 50 MHz. At the higher frequencies, solid materials, particularly those with high indexes of refraction, are used. Some crystalline materials such as sapphire and lithium tantalate have been used well into the microwave region, even at room temperature.

Types of effect. In the interaction of a light and a sound beam, where each is reasonably well collimated and monochromatic, and they cross each other at right angles in an acoustooptical material, let the diameter of the light beam be small compared to the wavelength of the sound (its velocity v divided by its frequency). Under these conditions the light beam is alternately bent (refracted) back and forth from its original direction as the peaks and valleys of the acoustic pressure wave pass through the light beam. The acoustic wave acts as a time-varying prism for the light passing through it. This effect has been used in optical deflectors and control devices for lasers. It is useful only at relatively low frequencies, less than about 500 kHz. As the frequency increases and the wavelength becomes comparable to the diameter, the acoustic disturbance appears as a time-varying lens, alternately converging and diverging the light beam. This effect has also been used to control certain laser systems. *See* Refraction of waves.

When the sound wavelength has decreased well below the diameter of the light beam, the light now must interact simultaneously with a series of periodic variations in the index of refraction as the sound wave propagates through it. The light beam is diffracted by this "acoustic diffraction grating" into one or more additional light beams propagating in new directions angularly displaced from the initial direction. An analogy can be made to the usual type of optical diffraction grating (consisting of a periodic array of slits), in which the angular displacement of the diffracted beams is proportional to the optical wavelength divided by the slit spacing. For acoustically diffracted light beams, the angular deflection is proportional to the optical wavelength divided by the acoustic wavelength. In this case, however, the "grating" is moving, and therefore causes a Doppler shift in the optical frequency of the diffracted light beams. This shift in optical frequency is, in most cases, exactly equal to the acoustic frequency. The intensity of the diffracted light beams is usually proportional to the intensity of the sound. *See* Diffraction.

Applications. The ability of sound to control the amplitude, direction, and frequency of a light beam (particularly a laser beam) has led to a great variety of practical devices. The well defined frequency and direction of a laser are used to full advantage in acoustooptical devices and these devices are being increasingly used in image scanners, recorders, printers, and projectors, as well as in a large number of signal-processing devices and laser controllers. *See* Laser.

The use of acoustooptic devices for real-time signal processing is of considerable contemporary importance because of their ability to handle wide-bandwidth, high-frequency signals, such as might appear in a radar environment. Contrary to digital processing systems, acoustooptic devices operate in real time with continuous readout. The special capabilities include radio-frequency spectral analysis, signal correlations and convolvers, and hybrid multiparameter processors. Both spatial and temporal integration is possible along with real-time Fourier transformation and electronically controllable filtering, a task which is usually not possible with digital computational techniques in real time. However, very high-speed integrated circuit (VHSIC) technology is permitting more and more of these tasks to be performed digitally in very small physical configuration.

Production of sound. In addition to the effect of acoustic disturbances on optical beams, the opposite effect occurs. That is, two light beams can be made to interact in an acoustooptical material to produce a sound wave. This experiment has been made possible only by the availability of very high-power, monochromatic laser beams.

Sound can also be induced in a material by the periodic irradiation of the material with an optical beam. The sound is produced by the periodic heating effects of the absorbed radiation. This technique is often used to study the properties of materials which cannot be prepared for other, more usual optical or acoustic investigation techniques. For example, the optical absorption properties of a powdered material can be investigated by using a microphone to sense the sound generated by the periodic irradiance of the powder as properties of the incident optical beam are varied.

Bibliography. R. Adler, Interaction between light and sound, *IEEE Spectrum*, 4:42–54, 1967; N. J. Berg and J. N. Lee (eds.), *Acousto-Optic Signal Processing: Theory and Implementation*, 1983; P. Debye and F. W. Sears, On the scattering of light by supersonic waves, *Proc. Nat. Acad. Sci. USA*, 18:409–421, 1932; E. I. Gordon, A review of acousto-optical deflection and modulation devices, *Proc. IEEE*, 54:1391–1401, 1966; C. F. Quate, C. D. W. Wilkenson, and D. K. Winslow, Interaction of light and microwave sound, *Proc. IEEE*, 53:1604–1623, 1965; Special Issue on Acousto-Optic Signal Processing, *Proc. IEEE*, 69:48–118, January 1981.

OPTICAL MODULATORS
Ivan P. Kaminow

Devices that serve to vary some property of a light beam. The direction of the beam may be scanned as in an optical deflector, or the phase or frequency of an optical wave may be modulated. Most often, however, the intensity of the light is modulated.

Rotating or oscillating mirrors and mechanical shutters can be used at relatively low frequencies (less than 10^5 Hz). However, these devices have too much inertia to operate at much higher frequencies. At higher frequencies it is necessary to take advantage of the motions of the low-mass electrons and atoms in liquids or solids. These motions are controlled by modulating the applied electric fields, magnetic fields, or acoustic waves in phenomena known as the electrooptic, magnetooptic, or acoustooptic effect, respectively. *See* Acoustooptics; Electrooptics; Magnetooptics.

For the most part, it will be assumed that the light to be modulated is nearly monochromatic—either a beam from a laser or a narrow-band incoherent source.

Electrooptic effect. The quadratic or Kerr electrooptic effect is present in all substances and refers to a change in refractive index, Δn, proportional to the square of the applied electric field E. The liquids nitrobenzene and carbon disulfide and the solid strontium titanate exhibit a large Kerr effect. *See* Kerr effect.

Much larger index changes can be realized in single crystals that exhibit the linear or Pockels electrooptic effect. In this case, Δn is directly proportional to E. The effect is present only in noncentrosymmetric single crystals, and the induced index change depends upon the orientations of E and the polarization of the light beam. Well-known linear electrooptic materials include potassium dihydrogen phosphate (KDP) and its deuterated isomorph (DKDP or KD*P), lithium niobate ($LiNbO_3$) and lithium tantalate ($LiTaO_3$), and semiconductors such as gallium arsenide (GaAs) and cadmium telluride (CdTe). The last two compounds are useful in the infrared (1–10 micrometers), while the others are used in the near-ultraviolet and visible regions (0.3–3 μm). *See* Crystal optics.

The phase increment Φ of an optical wave of wavelength λ that passes through a length L of material with refractive index n is given by Eq. (1). Thus phase modulation can be achieved by

$$\Phi = \frac{2\pi}{\lambda} nL \tag{1}$$

varying n electrooptically. Since the optical frequency of the wave is the time derivative of Φ, the frequency is also shifted by a time-varying Φ, yielding optical frequency modulation.

The refractive index change is given in terms of an electrooptic coefficient r by Eq. (2),

$$n(E) = n(0) \frac{-n^3 rE}{2} \tag{2}$$

where typical values for n and r are 2 and 3×10^{-11} m/V, respectively.

Electrooptic intensity modulation. Intensity modulation can be achieved by interfering two phase-modulated waves as shown in **Fig. 1**. The electrooptically induced index change is different for light polarized parallel (p) and perpendicular (s) to the modulating field; that is, $I_p \neq I_s$. The phase difference for the two polarizations is the retardation Γ given by Eq. (3), which

$$\Gamma(V) = \Phi_p - \Phi_s = \frac{2\pi}{\lambda}(n_p - n_s)L \tag{3}$$

is proportional to the applied voltage V. If it is assumed that $n_p(0) = n_s(0)$ for $V = 0$, then $\Gamma(0) = 0$. Further assume that $\Gamma(V') = \pi$. Then incident light polarized at 45° to the field may be resolved into two equal, in-phase components parallel and perpendicular to the field. For $V = 0$, they will

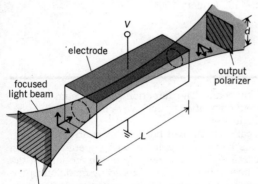

Fig. 1. Electrooptic intensity modulator.

still be in phase at the output end of the electrooptic crystal and will recombine to give a polarization at +45° which will not pass through the output polarizer in Fig. 1. For $V = V'$, however, the two components will be out of phase and will recombine to give a polarization angle of −45° which will pass through the output polarizer. The switching voltage V' is called the half-wave voltage and is given by Eq. (4), with d the thickness of the crystal and r_c an effective electrooptic

$$V' = \frac{\lambda d}{n^3 r_c L} \tag{4}$$

coefficient. A typical value for $n^3 r_c$ is 2×10^{-10} m/V. SEE POLARIZED LIGHT.

Electrooptic devices can operate at speeds up to several GHz.

Acoustooptic modulation and deflection. All transparent substances exhibit an acoustooptic effect—a change in n proportional to strain. Typical materials with a substantial effect are water, glass, arsenic selenide (As_2Se_3), and crystalline tellurium dioxide (TeO_2).

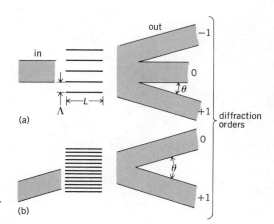

Fig. 2. Acoustooptic modulator-deflector. (a) Raman-Nath regime. (b) Bragg regime.

An acoustic wave of frequency F has a wavelength given by Eq. (5), where C is the acous-

$$\Lambda = \frac{C}{F} \tag{5}$$

tic velocity, which is typically 3×10^3 m/s. Such a wave produces a spatially periodic strain and corresponding optical phase grating with period Λ that diffracts an optical beam as shown in **Fig. 2**. A short grating ($Q < 1$) produces many diffraction orders (Raman-Nath regime), whereas a long grating ($Q > 10$) produces only one diffracted beam (Bragg regime). The quantity Q is defined by Eq. (6). If one detects the Bragg-diffracted beam (diffraction order $+ 1$), its intensity can be

$$Q = \frac{2\pi L \lambda}{n \Lambda^2} \tag{6}$$

switched on and off by turning the acoustic power on and off.

The Bragg angle θ through which the incident beam is diffracted is given by Eq. (7). Thus the beam angle can be scanned by varying F.

$$\sin \frac{\theta}{2} = \frac{\lambda}{2\Lambda} \tag{7}$$

Acoustooptic devices operate satisfactorily at speeds up to several hundred megahertz.

Optical waveguide devices. The efficiency of the modulators described above can be put in terms of the electrical modulating power required per unit bandwidth to achieve, say, 70% intensity modulation or one radian of phase modulation. For very good bulk electrooptic or acoustooptic modulators, approximately 1 to 10 mW is required for a 1-MHz bandwidth; that is, a figure of merit of 1 to 10 mW/MHz. This figure of merit can be improved by employing a dielectric waveguide to confine the light to the same small volume occupied by the electric or acoustic field, without the diffraction that is characteristic of a focused optical beam. This optical waveguide geometry is also compatible with optical fibers and lends itself to integrated optical circuits. SEE INTEGRATED OPTICS; OPTICAL COMMUNICATIONS; OPTICAL FIBERS.

An optical wave can be guided by a region of high refractive index surrounded by regions of lower index. A planar guide confines light to a plane, but permits diffraction within the plane; a strip or channel guide confines light in both transverse dimensions without any diffraction.

Planar guides have been formed in semiconductor materials such as $Al_xGa_{1-x}As$ by growing epitaxial crystal layers with differing values of x. Since the refractive index decreases as x increases, a thin GaAs layer surrounded by layers of $Al_{0.3}Ga_{0.7}As$ will guide light. If one of the outer layers is p-type and the other two layers are n-type, a reverse bias may be applied to the pn junction to produce a large electric field in the GaAs layer. Very efficient electrooptic modula-

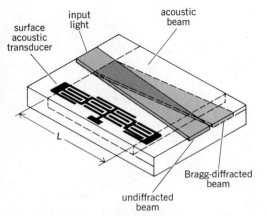

Fig. 3. Waveguide acoustooptic modulator-deflector.

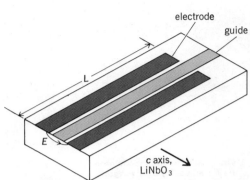

Fig. 4. Titanium-diffused LiNbO₃ strip waveguide with coplanar electrodes for electrooptic phase modulation.

tors requiring about 0.1 mW/MHz at $\lambda = 1$ μm have been realized in such junctions. Further improvement has been realized by loading the planar guide with a rib or ridge structure to provide lateral guiding within the plane. Ridge devices 1 mm long require about 5 V to switch light on or off, with a figure of merit of 10 μW/MHz.

Both planar and strip guides have been formed in LiNbO₃ by diffusing titanium metal into the surface in photolithographically defined patterns. Planar guides with surface acoustic wave transducers (**Fig. 3**) have been employed in waveguide acoustooptic devices. Strip guides (**Fig. 4**) have been used to make electrooptic phase modulators 3 cm long requiring 1 V to produce π radians of phase shift at $\lambda = 0.63$ μm; the figure of merit is 2 μW/MHz. In order to make an intensity modulator or a 2 × 2 switch, a directional coupler waveguide pattern is formed, and electrodes are applied as shown in **Fig. 5**. With no voltage applied and L the proper length, light

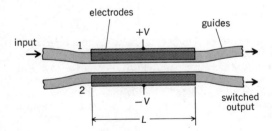

Fig. 5. Directional coupler waveguide modulator switch.

entering guide 1 will be totally coupled to and exit from guide 2; with the proper applied voltage, the phase match between the two guides will be destroyed, and no light will be coupled to guide 2, but will all exit from guide 1. Switches and modulators operating at speeds up to 5 GHz have been realized in this way.

Bibliography. R. C. Alferness, Guided wave devices for optical communication, *IEEE J. Quart. Elec.*, 17(6), June 1981; H. Elion and V. N. Morozov, *Optoelectronic Switching Systems in Telecommunications and Computers*, 1984; I. P. Kaminow, *An Introduction to Electrooptic Devices*, 1974; J. F. Nye, *Physical Properties of Crystals*, 1960; T. Tamir, *Integrated Optics*, 1979.

PHOTOELASTICITY
William Zuk

An experimental technique for the measurement of stresses and strains in material objects by means of the phenomenon of mechanical birefringence. Photoelasticity is especially useful for the study of objects with irregular boundaries and stress concentrations, such as pieces of machinery with notches or curves, structural components with slits or holes, and materials with cracks. The method provides a visual means of observing overall stress characteristics of an object by means of light patterns projected on a screen or photographic film. Regions of stress concentrations can be determined in general by simple observation. However, precise analysis of tension, compression, and shear stresses and strains at any point in an object requires more involved techniques. Photoelasticity is generally used to study objects stressed in two planar directions (biaxial), but with refinements it can be used for objects stressed in three spatial directions (triaxial).

For biaxial studies, a model geometrically similar to the object to be analyzed is prepared from a sheet of special transparent material and loaded as the object would be loaded.

Use of birefringent phenomenon. Model materials commonly used for photoelasticity are Bakelite, celluloid, gelatin, synthetic resins, glass, and other commercial products that are optically sensitive to stress and strain. The materials must have the optical properties of polarizing light when under stress (optical sensitivity) and of transmitting it on the principal stress planes with velocities dependent on the stresses (birefringence or double refraction). In addition, the material should be clear, elastic, homogeneous, optically isotropic when under no stress or strain, and reasonably free from creep, aging, and edge disturbances. *See* BIREFRINGENCE.

When the stressed model is subjected to monochromatic polarized light in a polariscope, the birefringence of the model causes the light to emerge refracted into two orthogonal planes. Because the velocities of light propagation are different in each direction, there occurs a phase shifting of the light waves. When the waves are recombined with the polariscope, regions of stress where the wave phases cancel appear black, and regions of stress where the wave phases combine appear light. Therefore, in models of complex stress distribution, light and dark fringe patterns (isochromatic fringes) are projected from the model. These fringes are related to the stresses. *See* POLARIZED LIGHT.

When white light is used in place of monochromatic light, the relative retardation of the model causes the fringes to appear in colors of the spectrum. White light is often used for demonstration, and monochromatic light is used for precise measurements.

Polariscope. A basic polariscope used in photoelasticity has a light source (generally monochromatic), a collimating lens, a polarizer, and a quarter-wave plate (**Fig. 1**). This plate is a birefringent material that causes the relative retardation of light to be exactly one-quarter the wavelength of the light. Next in the optical path is a planar model of the object under test and stressed in the direction of the plane. Finally there are a second quarter-wave plate, a polarizer called the analyzer, a focusing lens, and a viewing screen or film. Many variations of this basic transmission-type apparatus are in use. Other lenses may be added and components rearranged. If appropriate mirrors are added, the polariscope converts to a doubling type which is useful for the study of thin models under low stress, as the number of fringes doubles.

A typical isochromatic fringe pattern shows the effect on a flat plate with a central hole, pulled at the upper and lower ends (**Fig. 2**). The congestion of fringes at the boundary of the hole indicates a region of stress concentration, typical of stress behavior at cutouts. To study the exact stress at a given point in the model, the model is gradually loaded (from a condition of no load) and the number of fringe changes (fringe order) at that point is observed. Special equipment is sometimes employed to obtain partial fringe orders and to sharpen vague fringe boundaries. The fringe order is directly related by a calibrated constant to the difference of the principal stresses at that point.

Determination of principal stresses. Shear stresses can be related mathematically to the difference of the principal stresses, thereby relating shear stresses directly to fringe order. High shear stresses often cause the material to yield or fail, so that a point of large fringe order indicates a point of potential failure. In many applications of photoelasticity a knowledge of shear stress is all that is needed. This fact makes photoelasticity a simple and direct tool for investiga-

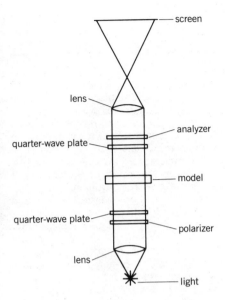

Fig. 1. Basic photoelastic polariscope.

tion of complex stress systems. However, if principal stresses and their directions are required, additional experimentation is necessary, as described later. SEE FRINGE.

Isoclinic fringes are a different set of interference patterns made by using white light, removing the quarter-wave plates and rotating the polarizer and analyzer a fixed number of degrees. These fringes represent lines making known angles with the principal planes of stress.

Stress trajectories are lines of principal stress directions over the model, obtained graphically from the isoclinic fringes. Stress trajectories are not lines of constant stress.

The determination of principal stresses requires additional information, which may be obtained in several ways. Principal stresses are determined analytically by differences of the shear stresses based on equilibrium equations. This procedure requires a numerical point by point study of the model, utilizing the shear stresses and stress trajectories. Principal stresses can be found analytically or experimentally by solution of Laplace's equation of elasticity. In principle, this procedure supplies equations pertaining to the sum of the principal stresses at any point in the model. Utilizing equations for the difference of the principal stresses from the isochromatic fringe orders, the stresses may be found by solving the two equations for the two principal stresses. Principal stresses are found experimentally by measuring the thickness of the model under stress. Because thickness changes caused by the Poisson's ratio effect are minute, a sensitive measuring

Fig. 2. Isochromatic fringe pattern for plate with hole (*From M. M. Frocht, Photoelasticity, vol. 2, John Wiley and Sons, 1948*)

device such as an optical interferometer is needed, although direct-reading thickness gages are sometimes used. The interferometer produces fringe patterns called isopachic fringes. This method essentially provides information regarding the sum of the principal stresses as with Laplace's equation. Another experimental method is to pass polarized light obliquely to the surface of the model. The relative retardations of the light produce interference fringes. These oblique fringe orders can be related to the principal stresses differently from those obtained by isochromatic fringes. Using the information on stresses from the isochromatics in conjunction with the oblique relations, the principal stresses may be obtained.

With care, stresses determined by photoelasticity are 98% accurate. With stresses determined, strains may be computed by elastic relations.

Three-dimensional measurements. Three-dimensional photoelasticity is also possible, although the techniques and stress-strain relationships are more involved than for planar objects.

The frozen stress method is well suited for three-dimension studies. Certain optically sensitive materials, such as Bakelite, when annealed in a stressed condition retain the deformation and birefringent characteristics of the initially stressed state when the load is removed. A three-dimensional model may therefore be cut into slices that may then be analyzed individually on somewhat the same principles as are the planar models.

The scattered light technique may also be used for three-dimensional models, such as torsion bars. The scattered light principle is based on the fact that polarized light passing through birefringent materials scatters in a predictable manner, acting as an optical analyzer in a polariscope.

Measurements on actual objects. The reflective polariscope method is essentially a variation of the normal polariscope. A sheet of birefringent material is bonded to the polished surface of the actual object to be studied, with a polarizer and quarter-wave plate interposed between the light and the object. The light (usually white light) passes through the stressed birefringent material and is reflected back through the quarter-wave plate and the polarizer (which now acts as an analyzer). Isochromatic fringes are projected as in a normal polariscope. Dependent on good bonding between the birefringent material and the stressed object, the reflective polariscope has the advantage that it can be used on actual structures under service loads without being reduced to model form. It may be used to find surface stresses and strains on curved or three-dimensional objects. Although photoelasticity is generally limited to elastic behavior, reflective techniques may be used to determine the nonelastic strains (but not stresses) of objects, provided the bonded optical material remains elastic. The technique may also be used with non-heterogeneous materials such as wood and concrete.

Dynamically induced stresses and strains may also be studied by photoelasticity when high-speed motion picture cameras are used to photograph the fringe patterns. Another development in photoelasticity is the use of coherent light to produce holographic interference patterns, instead of the normal fringe patterns. SEE HOLOGRAPHY.

Bibliography. H. Aben, *Integrated Photoelasticity*, 1979; J. W. Dally and W. F. Reiley, *Experimental Stress Analysis*, 2d ed., 1977; P. S. Theocaris and E. E. Gdoutos, *Matrix Theory of Photoelasticity*, 1979; T. S. Narasimhamurty, *Photoelastic and Electro-optic Properties of Crystals*, 1981; M. Nisida, *Photoelasticity '86*, 1986; F. Zandman et al., *Photoelastic Coatings*, 1977.

7

LASERS: TECHNOLOGY AND APPLICATIONS

Laser	262
Optical pumping	268
Quantum electronics	269
Coherence	270
Holography	276
Speckle	279
Optical fibers	282
Optical communications	284
Integrated optics	287
Optical information systems	291
Nonlinear optical devices	294
Optical phase conjugation	297
Optical bistability	300
Optical pulses	304

LASER

STEPHEN F. JACOBS AND ARTHUR L. SCHAWLOW

A device that uses the principle of amplification of electromagnetic waves by stimulated emission of radiation and operates in the infrared, visible, or ultraviolet region. The term laser is an acronym for light amplification by stimulated emission of radiation, or a light amplifier. However, just as an electronic amplifier can be made into an oscillator by feeding appropriately phased output back into the input, so the laser light amplifier can be made into a laser oscillator, which is really a light source. Laser oscillators are so much more common than laser amplifiers that the unmodified word "laser" has come to mean the oscillator, while the modifier "amplifier" is generally used when the oscillator is not intended.

The process of stimulated emission can be described as follows: When atoms, ions, or molecules absorb energy, they can emit light spontaneously (as in an incandescent lamp) or they can be stimulated to emit by a light wave. This stimulated emission is the opposite of (stimulated) absorption, where unexcited matter is stimulated into an excited state by a light wave. If a collection of atoms is prepared (pumped) so that more are initially excited than unexcited, then an incident light wave will stimulate more emission than absorption, and there is net amplification of the incident light beam. This is the way the laser amplifier works.

A laser amplifier can be made into a laser oscillator by arranging suitable mirrors on either end of the amplifier. These are called the resonator. Thus the essential parts of a laser oscillator are an amplifying medium, a source of pump power, and a resonator. Radiation that is directed straight along the axis bounces back and forth between the mirrors and can remain in the resonator long enough to build up a strong oscillation. (Waves oriented in other directions soon pass off the edge of the mirrors and are lost before they are much amplified.) Radiation may be coupled out as shown in **Fig. 1** by making one mirror partially transparent so that part of the amplified light can emerge through it. The output wave, like most of the waves being amplified between the mirrors, travels along the axis and is thus very nearly a plane wave. *See* OPTICAL PUMPING.

Comparison with other sources. In contrast to lasers, all conventional light sources are basically hot bodies which radiate by spontaneous emission. The electrons in the tungsten filament of an incandescent lamp are agitated by, and acquire excitation from, the high temperature of the filament. Once excited, they emit light in all directions and revert to a lower energy state. Similarly, in a gas lamp, the electron current excites the atoms to high-energy quantum states, and they soon give up this excitation energy by radiating it as light. In all the above, spontaneous emission from each excited electron or atom takes place independently of emission from the others. Thus the overall wave produced by a conventional light source is a jumble of waves from the numerous individual atoms. The phase of the wave emitted by one atom has no relation to the phase emitted by any other atom, so that the overall phase of the light fluctuates randomly from moment to moment and place to place. The lack of correlation is called incoherence.

Hot bodies emit more or less equally in all directions radiation whose wavelength distribution is dictated by the Planck blackbody radiation curve. For example, the surface of the Sun radiates like a blackbody at a temperature of about 6000 K, and emits a total of 7 kW/cm^2, spread

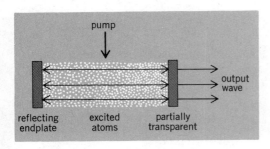

Fig. 1. Structure of a parallel-plate laser.

out over all wavelengths and directions. Light from gas lamps can be more monochromatic (wavelengths radiated are restricted by the quantized energies allowed in atoms), but radiation still occurs in all directions. In contrast to this, an ideal plane wave would have the same phase all across any wavefront, and the time fluctuations would be highly predictable (coherent). The output of the parallel-plate laser described above is very nearly such a plane wave and is therefore highly directional. This arises because in the laser oscillator atoms are stimulated to emit in phase with the stimulating wave, rather than independently, and the wave that builds up between the mirrors matches very closely the mirrors' surfaces. The output is powerful because atoms can be stimulated to emit much faster than they would spontaneously. It is highly monochromatic largely because stimulated emission is a resonance process that occurs most rapidly at the center of the range of wavelengths that would be emitted spontaneously. Since atoms are stimulated to emit in phase with the existing wave, the phase is preserved over many cycles, resulting in the high degree of time coherence of laser radiation. SEE COHERENCE.

The various types of lasers are discussed below, classified according to their pumping (or excitation) scheme. The function of the pumping system is to maintain more atoms in the upper than in the lower state, thereby assuring that stimulated emission (gain transitions) will exceed (stimulated) absorption (loss transitions). This so-called population inversion ensures net gain (amplification greater than unity).

Optically pumped lasers. One way to achieve population inversion is by concentrating light (for example, from a flash lamp or the Sun) onto the amplifying medium. Alternatively, lasers may be used to optically pump other lasers. For example, powerful continuous-wave (cw) ion lasers can pump liquid dyes to lase, yielding watts of tunable visible and near-visible radiation. Molecular lasers, like carbon dioxide (CO_2), can pump gases, like deuterium cyanide (DCN), to lase powerfully in the far infrared, operating out to several hundred micrometers wavelength. Optical pumping can be employed to pump gases at very high pressures (for example, 42 atm or 4.3 megapascals) to obtain tunability where other excitation methods would be difficult if not impossible to implement.

Many lasers are three-level lasers; that is, ground-state atoms are excited by absorption of light to a broad upper-energy state, from which they quickly relax to the emitting state. Laser action occurs as they are stimulated to emit radiation and so return to the original ground state. Crystalline, glass, liquid, and gaseous systems have been found suitable, but many possible materials remain to be explored. Solid three-level lasers usually make use of ions of a rare-earth element, such as neodymium, or of a transition metal, such as chromium, dispersed in a transparent crystal or glass. For example, ruby, which is crystalline aluminum oxide containing a fraction of a percent of trivalent chromium ions in place of aluminum ions, has been used for lasers to produce red light with wavelengths of 693 to 705 nm. The chromium ions have broad absorption bands in which pumping radiations can be absorbed. Thus broad-band white light can be used to excite the atoms.

Many rare-earth ion lasers use a fourth level above the ground state. This level serves as a terminal level for the laser transition, and is kept empty by rapid nonradiative relaxation to the ground state. This means that, relative to three-level systems, a population inversion is easier to maintain, and therefore such materials require relatively low pumping light intensity for laser action. Neodymium ions can provide laser action in many host materials, producing outputs in the infrared, around 1 micrometer. In glass, which can be made in large sizes, neodymium ions can generate high-energy pulses or very high peak powers (**Fig. 2**). Lasers using neodymium ions in crystals such as yttrium aluminum garnet (YAG) can provide continuous output powers up to a kilowatt. The output of either type (in glass or in crystals) can be converted to visible light, near 500 nm, by a harmonic generator crystal, as discussed below. Optically pumped solid-state lasers provide relatively high peak output powers. Tens of kilowatts can easily be obtained in a pulse lasting 100 µs. Much higher peak powers can be obtained by special techniques described below.

The structure of an optically pumped solid-state laser can be as simple as a rod of the light-amplifying material with parallel ends polished and coated to reflect light. Pumping radiation can enter either through the transparent sides or the ends. Other structures can be used. The mirror ends can be spherical rather than plane, with the common focal point of the two mirrors lying halfway between them. Still other structures make use of internal reflection of light rays that strike the surface of a crystal at a high angle.

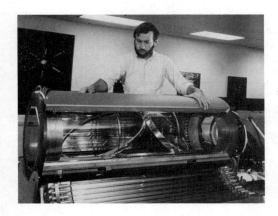

Fig. 2. High-power laser amplifier stage, with a 30-cm (1 ft) aperture, using liquid-cooled slabs of neodynium glass. (*Lawrence Livermore National Laboratory*)

Liquid lasers have structures generally like those of optically pumped solid-state lasers, except that the liquid is generally contained in a transparent cell. Some liquid lasers make use of rare-earth ions in suitable dissolved molecules, while others make use of organic dye solutions. The dyes can lase over a wide range of wavelengths, depending upon the composition and concentration of the dye or solvent. Thus tunability is obtained throughout the visible, and out to a wavelength of about 0.9 μm. Fine adjustment of the output wavelength can be provided by using a diffraction grating or other dispersive element in place of one of the laser mirrors. The grating acts as a good mirror for only one wavelength, which depends on the angle at which it is set. With further refinements, liquid-dye lasers can be made extremely monochromatic as well as broadly tunable. They may be pumped by various lasers to generate either short, intense pulses or continuous output. Dyes may also be incorporated into solid media such as plastics or gelatin to provide tunable laser action. Then the tuning may be controlled by a regular corrugation in the refractive index of the host medium, which acts as a distributed Bragg reflector. The reflection from any one layer is small, but when the successive alternations of refractive index provide reflections in phase, the effect is that of a strong, sharply tuned reflection.

Certain color centers in the alkali halides can be optically pumped to make efficient, tunable lasers for the region 0.8 to 3.3 μm, thus taking over just where the organic dyes fail. Both pulsed (10 kW) and continuous (watts) operation have been achieved.

In several infrared regions, tunable laser action can be obtained by using an infrared gas laser to pump a semiconductor crystal in a magnetic field, giving amplification by stimulating spin-flip Raman scattering from the electrons in the semiconductor. Tuning is achieved by varying the magnetic field.

Gas-discharge lasers. Another large class of lasers makes use of nonequilibrium processes in a gas discharge. At moderately low pressures (on the order of 1 torr or 10^2 Pa) and fairly high currents, the population of energy levels is far from an equilibrium distribution. Some levels are populated especially rapidly by the fast electrons in the discharge. Other levels empty particularly slowly and so accumulate large numbers of excited atoms. Thus laser action can occur at many wavelengths in any of a large number of gases under suitable discharge conditions. For some gases, a continuous discharge, with the use of either direct or radio-frequency current, gives continuous laser action. Output powers of continuous gas lasers range from less than 1 μW up to about 100 W in the visible region. Wavelengths generated span the ultraviolet and visible regions and extend out beyond 700 μm in the infrared. They thus provide the first intense sources of radiation in much of the far-infrared region of the spectrum.

The earliest, and still most widely used, gas-discharge laser utilizes a mixture of helium and neon. Various infrared and visible wavelengths can be generated, but most commonly they produce red light at a 632.8-nm wavelength, with power outputs of a few milliwatts or less, although they can be as high as about a watt. Helium-neon lasers can be small and inexpensive. Argon and krypton ion gas–discharge lasers provide a number of visible and near-ultraviolet wavelengths with continuous powers typically about 1 to 10 W, but ranging up to more than 100 W. Unfortunately, efficiencies are low.

Many molecular gases, such as hydrogen cyanide, carbon monoxide, and carbon dioxide, can provide infrared laser action. Carbon dioxide lasers can be operated at a number of wavelengths near 10 μm on various vibration-rotation spectral lines of the molecule. They can be relatively efficient, up to about 30%, and have been made large enough to give continuous power outputs exceeding tens of thousands of watts.

Many gas-discharge lasers, for example, helium-neon, produce only very small optical gain; thus losses must be kept low. Consequently, mirrors with very high reflectivity (greater than 99%) must generally be used. Diffraction losses can be kept low by using curved mirrors. One common arrangement, which combines relatively low diffraction losses with good mode selection, uses a flat mirror at one end and a spherically concave mirror at the other. The spacing between mirrors is made equal to the radius of curvature of the spherical mirror. On the other hand, in some of the higher-power, higher-gain lasers even the plane-parallel mirror structure does not give sufficient discrimination against those undesired modes of oscillation which cause the beam to be excessively divergent. It is then helpful to use "unstable" resonators, with at least one of the mirrors convex toward the other.

A smooth, small-bore dielectric tube can guide a light wave in its interior with little loss. Thus a light wave can be amplified by a long, narrow medium without spreading or diffraction. A gas discharge in a hollow optical waveguide can be run at high pressure and benefits from cooling by the nearby walls, so that relatively high-power outputs (watts) can be obtained from a small volume. Wave-guide structures are also useful when a medium is pumped optically by another laser, whose narrow beam can be confined within the bore of the tube. For example, pumping of various molecules, such as methyl fluoride, by a carbon dioxide laser has been used to generate coherent light in the very far-infrared (submillimeter-wavelength) region.

Pulsed gas lasers. Pulsed gas discharges permit a further departure from equilibrium. Thus pulsed laser action can be obtained in some additional gases which could not be made to lase continuously. In some of them the length of the laser pulse is limited when the lower state is filled by stimulated transitions from the upper level, and so introduces absorption at the laser wavelength. An example of such a self-terminating laser is the nitrogen laser, which gives pulses of several nanoseconds duration with peak powers from tens of kilowatts up to a few megawatts at a wavelength of 337 nm in the ultraviolet. They are easy to construct and are much used for pumping tunable dye lasers throughout the wavelength range from about 350 to 1000 nm. Very powerful laser radiation in the vacuum ultraviolet region, between 100 and 200 nm, can be obtained from short-pulse discharges in hydrogen, and in rare gases such as xenon at high pressure. (High pressure leads to higher power.) When the gas pressure is too high to permit an electric discharge, excitation may be provided by an intense burst of fast electrons from a small accelerator, the so-called E beam.

Some gases, notably carbon dioxide, which can provide continuous laser action, also can be used to generate intense pulses of microsecond duration. For this purpose, gas pressures of about 1 atm (10^5 Pa) are used, and the electrical discharge takes place across the diameter of the laser column, hence the name transverse-electrical-atmospheric (TEA) laser.

Chemical lasers. It is also possible to obtain laser action from the energy released in some fast chemical reactions. Atoms or molecules produced during the reaction are often in excited states. Under special circumstances there may be enough atoms or molecules excited to some particular state for amplification to occur by stimulated emission. Usually the reacting gases are mixed and then ignited by ultraviolet light or fast electrons. Both continuous infrared output and pulses up to several thousand joules of energy have been obtained in reactions which produce excited hydrogen fluoride molecules.

Pulsed laser action in the ultraviolet (193 to 353 nm) has been obtained from excimer states of rare gas monohalides (for example, KrF, XeF, KrCl, XeCl, XeBr). Such molecules have ground states which are unstable, thereby making a population inversion easy to achieve. Although these lasers require an electrical source for initial gas reaction, laser pumping is dependent on chemical reactions.

Photodissociation lasers. Intense pulses of ultraviolet light can dissociate molecules in such a way as to leave one constituent in an excited state capable of sustaining laser action. The most notable examples are iodine compounds, which have given peak 1.3-μm pulse powers above 10^9 W from the excited iodide atoms.

Nuclear lasers. Laser action in several gases has also been excited by the fast-moving ions produced in nuclear fusion. These fusion products excite and ionize the gas atoms, and make it possible to convert directly from nuclear to optical energy.

Gas-dynamic lasers. When a hot molecular gas is allowed to expand suddenly through a nozzle, it cools quickly, but different excited states lose energy at different rates. It can happen that, just after cooling, some particular upper state has more molecules than some lower one, so that amplification by stimulated emission can occur. Very high continuous power outputs have been generated from carbon dioxide in large gas-dynamic lasers.

Semiconductor lasers. Another method for providing excitation of lasers can be used with certain semiconducting materials. Laser action takes place when free electrons in the conduction band are stimulated to recombine with holes in the valence band. In recombining, the electrons give up energy corresponding nearly to the band gap. This energy is radiated as a light quantum. Suitable materials are the direct-gap semiconductors, such as gallium arsenide. In them, recombination occurs directly without the emission or absorption of a quantum of lattice vibrations. A flat junction between *p*-type and *n*-type material may be used. When a current is passed through this junction in the forward direction, a large number of holes and electrons are brought together. This is called recombination and is accompanied by emission of radiation. A light wave passing along the plane of the junction can be amplified by stimulating such recombination of electrons and holes. The ends of the semiconducting crystal provide the mirrors to complete the laser structure. In indirect-gap semiconductors, such as germanium and silicon, only a small amplification by stimulated emission is possible because of their requirement for interaction with the lattice vibrations.

Semiconductor lasers can be very small, less than 1 mm in any direction. They can have efficiencies higher than 50% (electricity to light). Power densities are high, but the thinness of the active layer tends to limit the total output power. Even so, maximum continuous powers are comparable with those of other moderate-size lasers. Since semiconductor lasers are so small, they can be assembled into compact arrays of many units, so as to generate higher peak powers. An alternative excitation method, bombardment of the semiconductor by a high-voltage beam of electrons, may provide laser action in larger crystal volumes, but it is likely to cause damage to the crystal.

Early semiconductor lasers required cooling and pulsed operation, emitting light only in the infrared. Great progress has been made, making possible room temperature, continuous operation. Monochromaticity has been improved to degrees almost comparable with gas lasers.

Free-electron lasers. Free-electron lasers are of interest because of their potential for efficiently producing very high-power radiation, tunable from the millimeter to the x-ray region. The principle of operation, so far demonstrated only at 3.4 μm and at very low-power levels, involves passage of electrons through a spatially varying magnetic field which causes the electron beam to "wiggle" and hence to radiate. The large Doppler upshift due to relativistic electron velocities can be adjusted, resulting in tunable emission at optical frequencies.

High-power, short-pulse lasers. The output power of pulsed lasers can be greatly increased, with correspondingly shorter pulse durations, by the Q-switch technique. In this method, the optical path between one mirror and the amplification medium is blocked by a shutter. The medium is then excited beyond the degree ordinarily needed, but the shutter prevents laser action. At this time the shutter is abruptly opened and the stored energy is released in a giant pulse (1–100 MW peak power, lasting 1–30 ns for optically pumped solid-state lasers). Still higher peak powers can be obtained by passing this output through a traveling-wave laser amplifier (without mirrors). Peak powers in excess of 100 MW have been obtained in this way.

Still shorter, and higher-power, pulses can be generated by mode-locking techniques. A typical laser without mode locking usually oscillates simultaneously and independently at several closely spaced wavelengths. These modes of oscillation can be synchronized so that the peaks of their waves occur simultaneously at some instant. The result is a very short, intense pulse which quickly ends as the waves of different frequency get out of step. Mode-locked lasers have generated pulses shorter than 1 ps. Since such brief pulses tend to produce somewhat less damage to materials, they can be amplified to very high peak intensities. Power outputs of picosecond pulses

as high as 10–100 MW have been obtained, limited as in the Q-switch case by material damage.
SEE OPTICAL PULSES.

For the highest peak power, the output may be further intensified by additional stages of laser amplification. The beam diameter is increased by some optical arrangement, such as a telescope, so as to expand (dilute) the beam and thereby prevent damage to the laser material and optics. Sometimes the amplifying medium is divided into flat slabs separated by cooling liquid (Fig. 2). Here the open faces of the light-amplifying slabs present a large area to receive pumping light and liquid cooling.

Development of very large multistage lasers has been undertaken for research on thermonuclear fusion. In a particularly large one, at Lawrence Livermore Laboratory, a single neodymium-glass laser oscillator is designed to drive 20 amplifier chains of glass rods and disks, each delivering a pulse of more than 10^{12} W. Focusing all these pulses onto the surface of a small pellet of heavy hydrogen is designed to heat and compress the pellet by ablation, until it is so hot that the hydrogen nuclei fuse to produce helium and release large amounts of nuclear energy. Ultimately this type of controlled laser fusion may become an important source of thermal, electrical, and chemical energy.

The technology of scaling up lasers to higher and higher powers has been so successful that the limitations are often set by material damage thresholds of the laser medium or associated optics. At the high-power densities attainable by focusing laser beams, the electric fields of the light can be large. Thus when the light intensity is 10^{12} W/cm^2, the corresponding electric field is 10^7 V/cm. To such large fields, many transparent materials have a nonlinear dielectric response. This nonlinearity can be large enough to permit the generation of optical harmonics. It is possible, with careful design and good nonlinear materials, to obtain substantially complete conversion of a laser's output to the second harmonic, at twice the frequency, even for continuous lasers near 1-W output power. Nonlinear dielectrics can be used as mixers to give the sum or difference of

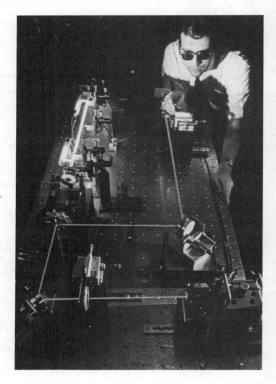

Fig. 3. Experimental arrangement to impress pulse-code modulation on a laser beam for optical communication. (*Bell Laboratories*)

two laser frequencies. They also permit the construction of optical parametric oscillators which, when pumped by a laser, can generate coherent light tunable over a wide range of wavelengths. SEE NONLINEAR OPTICS.

Applications. The variety of technological uses for lasers has increased steadily since their appearance in 1960. Among the noticeable applications are those that utilize high-speed controllability of the tiny focal spot of a laser beam. For example, high-speed automatic scanners identify library cards, ski passes, and supermarket purchases and perform a variety of functions known as optical processing. Other uses for the laser beam's programmable control include information storage and retrieval (including three-dimensional holography and video disk reading), laser printing, micromachining, and automated cutting. Further applications involving high power include weaponry, laser welding, laser surgery (self-cauterizing), laser fusion, and materials processing. Optical communications utilize the laser's high frequency, which makes possible high information capacity (**Fig. 3**). Except for space applications, laser light communication is primarily done through glass fibers. Fiberoptic laser telephone systems are presently in operation in many cities around the world. Low-loss optical fibers are far more compact, light weight, and economical than copper wires. Bright laser beams are used by the construction industry to align straight excavations and for surveying. Some applications include rotation rate sensing (laser gyros), laser velocity sensing (laser radar or lidar), optical testing, metrology (including the distance to the Moon), laser spectroscopy (including pollution monitoring), pumping other lasers to produce beams of coherent light at other wavelengths, from millimeters to x-rays, and exploration of ultrafast phenomena (by using picosecond laser pulses). SEE HOLOGRAPHY; INTEGRATED OPTICS; OPTICAL COMMUNICATIONS; OPTICAL FIBERS.

Bibliography. S. F. Jacobs et al. (eds.), *Free Electron Generators of Coherent Radiation*, vols. 7–9, 1980–1982; S. F. Jacobs et al., *Laser Photochemistry, Tunable Lasers, and Other Topics*, 1976; L. Marton and C. Marton, *Methods of Experimental Physics*, vol. 15, 1979; D. C. O'Shea, W. R. Callen, and W. T. Rhodes, *An Introduction to Lasers and Their Applications*, 1977; A. E. Siegman, *Introduction to Lasers and Masers*, 1971; D. C. Sinclair and W. E. Bell, *Gas Laser Technology*, 1969; O. Svelto, *Principles of Lasers*, 2d ed., 1982; A. Yariv, *Optical Electronics*, 3d ed., 1985.

OPTICAL PUMPING
WILLIAM WEST

The process of causing strong deviations from thermal equilibrium populations of selected quantized states of different energy in atomic or molecular systems by the use of optical radiation (that is, light of wavelengths in or near the visible spectrum), called the pumping radiation.

At thermal equilibrium at temperature TK, the relative numbers of atoms, N_2/N_1, in quantized levels E_2 and E_1, respectively, where E_2 is the higher, are given by $N_2/N_1 = e^{-(E_2 - E_1)/kT}$, where k is Boltzmann's constant. The number of atoms in the higher level is, at equilibrium, always less than that in the lower, and as the energy difference between the two levels increases the number in the higher level becomes very small indeed. By exposing a suitable system to optical radiation, one can, so to speak, pump atoms from a lower state to an upper state so as greatly to increase the number of atoms in the upper state above the equilibrium value.

In an early application of the principle, the levels E_2 and E_1 were not far apart, so that the equilibrium populations of the atoms in the two levels were not greatly different. The system was chosen to possess a third level E_3, accessible from E_1 but not from E_2 by the absorption of monochromatic visible light. (The states involved were the paramagnetic Zeeman components of atomic states, and the necessary selectivity of the transitions excitable by the visible light was secured by appropriate choice of the state of polarization of this light.) The visible light excites atoms from E_1 to E_3, from which they return, with spontaneous emission, with about equal probability, to the lower states E_2 and E_1. After a period of time, provided there has been sufficiently intense excitation by the visible light, most of the atoms are in state E_2 and few are in the lower state E_1—atoms have been pumped from E_1 to E_2 by way of the highly excited state E_3.

Optical pumping is vital for light amplification by stimulated emission in an important class of lasers. For example, the action of the ruby laser involves the fluorescent emission of red light by a transition from an excited level E_2 to the ground level E_1. In this case E_2 is relatively high above E_1 and the equilibrium population of E_2 is practically zero. Amplification of the red light by laser action requires that N_2 exceed N_1 (population inversion). The inversion is accomplished by intense green and violet light from an external source which excites the chromium ion in the ruby to a band of levels, E_3, above E_2. From E_3 the ion rapidly drops without radiation to E_2, in which its lifetime is relatively long for an excited state. Sufficiently intense pumping forces more luminescent ions into E_2 by way of the E_3 levels than remain in the ground state E_1, and amplification of the red emission of the ruby by stimulated emission can then occur. SEE LASER.

Bibliography. D. C. O'Shea, W. R. Callen, and W. T. Rhodes, *An Introduction to Lasers and Their Applications*, 1977; O. Svelto, *Principles of Lasers*, 1976; A. Yariv, *Optical Electronics*, 3d ed., 1985.

QUANTUM ELECTRONICS
A. E. SIEGMAN

A loosely defined field concerned with the interaction of radiation and matter, particularly those interactions involving quantum energy levels and resonance phenomena, and especially those involving lasers and masers. Quantum electronics encompasses useful devices such as lasers and masers and their practical applications; related phenomena and techniques, such as nonlinear optics and light modulation and detection; and related scientific problems and applications, such as quantum noise processes, laser spectroscopy, picosecond spectroscopy, and laser-induced optical breakdown.

In one sense any electronic device, even one as thoroughly classical in nature as a vacuum tube, may be considered a quantum electronic device, even one as thoroughly classical in nature as a vacuum tube, may be considered a quantum electronic device, since quantum theory is presently accepted to be the basic theory underlying all physical devices. In practice, however, quantum electronics is usually understood to refer to only those devices such as lasers and atomic clocks in which stimulated transitions between discrete quantum energy levels are important, together with related devices and physical phenomena which are excited or explored using lasers. Other devices such as transistors or superconducting devices which may be equally quantum-mechanical in nature are not usually included in the domain of quantum electronics.

Stimulated emission and amplification. Quantum electronics thus centers on stimulated-emission devices, primarily lasers and masers. Atoms, molecules, and other small isolated quantum systems typically have discrete, well-resolved quantum-mechanical energy levels. In a collection of a large number of identical such atoms or molecules, one can speak of the number of atoms, or the population, residing in each such energy level. Under normal thermal equilibrium conditions, lower energy levels always have larger populations. One may apply electromagnetic radiation (radio waves or light waves, as appropriate) to such a collection of atoms at a frequency (or wavelength) corresponding through Planck's law to the quantum-mechanical transition frequency between any two levels. This radiation will then be absorbed by the atoms, and some of the atoms will be correspondingly lifted from the lower energy level to the upper. Measurement of the strength of this absorption versus the probing frequency or wavelength provides a powerful way of studying the quantum energy levels, and provides the basic approach of spectroscopy or of resonance physics.

Under suitable conditions it is also possible, in a wide variety of atomic systems, to create a condition of population inversion, in which more atoms are temporarily placed in some upper energy level than in a lower energy level. Population inversion is created by "pumping" the atomic system, by using a wide variety of techniques. Application of a signal at the appropriate transition frequency then produces coherent amplification rather than absorption of the applied signal, with the amplification energy being supplied by a net flow of atoms from the upper to the lower energy level through stimulated transitions. The addition of electromagnetic feedback to the atomic system by means of lumped electrical circuitry, microwave circuity, or optical mirrors, as appropriate to the frequency range involved, can convert this amplification into coherent oscillation at the atomic transition frequency. While this type of stimulated emission and amplification has found useful but limited application at ordinary radio-wave and microwave frequencies (primarily in ultrastable atomic clocks and very low-noise microwave maser amplifiers), its overwhelming importance is at optical frequencies, where it makes possible for the first time the coherent amplification and oscillation of optical signals in laser devices. SEE OPTICAL PUMPING.

Nonlinear optical phenomena. Soon after the invention of the laser, it was discovered that a wide assortment of interesting and useful nonlinear optical phenomena could be produced using coherent beams from lasers because of their unprecedentedly high power and coherence. By using nonlinear optical phenomena such as harmonic generation, subharmonic generation, parametric amplification and oscillation, as well as stimulated Raman and Brillouin scattering, it becomes possible to produce a wide range of new infrared, optical, and ultraviolet wavelengths, some of them widely tunable, in addition to obtaining important basic physical information about the nonlinear materials involved. Although the elementary description of these nonlinear optical devices is thoroughly classical in nature, they have become a major portion of the field of quantum electronics. SEE NONLINEAR OPTICS; RAMAN EFFECT.

Applications of lasers. Lasers have found an enormous variety of important practical applications in engineering, technology, and medicine, ranging from highway surveying and supermarket checkout counters to automobile production lines and retinal surgery. As these applications have become established as routine techniques, they have generally moved out of the domain of quantum electronics considered as a scientific discipline. At the same time, the laser has been applied as the primary tool for many fundamental as well as exotic measurement techniques in science, including particularly high-resolution spectroscopy, such as saturated-absorption, tunable-laser, and picosecond spectroscopy. Many of these more complex techniques have been retained within the domain of quantum electronics, along with certain exotic engineering applications such as laser isotope separation and the study of laser interactions with plasmas, particularly for laser-induced fusion. SEE LASER; OPTICAL DETECTORS; OPTICAL MODULATORS.

Bibliography. F. T. Arecchi and E. O. Schulz-DuBois (eds.), *Laser Handbook*, 1972; D. Marcuse, *Principles of Quantum Electronics*, 1980; A. E. Siegman, *An Introduction to Masers and Lasers*, 1971; A. E. Siegman, *Lasers*, 1986; A. Yariv, *Quantum Electronics*, 1975.

COHERENCE
ROLF G. WINTER

The attribute of two or more waves, or parts of a wave, whose relative phase is nearly constant during the resolving time of the observer. The concept has been developed most extensively in optics, but is applicable to all wave phenomena.

Coherence of two beams. Consider two waves, with the same mean angular frequency ω, given by Eqs. (1) and (2). These expressions as they stand could describe de Broglie waves in

$$\Psi_A(x,t) = A \exp i\,[k(\omega)x - \omega t - \delta_A(t)] \quad (1) \qquad \Psi_B(x,t) = B \exp i\,[k(\omega)x - \omega t - \delta_B(t)] \quad (2)$$

quantum mechanics. For real waves, such as components of the electric field in light or radio beams, or the pressure oscillations in sound, it is necessary to retain only the real parts of these and subsequent expressions. Assume that the frequency distribution is narrow, in the sense that a Fourier analysis of expressions (1) and (2) gives appreciable contributions only for angular frequencies close to ω. This assumption means that, on the average, $\delta_A(t)$ and $\delta_B(t)$ do not change much per period. For any actual wave, however, they must undergo some change; only waves that have existed forever and that fill all of space can have absolutely fixed frequency and phase. SEE ELECTROMAGNETIC RADIATION.

Suppose that the waves are detected by an apparatus with resolving time T; that is, T is the shortest interval between two events for which the events do not seem to be simultaneous. For the human eye and ear, T is about 0.1 s, while a fast electronic device might have a T of 10^{-10} s. If the relative phase $\delta(t)$, given by Eq. (3), does not, on the average, change noticeably

$$\delta(t) = \delta_B(t) - \delta_A(t) \qquad (3)$$

during T, then the waves are coherent. If during T there are sufficient random fluctuations for all values of $\delta(t)$, modulus 2π, to be equally probable, then the waves are incoherent. If during T the fluctuations in $\delta(t)$ are noticeable, but not enough to make the waves completely incoherent, then the waves are partially coherent. These distinctions are not useful unless T is specified. Two independent sound waves in which the phases change appreciably in 0.01 s would seem incoherent to unaided human perception, but would seem highly coherent to a fast electronic device.

The degree of coherence is related to the interference patterns that can be observed when the two beams are combined. The following variations of Young's two-slit interference experiment illustrate several possibilities. *See* INTERFERENCE OF WAVES.

Young's two-slit experiment. **Figure 1** shows the usual Young's experiment. Near P, for linear wave phenomena, the resultant wave $\Psi(x,t)$ is the sum of $\Psi_A(x,t)$ and $\Psi_B(x,t)$. The observable intensity is proportional to $|\Psi|^2$, the square of the magnitude of Ψ. This observable intensity is the energy density or the mean photon density for electromagnetic waves, the energy density for acoustic waves, and the mean particle density for the wave functions of quantum mechanics. With the wave forms of Eqs. (1) and (2), for real A and B, it is given by Eq. (4). The

$$|\Psi|^2 = |\Psi_A + \Psi_B|^2 = A^2 + B^2 + 2AB \cos \delta(t) \qquad (4)$$

first term gives the intensity of beam A alone, the second gives the intensity of B alone, and the third term depends on the relative phase, given by Eq. (3).

The mean life of a typical excited atomic state is about 10^{-8} s. Collisions and thermal motion reduce the effective time of undisturbed emission to around 10^{-10} or 10^{-11} s in standard discharge tubes. The phase $\delta_A(t)$ of beam A therefore dances about erratically, with substantial

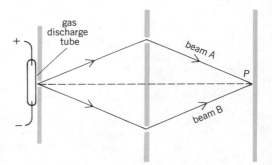

Fig. 1. Young's experiment. If beams A and B come from the same point source and traverse the same distance, they are coherent around P and produce there an interference pattern that has high visibility.

changes occurring perhaps 10^{11} times per second. However beam B comes from the same atoms and travels nearly the same distance, and the changes in $\delta_B(t)$ are the same as those in $\delta_A(t)$. The relative phase is always zero at the exact midpoint P and takes on other time-independent values in the neighborhood of P. For each point in that neighborhood, $|\Psi|^2$ has a constant value that satisfies inequality (5). A clear interference pattern is observed even with a large-T detector such as a photographic plate.

$$|A - B|^2 \leq |\Psi|^2 \leq |A + B|^2 \qquad (5)$$

Independent sources. **Figure 2** shows an arrangement that uses two independent sources to produce the two beams. If the two sources are standard discharge tubes, $\delta_A(t)$ and $\delta_B(t)$ independently change erratically around 10^{11} times per second. Equation (4) is still valid, but the observed quantity is $|\Psi|^2$ averaged over the resolving time T, denoted by $\langle|\Psi|^2\rangle_T$. When the average of Eq. (4) is formed, the last term on the right contributes nothing because $\cos \delta(t)$ randomly takes on values between $+1$ and -1 and thus obeys Eq. (6). The two beams are incoherent, the

$$\langle \cos \delta(t) \rangle_T = 0 \qquad (6)$$

observed intensity is simply the sum of the separate intensities, and the superposition does not produce a visible interference pattern. In general, if incoherent waves Ψ_A, Ψ_B, Ψ_C, . . . , are

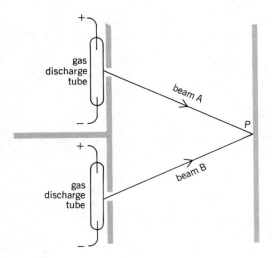

Fig 2. Independent sources that change their phases frequently and randomly during the resolving time T are incoherent, and their superposition does not produce a visible interference pattern.

combined, much the same argument applies because the average over T of each cross term is zero, so that the observed intensity is given by Eq. (7).

$$\langle |\Psi|^2 \rangle_T = \langle |\Psi_A|^2 \rangle_T + \langle |\Psi_B|^2 \rangle_T + \langle |\Psi_C|^2 \rangle_T + \cdots \qquad (7)$$

The argument that results in Eq. (6) depends on there being random independent changes in the separate phases during times much shorter than T. If the discharge tubes in Fig. 2 are replaced by very well stabilized lasers, or by loudspeakers or radio transmitters, then the separate phases $\delta_A(t)$ and $\delta_B(t)$ can be made nearly constant during 0.1 s. The relative phase $\delta(t)$ is then also nearly constant, an interference pattern is visible, and the two waves can be called coherent. SEE LASER.

Effect of path length difference. Figure 3, like Fig. 1, shows two beams that are emitted by the same point source and are therefore coherent at their origin. Here, however, the distance from the source to the point P' of wave addition is larger along beam B than along beam A by an amount l. In Fig. 3, the path length difference is due to the observation point P' not being near the central axis, but it could equally well be caused by the insertion of different optical devices in the two beams.

The waves that arrive at P' via path B at any instant must have left the source earlier by a time $\tau = l/c$, where c is the wave speed, than those that arrive via path A. Suppose again that the source is a standard discharge tube, so that the phase changes randomly about 10^{11} times per

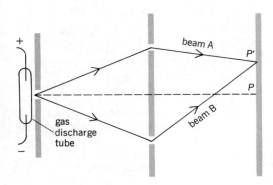

Fig. 3. If the difference between the distances traversed by beams A and B is of the order of the coherence length, then the beams are partially coherent at P'.

second. For $\tau \ll 10^{-11}$ s, the coherence between A and B at P' is not spoiled by the path length difference, and there is a visible interference pattern around P'. For $\tau \gg 10^{-11}$ s, the light that arrives via A left the source so much later than that which arrives via B at the same instant that the phase changed many times in the interim. The two beams are then incoherent, and their superposition does not produce a visible interference pattern around P'. For $\tau \simeq 10^{-11}$ s, perfect coherence is spoiled but there less than complete incoherence; there is partial coherence and an interference pattern that is visible but not very sharp. The time difference that gives partial coherence is the coherence time Δt. The coherence time multiplied by the speed c is the coherence length Δl. In this example, Δl is given by Eq. (8).

$$\Delta l \simeq 10^{-11} \text{ s} \times 3.0 \times 10^{10} \text{ cm/s} = 0.3 \text{ cm} \tag{8}$$

Effect of extended source. **Figure 4** differs from Fig. 1 in that it shows a large hole in the collimator in front of the discharge tube, so that there is an extended source rather than a point source. Consider two regions of the source, S near the center and S' near one end, separated by a distance y. During times longer than the coherence time, the difference between the phases of the radiation from the two regions changes randomly. Suppose that, at one moment, the radia-

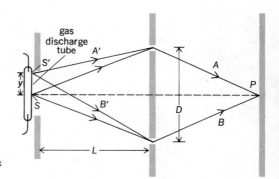

Fig. 4. If the lateral extent y of the source is of the order of $L\lambda/D$, then the beams are partially coherent at P.

tion from S happens to have a large amplitude while that from S' happens to be negligible. The waves at the two slits are then in phase because S is equidistant from them, and interference at P gives a maximum. Suppose that a few coherence times later the radiation from S happens to be negligible while that from S' happens to have a large amplitude. If the path length along A' differs from that along B' by half of the wavelength λ, then the waves at the two slits are out of phase. Now interference at P gives a minimum, and the pattern has shifted in a few coherence times. If the resolving time T is much larger than the coherence time Δt, the consequence is that beams A and B are not coherent at P.

The condition that there be a path length difference of $\lambda/2$ between B' and A' is given by Eq. (9), and for D and y both much smaller than L, expansion of the square roots gives Eq. (10).

$$\sqrt{L^2 + (y + D/2)^2} - \sqrt{L^2 + (y - D/2)^2} = \lambda/2 \quad (9) \qquad 2Dy = L\lambda \tag{10}$$

This argument is rather rough, but it is clear that a source that satisfies Eq. (11) is too large to

$$y \simeq \frac{L\lambda}{D} \tag{11}$$

give coherence and too small to give complete incoherence; it gives partial coherence.

Coherence of a single beam. Coherence is also used to describe relations between phases within the same beam. Suppose that a wave represented by Eq. (1) is passing a fixed observer characterized by a resolving time T. The phase δ_A may fluctuate, perhaps because the source of the wave contains many independent radiators. The coherence time Δt_W of the wave is defined to be the average time required for $\delta_A(t)$ to fluctuate appreciably at the position of the observer. If Δt_W is much greater than T, the wave is coherent; if Δt_W is of the order of T, the wave is partially coherent; and if Δt_W is much less than T, the wave is incoherent. These concepts are very close to those developed above. The two beams in Fig. 2 are incoherent with respect to each other if one or both have $\Delta t_W \ll T$, but are coherent with respect to each other if both have $\Delta t_W \gg T$.

The degree of coherence of a beam is of course determined by its source. Discharge tubes that emit beams with Δt_W small compared to any usual resolving time are therefore called incoherent sources, while well-stabilized lasers that give long coherence times are called coherent sources.

Consider two observers that measure the phase of a single wave at the same time. If the observers are close to each other, they will usually measure the same phase. If they are far from each other, the phase difference between them may be entirely random. The observer separation that shows the onset of randomness is defined to be the coherence length of the wave, Δx_W. If this separation is in the direction of propagation, then Δx_W is given by Eq (12), where c is the

$$\Delta x_W = c \Delta t_W \tag{12}$$

speed of the wave. Of course, the definition of Δx_W is also applicable for observer separations perpendicular to the propagation direction.

The coherence time Δt_W is related to the spectral purity of the beam, as is shown by the following argument. The wave can be viewed as a sequence of packets with spatial lengths around Δx_W. A typical packet requires Δt_W to pass a fixed point. Suppose that n periods of duration $2\pi/\omega$ occur during Δt_W, as expressed in Eq. (13). The packet can be viewed as a superposition

$$\Delta t_W \frac{\omega}{2\pi} = n \tag{13}$$

of plane waves arranged to cancel each other outside the boundaries of the packet. In order to produce cancellation, waves must be mixed in which are in phase with the wave of angular frequency ω in the middle of the packet and which are half a period out of phase at both ends. In other words, waves must be mixed in with periods around $2\pi/\omega'$, where Eq. (14) holds. The

$$\Delta t_W \frac{\omega'}{2\pi} = n \pm 1 \tag{14}$$

difference of Eqs. (13) and (14) gives Eq. (15), where the spread in angular frequencies $|\omega - \omega'|$

$$\Delta t_W \Delta \omega = 2\pi \tag{15}$$

is called $\Delta \omega$. This result should be stated as an approximate inequality (16). A large Δt_W permits

$$\Delta t_W \Delta \omega \gtrsim 1 \tag{16}$$

a small $\Delta \omega$ and therefore a well-defined frequency, while a small Δt_W implies a large $\Delta \omega$ and a poorly defined frequency.

Quantitative definitions. To go beyond qualitative descriptions and order-of-magnitude relations, it is useful to define the fundamental quantities in terms of correlation functions.

Self-coherence function. As discussed above, there are two ways to view the random fluctuations in the phase of a wave. One can discuss splitting the wave into two beams that interfere after a difference τ in traversal time, or one can discuss one wave that is passing over a fixed observer who determines the phase fluctuations. Both approaches concern the correlation between $\Psi(x,t)$ and $\Psi(x, t + \tau)$. This correlation is described by the normalized self-coherence function given by Eq. (17). The brackets $\langle \; \rangle$ indicate the average of the enclosed quantity over a

$$\gamma(\tau) \equiv \frac{\langle \Psi(x, t + \tau) \Psi^*(x,t) \rangle}{\langle \Psi(x,t) \Psi^*(x,t) \rangle} \tag{17}$$

time which is long compared to the resolving time of the observer. It is assumed here that the statistical character of the wave does not change during the time of interest, so that $\gamma(\tau)$ is not a function of t.

With the aid of the self-coherence function, the coherence time of a wave can be defined quantitatively by Eq. (18). That is, Δt_W is the root-mean-squared width of $|\gamma(\tau)|^2$. The approximate

$$(\Delta t_W)^2 \equiv \frac{\int_{-\infty}^{+\infty} \tau^2 |\gamma(\tau)|^2 \, d\tau}{\int_{-\infty}^{+\infty} |\gamma(\tau)|^2 \, d\tau} \tag{18}$$

inequality (16) can also be made precise. The quantity $|g(\omega')|^2$ is proportional to the contribution to Ψ at the angular frequency ω', where $g(\omega')$ is the Fourier transform of Ψ, defined by Eq. (19). The root-mean-squared width $\Delta\omega$ of this distribution is given by Eq. (20), where ω is again the

$$g(\omega') \equiv \frac{1}{\sqrt{2\pi}} \int_{-\infty}^{+\infty} \Psi(0,t) e^{-i\omega' t} dt \tag{19} \qquad (\Delta\omega)^2 = \frac{\int_0^\infty (\omega' - \omega)^2 |g(\omega')|^2 d\omega'}{\int_0^\infty |g(\omega')|^2 d\omega'} \tag{20}$$

mean angular frequency. With these definitions, one can show that inequality (21) is satisfied. The

$$\Delta t_W \Delta\omega \geq 1/2 \tag{21}$$

symbol $\gtrsim$ has become $\geq$; no situation can lead to a value of $\Delta t_W \Delta\omega$ that is less than $1/2$. Relation (21) multiplied by Planck's constant is the Heisenberg uncertainty principle of quantum physics, and can in fact be proved in virtually the same way.

Complex degree of coherence. To describe the coherence between two different waves Ψ_A and Ψ_B, one can use the complex degree of coherence $\gamma_{AB}(\tau)$, defined by Eq. (22), where the

$$\gamma_{AB}(\tau) \equiv \frac{\langle \Psi_A(t+\tau) \Psi_B^*(t) \rangle}{[\langle \Psi_A^*(t) \Psi_A(t) \rangle \langle \Psi_B^*(t) \Psi_B(t) \rangle]^{1/2}} \tag{22}$$

averaging process $\langle \ \rangle$ and the assumptions are the same as for the definition of $\gamma(\tau)$. If beams A and B are brought together, interference fringes may be formed where the time-averaged resultant intensity $\langle |\Psi|^2 \rangle$ has maxima and minima as a function of position. The visibility of these fringes, defined by Eq. (23), can be shown, for the usual interference studies, to be proportional to $|\gamma_{AB}(\tau)|$.

$$V = \frac{\langle |\Psi|^2 \rangle_{\max} - \langle |\Psi|^2 \rangle_{\min}}{\langle |\Psi|^2 \rangle_{\max} + \langle |\Psi|^2 \rangle_{\min}} \tag{23}$$

Examples. The concept of coherence occurs in a great variety of areas. The following are some application and illustrations.

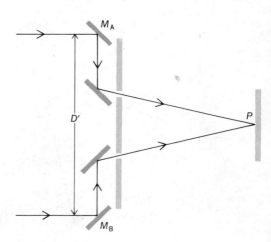

Fig. 5. Michelson's stellar interferometer. The effective slit separation D' can be varied with the movable mirrors M_A and M_B. The angular size of the source can be calculated from the resulting change in the visibility of the interference fringes around P.

Astronomical applications. As is shown in the discussion of Fig. 4, extended sources give partial coherence and produce interference fringes with visibility V [Eq. (23)] less than unity. A. A. Michelson exploited this fact with his stellar interferometer (**Fig. 5**), a modified double-slit arrangement with movable mirrors that permit adjustment of the effective separation D' of the slits. According to Eq. (11), there is partial coherence if the angular size of the source is about λ/D'. It can be shown that if the source is a uniform disk of angular diameter Θ, then the smallest value of D' that gives zero V is $1.22\lambda/\Theta$. The interferometer can therefore be used in favorable cases to measure stellar diameters. The same approach has also been applied in radio astronomy. A different technique, developed by R. Hanbury Brown and R. Q. Twiss, measures the correlation between the intensities received by separated detectors with fast electronics. SEE INTERFEROMETRY.

Lasers and masers. Because they are highly coherent sources, lasers and masers provide very large intensities per unit frequency, and double-photon absorption experiments have become possible. As a consequence, atomic and molecular spectroscopy were revolutionized. One can selectively excite molecules that have negligible velocity components in the beam direction and thus improve resolution by reduction of Doppler broadening. States that are virtually impossible to excite with single-photon absorption can now be reached.

Bibliography. M. Born and E. Wolf, *Principles of Optics*, 6th ed., 1980; L. Mandel and E. Wolf, Coherence properties of optical fields, *Rev. Mod. Phys.*, 37:231–287, 1965; J. I. Steinfeld (ed.), *Laser and Coherence Spectroscopy*, 1978; K. Thyagarajan and A. K. Ghatak, *Lasers, Theory and Applications*, 1981; R. G. Winter, *Quantum Physics*, 1979.

HOLOGRAPHY
JOSEPH W. GOODMAN

A technique for recording, and later reconstructing, the amplitude and phase distributions of a coherent wave disturbance. Invented by Dennis Gabor in 1948, the process was originally envisioned as a possible method for improving the resolution of electron microscopes. While this original application has not proved feasible, the technique is widely used as a method for optical image formation, and in addition has been successfully used with acoustical and radio waves.

Fundamentals of the technique. The technique is accomplished by recording the pattern of interference between the unknown object wave of interest and a known reference wave (**Fig. 1**). In general, the object wave is generated by illuminating the (possibly three-dimensional) subject of concern with the coherent light beam. The waves reflected from the object strike a light-sensitive recording medium, such as photographic film or plate. Simultaneously a portion of the light is allowed to bypass the object, and is sent directly to the recording plane, typically by means of a mirror placed next to the object. Thus incident on the recording medium is the sum of the light from the object and a mutually coherent reference wave.

While all light-sensitive recording media respond only to light intensity (that is, power), nonetheless in the pattern of interference between reference and object waves there is preserved

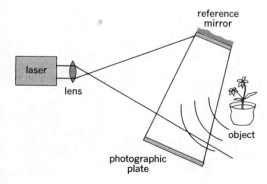

Fig. 1. Recording a hologram.

a complete record of both the amplitude and the phase distributions of the object wave. Amplitude information is preserved as a modulation of the depth of the interference fringes, while phase information is preserved as variations of the position of the fringes. SEE INTERFERENCE OF WAVES.

The photographic recording obtained is known as a hologram (meaning a total recording); this record generally bears no resemblance to the original object, but rather is a collection of many fine fringes which appear in rather irregular patterns (**Fig. 2**). Nonetheless, when this photographic transparency is illuminated by coherent light, one of the transmitted wave components is

Fig. 2. Typical appearance of a hologram (under magnification).

an exact duplication of the original object wave (**Fig. 3**). This wave component therefore appears to originate from the object (although the object has long since been removed) and accordingly generates a virtual image of it, which appears to an observer to exist in three-dimensional space behind the transparency. The image is truly three-dimensional in the sense that the observer's eyes must refocus to examine foreground and background, and indeed can "look behind" objects in the foreground simply by moving the head laterally.

Also generated are several other wave components, some of which are extraneous, but one of which focuses of its own accord to form a real image in space between the observer and the transparency. This image is generally of less utility than the virtual image because its parallax relations are opposite to those of the original object.

Applications. The holographic technique has a number of unique properties which make it of great value as a scientific tool. Although the field is young, and new applications are continually emerging, certain important areas can be identified.

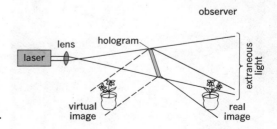

Fig. 3. Obtaining images from a hologram.

Microscopy. Historically, microscopy is the potential application of holography that has motivated much of the early work, including the original work of Gabor. The use of holography for optical microscopy has been amply demonstrated, but it is generally agreed that these techniques are not serious competitors with more conventional microscopes in ordinary microscopy.

Nonetheless, there is one area in which holography offers a unique potential for optical microscopy. This area might be called high-resolution volume imagery. In conventional microscopy, high transverse resolution is achieved only at the price of a very limited depth of focus; that is, only a limited portion of the object volume can be brought into focus at one time. It is possible, of course, to explore a large volume in sequence by continuously refocusing to examine new regions of the object volume, but such an approach is often unsatisfactory, particularly if the object is a dynamic one, continuously in motion. A solution to this problem is to record a hologram of the object by using a pulsed laser. The dynamic object is then "frozen" in time, but the recording contains all information necessary to explore the full object volume with an auxiliary optical system. Sequential observation is acceptable because the object (that is, the holographic image) is no longer dynamic. This approach has been fruitfully applied to the microscopy of three-dimensional volumes of living biological specimens and to the measurement of particle-size distributions in aerosols.

Interferometry. Holography has been demonstrated to offer the capability of several unique kinds of interferometry. This capability is a consequence of the fact that holographic images are coherent; that is, they have well-defined amplitude and phase distributions. Any use of holography to achieve the superposition of two coherent images will result in a potential method of interferometry.

The most powerful holographic interferometry techniques are based on the following property: When a photographic emulsion is multiply exposed to form several superimposed holograms, upon reconstruction the several corresponding virtual images are formed simultaneously and therefore interfere. Likewise the various real images interfere.

The most dramatic demonstrations of this type of interferometry were performed by R. E. Brooks, L. O. Heflinger, and R. F. Wuerker, using a pulsed ruby laser. Two laser pulses were used to record two separate holograms on the same transparency. Any changes of the object between pulses resulted in well-defined fringes of the interference in the reconstructed image (**Fig. 4**). The technique is particularly well suited for performing interferometry through imperfect optical elements (for example, windows of poor quality), thus making possible certain kinds of interferometry that could not be achieved by any classical means. *See* INTERFEROMETRY.

Memories. Optical memories for storing large volumes of binary data in the form of holograms have been intensively studied. Such a memory consists of an array of small holograms, each capable of reconstructing a different "page" of binary data. When one of these holograms is illuminated by coherent light, it generates a real image consisting of an array of bright or dark spots, each spot representing a binary digit. This image falls on a detector array, with one detector element for each binary digit. Thus to read a single binary digit at a specific location in the memory, a beam deflector causes light to illuminate the appropriate hologram page, and the out-

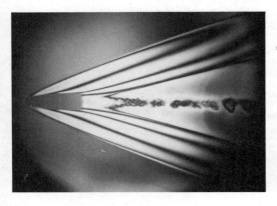

Fig. 4. Image taken by the technique of holographic interferometry, showing the compressional waves generated by a high-speed rifle bullet. (*Courtesy of R. E. Brooks, L. O. Heflinger, and R. F. Wuerker*)

put of the proper detector element is interrogated to determine whether a bright spot of light exists at that particular location in the image.

In spite of several identifiable advantages of holographic memories over other methods of optical storage, the holographic technique is not regarded as a viable commercial alternative to bit-by-bit optical storage in ablative media, as practiced, for example, with digital audio disks.

Display. There has been interest in the use of holography for purposes of display of three-dimensional images. Applications have been found in the field of advertising, and there is increased use of holography as a medium for artistic expression. A significant technical development in this area has been the perfection of a type of recording known as a multiplex hologram. Such a recording typically consists of a large number of separate holograms, all in the form of thin, contiguous, vertical strips on a single piece of film. Each of these holograms produces a virtual image of a different ordinary photograph of the subject of interest. In turn, each such photograph was originally taken from a slightly different angle. Thus when the observer examines the virtual image produced by the entire set of holograms, each eye looks through a different hologram and sees the subject from a different angle. The resulting stereo effect produces a nearly perfect illusion of three-dimensionality. Furthermore, as an observer moves the head horizontally, or as the collection of holograms is rotated, the observer's two eyes continuously see a changing pair of images. If the original set of photographs is properly chosen, the image can be made to move or dance about in nearly any desired fashion. Very dramatic three-dimensional displays of animated subjects can thus be constructed from a series of ordinary photographs. Such displays do not require a laser for viewing, but rather can be utilized with white-light sources.

Holographic optical elements. A hologram consisting of the interference of a plane reference wave and a diverging spherical wave, upon illumination by a reconstruction plane wave, will generate a diverging spherical wave (the virtual image) and a converging spherical wave (the real image), each traveling in a different angular direction. Thus such a hologram behaves as an optical focusing element, with properties similar to those of a lens (or, more accurately, a pair of lenses). More complex holograms can generate a multitude of foci, in virtually any pattern desired. Alternatively, by varying the periodicity of the grating-like structure of the hologram, a small laser beam can be deflected through an angle that is controlled by the local period of the structure. Holograms which are used to control transmitted light beams, rather than to display images, are called holographic optical elements. Interest in such elements has grown substantially, and commercial applications have been found. Most notable is the use of holograms in supermarket scanners at check-out stands. Light from a helium-neon laser falls on a small region of a holographic optical element, which was recorded on a disk and is rotating continuously. As the hologram rotates, different portions of the hologram containing different grating periods are illuminated, and the angle of deflection of the laser beam sweeps through a pattern that was predetermined when the hologram was recorded. In this way the laser beam is caused to follow a complicated scan pattern, which ultimately allows the reading of information from the bar-code patterns recorded on each product. *See* GEOMETRICAL OPTICS; OPTICAL IMAGE.

Other applications. A variety of other applications of holography has been proposed and demonstrated, including the analysis of modes of vibration of complicated objects, measurement of strain of objects under stress, generation of very precise depth contours on three-dimensional objects, and high-resolution imagery through aberrating media. These and other applications of holography will be useful in future scientific and engineering problems.

Bibliography. N. Abramson, *The Making and Evaluation of Holograms*, 1981; R. J. Collier et al. (eds.), *Principles of Holography*, 2d ed., 1977; J. W. Goodman, *Introduction to Fourier Optics*, 1968; E. N. Leith and J. Upatnieks, Photography by laser, *Sci. Amer.*, 212(6):24–35, 1965; H. M. Smith, *Principles of Holography*, 2d ed., 1975; C. M. Vest, *Holographic Interferometry*, 1979.

SPECKLE
JAMES C. WYANT

The generation of a random intensity distribution, called a speckle pattern, when light from a highly coherent source, such as a laser, is scattered by a rough surface or inhomogeneous medium. Although the speckle phenomenon has been known since the time of Isaac Newton, the

Fig. 1. Photograph of speckle pattern generated by illuminating rough surface with laser radiation.

development of the laser is responsible for the present-day interest in speckle. Speckle has proved to be a universal nuisance as far as most laser applications are concerned, and it was only in the mid-1970s that investigators turned from the unwanted aspects of speckle toward the uses of speckle patterns, in a wide variety of applications. *See* Laser.

Basic phenomenon. Objects viewed in coherent light acquire a granular appearance (**Fig. 1**). The detailed irradiance distribution of this granularity appears to have no obvious relationship to the microscopic properties of the illuminated object, but rather it is an irregular pattern that is best described by the methods of probability theory and statistics. Although the mathematical description of the observed granularity is rather complex, the physical origin of the observed speckle pattern is easily described. The surfaces of most materials are extremely rough on the scale of an optical wavelength (approximately 5×10^{-7} m). When nearly monochromatic light is reflected from such a surface, the optical wave resulting at any moderately distant point consists of many coherent wavelets, each arising from a different microscopic element of the surface. Since the distances traveled by these various wavelets may differ by several wavelengths if the surface is truly rough, the interference of the wavelets of various phases results in the granular pattern of intensity called speckle. If a surface is imaged with a perfectly corrected optical system as shown in **Fig. 2**, diffraction causes a spread of the light at an image point, so that the intensity at a given image point results from the coherent addition of contributions from many independent surface areas. As long as the diffraction-limited point-spread function of the imaging system is broad by comparison with the microscopic surface variations, many dephased coherent contributions add at each image point to give a speckle pattern.

The basic random interference phenomenon underlying laser speckle exists for sources other than lasers. For example, it explains radar "clutter," results for scattering of x-rays by liquids, and electron scattering by amorphous carbon films. Speckle theory also explains why twinkling may be observed for stars, but not for planets. *See* Coherence; Diffraction; Interference of waves.

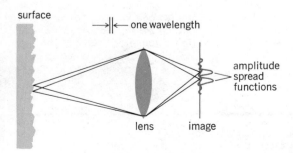

Fig. 2. Physical origin of speckle for an imaging system.

Applications. The principal applications for speckle patterns fall into two areas: metrology and stellar speckle interferometry.

Metrology. In the metrology area, the most obvious application of speckle is to the measurement of surface roughness. If a speckle pattern is produced by coherent light incident on a rough surface, then surely the speckle pattern, or at least the statistics of the speckle pattern, must depend upon the detailed surface properties. If the surface root-mean-square roughness is small compared to one wavelength of the fully coherent radiation used to illuminate the surface, the roughness can be determined by measuring the speckle contrast. If the root-mean-square roughness is large compared to one wavelength, the radiation should be spatially coherent polychromatic, instead of monochromatic; the roughness is again determined by measuring the speckle contrast.

An application of growing importance in engineering is the use of speckle patterns in the study of object displacements, vibration, and distortion that arise in nondestructive testing of mechanical components. The key advantage of speckle methods in this case is that the speckle size can be adjusted to suit the resolution of the most convenient detector, whether it be film or a television camera, while still retaining information about displacements with an accuracy of a small fraction of a micrometer. Although several techniques for using laser speckle in metrology are available, the basic idea can be explained by the following example for measuring vibration. A vibrating surface illuminated with a laser beam is imaged as shown in Fig. 2. A portion of the laser beam is also superimposed on the image of the vibrating surface. If the object surface is moving in and out with a travel of one-quarter of a wavelength or more, the speckle pattern will become blurred out. However, for areas of the surface that are not vibrating (that is, the nodal regions) the eye will be able to distinguish the fully developed high-contrast speckle pattern.

Electronic speckle interferometry (ESPI) is finding much use in industrial applications involving measurement of the vibrational properties of structures. In ESPI a television camera is used to view a speckle pattern. The output of the television camera is processed electronically to emphasize the high spatial frequency structure. The result is viewed on a television monitor (**Fig. 3**). The bright fringes correspond to the region of the object which are at the nodes of the vibration pattern. By knowing the angles at which the object is illuminated and viewed, it is possible to calculate the amplitude of the vibration from the number of dark fringes present. Thus, simply by looking at the television monitor it is possible to determine the vibrational pattern over the entire object. The amplitude and frequency of the drive source can be changed and instantly the new vibrational pattern can be observed.

Speckle patterns can also be used to examine the state of refraction of the eye. If a diffusing surface illuminated by laser light moves perpendicular to the line of sight of an observer, the speckles may appear to move with respect to the surface. For a normal eye, movement in a direction opposite to that of the surface indicates underaccommodation, and movement with the surface indicates overaccommodation. If the speckles do not move, but just seem to "boil," the observed surface is imaged on the retina.

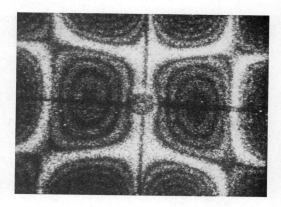

Fig. 3. Electronic speckle interferogram of vibrating object.

Stellar speckle interferometry. Stellar speckle interferometry, which has many similarities with the laser speckle methods used in metrology, is a technique for obtaining diffraction-limited resolution of stellar objects despite the presence of the turbulent atmosphere which limits the resolution of conventional pictures to approximately 1 arc-second. If a short-exposure photograph is taken of a magnified image of an unresolved star and a narrow-bandwidth spectral filter is used, the picture has a specklelike structure. The size of the speckles is equal to the diffraction-limited resolution limit of the telescope, regardless of the resolution limit determined by the turbulent atmosphere. This means that the short-exposure photograph of a resolvable object, for example a binary star, contains information about the object down to the diffraction limit of the telescope, which is approximately 0.02 arc-second for the 200-in. (5-m) Palomar Mountain telescope, one-fiftieth the resolution limit set by the atmosphere. Hence, by extracting correctly the information in short-exposure pictures of objects with more than one resolvable element, detail down to the diffraction limit of the telescope can be observed. The technique has proved to be of enormous value in the study of binary stars and centrosymmetric resolvable stars, and research is directed at making the technique useful for observing objects having a more general shape. Image intensifiers and modern electronics have made it possible to look at much dimmer astronomical objects. The use of solid-state detector arrays and computers utilizing special processors such as array processors makes it possible to eliminate much of the optical analog processing and to do complete digital processing and obtain higher accuracy in the measurements. SEE INTERFEROMETRY; OPTICAL TELESCOPE.

Bibliography. J. C. Dainty (ed.), *Laser Speckle and Related Phenomena*, 1975; R. K. Erf (ed.), *Speckle Metrology*, 1978; M. Francon, *Laser Speckle and Applications in Optics*, 1979; G. A. Slettemoen, Electronic speckle pattern interferometric system based on a speckle reference beam, *Appl. Opt.*, 19:616–623, 1980.

OPTICAL FIBERS
SUZANNE R. NAGEL

Flexible transparent fiber devices, used for either image or information transmission, in which light is propagated by total internal reflection. In simplest form, the optical fiber or lightguide consists of a core of material with a refractive index higher than the surrounding cladding. The optical fiber properties and requirements for image transfer, in which information is continuously transmitted over relatively short distances, are quite different than those for information transmission, where typically digital encoding of on-off pulses of light is used to transmit audio, video, or data over much longer distances at high bit rates. SEE OPTICAL COMMUNICATIONS; REFLECTION OF ELECTROMAGNETIC RADIATION; REFRACTION OF WAVES.

Fiber designs. There are three basic types of optical fibers (**Fig. 1**). Propagation in these lightguides is most easily understood by ray optics, although the wave or modal description must be used for an exact description. In a multimode, stepped-refractive-index-profile fiber (Fig. 1a),

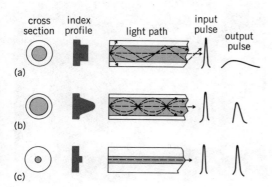

Fig. 1. Types of optical fiber designs. (*a*) Multimode, stepped-refractive-index-profile. (*b*) Multimode, graded-index-profile. (*c*) Single-mode, stepped-index. Graded-index is possible.

the number of rays or modes of light which are guided, and thus the amount of light power coupled into the lightguide is determined by the core size and the core-cladding refractive index difference. Such fibers, used for conventional image transfer, are limited to short distances for information transmission due to pulse broadening. An initially sharp pulse made up of many modes broadens as it travels long distances in the fiber, since high-angle modes have a longer distance to travel relative to the low-angle modes. This limits the bit rate and distance because it determines how closely input pulses can be spaced without overlap at the output end.

A graded-index multimode fiber (Fig. 1b), where the core refractive index varies across the core diameter, is used to minimize pulse broadening due to intermodal dispersion. Since light travels more slowly in the high index region of the fiber relative to the low index region, significant equalization of the transit time for the various modes can be achieved to reduce pulse broadening. This type of fiber is suitable for intermediate-distance, intermediate-bit-rate transmission systems. For both fiber types, light from a laser or light-emitting diode can be effectively coupled into the fiber. SEE LASER.

A single-mode fiber (Fig. 1c) is designed with a core diameter and refractive index distribution such that only one mode is guided, thus eliminating intermodal pulse-broadening effects. Material and waveguide dispersion effects cause some pulse broadening, which increases with the spectral width of the light source. These fibers are best suited for use with a laser source in order to efficiently couple light into the small core of the lightguide, and to enable information transmission over long distances at very high bit rates. The specific fiber design and the ability to manufacture it with controlled refractive index and dimensions determine its ultimate bandwidth or information-carrying capacity.

Attenuation. The attenuation or loss of light intensity is an important property of the lightguide since it limits the achievable transmission distance, and is caused by light absorption and scattering. Every material has some fundamental absorption due to the atoms or molecules composing it. In addition, the presence of other elements as impurities can cause strong absorption of light at specific wavelengths. Fluctuations in a material on a molecular scale cause intrinsic Rayleigh scattering of light. In actual fiber devices, fiber-core-diameter variations or the presence of defects such as bubbles can cause additional scattering light loss. The light loss of a material, after traveling a length L, is related to the initial power coupled into the fiber, P_0, versus the power at the output end, P, by the equation below.

$$\text{Loss (dB/km)} = \frac{10}{L(\text{km})} \log \left(\frac{P_0}{P}\right)$$

SEE ABSORPTION OF ELECTROMAGNETIC RADIATION; SCATTERING OF ELECTROMAGNETIC RADIATION.

Optical fibers based on silica glass have an intrinsic transmission window at near-infrared wavelengths with extremely low losses (**Fig. 2**). Very special glassmaking techniques are required to reduce iron and water (OH) to the parts-per-billion level, and have resulted in losses as low as 0.16 dB/km (0.26 dB/mi). Such fibers are used with solid-state lasers and light-emitting diodes for

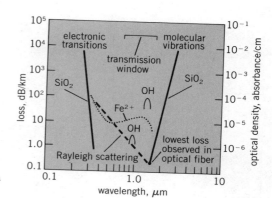

Fig. 2. Loss mechanisms in silica-based fibers.

information transmission, especially for long distances (greater than 1 km or 0.6 mi). Plastic fibers exhibit much higher intrinsic as well as total losses, and are more commonly used for image transmission, illumination, or very short-distance data links. SEE OPTICAL MATERIALS.

Many other fiber properties are also important, and their specification and control are dictated by the particular application. Good mechanical properties are essential for handling: plastic fibers are ductile, while glass fibers, intrinsically brittle, are coated with a protective plastic to preserve their strength. Glass fibers have much better chemical durability and can operate at higher temperatures than plastics. Very tight tolerances on core and outer-diameter control are essential for information transmission fibers, especially to allow long lengths to be assembled with low-loss joining or splicing.

Bibliography. A. H. Cherin, *An Introduction to Optical Fibers*, 1983; N. S. Kapany, *Fiber Optics Principles and Applications*, 1967; S. E. Miller and A. G. Chynoweth (eds.), *Optical Fiber Telecommunications*, 1979; S. D. Personick, *Fiber Optics: Technology and Applications*, 1985.

OPTICAL COMMUNICATIONS
W. M. HUBBARD

The transmission of speech, data, video, and other information by means of the visible and the infrared portion of the electromagnetic spectrum. A communication system consists of a transmitter, a transmission medium, and a receiver (see **illus.**).

Optical communication is one of the newest and most advanced forms of communication by electromagnetic waves. In one sense, it differs from radio and microwave communication only in that the wavelengths employed are shorter (or equivalently, the frequencies employed are higher). However, in another very real sense it differs markedly from these older technologies because, for the first time, the wavelengths involved are much shorter than the dimensions of the devices which are used to transmit, receive, and otherwise handle the signals.

The advantages of optical communication are threefold. First, the high frequency of the optical carrier (typically of the order of 300,000 GHz) permits much more information to be transmitted over a single channel than is possible with a conventional radio or microwave system. Second, the very short wavelength of the optical carrier (typically of the order of 1 micrometer) permits the realization of very small, compact components. Third, the highest transparency for electromagnetic radiation yet achieved in any solid material is that of silica glass in the wavelength region 1–1.5 µm. This transparency is orders of magnitude higher than that of any other solid material in any other part of the spectrum. SEE ELECTROMAGNETIC RADIATION; LIGHT.

Communication by means of light is not a new concept. A. G. Bell patented the "photophone" in 1880. In this device, the sound waves from a speaker's voice caused a mirror to vibrate and this in turn caused a beam of sunlight to be modulated. At the receiver, a selenium detector converted the sunlight into electronic current to recreate the speech. Nevertheless, optical com-

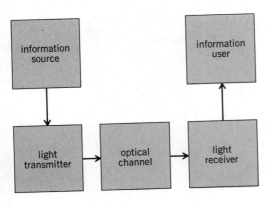

Block diagram of a simplified optical communication system.

munication in the modern sense of the term dates from about 1960, when the advent of lasers and light-emitting diodes (LEDs) made the exploitation of the wide-bandwidth capabilities of the light wave practical. During the 1960s much effort was devoted to utilizing light waves which were propagating through the atmosphere. But in 1970 an optical fiber which was orders of magnitude more transparent than any of its predecessors was made. The advent of this low-loss fiber stimulated a worldwide effort in what is now known as fiber optics. The result has been a tremendous increase in the transparency of silica fibers. (In the best fibers available prior to 1970, light waves retained only about 1% of their energy after traveling a distance of 65 ft or 20 m. By 1984 light waves in the best fibers retained about 10% of their energy after traveling over 12 mi or 20 km.) *See Laser; Optical fibers.*

Although the majority of the effort in optical communication is now in the area of optical fiber systems, they were not the only or the first systems considered or used. Other optical channels which have been considered include light beams in free space (between satellites), light beams through the atmosphere, and light beams through tubes which control the atmosphere.

In addition to categorizing optical communication systems by the type of transmission medium, it is possible to categorize them by wavelength, by the characteristics of the signal transmitted, and by type of source.

Free-space optical communications. A free-space optical channel exists, for example, between orbiting satellites. A free-space channel between satellites is, in some sense, ideal; it does not distort or attenuate the light beam. For this application the laser is the best source because its spatially coherent radiation can be confined to a much smaller angle of divergence than can the incoherent radiation from other sources. Lasers particularly well suited to this application are the carbon dioxide gas laser and the neodymium yttrium-aluminum-garnet (Nd:YAG) solid-state laser. The beam can be launched at the transmitter and picked up at the receiver by telescopes with apertures limited only by the weight and by the precision with which they can be pointed at each other. Pointing is a severe limitation since light beams less than 1 second of arc wide may be required, a beam width attainable with a 4-in. (10-cm) telescope at visible wavelengths. For this reason, lightwave systems have had difficulty competing with millimeter-wave systems for satellite-to-satellite communication.

Atmospheric optical communications. For satellite-to-Earth communications and terrestrial communications through the air, the Earth's atmosphere strongly influences the light transmission. In the visible wavelength band and in a few narrow windows in the near-infrared, transmission losses are low in clear weather. However, minute temperature gradients along the path of the light beam cause the beam to broaden and bend so that even in clear air the degradation can be severe over longer paths. Rain, fog, and snow cause even more severe transmission degradations. Attenuation of the light beam power to less than 1/1000 the clear weather value can occur in a 2-mi (3-km) path.

The primary source of excess attenuation in an Earth-to-satellite path is clouds. Fog and snow, rather than rain, are the most serious offenders on terrestrial paths. In neither case would the transmission reliability be considered satisfactory for most communication purposes. But for very short transmission paths the probability of outage may be low enough to be acceptable. For example, it is sometimes feasible to provide data links between nearby buildings over optical beams.

In a shielded atmosphere (such as in a room or in a pipe) the effects of precipitation can be avoided so that attenuation is quite low. Optical communication systems which guide light beams along a path inside a pipe with the help of lenses and mirrors have been studied. Even in a confined atmosphere, however, thermal effects must be very well controlled to prevent degradation of the light beams. The information-carrying capacity of a pipe carrying many such light beams could be very high, perhaps a few million telephone conversations. But the installation and maintenance costs would be very high, and the demand for such a system may never be sufficient to justify the expense of its development.

Optical fiber communications. Optical fiber bundles had been used to transmit light and even images for many years when the use of silica-based optical fibers to transmit data was proposed in 1966. With the development of extremely low-loss optical fibers during the 1970s, optical fiber communication became a very important form of telecommunication almost instantaneously. For fibers to become useful as light waveguides (or light guides) for communications applications, transparency and control of signal distortion had to be improved dramatically and a method had to be found to connect separate lengths of fiber together.

The transparency objective was achieved by making glass rods almost entirely of silica. These rods could be pulled into fibers at temperatures approaching 3600°F (2000°C).

Reducing distortion over long distances required modification of the method of guidance employed in early fibers. These early fibers (called step-index fibers) consisted of two coaxial cylinders (called core and cladding) which were made of two slightly different glasses so that the core glass had a slightly higher index of refraction than the cladding glass. Light rays that strike the core-cladding interface at a grazing angle are reflected into the core by means of a theoretically lossless process called total internal reflection, and thus are confined there. Depending on the angle of incidence, therefore, the rays follow different paths and hence require different lengths of time to travel down the fiber. This difference in propagation time results in distortion of the signal—a broadening and overlapping of the pulses in a digital signal, for example. Because the propagation time increases with path length, this effect places a limit on the transmission distance for a given pulse rate. SEE REFLECTION OF ELECTROMAGNETIC RADIATION; REFRACTION OF WAVES.

Two modifications of the step-index fiber have led to large improvements in the transmission of high data rates over large distances: (1) By causing the index of refraction to decrease continuously according to a particular formula, the effect of this delay distortion can be greatly reduced. This happens because in a properly designed fiber the light which travels the greatest distance spends more time farther from the axis in a region of lower index and therefore travels at a greater velocity. These graded-index fibers were widely used in first-generation optical fiber transmission systems, and they will continue to be used in certain applications in the future. However, an alternative approach is more attractive and is the method of choice in most long-distance systems. (2) By reducing the core size and the index difference in a step-index fiber, it is possible to reach a point at which only axial propagation is possible. In this condition, only one mode of propagation exists and determines the travel time. These single-mode fibers are capable of transmitting rates in excess of 10^9 pulses per second over distances of 75 mi (120 km). In fact, it will become feasible in the near future to provide individual offices and even individual homes with the capability of receiving and sending data at rates in excess of 2×10^9 bits per second.

The problem of joining fibers together was solved in two ways. For permanent connections, fibers can be spliced together by carefully aligning the individual fibers and then epoxying them together or fusing them together. In fact, permanent connection of fiber ribbons—linear arrays of several fibers—can be achieved by splicing the entire ribbon as a single unit. Fusion splices have been made with losses of less than 3% while epoxied splices typically have losses of about 12%.

For temporary connections, or for applications in which it is not desirable to make splices, fiber connectors have been developed. Connector losses for good connectors are typically less than 12%.

Optical transmitters. In principle any light source could be used as an optical transmitter. In modern optical communication systems, however, only lasers and light-emitting diodes are generally considered for use. The most simple device is the light-emitting diode which emits in all directions from a fluorescent area located in the diode junction. Since optical communication systems usually require well-collimated beams of light, light-emitting diodes are relatively inefficient. In particular, since optical fibers accept only light entering the core in a relatively narrow solid angle about the axis, only a small portion of the emitted light is captured and transmitted by the fiber. In fact, the acceptance angle for single-mode fibers is so small that light-emitting diodes are not well suited for use with them unless a high loss can be tolerated. On the other hand, light-emitting diodes are less expensive than lasers and, at least until recently, have exhibited longer lifetimes.

Another device, the semiconductor laser, provides comparatively well-collimated light. In this device, two ends of the junction plane are furnished with partially reflecting mirror surfaces which form an optical resonator. (In practical semiconductor lasers the partially reflecting mirrors are formed by simply cleaving the two sides of the junction.) The device enhances the light bouncing back and forth between these mirrors by means of stimulated emission. As a result of cavity resonances, the light emitted through the partially reflecting mirrors is well collimated within a narrow solid angle, and a large fraction of it can be captured and transmitted by an optical fiber.

Both light-emitting diodes and laser diodes can be modulated by varying the forward diode current. Typically, the message is a digital sequence of pulses which are used to turn the diode on and off. The light injected into the optical channel is a faithful representation of the information sequence.

It is also possible to fabricate a laser directly in an optical fiber by doping the core glass with ions of an element such as neodymium and pumping this active medium with light-emitting diodes of the proper wavelength. Lasers of this type require external modulators which make them less attractive than the diode lasers described above. Such lasers are still in the early experimental stage, but they show potential for use as amplifiers in long-haul systems.

Optical receivers. Semiconductor photodiodes are used for the receivers in virtually all optical communication systems. There are two basic types of photodiodes in use. The most simple comprises a reverse-biased junction in which the received light creates electron-hole pairs. These carriers are swept out by the electric field and induce a photocurrent in the external circuit. The minimum amount of light needed for correct reconstruction of the received signal is limited by noise superimposed on the signal by the following circuits. Such a photodiode must collect more than 100 nanowatts of signal in order to receive a signal of 10^8 pulses per second with sufficient fidelity.

Avalanche photodiodes provide some increase in the level of the received signal before it reaches the external circuits. They achieve greater sensitivity by multiplying the photogenerated carriers in the diode junction. This is done by creating an internal electric field sufficiently strong to cause avalanche multiplication of the free carriers. Unfortunately, the avalanche gain process is a random phenomenon and introduces additional noise into the signal. This excess noise becomes the limiting factor in the amount of avalanche gain which can be beneficially obtained. Nevertheless, the improvement available from avalanche detectors is significant. A reduction in required signal power of between 10 and 100 times (depending on the type of devices used) is routinely realizable. Avalanche photodiodes are used only when the extra sensitivity which they provide is required because they are more expensive than simple photodiodes, they require much more expensive power supplies than simple photodiodes, and they require very careful temperature stabilization. SEE OPTICAL DETECTORS.

Bibliography. A. H. Cherin, *An Introduction to Optical Fibers*, 1983; C. K. Kao, *Optical Fiber Systems: Technology, Design, and Applications*, 1982; S. E. Miller and A. G. Chynoweth (eds.), *Optical Fiber Telecommunications*, 1979; S. D. Personick, *Fiber Optics: Technology and Applications*, 1985.

INTEGRATED OPTICS
WILLIAM STREIFER

The study of optical devices that are based on light transmission in planar waveguides, that is, dielectric structures that confine the propagating light to a region with one or two very small dimensions, on the order of the optical wavelength. The principal motivation for these studies is to combine miniaturized individual devices through waveguides or other means into a functional optical system incorporated into a small substrate. The resulting system is called an integrated optical circuit (IOC) by analogy with the semiconductor type of integrated circuit. An integrated optical circuit could include lasers, integrated lenses, switches, interferometers, polarizers, modulators, detectors, and so forth. Important uses envisioned for integrated optical circuits include signal processing (for example, spectrum analysis and analog-to-digital conversion) and optical communications through glass fibers, which are themselves circular (or elliptical) waveguides. Integrated optical circuits could be used in such systems as optical transmitters, switches, repeaters, and receivers. SEE OPTICAL COMMUNICATIONS; OPTICAL FIBERS.

The advantages of having an optical system in the form of an integrated optical circuit rather than a conventional series of components include (apart from miniaturization) reduced sensitivity to air currents and to mechanical vibrations of the separately mounted parts, low driving voltages and high efficiency, robustness, and (potentially) reproducibility and economy. As in the case of semiconductor integrated circuits, an integrated optical circuit might be fabricated on or just within the surface of one material (the substrate) modified for the different components by shaping structures (using etching, for example) or incorporating suitable substitutes or dopants, or alternatively, by depositing or expitaxially growing additional layers. It is also possible to construct independent components which are then attached to form the integrated optical circuit. This option, called hybrid, has the advantage that each component could be optimized, for example, by using gallium aluminum arsenide lasers as sources for an integrated optical circuit and silicon detectors. In the former case, the integrated optical circuit is called monolithic, and is expected to have the advantage of ease of processing, similar to the situation for monolithic semiconductor integrated circuits. Perhaps the most promising materials for monolithic integrated optical circuits are direct band-gap semiconductors composed of III–V materials such as gallium aluminum arsenide [(GaAl)As] and indium gallium arsenide phosphide [(InGa)(AsP)] since with suitable processing they may perform almost all necessary operations as lasers, switches, modulators, detectors, and so forth.

Guided waves. The simplest optical waveguide is a three-layer or sandwich structure with the index of refraction largest in the middle or waveguiding layer (**Fig. 1**). The lower and top layers are usually the substrate and superstrate, respectively. Often, the top layer is air and the waveguide layer is referred to as a film. Sometimes, too, the outer regions are called cladding layers. A guided wave does not have light distributed uniformly across the waveguide, but is a pattern that depends on the indices of refraction of all three layers and the guide thickness. The waveguide is usually designed by selecting its refractive index and thickness, so that only one such characteristic pattern propagates with no change in shape. This pattern, referred to as the fundamental or lowest-order mode, travels down the guide with a characteristic velocity.

A mode pattern such as that of Fig. 1a may be viewed as the superposition of two progressive waves moving at a particular slight angle θ to the propagation direction as shown in Fig. 1b. If the ray directions are sufficiently near grazing incidence, the large refractive index of the waveguide layer will cause total internal reflection at the interfaces and the mode will propagate along the waveguide without loss.

Materials and fabrication. Waveguides have been made of many different materials, most of which may be categorized as ferroelectric, semiconductor, or amorphous. Examples of these classes are lithium niobate, gallium arsenide/gallium aluminum arsenide [GaAs/(GaAl)As], and glass, respectively. Methods for fabricating a waveguide layer at the surface of lithium niobate include heating the crystal in a vacuum to drive off lithium oxide or diffusing titanium metal into the crystal. Both processes create a region of high refractive index near the surface; air is the superstrate. Although the refractive index of the waveguide layer decreases with depth rather than being constant, the guided mode is similar to that shown in Fig. 1. A semiconductor waveguide is fabricated, for example, by growing successively crystalline layers of $(Ga_{0.7}Al_{0.3})As$, GaAs, and $(Ga_{0.7}Al_{0.3})As$. The thin GaAs waveguide layer of high refractive index is thus interposed between thicker cladding regions of the lower-index $(Ga_{0.7}Al_{0.3})As$. Glass waveguides may be formed, for example, by sputter deposition of a relatively high-refractive-index glass on a lower-index glass substrate.

Waveguides that confine light in two dimensions, rather than one as shown in Fig. 1, utilize refractive index differences in both transverse directions. Examples of these are the rib guide (**Fig. 2a**) and the titanium-diffused channel guide (Fig. 2b). In fabricating a rib guide, photolithography is employed to delineate the stripe, followed by chemical or dry etching to remove the undesired material. The channel guide is produced by etching away all but a strip of metal prior to diffusion.

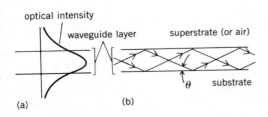

Fig. 1. Optical waveguide. (a) Optical intensity of the fundamental guided mode. (b) Ray picture of the guided mode.

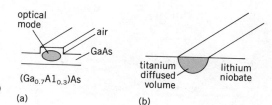

Fig. 2. Two-dimensional transverse waveguiding. (a) Rib waveguide. (b) Diffused channel guide.

Coupling of external light beams. An external light beam may be coupled into a waveguide by introducing the light at the end of the guide at an edge of the substrate or through the surface of the waveguide. The former approach may employ a lens to focus the light beam onto the guide end. Alternatively, the laser or an optical fiber is placed against or in close proximity to the guide end, which is referred to as butt coupling. Light may also be injected through the guide surface with an auxiliary element. Simply illuminating that surface will not suffice since the refractive index difference needed for effective waveguiding prevents light from entering at the required angle. The necessary angle, which is exactly that of the propagating waves shown in Fig. 1b, may be obtained by passing the light through a high-refractive-index prism whose surface is so close to the guide layer that waveguiding is disturbed in that region. The prism not only permits the introduction of light, but also acts to couple light from the waveguide. For this reason, the light must illuminate the guide near the end of the prism (**Fig. 3**). The angle of light incident on the surface of the waveguide layer can also be modified so as to coincide with the internal angle of the propagating wave by a periodic structure or grating on the waveguide surface. Such a grating may be made by a multiple-step process which begins by depositing a light-sensitive material called photoresist and concludes with etching and cleaning. The grating also functions as an output coupler. By reversing the incident and output beam directions in Fig. 3, the roles of the prism and grating couplers are interchanged.

Lasers and distributed feedback. The diode laser is already an integrated optics device in a sense since the lasing medium is a waveguide laser interposed between two cladding layers. The waveguide layer may be GaAs, in which case the outer regions are composed of $(Ga_{0.7}Al_{0.3})As$, for example. When used as a separate source, the crystal facets act as end reflectors. The diode laser's high efficiency, low-voltage operation, small size, and physical integrity cause it to be the laser of choice in many hybrid integrated optics applications. Moreover, it lends itself to integration in an integrated optical circuit where reflectivity may be provided by introducing an appropriate periodic structure. This could be a thickness variation in the waveguide layer. Devices utilizing such structures are called distributed feedback or distributed Bragg reflector lasers. *See* LASER.

Switches and modulators. Both lithium niobate and gallium arsenide belong to the family of electro-optically active crystals. When an electric field is applied to these materials, their refractive indices are modified. This effect is employed in integrated optical circuit switching and modulation applications. To construct a switch, gold or other conducting electrodes are deposited on a lithium niobate integrated optical circuit surface parallel to two closely spaced waveguides.

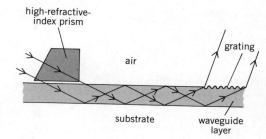

Fig. 3. Prism input coupling and grating output coupling. With the ray directions reversed, the grating acts as an input coupler and the prism as an output coupler.

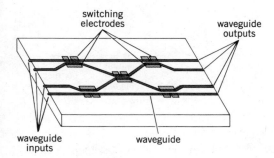

Fig. 4. A "4 by 4" directional coupler switch.

If the electrodes and waveguides are suitably designed, the applications of specific small voltages to the electrodes will cause the transfer of optical power from one waveguide to its neighbor with high efficiency and little residual power in the initial guide. Shown in **Fig. 4** is a "4 by 4" switching network in which each of four input optical signals, possibly from optical fibers butt-coupled to the waveguide inputs, may be switched to any one of the four output ports. Such an integrated optical circuit serves to interconnect four computers through optical fibers, for example. SEE ELECTROOPTICS.

A switch is in effect a modulator as well. Modulation is a process in which information is encoded onto an optical or other electromagnetic wave. Pulse modulation results simply by interrupting or connecting a light wave in a manner intelligible to the receiver. (Morse code is an unsophisticated example.) By transferring light into or out of a waveguide in response to an electric signal at a switching electrode, the output optical wave becomes modulated; that is, the switch operates as a modulator. SEE OPTICAL MODULATORS.

Spectrum analyzer. Light propagating in an integrated optical circuit within a one-dimensional waveguide may be deflected by interacting with an acoustic wave. The deflection angle is a function of the acoustic frequency. This interaction forms the basis of an integrated optics spectrum analyzer (**Fig. 5**). An acoustic wave is generated by an electric signal whose frequencies are being analyzed. The acoustic wave interacts with a guided optical wave and the

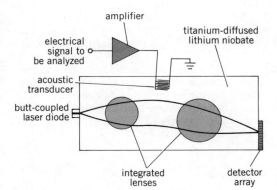

Fig. 5. Top view of an integrated optics spectrum analyzer.

deflected light is focused onto a detector array. The one or more detectors which respond determines the previously unknown frequencies. The integrated optical circuit incorporates lenses which focus the light and a transducer to generate the acoustic wave. The lenses may be fabricated by depositing additional high-refractive-index material in an area so as to decrease locally the propagation velocity of the guided wave. The transducer is made of a piezoelectric material (such as lithium niobate) which has the property of converting an electric field into a strain field. If lithium niobate is used for the integrated optical circuit, the transducer consists of a sophisticated interdigital electrode structure deposited on the waveguide surface. SEE ACOUSTOOPTICS.

Bibliography. F. J. Leonberger, Applications of guided wave interferometers, *Laser Focus Fiberopt. Tech.*, 18:125–129, 1982; Optical Society of America, Integrated and guided wave optics, *Tech. Dig.*, January 6–8, 1982; T. Tamir (ed.), *Integrated Optics*, 1979; P. K. Tien, Integrated optics and new wave phenomena in optical waveguides, *Rev. Mod. Phys.*, 49(2):361–420, 1977.

OPTICAL INFORMATION SYSTEMS
DAVID CASASENT

Devices that use light to process information. Optical information systems or processors consist of one or several light sources; one- or two-dimensional planes of data such as film transparencies, various lenses, and other optical components; and detectors. These elements can be arranged in various configurations to achieve different data processing functions. As light passes through various data planes, the light distribution is spatially modulated proportional to the information present in each plane. This modulation occurs in parallel in one or two dimensions, and the processing is performed at the speed of light. The high-speed and parallel-processing features of optical information processors, together with their small size and low power dissipation, are their attractive features. The processing speed is limited in practice by the rate at which new planes of data can be introduced into the system and the rate at which processed data can be removed from the output detectors.

Practical systems employ real-time and reusable spatial light modulators rather than film. These devices include various liquid, ferroelectric, magnetooptic, electrooptic, and acoustooptic crystals to convert optical data (for example, an ambient scene) or electrical data into a form suitable for spatially modulating the light passing through the system. (Usually, coherent laser light is employed.) Lenses, mirrors, computer-generated holograms, holographic optical elements, and fiber optics are used to manipulate and control the light as it passes through the system and thus to provide a quite versatile collection of architectures for diverse applications. Optical information processors can be grouped into three major classes: optical image processors, optical signal processors, and optical computers. The architecture of the optical system and the type of processing functions performed in each case differ considerably. SEE ACOUSTOOPTICS; ELECTROOPTICS; HOLOGRAPHY; LASER; MAGNETOOPTICS; OPTICAL FIBERS.

Optical image processing. The basic optical image processing functions are shown in **Fig. 1**. An image $g(x,y)$ is placed in plane P_1 and illuminated with laser light. The light distribution incident on P_2 is the two-dimensional Fourier transform $G(u,v)$ of the input image $g(x,y)$. This Fourier-transform distribution is a representation of the spatial frequencies (u,v) present in the input image. Lower spatial frequencies (corresponding to larger input shapes) lie closer to the center of P_2, and higher spatial frequencies (corresponding to smaller input image regions) lie

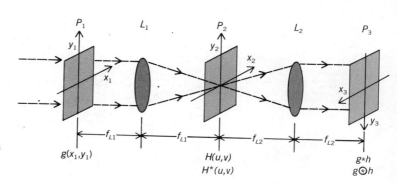

Fig. 1. Optical Fourier transform and correlation image processor. Lenses L_1 and L_2 have focal lengths f_{L1} and f_{L2}.

further from the center of P_2. The orientation of each input object is reflected in the angular location of its corresponding spatial frequency distribution in P_2. Detectors placed in P_2 can feed this Fourier-coefficient information to a digital postprocessor for subsequent analysis and classification of the type of input object and its size and orientation. This is one form of optical extraction for pattern recognition and precise measurement of object size and orientation.

Other architectures can optically compute other features of an input object such as its geometrical moments and its chord histogram distribution (the length and angle of all chords that define the object), which a digital postprocessor can analyze to determine the final class and orientation of the input object. Such feature extraction systems are the simplest form of optical pattern recognition image processors.

However, the full optical system of Fig. 1 can also perform the correlation operation. In this case, the transmittance of P_2 is made proportional to $H^*(u,v)$, the conjugate Fourier transform of a reference function $h(x,y)$. The $H^*(u,v)$ filter function at P_2 is formed by holographic techniques. Then, the P_3 pattern is the correlation of the two space functions g and h, written $g \circledast h$. The correlation of two functions is equivalent to translating h spatially over g and, for each translation, forming the product and sum of h and the associated region of g. The regions of an input scene g that most closely match the reference function h being searched for yield the largest output. This template matching operation can thus locate the presence and positions of reference objects in an input image. This object pattern recognition operation is achieved with no moving parts and is quite effective for extracting objects in clutter or high noise and for locating multiple occurrences of an object. Advanced filter synthesis techniques allow such systems to operate independent of scale, rotation, and other geometrical distortions between the input and reference object. Character recognition, robotics, missile guidance, and industrial inspection are the major applications that have been pursued for such architectures. If $H(u,v)$ rather than $H^*(u,v)$ is recorded at P_2, the P_3 output is an image, $g*h$, that is a filtered version of the input. The filter function of the system can be controlled by spatially varying the contents of P_2 to achieve image enhancement and restoration of blurred and degraded images. SEE IMAGE PROCESSING.

Optical signal processing. Signal processing systems are used in radar, sonar, electronic warfare, and communications to determine the frequency and direction of arrival of input signals, and for correlation applications. These systems generally employ acoustooptic transducers to input electrical signals to the optical processor. The electronic input to an acoustooptic cell is converted to a sound wave which travels the length of the cell. If the cell is illuminated with light, the amplitude of the light leaving the cell will be proportional to the strength of the input signal, and the angle at which the light leaves will be proportional to the frequency of the input signal. Since the device and system are linear, if N input signals are simultaneously present, N light waves leave the cell at N angles and strengths. A lens behind the cell focuses each light wave to a different spatial location in an output detector plane. Thus, an optical spectrum analyzer results (**Fig. 2**). Compact systems of this type exist, and bandwidths in excess of 2 GHz can be achieved and up to 1000 signals can be simultaneously processed. If a multichannel acoustooptic cell is used with input signals from different antennas or antenna elements, the two-dimensional Fourier

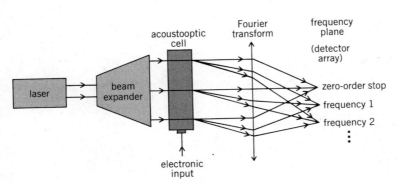

Fig. 2. Acoustooptic signal spectrum analyzer.

transform of its output provides a two-dimensional display of the simultaneous frequency and direction of arrival of all received signals within the bandwidth of the cell. The angle at which light is deflected can be varied by varying the frequency of the input signal to the acoustooptic cell; a beam deflector or scanner based on this principle is used in most xerographic reproduction systems and laser printers. Other architectures using multiple cells can correlate two or more signals. These systems are used for simultaneous range and Doppler processing and for demodulation and synchronization of signals for communication applications.

Optical computing. The optical processors described above are special-purpose architectures. The parallel and high-speed features of optical systems can also be used to fabricate rather general-purpose optical computers (which can be used to perform general functions). Much research has been done on the optical realization of parallel processors for multiplication and addition. A parallel two-dimensional multiplication occurs when a two-dimensional spatially modulated light beam is passed through a two-dimensional transparency. Certain two-dimensional spatial light modulators can also produce an output that is the difference between two successive two-dimensional inputs or contains only the moving objects within a two-dimensional input scene. The addition of two two-dimensional data planes occurs if two spatially modulated beams are incident on a detector array simultaneously or sequentially. Various number representations have been used (such as residue arithmetic) to design efficient parallel optical and numerical computers.

The most attractive class of general-purpose optical computers is presently optical linear algebra processors. These systems perform matrix-vector operations and similar linear algebra algorithms, often in a systolic form. The architecture for one rather general form of such a processor (**Fig. 3**) exemplifies the basic concepts. This system consists of a linear array of input point modulators (for example, laser diodes), each of which is imaged through a different spatial region of an acoustooptic cell. The Fourier transform of the light distribution leaving the cell is formed

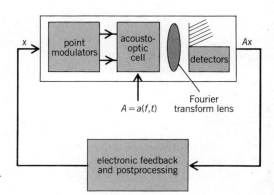

Fig. 3. Optical linear algebra processor.

on a linear output detector array. If the elements of N input vectors are fed simultaneously at time t to the acoustooptic cell, each vector on a different frequency f, then the transmittance of the cell will be N vectors (that is, a matrix A). If the input point modulators are fed in parallel with the elements of another vector x, then the product of the input vector and the N vectors in the cell is formed, and each vector inner product is produced (by the Fourier-transform lens) on a separate output detector. Thus, the system performs N vector inner products or a matrix-vector multiplication in parallel. With various frequency, time, and space encoding techniques, different electrical postprocessing, and various feedback configurations, all of the major linear algebra operations required in modern signal processing can be achieved on this one system. Use of multichannel acoustooptic cells and related architectures can increase the processing capacity of the system and allow optical processing of data to the accuracy possible on digital computers.

Bibliography. N. J. Berg and J. N. Lee (eds.), *Acousto-Optic Signal Processing: Theory and Practice*, 1983; D. Casasent, Coherent optical pattern recognition, *Opt. Eng.*, 24:26–32, January 1985; Special issue on acousto-optic signal processing, *Proc. IEEE*, vol. 69, January 1981; Special issue on optical computing, *Proc. IEEE*, vol. 72, July 1984.

NONLINEAR OPTICAL DEVICES
Nasser Peyghambarian and Hyatt M. Gibbs

Devices based on a class of optical effects that result from the interaction of electromagnetic radiation from lasers with nonlinear materials. Nonlinear means that the effect depends on the intensity of the light. Nonlinear effects are due to the nonlinear contribution to the polarization of the medium, which can be expressed as a power series expansion in the incident electric field E by Eq. (1), where $\chi^{(1)}$ is the linear, and $\chi^{(2)}$ and $\chi^{(3)}$ are second- and third-order susceptibilities,

$$P = \epsilon_0(\chi^{(1)} E + \chi^{(2)} E^2 + \chi^{(3)} E^3 + \cdots) \tag{1}$$

respectively. $\chi^{(2)}$ has a nonzero value only in materials that do not possess inversion symmetry, but $\chi^{(3)}$ is usually nonzero in all materials. Second-order nonlinearities, which arise from $\chi^{(2)}$, can cause second-harmonic generation, optical parametric amplification, and oscillation. Four-wave mixing and phase conjugation are examples of third-order, $\chi^{(3)}$ nonlinear effects. Optical bistability, which can also be expressed as a $\chi^{(3)}$ mechanism, occurs when the nonlinearity is coupled with feedback.

Nonlinear optical devices can be classified roughly into two categories: $\chi^{(2)}$ devices that generate light at new frequencies and $\chi^{(3)}$ devices that process optical signals. The first category covers second-harmonic generators and optical parametric oscillators, while the second category contains four-wave-mixing beam deflectors, phase-conjugate mirrors, etalon switches and logic devices, and waveguide couplers.

Devices for frequency generation. Generation of laser frequencies in the ultraviolet region of the spectrum is of considerable importance in many applications such as laser fusion and laser isotope separation. Second-harmonic generators and optical parametric oscillators have produced tunable coherent radiation for these and other applications. *See* Laser.

Optical parametric oscillators. When an intense pump beam of frequency ω_3 and a weak "signal" beam of frequency ω_1 are simultaneously incident on a nonlinear crystal, an "idler" wave at frequency $\omega_2 = \omega_3 - \omega_1$ is generated. Also, the "signal" beam at ω_1 is amplified at the expense of the pump photons. This process is called optical parametric amplification. If only the pump frequency ω_3 is present and the crystal is placed inside an optical cavity, then under appropriate conditions, frequencies ω_1 and ω_2 will build up from noise and oscillate inside the resonator. This so-called optical parametric oscillation occurs when the conservation of momentum (phase-matching condition) is satisfied, namely, Eq. (2), where the **k**'s are propagation constants. For a collinear geometry, each wave satisfies Eq. (3), where n is the index of refraction of the crystal

$$\mathbf{k}_1 + \mathbf{k}_2 = \mathbf{k}_3 \tag{2} \qquad k = \frac{n\omega}{c} \tag{3}$$

and c is the speed of light. The phase-matching condition can then be expressed in terms of the indices of refraction of the crystal for the three waves as Eq. (4). The conservation of energy requires Eq. (5), and efficient parametric generation therefore requires that Eqs. (4) and (5) be

$$\omega_3 n_3 = \omega_1 n_1 + \omega_2 n_2 \tag{4} \qquad \omega_3 = \omega_1 + \omega_2 \tag{5}$$

satisfied simultaneously. A variation in the temperature of the crystal or a change in crystal orientation will change the three indices of refraction and will tune ω_1 and ω_2. The tuning of the "signal" and "idler" frequencies over a large range results from the nonresonant nature of the nonlinear mechanism in materials such as lithium niobate ($LiNbO_3$). *See* Crystal optics.

Second-harmonic generators. Second-harmonic generation is a special case of parametric generation where $\omega_1 = \omega_2 = \omega$. The interaction of a beam of frequency ω with the nonlinear crystal causes the generation of a wave at twice the frequency, 2ω. Second-harmonic generators are widely used with neodymium:yttrium-aluminum-garnet (Nd:YAG) lasers and in correlation techniques for measurement of the time profile of fast (picosecond and subpicosecond) optical pulses. Quartz, potassium dihydrogen phosphate (KDP), and ammonium dihydrogen phosphate (ADP) are among the crystals used for second-harmonic generation. Rare gases are often used for second- and higher-harmonic generation of ultraviolet light because they are more transparent at shorter wavelengths than most materials. SEE OPTICAL PULSES.

Devices for optical signal processing. Nonlinear optical devices may be used in digital signal processing and optical computing. The parallel nature of light, which may make the simultaneous operation of many devices possible, together with the high speed of optical components, promises very large bandwidths for number crunching computation and information processing. Many counterparts of electronic circuit elements such as transistors and gates have already been demonstrated by using optical technology. Methods for optical interconnections within electronic computers have been considered.

Etalon devices. When a nonlinear material is sandwiched between two partially reflecting mirrors, a nonlinear Fabry-Perot etalon is formed. Normally, a thin semiconductor slab is used to provide the nonlinearity. Such a semiconductor etalon is capable of performing logic operations, switching, and controlling one light beam with another. The principle of operation of such optical transistors can be described as follows. The incident laser frequency ω_L is initially detuned from the resonances (transmission peaks) of the Fabry-Perot etalon, resulting in low transmission; the device is in its "off" state (**Fig. 1***a*). To switch the device to its "on" state, a second laser beam is directed toward the same spot on the device. The primary function of the second laser is to change the index of refraction of the semiconductor by any of several physical mechanisms: heating the etalon, creating free carriers that screen an exciton feature or partially fill the conduction band, and so forth. The index change causes the nearest etalon resonance to shift toward ν_L, resulting in an abrupt increase in the transmission and the consequent turn "on" of the switch (Fig. 1*b*). Optical switching can be accomplished with a single laser beam also. In this case, when the device is initially "off," increasing the laser intensity beyond a critical value results in an abrupt turn "on" of the device while turning "off" can be performed by decreasing the intensity. Optical bistability refers to situations where the intensities for turning "on" and "off" are not the same. A typical optical bistability hysteresis curve is shown in **Fig. 2**. SEE INTERFEROMETRY; OPTICAL BISTABILITY.

Operations of bistable devices have been realized in many semiconductors, including bulk gallium arsenide (GaAs), gallium arsenide–aluminum gallium arsenide (GaAs-AlGaAs) superlattices, copper chloride (CuCl), indium arsenide (InAs), indium antimonide (InSb), zinc sulfide (ZnS),

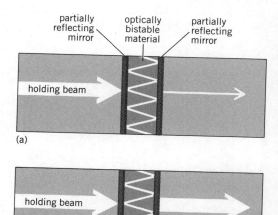

Fig. 1. Operations of a nonlinear optical switch. (*a*) Switch in its "off" state. (*b*) Switch abruptly turns "on" by the application of a second switching beam.

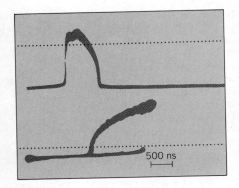

Fig. 2. Room-temperature optical bistability in a gallium arsenide–aluminum gallium arsenide (GaAs–AlGaAs) nonlinear optical device. The top trace is the output transmission of the device versus time. The bottom trace is the output intensity versus input intensity, showing hysteresis. The input pulse versus time has a triangular shape and is not shown.

zinc selenide (ZnSe), cadmium sulfide (CdS), and cadmium mercury telluride (CdHgTe). The ideal practical device should be small (with micrometer or submicrometer sizes) and fast (with picosecond switching times), require very little switching energy (of the order of 10^{-15} joule) and low holding power (submilliwatts), and operate at a convenient temperature and wavelength. Gallium arsenide is the most promising candidate: It can operate with a few milliwatts of power at room temperature; and switch-on times of a few picoseconds and switch-off times of a few nanoseconds have been demonstrated by using a few-micrometers-square spot on a gallium arsenide etalon. Operation at submilliwatt power levels at $-307°F$ (85 K) has been accomplished. The response time of a gallium arsenide NOR gate was measured to be approximately 1 picosecond (**Fig. 3**). Even though the NOR gate responds in 1 picosecond, the next NOR gate operation cannot be performed for a few nanoseconds because it takes that long for the free carriers to recombine or diffuse out of the illuminated region and to allow the index of refraction to return to its initial value. For the NOR gate operation, the device is initially made totally transmitting by tuning one of the etalon resonances to the laser frequency. The application of one or two input beams shifts the etalon peak away from the laser frequency, resulting in a decrease in transmission.

Etalon devices are most suitable for parallel processing because many beams could be focused on a single etalon, thereby defining a two-dimensional array of pixels. Employing parallel processors in optical computing could considerably enhance the speed of computations. For example, simultaneous operation of 10^6 pixels at a rate of 100 MHz per pixel results in an effective bandwidth of 10^{14} operations per second. Computer architecture approaches are based on parallel computations, and fifth-generation supercomputers will surely employ parallel architecture techniques. SEE OPTICAL INFORMATION SYSTEMS.

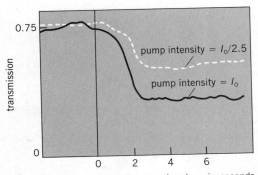

Fig. 3. Transmission of a gallium arsenide–aluminum gallium arsenide (GaAs–AlGaAs) optical NOR gate versus time at two pump intensities, I_0 and $I_0/2.5$, showing a 1-picosecond response time for the device.

Waveguide devices. When light travels in a medium with an index of refraction larger than the surrounding material, light traveling at a shallow enough angle is trapped in the higher-index medium by total internal reflection. Such waveguiding can occur in one dimension only (as in many multimode diode lasers) or in both transverse dimensions (as in rib waveguides, single-transverse-mode diode lasers, or optical fibers). The guided-wave devices confine the light over a long interaction length (a fraction of an inch or few millimeters), thereby permitting high intensities in the guided region. (The intensity I is related to power P by $I = P/A$, where A is the focused light area; therefore, for a small A, intensity can be high.) Also, the long interaction length permits the use of materials with weaker nonlinearities to obtain the phase shift (of the order of 180°) needed for logic devices. However, absorption in the guided material usually limits the extent of the interaction length. SEE OPTICAL FIBERS; REFLECTION OF ELECTROMAGNETIC RADIATION.

Four-wave-mixing beam deflectors. Optical techniques for intracomputer communications have become increasingly popular because of the limit imposed by the electronic interconnections on the scale-downs of computer circuits. One method that has been assessed for optical interconnection is the use of holographic elements to diffract the light to the desired locations. Holographic arrays may someday replace the huge number of wires or buses that are used inside the computer to transfer information. However, conventional holograms are not programmable and would have to be replaced every time a different set of pixels was to be accessed. Four-wave mixing is one of the avenues that has been investigated to perform real-time programmable holography. Dynamic gratings that are generated by four-wave mixing may be able to deflect light that has originated from a source on a very large-scale integrated circuit (VLSI) chip to a spot on the same chip or to another VLSI chip. If the light is desired to illuminate a different spot, the direction of the grating must be changed; this can be accomplished by changing the angle at which the two "write" beams intersect. SEE HOLOGRAPHY; INTEGRATED OPTICS; NONLINEAR OPTICS.

Bibliography. H. M. Gibbs, *Optical Bistability: Controlling Light with Light*, 1985; A. Lattes et al., An ultrafast all-optical gate, *IEEE J. Quant. Electr.*, QE-19:1718–1723, 1983; A. Migus et al., One-picosecond optical NOR gate at room-temperature with a GaAs-AlGaAs multiple-quantum-well Fabry-Perot etalon, *Appl. Phys. Lett.*, 46:70, 1985; D. A. B. Miller and T. H. Wood, Quantum-well optical modulators and SEED all-optical switches, *Opt. News*, 10(6):19–20, 1984; N. Peyghambarian and H. M. Gibbs, Optical nonlinearity, bistability and signal processing in semiconductors, *J. Opt. Soc. Amer. B*, 2:1215, 1985; N. Peyghambarian and H. M. Gibbs, Optical bistability for optical signal processing and computing, *Opt. Eng.*, 24:68–73, 1975.

OPTICAL PHASE CONJUGATION
ROBERT A. FISHER AND BARRY J. FELDMAN

A process that involves the use of nonlinear optical effects to precisely reverse the direction of propagation of each plane wave in an arbitrary beam of light, thereby causing the return beam to exactly retrace the path of the incident beam. The process is also known as wavefront reversal or time-reversal reflection. The unique features of this phenomenon suggest widespread application to the problems of optical beam transport through distorting or inhomogeneous media. Although closely related, the field of adaptive optics will not be discussed here. SEE ADAPTIVE OPTICS.

Fundamental properties. Optical phase conjugation is a process by which a light beam interacting in a nonlinear material is reflected in such a manner as to retrace its optical path. As **Fig. 1** shows, the image-transformation properties of this reflection are radically different from those of a conventional mirror. The incoming rays and those reflected by a conventional mirror (Fig. 1a) are related by reversal of the component of the wave vector $\vec{k}$ which is normal to the mirror surface. Thus a light beam can be arbitrarily redirected by adjusting the orientation of a conventional mirror. In contrast, a phase-conjugate reflector (Fig. 1b) inverts the vector quantity $\vec{k}$ so that, regardless of the orientation of the device, the reflected conjugate light beam exactly retraces the path of the incident beam. This retracing occurs even though an aberrator (such as a piece of broken glass) may be in the path of the incident beam. Looking into a conventional mirror, one would see one's own face, whereas looking into a phase-conjugate mirror, one would

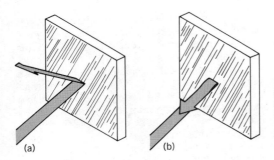

Fig. 1. Comparison of reflections (a) from a conventional mirror and (b) from an optical phase conjugator. (*After V. J. Corcoran, ed., Proceedings for the International Conference on Laser '78 for Optical and Quantum Electronics, STS Press, McLean, Virginia, 1979*)

see only the pupil of the eye. This is because any light emanating from, say, one's chin would be reversed by the phase conjugator and would return to the chin, thereby missing the viewer's eye. A simple extension of the arrangement in Fig. 1b indicates that the phase conjugator will reflect a diverging beam as a converging one, and vice versa. These new and remarkable image-transformation properties (even in the presence of a distorting optical element) open the door to many potential applications in areas such as laser fusion, atmospheric propagation, fiber-optic propagation, image restoration, real-time holography, optical data processing, nonlinear microscopy, laser resonator design, and high-resolution nonlinear spectroscopy. *See* Mirror optics.

Optical phase conjugation techniques. Optical phase conjugation can be obtained in many nonlinear materials (materials whose optical properties are affected by strong applied optical fields). The response of the material may permit beams to combine in such a way as to generate a new beam that is the phase-conjugate of one of the input beams. Processes associated with degenerate four-wave mixing, scattering from saturated resonances, stimulated Brillouin scattering, stimulated Raman scattering, photoreactive phenomena, surface phenomena, and thermal scattering have all been utilized to generate optical phase-conjugate reflections.

Kerr-like degenerate four-wave mixing. As shown in **Fig. 2**, two strong counterpropagating (pump) beams with k-vectors $\vec{k}_1$ and $\vec{k}_2$ (at frequency ω) are directed to set up a standing wave in a clear material whose index of refraction varies linearly with intensity. This arrangement provides the conditions in which a third (probe) beam with k-vector $\vec{k}_p$, also at frequency ω, incident upon the material from any direction would result in a fourth beam with k-vector $\vec{k}_c$ being emitted in the sample precisely retracing the third one. (The term degenerate indicates that all beams have exactly the same frequency.) In this case, phase matching (even in the birefringent materials) is obtained independent of the angle between $\vec{k}_p$ and $\vec{k}_1$. The electric field of the conjugate wave E_c is given by the equation below, where δn is the change in the

$$E_c = E_p^* \tan\left(\frac{2\pi}{\lambda_0} \delta n \ell\right)$$

index of refraction induced by one strong counterpropagating wave, λ_0 is the free-space optical wavelength, and ℓ is the length over which the probe beam overlaps the conjugation region. The conjugate reflectivity is defined as the ratio of reflected and incident intensities, which is the square of the above tangent function. The essential feature of phase conjugation is that E_c is proportional to the complex conjugate of E_p. Although degenerate four-wave mixing is a nonlinear

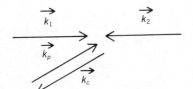

Fig. 2. Geometry of k-vectors for optical phase conjugation using degenerate four-wave mixing.

optical effect, it is a linear function of the field one wishes to conjugate. This means that a superposition of E_p's will generate a corresponding superposition of E_c's; therefore faithful image reconstruction is possible. SEE KERR EFFECT.

To visualize the degenerate four-wave mixing effect, consider first the interaction of the weak probe wave with pump wave number two. The amount by which the index of refraction changes is proportional to the quantity $(E_p + E_2)^2$, and the cross term corresponds to a phase grating (periodic phase disturbance) appropriately oriented to scatter the pump wave number one into the $\vec{k}_c$ direction. Similarly, the very same scattering process occurs with the roles of two pump waves reversed. Creating these phase gratings can be thought of as "writing" of a hologram, and the subsequent scattering can be thought of as "reading" the hologram. Thus the four-wave mixing process is equivalent to volume holography in which the writing and reading are done simultaneously. SEE DIFFRACTION GRATING; HOLOGRAPHY.

Conjugation using saturated resonances. Instead of using a clear material, as outlined above, the same four-wave mixing beam geometry can be set up in an absorbing or amplifying medium partially (or totally) saturated by the pump waves. When the frequency of the light is equal to the resonance frequency of the transition, the induced disturbance corresponds to amplitude gratings which couple the four waves. When the single frequency of the four waves differs slightly from the resonance frequency of the transition, both amplitude gratings and index gratings become involved. Because of the complex nature of the resonant saturation process, the simple $\tan^2 [(2\pi/\lambda_0)(\delta n \ell)]$ expression for the conjugate reflectivity is no longer valid. Instead, this effect is maximized when the intensities of the pump waves are about equal to the intensity which saturates the transition.

Stimulated Brillouin and Raman scattering. Earliest demonstrations of optical phase conjugation were performed by focusing an intense optical beam into a waveguide containing materials that exhibit backward stimulated Brillouin scattering. More recently, this technique has been extended to include the backward stimulated Raman effect. In both cases, the conjugate is shifted by the frequencies characteristic of the effect. SEE RAMAN EFFECT.

Backward stimulated Brillouin scattering involves an inelastic process where an intense laser beam in a clear material is backscattered through the production of optical phonons. Here, the incoming and outgoing light waves interfere to beat at the frequency and propagation vector of the sound wave, and the incoming light wave and the sound wave interfere to produce a radiating nonlinear polarization at the frequency and propagation vector of the outgoing light wave. The production of the sound wave involves the electrostrictive effect in which the Brillouin scattering medium is forced into regions of high electric field. The scattering of the light by the sound wave can be viewed as Bragg scattering from the sound wave grating with a Doppler shift arising because the sound wave is moving. The coupled process of stimulated backward Brillouin scattering can then be understood by considering that the forward-going beam scatters off of a small amount of sound wave to make more backward-going light, and at the same time the forward-going beam and a small amount of backward-going beam interfere to promote the growth of the sound wave. This doubly coupled effect rapidly depletes the incoming laser beam while producing the backward beam and while leaving a strong sound wave propagating in the medium. SEE SCATTERING OF ELECTROMAGNETIC RADIATION.

Photorefractive effects. A degenerate four-wave mixing geometry can be set up in a nonlinear crystal which exhibits the photorefractive effect. In such a material, the presence of light releases charges which migrate until they are trapped. If the light is in the form of an interference pattern, those migrating charges which get trapped in the peaks are more likely to be re-released than would those which become recombined in the valleys. This process leads to the charges being forced into the valleys of the interference pattern. This periodic charge separation sets up large periodic electric fields which, through the Pockels effect, produce a spatially periodic index of refraction. Hence a light-wave grating is converted into an index grating which can then scatter a pump wave into the conjugate direction. This process requires very little power, and therefore can be studied without the use of high-power lasers. SEE ELECTROOPTICS; INTERFERENCE OF WAVES.

Surface phase conjugation. Here a strong pump wave is directed normal to a surface which exhibits nonlinear behavior. Phenomena which can couple the light waves include electrostriction, heating, damage, phase changes, surface charge production, and liquid crystal effects. In general, the surface phase-conjugation process is equivalent to two-dimensional (thin) holography.

Thermal scattering effects. Here an optical interference pattern in a slightly absorbing material will be converted to an index grating if the material has a temperature-dependent index of refraction. In the four-wave mixing geometry, these thermal gratings can then scatter pump waves into the conjugate wave direction.

Practical applications. Many practical applications of optical conjugators utilize their unusual image-transformation properties. Because the conjugation effect is not impaired by interposition of an aberrating material in the beam, the effect can be used to repair the damage done to the beam by otherwise unavoidable aberrations. This technique can be applied to improving the output beam quality of laser systems which contain optical phase inhomogeneities or imperfect optical components. In a laser, one of the two mirrors could be replaced by a phase-conjugating mirror, or in laser amplifier systems, a phase conjugate reflector could be used to reflect the beam back through the amplifier in a double-pass configuration. In both cases, the optical-beam quality would not be degraded by inhomogeneities in the amplifying medium, by deformations or imperfections in optical elements, windows, mirrors, and so forth, or by accidental misalignment of optical elements. SEE IMAGE PROCESSING.

Aiming a laser beam through an imperfect medium to strike a distant target may be another application. The imperfect medium may be turbulent air, an air-water interface, or the focusing mirror in a laser-fusion experiment. Instead of conventional aiming approaches, one could envision a phase-conjugation approach in which the target would be irradiated first with a weak diffuse probe beam. The glint returning from the target would pass through the imperfect medium, and through the laser amplifier system, and would then strike a conjugator. The conjugate beam would essentially contain all the information needed to strike the target after passing through both the amplifier and the imperfect medium a second time. Just as imperfections between the laser and the target would not impair the results, neither would problems associated with imperfections in the elements within the laser amplifier.

Other applications are based upon the fact that four-wave mixing conjugation is a narrowband mirror that is tunable by varying the frequency of the pump waves. There are also applications to fiber-optic communications. For example, a spatial image could be reconstructed by the conjugation process after having been "scrambled" during passage through a multimode fiber. Also, a fiberoptic communication network is limited in bandwidth by the available pulse rate; this rate is determined to a large extent by the dispersive temporal spreading of each pulse as it propagates down the fiber. The time-reversal aspect of phase conjugation could undo the spreading associated with linear dispersion and could therefore increase the possible data rate. SEE LASER; OPTICAL COMMUNICATIONS; OPTICAL FIBERS; OPTICAL INFORMATION SYSTEMS.

Bibliography. R. A. Fisher (ed.), *Optical Phase Conjugation*, 1983; R. A. Fisher (feature ed.), Special issue on optical phase conjugation, *J. Opt. Soc. Amer.*, 73(5):524–660, May 1983; D. M. Pepper (guest ed.), Special issue on nonlinear optical phase conjugation, *Opt. Eng.*, 21(2):156–283, March/April 1982; A. Yariv, Phase conjugate optics and real-time holography, *IEEE J. Quantum Electron.*, QE-14:650–660, 1978.

OPTICAL BISTABILITY
HYATT M. GIBBS

A phenomenon whereby an optical device can have either of two stable output states, labeled 0 and 1, for the same input (**Fig. 1a**). Optical bistability is an expanding field of research because of its potential application to all-optical logic and because of the interesting physical phenomena it encompasses. A bistable optical device can function as a variety of logic devices or as an optical memory element. Under slightly modified operating conditions, the same device can exhibit the optical transistor characteristic of Fig. 1b. For input intensities close to I_{gain}, small variations in the input light are amplified in much the same way that a vacuum tube triode or transistor amplifies electrical signals. The characteristic of Fig. 1b can also be used as a discriminator; inputs above I_{gain} are transmitted with far less attenuation than those below. Finally, there is limiting

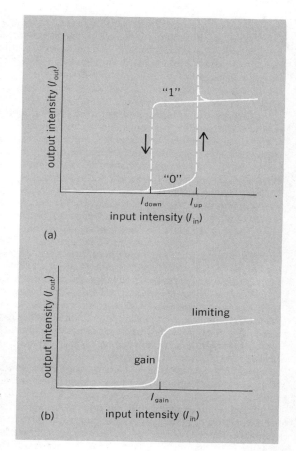

Fig. 1. Transmission of a typical bistable optical device, under conditions of (a) bistability (memory) and (b) high ac gain (optical transistor, discriminator, or limiter).

action above I_{gain}: large changes in the input hardly change the output. There is hope that bistable devices will revolutionize optical processing, switching, and computing. Since the first observation of optical bistability in a passive, unexcited medium in 1974, bistability has been observed in many different materials, including tiny semiconductor etalons. Research has focused on optimizing these devices and developing smaller devices of better materials which operate faster, at higher temperatures, and with less power.

Optical bistability has also attracted the attention of physicists interested in fundamental phenomena. In fact, a bistable device often consists of a nonlinear medium within an optical resonator; it is thus similar to a laser except that the medium is unexcited (except by the coherent light incident on the resonator). Such a device constitutes a simple example of a strongly coupled system of matter and radiation. The counterparts of many of the phenomena studied in lasers such as fluctuations, regenerative pulsations, and optical turbulence can be observed under better-controlled conditions in passive bistable systems. SEE LASER.

All-optical systems. The transmission of information as signals impressed on light beams traveling through optical fibers is replacing electrical transmission over wires. The low cost and inertness of the basic materials of fibers and the small size and low loss of the finished fibers are important factors in this evolution. Furthermore, for very fast transmission systems, for example, for transmitting a multiplexed composite of many slow signals, optical pulses are best. This is because it is far easier to generate and transmit picosecond optical pulses than electrical pulses. With optical pulses and optical transmission, the missing component of an all-optical signal processing system is an optical logic element in which one light beam or pulse controls another. Because of the high frequencies of optical electromagnetic radiation, such all-optical systems have the potential for subpicosecond switching and room-temperature operation. Although any information processing and transmitting system is likely to have electrical parts, especially for powering the lasers and interfacing to humans, the capability of subpicosecond switching appears unique to the optical part of such a system. SEE OPTICAL COMMUNICATIONS; OPTICAL FIBERS; OPTICAL INFORMATION SYSTEMS.

Bistable optical devices. Bistable optical devices have been constructed which have many of the desirable properties of an all-optical logic element. A device is said to be bistable if it has two output states for the same value of the input over some range of input values. Thus a device having the transmission curve of Fig. 1a is said to be bistable between I_{down} and I_{up}. This device is clearly nonlinear; that is, I_{out} is not just a multiplicative constant times I_{in}. In fact, if I_{in} is between I_{down} and I_{up}, knowing I_{in} does not reveal I_{out}. To accomplish this behavior, an all-optical bistable device requires feedback. Even though a nonlinear medium is essential, the nonlinearity alone only means that I_{out} versus I_{in} is not a straight line. The feedback is what permits the nonlinear transmission to be multivalued, that is, bistable.

Fabry-Perot interferometer. The most nearly practical devices so far are tiny semiconductor etalons, that is, tiny Fabry-Perot interferometers consisting of a gallium arsenide (GaAs) or indium antimonide (InSb) crystal with flat parallel faces sometimes coated with dielectrics to increase the reflectivity to about 90% (**Fig. 2**). In these etalons the Fabry-Perot cavity provides the optical feedback, and the nonlinear index of refraction n_2 is the intensity-dependent parameter. The bistable operation of such a Fabry-Perot cavity containing a medium with a nonlinear refraction can be pictured as follows. In the off or low transmission state, the laser is detuned from one of the approximately equally spaced transmission peaks of the etalon and most of I_{in} is reflected (**Fig. 3**a). The refractive index is approximately n_0, its value for weak light intensity. In the on state the index is approximately $n_0 + n_2 I_c$, where the intensity inside the cavity is I_c. This change in index shifts the etalon peak to near coincidence with the laser frequency, permitting a large transmission and a large I_c. Clearly there must be a consistency between the index and the laser frequency; each affects the other through the feedback. As the input is increased from low values, the frequency ν_{FP} of the Fabry-Perot peak begins to shift when $n_2 I_c$ becomes significant. But this shift increases I_c, which further increases $n_2 I_c$, and so on. This positive feedback continues until, at I_{up}, the effect runs away, the device turns on, and the transmission reaches a value on the negative feedback side of the etalon peak consistent with $n = n_0 + n_2 I_c$ (Fig. 3b). Once the device is turned on, I_c is larger than I_{in} because of the storage property of the cavity. Therefore I_{in} can now be lowered to a value below I_{up}, and the large I_c will keep the device on. Thus the

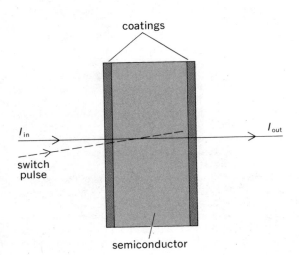

Fig. 2. Bistable optical etalon.

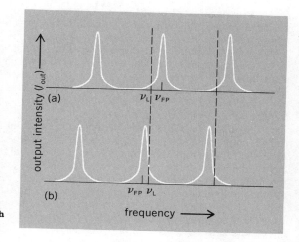

Fig. 3. Transmission function of a nonlinear Fabry-Perot interferometer: (a) device off, index $n \cong n_0$, and (b) device on, $n \cong n_0 + n_2 I_c$. The laser frequency ν_L is fixed, but the cavity peak frequency ν_{FP} changes with light intensity via n_2.

hysteresis of the bistability loop arises from the fact that, for the same input intensity, the intracavity intensity and index contribution are small in the off or detuned state and large in the on or in-tune state. *See* INTERFEROMETRY.

Clearly, a device with the transmission characteristic of Fig. 1 can serve as an optical memory. If I_{in} is maintained between I_{down} and I_{up}, the value of I_{out} reveals the state of the device. Light pulses can be used to switch the device on and off, just as for an electrical flip-flop. A simple analysis of a nonlinear Fabry-Perot cavity and experiments show that by changing the initial detuning this device can perform a whole host of optical operations. Bistable optical devices have been operated as an all-optical differential or ac amplifier (also called an optical transistor and a transphasor), limiter, discriminator, gate, oscillator, and pulse shaper.

Properties. A GaAs device has been fabricated that was only 5 micrometers thick, and the laser beam diameter on the etalon was only 10 μm—about one-tenth the diameter of a human hair. The device was switched on by a 10-ps pulse in a time shorter than the 200-ps detector response time, and a time of a few picoseconds is believed possible. The GaAs device operates at a convenient wavelength (0.83 μm) and is constructed of a material used for electronics and diode lasers, facilitating interfacing and integration. But all the properties of the demonstrated device are not ideal. The switch-down time is about 40 nanoseconds, presumably arising from the lifetime of the free carriers and excitons created by the intense light in the on state. The minimum input power is about 100 mW, and the minimum switch-on energy about 1 nanojoule, far more than a practical device can have. And the highest operating temperature is −244°F (120 K), because the free exciton resonance used to produce n_2 decreases, and the undesirable band-tail absorption increases, with increased temperature. However, measurements of the intrinsic properties of pure GaAs reveal that the switching times and energy should be greatly reduced in optimized etalons.

The InSb etalons are longer (a few hundred micrometers), operate at longer wavelengths (approximately 5 μm), and hence have larger transverse dimension limitations than GaAs, have demonstrated switching times of a few hundred nanoseconds, operate at least up to −321°F (77 K), and have functioned with as little as 8-mW input power. And, just as for GaAs, it is unlikely that the InSb devices are optimized.

The results of these first experiments seeking practical devices point to obvious areas currently under intense research. The physics of semiconductors is being challenged to identify giant nonlinearities at convenient wavelengths and at higher temperatures. In addition to the GaAs free exciton and InSb below-edge band-to-band saturation mechanisms already observed, electron-hole plasma and biexciton mechanisms have been proposed. Other bistable configurations are being sought to better utilize the nonlinearities and minimize the power required: thin etalons, nonlinear interfaces, self-focusing devices, and guided-wave structures.

Fundamental studies. In addition to work on the application of optical bistability to practical devices, a comparable effort is being directed toward the properties of these devices, whose operation depends on the nonlinear coupling between electronic material and light. Figure 1 shows a discontinuity in the transmission, reminiscent of a first-order phase transition. The input light maintains the system in equilibrium which is far from thermal. Fluctuations and transient behavior of such systems are of great interest. A sluggishness in response, called critical slowing down, has already been observed for I_{in} close to I_{up}. Regenerative pulsations have also been seen in which competing mechanisms cause the device to switch on and off repeatedly, forming an all-optical oscillator.

Hybrid devices. Many of these fundamental studies have been conducted with hybrid (mixed optical and electronic) bistable devices in which the intensity dependence of the intracavity index results from applying a voltage proportional to the transmitted intensity across an intracavity modulator. In fact, the cavity is not necessary, since electrical feedback is present. Placing the modulator between crossed polarizers provides the required nonlinearity. Integrated hybrids are being considered for practical devices, but the ultimate in shortening the detector-to-modulator wire length is to place the detector inside the cavity. The distinction between hybrid and intrinsic then fades. The best device may have a voltage across the semiconductor to increase its speed and sensitivity. Arrays of hybrid devices have been used to study image processing using bistable elements; eventually bistable arrays could be used for parallel computing, for example, for propagation problems with transverse effects. SEE IMAGE PROCESSING.

Optical turbulence. Hybrid devices have been considered to be completely analogous to intrinsic devices. It has been predicted that an intrinsic ring-cavity device subjected to a steady input will exhibit periodic oscillations and turbulence or chaos if the medium response time is short compared with the cavity round-trip time. The hybrid analogy is to delay the feedback by a time longer than the detector-feedback-modulator response time. Optical chaos, observed in such a hybrid, has been used to study the evolution, from a stable output, through periodic oscillations, to chaos as the input is increased. The optical turbulence studies interface to phenomena in many other disciplines, such as mathematics, genetics, and hydrodynamics, emphasizing the basic similarity of these seemingly diverse phenomena, in agreement with some recent mathematical findings. SEE NONLINEAR OPTICAL DEVICES; NONLINEAR OPTICS.

Bibliography. H. M. Gibbs, S. L. McCall, and T. N. C. Venkatesan, Optical bistable devices—The basic components of all-optical systems? *Opt. Eng.*, 19:463–468, 1980; H. M. Gibbs et al., Observation of chaos in optical bistability, *Phys. Rev. Lett.*, 46:474–477, 1981; S. D. Smith and D. A. B. Miller, Computing at the speed of light, *New Sci.*, 85:554–556, February 21, 1980; P. W. Smith and W. J. Tomlinson, Bistable optical devices promise subpicosecond switching, *IEEE Spectrum*, 18(6):26–33, 1981.

OPTICAL PULSES
ERICH P. IPPEN

Flashes of light used to study high-speed events or to transmit information.

Flash and strobe photography are time-honored methods for capturing fleeting moments and for studying physical motion too rapid for the eye to see. With pulses of less than 10^{-7}s it is possible to freeze even the fastest mechanical movement of macroscopic objects. Higher-speed flashlamps and electronic sparks have been developed to produce pulses as short as 10^{-10} s for scientific studies of fast photophysical and photochemical processes. SEE STROBOSCOPIC PHOTOGRAPHY.

With the advent of the laser it became possible to generate optical pulses that are dramatically shorter, more intense, and more rapidly repetitive than those achieved previously. In 1966, generation of the first pulses shorter than 10 picoseconds greatly stimulated interest in ultrashort pulses and their potential application to studies of ultrafast phenomena. There followed a rapid development of nonlinear optical techniques for detection and measurement with picosecond and subpicosecond resolution. Methods of generation also improved with the use of other laser systems, notably continuously operated dye lasers. By 1981, a reliable source of pulses shorter than 0.1 ps was in use in research laboratories, and interest in the femtosecond time domain had been kindled. With sophisticated pulse-compression techniques, pulses composed of fewer than eight optical wavelengths have been produced. SEE LASER.

Methods of generation. Pulsed electrical excitation of lasers and other light-emitting devices is the most direct means of producing pulses of 10^{-10} s or longer. Lasers can also be induced to emit much shorter pulses by a technique called mode locking, which requires a laser to operate simultaneously at a variety of different frequencies (modes). These frequencies are then locked together in the appropriate phase relationship to produce an output sequence of short pulses. Fourier theory implies that the larger bandwidth over which frequencies are locked, the shorter the pulses are in time. The shortest pulses have been achieved with well-controlled, continuously operated organic dye lasers and semiconductor diode lasers which have sufficient bandwidth to produce subpicosecond pulses directly. Picosecond pulses from other mode-locked lasers can be compressed to subpicosecond durations by nonlinear optical techniques.

Applications. Ultrashort optical pulses have found widespread application to scientific studies of ultrafast phenomena. Although macroscopic motion on a picosecond time scale is negligible, considerable activity occurs on the atomic and molecular level. Optical pulses are used both to excite such activity and then to probe the way in which the excitation energy is exchanged, transported, and dissipated. Important information has been obtained about molecular motion, internal vibrations, photodissociation, chemical reactions, and primary events in vision and photosynthesis. The dynamics of carrier interactions in semiconductors and the evolution of changes in states of matter are also topics of active investigation.

Optical fiber communication systems use optical pulses for the transmission of information. Some commercial systems already operate with pulses as short as 10^{-9} s, and fibers have the potential for still greater bandwidths. The propagation, dispersion, detection, and processing of picosecond pulses are subjects of much research. Diode laser transmitters have been made to produce subpicosecond pulses, integrated optic devices have demonstrated capability for switching picosecond pulses, and picosecond pulses have been transmitted without distortion over mile lengths of single-mode optical fiber. SEE INTEGRATED OPTICS; OPTICAL COMMUNICATIONS; OPTICAL FIBERS.

Bibliography. S. L. Shapiro (ed.), *Ultrashort Light Pulses*, 1977; K. Young (ed.), *Opto-electronics*, 1982.

LIGHT DETECTION AND PROCESSING

Optical detectors	**308**
Image tube (astronomy)	**309**
Photography	**313**
Latent image	**326**
Photographic materials	**327**
Stroboscopic photography	**331**
Cinematography	**336**
Image processing	**338**

OPTICAL DETECTORS
J. K. GALT

Devices that generate a signal when light is incident upon them. The signal may be observed visually in reflected or transmitted light, or it may be electrical.

The eyes of animals and the leaves of plants, for example, perform this function. Light, as well as other forms of electromagnetic energy, can also be detected by observing the temperature change which is produced in a body absorbing it. Furthermore, it can be detected by the color changes its absorption produces in certain materials. However, the need for high sensitivity or fast response or both usually leads, in optical technology, to detectors based on other mechanisms. SEE PHOTOMETER.

One of the most widely used optical detectors is photographic film. This method of detection is based on a photochemical process in particles embedded in the film. This fairly sensitive technique is more effective than others in detecting or recording, or both, the large amount of information usually contained in an image or a complicated optical spectrum. Disadvantages are its insensitivity to wavelengths longer than those in the optical range and the need to develop film after exposure. SEE PHOTOGRAPHY.

In many cases, fast detector response is needed to detect rapid changes in incident light; also, an electrical output signal may be desired. The optimum detector here would be based on external or internal photoemission.

In an externally photoemitting device, the light incident upon a surface causes an emission of electrons from that surface into a vacuum. These electrons can be amplified in number by using electric and magnetic fields to cause them to impinge on other surfaces, thereby producing secondary electrons. Detectors of this type can be fabricated in such a way as to detect images. These detectors are limited to wavelengths in the visible region or shorter, but are capable of a sensitivity and speed of response higher than any other type.

In an internally photoemitting device, incident light produces free charge carriers within a body and is detected through the effect of these charge carriers on the electrical impedance of the body. In its simplest and most widely used form, the body is a photoconductor whose impedance is high in the absence of incident light. Such detectors are less sensitive and slower than external photoemitters, but they require no vacuum envelope and operate at low voltages. They also have the advantage that some of them are sensitive to all wavelengths shorter than about 40 micrometers, far into the infrared. In general, however, operation at the longer wavelengths requires that the device be cooled. Detectors of this type can also be built in such a way as to detect images. This, however, is generally done with an electron beam as part of the read-out circuit and does require a vacuum envelope.

A development among internal photoemitting detectors is the reverse-biased semiconductor diode, which takes advantage of a large electric field at the reverse-biased semiconductor pn junction to give a response faster than that of the devices simply based on photoconductivity in bulk samples. The device, when made of germanium, is sensitive to all wavelengths shorter than 1.6 μm. Its advantages are further increased by operating it at voltages which cause avalanche processes to follow each absorption of a photon. This gives rise to internal gain, and thereby increases the figure of merit of the device if, as is often the case, other sources of noise are dominant. It is difficult to produce avalanche diodes so that the properties of successive diodes are identical. This problem has been solved for the nonavalanching devices, however. Thus it has been possible to fabricate arrays which have potential for use in vidicons as image detectors. SEE OPTICAL COMMUNICATIONS; OPTICAL MODULATORS.

Bibliography. L. K. Anderson and B. J. McMurty, High speed photodetectors, *Appl. Opt.*, 5:1573–1587, October 1966; E. L. Derniak and D. G. Crowe, *Optical Radiation Detectors*, 1984; H. Melchior, M. B. Fisher, and F. R. Arams, Photodetectors for optical communication systems, *Proc. IEEE*, 58:1466–1486, 1970; T. P. Pearsall, Long wavelength photodetectors, *Fiber Integr. Opt.*, 4:107–128, 1982; G. H. S. Rokos, Optical detection using photodiodes, *Opto-Electronics*, 5:351–366, 1973; G. E. Stillman and C. M. Wolfe, Avalanche photodiodes, in R. W. Willardson and A. C. Beer (eds.), *Semiconductors and Semimetals*, vol. 12, 1977; L. R. Tomasetta et al., High sensitivity optical receivers for 1.0–1.4 μm fiber-optic systems, *IEEE J. Quantum Electron.*, QE-14:800–804, 1978.

IMAGE TUBE (ASTRONOMY)
RICHARD G. ALLEN

A photoelectric device for intensifying faint astronomical images. Photographic emulsions on plates or film have several characteristics that have made them useful for recording astronomical images—high resolution, large two-dimensional formats with millions of sensitive elements, good dimensional stability and permanence, and a wide spectral response. Unfortunately, they lack one characteristic of great importance to the astronomer—a high quantum efficiency, or the ability to record a high percentage of the light quanta in the incident image. Even the most sensitive photographic emulsions are unable to record more than a few photons in each thousand that are collected and focused into an image by the telescope. All the rest, and the information they bring, are lost. The photographic process has other disadvantages as well. Because of a phenomenon called reciprocity failure, the effective speed of a photographic emulsion drops when the required exposure time is longer than a few seconds. A linear relationship between emulsion density and exposure also exists over only a relatively narrow range of exposures. Outside this range, the response is nonlinear. *See Photon.*

Photocathode. Modern image tubes are free of the drawbacks cited above. In these devices a photoemissive surface, called the photocathode, emits electrons through the photoelectric effect. In most tubes the photocathode is semitransparent and is deposited on the inside of a transparent window that is mounted on the end of an evacuated glass or ceramic cylinder. When light from a telescope or spectrograph is imaged on the photocathode, electrons are ejected into the vacuum inside the tube. Electric or electric and magnetic fields in combination then accelerate and direct the photoelectrons through the device. In most image tubes the photoelectrons are ultimately reimaged on a phosphor-coated output window that converts them back into a visible image. In other devices the photoelectrons are brought to focus directly on a photographic plate or solid-state array detector.

The quantum efficiency of an image tube is simply the probability that an incoming photon of light will release a detectable photoelectron inside the tube. Although a single photon of light is all that is necessary to create a photoelectron, some photons pass through the photocathode without being absorbed. Others are absorbed but do not produce a free electron. In some devices a significant number of photoelectrons are also lost after they are released. The effective quantum efficiency of such a device is therefore even less than that of its photocathode. In spite of these losses, many low-light imaging systems are capable of detecting about one-fifth of the photons incident upon them. Since the effects of individual photons are recorded by most of these systems, measurement uncertainties are primarily caused by statistical fluctuations in the photon arrival rate. There is no reciprocity failure, and the response of the system is linear as long as each photon is counted with the same efficiency as any other.

Two types of photocathode are commonly used in astronomical image tubes. Bialkali photocathodes are composed of antimony and two alkali metals. Photocathodes of this type are most sensitive to ultraviolet, blue, and green light at wavelengths below 600 nanometers. Multialkali photocathodes have three or more alkali metals in combination with antimony and have red responses that extend beyond 800 nm. Both types eventually lose sensitivity at long wavelengths because light photons at long wavelengths have very little energy to eject a photoelectron. Their responses at short wavelengths are usually limited only by the transparency of the photocathode window or by the Earth's atmosphere. The Earth's atmosphere becomes virtually opaque at about 320 nm. Satellite-borne image tubes with magnesium fluoride windows, however, can be used down to 120 nm in the far ultraviolet.

All photocathodes tend to emit photoelectrons at a modest rate even in total darkness. This spontaneous emission in the absence of incident light, or dark emission, makes observations of faint astronomical sources more difficult. Since bialkali photocathodes have somewhat lower dark emission rates than multialkali photocathodes, they are usually used whenever the extended red response of the multialkali photocathode is not required.

Image-intensifier tubes. The electron image produced at the photocathode is converted back into visible light in an image intensifier. In a single-stage intensifier the photoelectrons are accelerated away from the photocathode and eventually crash into a phosphor-coated output window at the other end of the tube. Electron optics are used to bring the photoelectrons back into focus on the phosphor screen of the output window. Optical gain is provided by the phosphor itself, and the output of the tube can be examined with the eye or recorded photographically. Solid-state detectors or television cameras may also be utilized.

Proximity-focused tubes. In a proximity-focused image intensifier, the photocathode and phosphor screen are both deposited on plane-parallel optical windows. The tube is very short, and the phosphor screen is only 0.06 to 0.14 in. (1.5–3.5 mm) from the photocathode. Electrons emitted from the photocathode encounter a very strong electric field and are rapidly accelerated across the gap between the photocathode and the phosphor screen. Although photoelectrons leave the photocathode at all angles to its surface, very little defocusing takes place. Unfortunately, most proximity-focused intensifiers have unacceptably high dark emission rates.

Electrostatically focused tubes. Electrostatically focused image intensifiers are usually constructed with fiber-optic input and output windows to optimize and simplify the electron imaging. The fiber-optic windows at both ends of such a tube are flat on the outside and spherical on the inside. Electrons emitted from the spherical surface of the photocathode are accelerated through a small hole at the tip of a conical-shaped electrode. They then come back to focus on the spherical surface of the output window. A phosphor screen on the output window converts the electron image back into light. Because electrostatically focused image tubes normally have fiber-optic photocathode windows made of glass, they cannot be used in the ultraviolet. *See* OPTICAL FIBERS.

A two-stage image intensifier containing a pair of electrostatically focused intensifiers is shown in **Fig. 1**. The output light of the first tube travels through the contacted fiber-optic windows and falls on the photocathode of the second. Light from the first tube is thus amplified and reimaged by the tube behind it. In normal practice, each tube has a 15,000-V potential difference between the photocathode and phosphor. At this voltage each photoelectron produces about 1000 photons; however, only about half of them are collected and transmitted through the fiber-optic windows. If both tubes happened to have photocathode quantum efficiencies of 10%, the photoelectron gain would be 50 per stage, and the overall light gain of the intensifier would be 2500. A similar four-stage intensifier would have a light gain of over a million. A single photon falling on such an intensifier could produce a pulse of light that is easily visible to the naked eye.

Magnetically focused tubes. A magnetically focused image intensifier uses both electric and magnetic fields for electron imaging. Electrons from the photocathode are accelerated by a series of ring-shaped electrodes that are spaced out at regular intervals down the length of the tube. A uniform magnetic field down the center of the intensifier forces the photoelectrons into a tight spiral as they are accelerated. By balancing the electric and magnetic fields properly, the

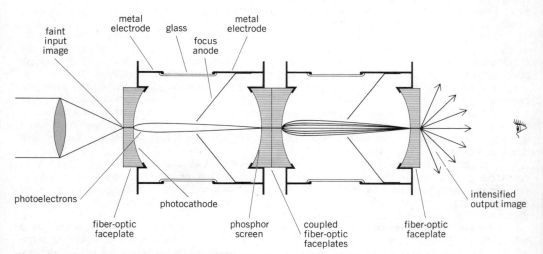

Fig. 1. Two-stage electrostatically focused image tube, fiber-optically coupled.

photoelectrons are brought to focus on a phosphor-coated output window at the opposite end of the tube. The input and output windows of a magnetically focused intensifier are flat on both sides. Since magnetically focused tubes can be made without fiber-optic input windows, they are useful in the ultraviolet.

A two-stage magnetically focused cascade intensifier is shown in **Fig. 2**. Additional gain is achieved in this tube by focusing the primary photoelectrons on a thin transparent membrane which has a phosphor screen on one side and a photocathode on the other. The electron gain with such a sandwich is usually somewhat higher than that obtained by coupling single-stage tubes. High gain is thus achieved without serious loss of resolution. Transmission secondary-emitting dynodes have also been used in magnetic intensifiers. Such dynodes emit secondary electrons directly when they are bombarded with energetic photoelectrons. Although such dynodes are simpler than phosphor-photocathode sandwiches, they have the disadvantage of producing only a very few secondary electrons for each incident primary. This means that additional stages of gain are required and that individual output pulses will vary significantly in brightness. Because of their inferior performance, such dynodes are no longer used in astronomical intensifiers.

Microchannel-plate tubes. An effective method of obtaining high electron gain within an image tube is to focus the photoelectrons on a microchannel plate. A microchannel plate (**Fig. 3**) consists of an array of closely spaced tubes, each of which is coated on the inside with a secondary-emitting material. When a primary electron enters a channel, it collides with the side wall of the tube and generates several secondary electrons. The secondary electrons travel farther down the tube and eventually produce secondary electrons of their own. Since this process is repeated several times down the channel, thousands of electrons exit the rear face of the microchannel plate. These electrons are then brought to focus on a phosphor screen on the output window of the tube.

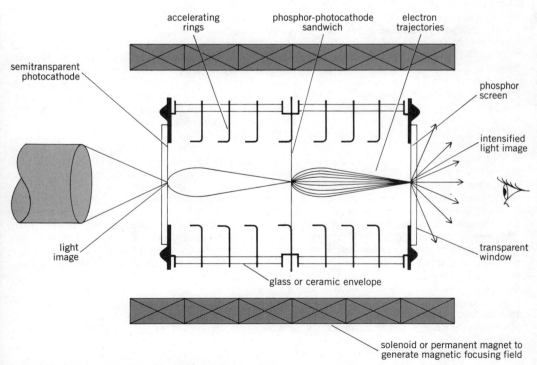

Fig. 2. Two-stage magnetically focused cascade image tube.

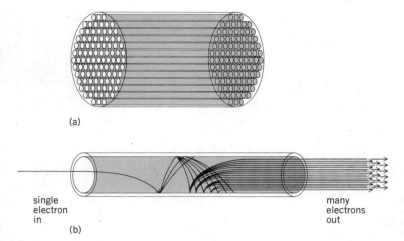

Fig. 3. Microchannel plate showing (a) array of tubes and (b) details of a single channel.

Microchannel plates are most often added to proximity-focused intensifiers. Such tubes are barely over 0.4 in. (1 cm) in length, but they can have approximately the same gain as conventional two-stage intensifiers. No permanent magnet or solenoid is required to operate one, and they are almost entirely free of distortion. Unfortunately, these devices also have disadvantages. Individual optical pulses from a microchannel-plate tube vary enormously in amplitude, and integration times often have to be increased to compensate for the added noise. Lifetimes of microchannel plates are also limited. In spite of these drawbacks, microchannel-plate tubes have been widely used.

Image-intensifier cameras. A separate two-dimensional detector is needed to permanently record the output of an image intensifier. Simple image-intensifier cameras often use a lens to relay the output of the intensifier to a television tube or photographic plate. Unfortunately, fast lenses of adequate resolution only collect about 5% of the light emitted from the intensifier. Tubes with transparent output windows must be used with relay lenses because the output image is formed on the inside of the window and is not accessible. Tubes with fiber-optic output windows, however, can be used with a detector in direct contact with the output window. This greatly increases the transfer efficiency and reduces the gain required from the image intensifier.

Image-intensifier cameras built with charge-coupled devices (CCDs) and other solid-state array detectors are used at many observatories. Originally, solid-state detectors were of limited use because individual detecting elements were fairly large and only a few of them could be fabricated on a single chip. Background noise was also a major problem. Solid-state linear arrays of hundreds of elements are commonplace now, however, and have almost totally displaced photography as a recording medium in astronomical spectroscopy. Charge-coupled devices with thousands of elements in a square or rectangular array are also now available in limited quantities for two-dimensional applications. Most solid-state detectors can be obtained with fiber-optic input windows which allow them to be directly coupled to the output windows of image tubes. Unfortunately, photon-counting detector systems with scanned arrays respond very nonlinearly if the photon arrival rate gets moderately high. Since each diode in a scanned array is typically read out only one thousand times a second, coincidence losses become severe whenever there are more than a few events a second in each diode.

Charge-coupled devices have been used without image tubes in the visible and infrared. In long integrations they have a high quantum efficiency and relatively low noise. At present, intensifiers are still required in the blue where charge-coupled devices are not as sensitive and for time-resolved applications in which the image must be read out at a high frame rate.

Electronographic cameras. Direct recording of electron images with photographic plates is called electronography, and image tubes which use this method are referred to as electronographic cameras. Electrons that have been accelerated through a several-thousand-volt potential are very efficient in sensitizing even very fine-grain, high-resolution photographic emulsions. These emulsions are generally too insensitive to light to be useful for direct astronomical photography, but some of them have a sufficiently good response to fast electrons to be of value in electronography. Furthermore, some of these emulsions have a very linear relationship between density and exposure to electrons. Unfortunately, the electronographic tube has a drawback that has never been fully surmounted. Photographic emulsions inside an image tube evolve gases that slowly destroy the photocathode. A few successful schemes have been employed to prevent these gases from reaching the photocathode. Because of the extreme complexity in using electronographic tubes, however, they have largely been phased out of use at most observatories.

Digicon cameras. A digicon camera is closely related to the electronographic camera. In this detector, however, the electron image is focused directly on a silicon diode array, and each incoming photoelectron produces an electrical pulse that is amplified and recorded. Since each diode is handled as a separate detector, digicons can record individual photoelectrons that arrive very close together in time. Digicons are expensive, and ones having more than 512 linear elements are not likely to be built. Active research, however, has been undertaken on a number of similar solid-state devices which show promise of being expandable into much larger one- and two-dimenstional formats.

PHOTOGRAPHY
ROBERT D. ANWYL AND VIVIAN K. WALWORTH

V. K. Walworth is coauthor of the section Color Photography.

The process of forming stable or permanent visible images directly or indirectly by the action of light or other forms of radiation on sensitive surfaces. Traditionally, photography utilizes the action of light to bring about changes in silver halide crystals. These changes may be invisible, necessitating a development step to reveal the image, or they may cause directly visible darkening (the print-out effect). Most photography is of the first kind, in which development converts exposed silver halide to (nonsensitive) metallic silver.

Since the bright parts of the subject normally lead to formation of the darkest parts of the reproduction, a negative image results. A positive image, in which the relation between light and dark areas corresponds to that of the subject, is obtained when the negative is printed onto a second negative-working material. Positive images may be achieved directly in the original material in two principal ways: (1) In reversal processing, a positive image results when the developed metallic silver is removed chemically and the remaining silver halide is then reexposed and developed. (2) In certain specialized materials, all of the silver halide crystals are rendered developable by chemical means during manufacture. Exposure to light alters the crystals so that development will not proceed. As a result, conventional "negative" development affects only the unexposed crystals and a positive image is formed directly.

The common materials of photography consist of an emulsion of finely dispersed silver halide crystals in gelatin coated in a thin layer (usually less than 25 micrometers) on glass, flexible transparent film, or paper (**Fig. 1**). The halides are chloride, bromide, or bromoiodide, depending upon the intended usage. The most sensitive materials, used for producing camera negatives, consist of silver bromide containing some silver iodide; the slow materials used for printing are usually of silver chloride; materials of intermediate speed are of silver bromide or silver bromide and chloride.

Following exposure of a sensitized recording material in a camera or other exposing device such as a line-scan recorder, oscillograph, plotter, or spectrograph, the film or plate is developed, fixed in a solution which dissolves the undeveloped silver halide, washed to remove the soluble salts, and dried. Printing from the original, if required, is done by contact or optical projection onto a second emulsion-coated material, and a similar sequence of processing steps is followed.

For about 100 years, the results of practical photography were almost exclusively in black and white or, more precisely, shades of gray. With the introduction of the Kodachrome process and materials in 1935 and of a variety of other systems over the succeeding years, a large and

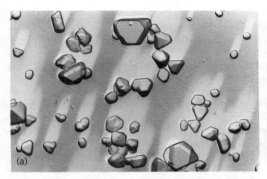 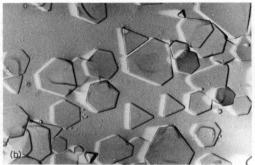

Fig. 1. Silver halide crystals, shown highly magnified. (a) Conventional pebblelike crystals. (b) Tablet-shaped crystals, used in some modern emulsions, that absorb light more efficiently, resulting in higher film speed.

increasing fraction of photography has been done in color. In most color photography, development is basically the same as in black-and-white photography, except that the chemical reactions result in the formation of dye images and subsequent removal of the image-forming silver. Both negative and positive (reversal) color systems are employed. A special class of print materials relies on bleachable dyes instead of conventional silver halide processes. SEE PHOTOGRAPHIC MATERIALS.

Infrared photography. Emulsions made with special sensitizing dyes can respond to radiation at wavelengths up to 1200 nanometers, though the most common infrared films exhibit little sensitivity beyond 900 nm. One specialized color film incorporates a layer sensitive in the 700–900-nm region and is developed to false colors to show infrared-reflecting subjects as bright red. SEE INFRARED RADIATION.

Photographs can thus be made of subjects which radiate in the near-infrared, such as stars, certain lasers and light-emitting diodes, and hot objects with surface temperatures greater than 500°F (260°C). Infrared films are more commonly used to photograph subjects which selectively transmit or reflect near-infrared radiation, especially in a manner different from visible radiation. Infrared photographs taken from long distances or high altitudes usually show improved clarity of detail because atmospheric scatter (haze) is diminished with increasing wavelength and because the contrast of ground objects may be higher as a result of their different reflectances in the near-infrared. Grass and foliage appear white because chlorophyll is transparent in the near-infrared, while water is rendered black because it is an efficient absorber of infrared radiation.

Infrared photography is used in a wide range of scientific and technical disciplines because it permits or greatly enhances the detection of features not obvious to the eye. For example, it has been used for camouflage detection because most green paints absorb infrared more strongly than the foliage they are meant to match, to distinguish between healthy and diseased trees or plants, to detect alterations in documents and paintings, to reveal subcutaneous veins (because the skin is somewhat transparent to infrared radiation), and to record the presence of certain substances which fluoresce in the near-infrared when exposed to visible radiation. While infrared photography can be used to map variations in the surface temperature of hot objects which are not quite incandescent, it cannot be used for detecting heat loss from buildings or other structures whose surface temperatures are usually close to that of the surrounding atmosphere.

One technique for detecting radiation at wavelengths near 10 micrometers relies on localized transient changes in the sensitivity of conventional photographic materials when exposed briefly to intense radiation from certain infrared lasers, and simultaneously or immediately thereafter to a general low-level flashing with light. Another technique for detecting intense infrared exposures of very short duration with films having no conventional sensitivity at the exposing wavelength is thought to rely on two-photon effects.

Ultraviolet photography. Two distinct classes of photography rely on ultraviolet radiation. In the first, the recording material is exposed directly with ultraviolet radiation emitted, reflected, or transmitted by the subject; in the other, exposure is made solely with visible radiation

LIGHT DETECTION AND PROCESSING

resulting from the fluorescence of certain materials when irradiated in the ultraviolet. In the direct case, the wavelength region is usually restricted by the camera lens and filtration to 350–400 nm, which is readily detected with conventional black-and-white films. Ultraviolet photography is accomplished at shorter wavelengths in spectrographs and cameras fitted with ultraviolet-transmitting or reflecting optics, usually with specialized films. In ultraviolet-fluorescence photography, ultraviolet radiation is blocked from the film by filtration over the camera lens and the fluorescing subject is recorded readily with conventional color or panchromatic films. Both forms of ultraviolet photography are used in close-up photography and photomicrography by mineralogists, museums, art galleries, and forensic photographers. SEE ULTRAVIOLET RADIATION.

Photographic radiometry. The intensity (brightness) and spectral distribution (color) of radiation emitted or reflected by a surface can be measured by photography by comparing the densities resulting from recording a test object with those obtained from photographing a standard source or surface, preferably on adjacent portions of the same film or plate. The method is capable of reasonable precision if the characteristics of the photographic material are accurately known, all of the other photographic conditions are closely or exactly matched, and the standard source is well calibrated. In practice, it is difficult to meet all of these conditions, so photographic radiometry is used mainly when direct radiometric or photometric methods are not practicable. It is even more difficult, though not impossible, when using color films. When photographic radiometry is applied to objects at elevated temperatures which self-radiate in the visible or near-infrared region and the emissivity of the materials is known, the surface temperature of the object may be determined and the process is known as photographic pyrometry. Reference objects at known temperatures must be included in the photograph. SEE PHOTOMETRY; RADIOMETRY.

THEORY OF THE PHOTOGRAPHIC PROCESS

The normal photographic image consists of a large number of small grains or clumps of silver. They are the end product of exposure and development of the original silver halide crystals in the emulsion and may be two to ten times larger than the crystals, which range in size up to a few micrometers (**Fig. 2**). The shape, size, and size distribution of the crystals are determined by the way the emulsions are made, and are closely related to their photographic properties. In general, high-speed, negative-type emulsions have crystals in a wide range of sizes, while emulsions with low speed and high contrast have smaller crystals that are more uniform in size.

Exposure of the crystals to light causes a normally invisible change called the latent image. The latent image is made visible by development in a chemical reducing solution which converts the exposed crystals to metallic silver, leaving unreduced virtually all of the crystals either unexposed or unaffected by exposure. The darkening which results is determined by the amount of

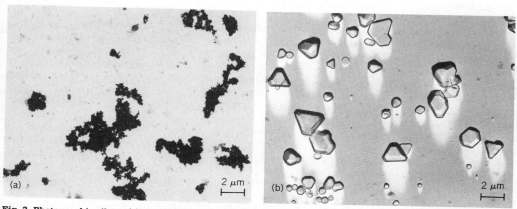

Fig. 2. Photographic silver. (a) Developed grains of a photographic emulsion. (b) Original silver halide crystals from which the silver grains were developed. Both views are highly magnified. (*Eastman Kodak Co.*)

exposure, the type of emulsion, and the extent of development. Density, a physical measure of the darkening, depends on the amount of silver developed in a particular area and the covering power of the silver. With very high exposures, darkening may occur directly, without development, as a result of photolysis. This is known as the print-out effect. Its use is confined mostly to making proof prints in black-and-white portraiture and papers for direct-trace recording instruments.

Latent image. As noted above, the term latent image refers to a change occurring in the individual crystals of photographic emulsions whereby they can be developed upon exposure to light. Early research indicated that development starts at localized aggregates of silver, mainly on the surfaces of the crystals. The existence of these discrete development centers suggested that the latent image is concentrated at specific sites. Subsequent research by numerous investigators relating to the chemical and physical properties of silver halide crystals during irradiation led eventually to a logical and comprehensive explanation for the process of latent image formation, formulated in 1938 by R. W. Gurney and N. F. Mott.

According to the Gurney-Mott theory, absorption of photon energy by a crystal liberates electrons which are able to move freely to sensitivity sites, presumably associated with localized imperfections in the crystal structure and concentrations of sensitizing impurities, where they are trapped and establish a negative charge. This charge attracts nearby positively charged silver ions, which are also mobile. The electrons and ions combine to form metallic silver, and the process continues until sufficient silver has accumulated to form a stable, developable aggregate. The evidence indicates that a minimum of three or four silver atoms is required.

At the exposure levels required to form a latent image, the absorption of one photon by a silver halide crystal can, in principle, form one atom of silver. However, many photons which reach a crystal are not absorbed, and many that are absorbed fail to contribute to the formation of a development center. The overall efficiency of latent image formation is dependent upon the structure of the crystal, the presence of sensitizing dyes and other addenda in the emulsion, temperature, and several other factors, but is essentially independent of grain size.

The efficiency of latent image formation is dependent on the rate of exposure. With low-intensity, long-duration exposure, the first silver atom formed may dissociate before succeeding ions are attracted to the site. Hence, exposing energy is wasted and the emulsion will exhibit low-intensity reciprocity failure. With very intense exposures of short duration, an abundance of sites compete for the available silver atoms with the result that many of the sites fail to grow sufficiently to become developable. In addition, such exposures frequently result in the formation of internal latent images which cannot be reached by conventional surface-active developers.

The Gurney-Mott theory was not universally accepted at first, and there is still uncertainty with regard to some details of the mechanism, but it prompted extensive research and continues to be the best available explanation of latent image formation and behavior.

Development. Development is of two kinds, physical and chemical. Both physical and chemical developers contain chemical reducing agents, but a physical developer also contains silver compounds in solution (directly added or derived from the silver halide by a solvent in the developer) and works by depositing silver on the latent image. Physical development as such is little used, although it usually plays some part in chemical development. A chemical developer contains no silver and is basically a source of reducing agents which distinguish between exposed and unexposed silver halide and convert the exposed halide to silver. Developers in general use are compounded from organic reducing compounds, an alkali to give desired activity, sodium sulfite which acts as a preservative, and potassium bromide or other compounds used as antifoggants (fog is the term used to indicate the development of unexposed crystals; it is usually desirable to suppress fog).

Most developing agents used in normal practice are phenols or amines, and a classical rule which still applies, although not exclusively, states that developers must contain at least two hydroxyl groups or two amino groups, or one hydroxyl and one amino group attached ortho or para to each other on a benzene nucleus. Some developing agents do not follow this rule.

Alkalies generally used are sodium carbonate, sodium hydroxide, and sodium metaborate. Sulfite in a developer lowers the tendency for oxidation by the air. Oxidation products of developers have an undesirable influence on the course of development and may result in stain.

Developing agents and formulas are selected for use with specific emulsions and purposes.

Modern color photography relies mainly on paraphenylene-diamine derivatives. So-called fine-grain developers are made to reduce the apparent graininess of negatives. They generally contain the conventional components but are adjusted to low activity and contain a solvent for silver bromide. One fine-grain developer is based on p-phenylenediamine, which itself is a silver halide solvent. Some developers are compounded for hardening the gelatin where development occurs, the unhardened areas being washed out to give relief images for photomechanical reproduction and imbibition color printing. In a number of products, the developing agent is incorporated in the emulsion, development being initiated by placing the film or paper in an alkaline solution called an activator. Hardening agents may be included in developers to permit processing at high temperatures.

Monobaths are developers containing agents which dissolve or form insoluble complexes with the undeveloped silver halide during developing.

When film is developed, there is usually a period during which no visible effect appears; after this the density increases rapidly at first and then more slowly, eventually reaching a maximum. In the simplest case, the relation between density and development time is given by Eq. (1), where D is the density attained in time t, D_x is the maximum developable density, and k is a

$$D = D_x (1 - e^{-kt}) \qquad (1)$$

constant called the velocity constant of development.

After processes. These include fixing, washing, drying, reduction, intensification, toning, and bleaching.

Fixing. After the image is developed, the unchanged halide is removed, usually in water solutions of sodium or ammonium thiosulfate (known as hypo and ammonium hypo, respectively). This procedure is called fixing. Prior to fixing, a dilute acid stop bath is often used to neutralize the alkali carried over from the developer. The rate of fixing depends on the concentration of the fixing agent (typically 20–40% for hypo) and its temperature (60–75°F or 16–24°C). Hardeners may be added to the fixer to toughen the gelatin, though they tend to reduce the fixing rate. In so-called stabilization processes, the undeveloped silver salts are converted into more or less stable complexes and retained in the emulsion to minimize the processing time associated with washing. Solutions of organic compounds containing sulfur are often used for this purpose.

Washing and drying. Negatives and prints are washed in water after fixing to remove the soluble silver halide-fixing agent complexes, which would render the images unstable or cause stain. The rate of removal declines exponentially and can be increased by raising temperature and increasing agitation. The rate can be accelerated with neutral salt solutions known as hypo clearing aids, thus reducing washing times. Print permanence can be enhanced with toning solutions (especially selenium toners) too dilute to cause a change in image coloration. After washing, the materials must be dried as uniformly as possible, preferably with dust-free moving warm air, though some papers are often dried on heated metal drums. Drying uniformity is aided by adding wetting agents to the final wash water.

Reduction and intensification. Reduction refers to methods of decreasing the density of images by chemically dissolving part of the silver with oxidizers. According to their composition, oxidizers may remove equal amounts of silver from all densities, remove silver in proportion to the amount of silver present, or remove substantially more silver from the higher densities than from the lower. Intensification refers to methods of increasing the density of an image, usually by deposition of silver, mercury, or other compound, the composition being selected according to the nature of the intensification required.

Toning. The materials and processes for black-and-white photography are normally designed to be neutral in color. The coloration can be modified by special treatments known as toning. This may be done with special developers or additions to conventional developers, but toning solutions used after fixing and washing are more common. These solutions convert the silver image into compounds such as silver sulfide or silver selenide, or precipitate colored metallic salts with the silver image. Dye images can be obtained by various methods.

Bleaching. In reduction and in color processes and other reversal processes where direct positives are formed, the negative silver is dissolved out by an oxidizing solution, such as acid permanganate or dichromate, which converts it to a soluble silver salt. If a soluble halide, ferricyanide, or similar agent is present, the corresponding silver salt is formed and can be fixed out or used as a basis for toning and intensifying processes. Bleach-fixing solutions (blixes) are used to dissolve out silver and fix out silver halide simultaneously.

SENSITOMETRY AND IMAGE STRUCTURE

Sensitometry refers to the measurement of the sensitivity or response to light of photographic materials. Conventional methods of measurement tend to mask the dependence of this response on factors which affect the structure of the photographic image. The simplest method of determining sensitivity is to give a graded series of exposures and to find the exposure required to produce the lowest visible density. Modern sensitometry relies on plotting curves showing the relation between the logarithm of the exposure H and the density of the silver image for specified conditions of exposure, development, and evaluation. Exposure is usually defined as the product of intensity and duration of illumination incident on the material, though it is more correct to define exposure as the time integral of illuminance I ($H = \int I dt$). Density D is defined as the logarithm of the opacity O, which in turn is defined as the reciprocal of transmittance T. If light of intensity I falls on a negative and intensity I' is transmitted, Eqs. (2) hold.

$$T = I'/I \qquad O = 1/T = I/I' \qquad D = \log I/I' \qquad (2)$$

Illuminance is the time rate at which light is directed at a unit area of film, exposure is the amount of energy per unit area received by the film, and density is a measure of the extent to which that energy is first converted to a latent image and then, with an infusion of chemical energy from the developer, is amplified to form a stable or permanent silver image. Since density is dependent mainly on the amount of silver per unit area of image, and since the amount of energy required to form a latent image in each crystal is independent of crystal size, it takes more exposure to obtain a specified density with a fine-grain emulsion than a coarse-grain one. However, through advances in emulsion and developer chemistry, modern picture-taking films are roughly 25,000 times more sensitive than the gelatin dry plates introduced in 1880.

Hurter and Driffield curve. The relationship between density and the logarithm of exposure is shown by the Hurter and Driffield characteristic curve, which has three recognizable sections, the largest corresponding usually to the straight line B to C in **Fig. 3**. In this region, density is directly proportional to the logarithm of exposure. In the lower and upper sections of the curve, this proportionality does not apply. These nonlinear regions, AB and CD, are commonly called the toe and shoulder, respectively.

The characteristic curve is used to determine the contrast, exposure latitude, and tone reproduction of a material as well as its sensitivity or speed. It is obtained under controlled conditions by using a light source of known intensity and spectral characteristics, a modulator which gives a series of graded exposures of known values, development under precise conditions, and an accurate means of measuring the resulting densities.

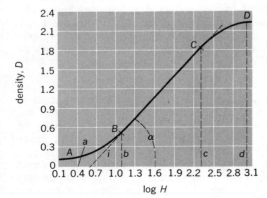

Fig. 3. Hurter and Driffield characteristic curve.

Sensitometric exposure. The internationally adopted light source is a tungsten-filament electric lamp operated at a color temperature close to 2850 K combined with a filter to give spectral quality corresponding to standard sensitometric daylight (approximately 5500 K). The light source and exposure modulator are combined to form a sensitometer, which gives a series of exposures increasing stepwise or in a continuous manner. Sensitometers are either intensity-scale or time-scale instruments, depending on whether the steps of exposure are of fixed duration and varying intensity or are made at constant intensity with varying duration. Special sensitometers are used to determine the response of materials to the short exposures encountered in electronic flash photography or in scanning recorders. The exposure may also be intermittent, that is, a series of short exposures adding up to the desired total, though continuous exposures are preferred to avoid intermittency effects. (Below a critical high frequency, the photographic material does not add up separate short exposures arithmetically.) The best sensitometers are of the continuous-exposure, intensity-scale variety, and the exposure time used should approximate that which would prevail in practice.

Development. Development is carried out in a standard developer or in a developer specifically recommended for use with the material under test, for the desired time at a prescribed standard temperature, and with agitation that will ensure uniform development and reproducibility. The procedure may be repeated for several development times.

Densitometers. Photographic density is measured with a densitometer. Early densitometers were based on visual photometers in which two beams are taken from a standard light source and brought together in a photometer head. One beam passes through the density to be measured, the other through a device for modulating the intensity so that the two fields can be matched. Visual instruments are seldom used now except as primary reference standards. Most densitometers are photoelectric devices which rely on one or more standard densities which have been calibrated on another instrument.

The simplest photoelectric transmission densitometers are of the deflection type, consisting of a light source, a photosensitive cell, and a microammeter or digital readout, the density being obtained from the relative meter deflections or from readings with and without the sample in place. Other photographic densitometers employ a null system in which the current resulting from the transmission of the test sample is balanced by an equivalent current derived from light transmitted through a calibrated modulator. In another null method, the sample and modulator are placed in one beam and the modulator controlled so that the net transmittance of the combination is constant. Some densitometers are equipped with devices for plotting the characteristic curves automatically. The densitometry of colored images is done through a suitable set of color filters and presents special problems in interpretation.

When light passes through a silver image, some is specularly transmitted and some is scattered. If all the transmitted light is used in densitometry, the result is called the diffuse density. If only the light passing directly through is measured, the density is called specular. The specular density is higher and related to diffuse density by a ratio known as the Callier Q factor. The Q of color films is negligible.

The density of paper prints or other images on reflective supports is measured by reflection. The reflection density is $D_R = \log(1/R)$, where R is the ratio of light reflected by the image to that reflected by an unexposed area of the paper itself or, better, by a reference diffuse light reflector. As with transmission densitometry, the angular relationships between incident and reflected light must be standardized.

Contrast. If extended, the straight-line portion of the characteristic curve BC intersects the exposure axis at the inertia point i on Fig. 3, and the angle α is a measure of contrast. The tangent of α, that is, the slope of the line BC, is known as gamma (γ). As development time or temperature increase, gamma increases, eventually reaching a maximum (**Figs. 4** and **5**). The projection of the straight line BC of Fig. 3 onto the exposure axis at bc is a measure of latitude, and the distance ad represents the total scale, that is, the whole exposure range in which brightness differences in the subject can be recorded.

Another measure of contrast, contrast index, is more useful than gamma in pictorial photography because it is the average gradient over the part of the characteristic curve used in practice. It is determined by measuring the slope of the line joining a point in the toe of the curve

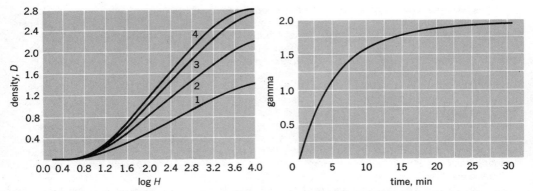

Fig. 4. Characteristic curves for development times increasing in the order 1, 2, 3, 4; horizontal scale is in lux seconds.

Fig. 5. Gamma versus time of development curve.

and one at a higher density that are separated by a distance related to a specified difference in exposure or density.

Sensitivity. The sensitivity or speed of a photographic material is not an intrinsic or uniquely defined property. However, it is possible to define a speed value which is a numerical expression of the material's sensitivity and is useful for controlling camera settings when the average scene brightness is known. Speed values are usually determined from an equation of the form $S = k/H$, where S is the speed, k is an arbitrary constant, and H is the exposure corresponding to a defined point on the characteristic curve.

Historically, Hurter and Driffield speeds were obtained by dividing 34 by the exposure corresponding to the inertia point, but most modern definitions use much smaller constants and the exposure corresponds to a point in the toe region which is at some arbitrary density above base plus fog or where the gradient of the curve has reached some specified value. The goal is always to establish speed values which lead either to correctly exposed transparencies or to negatives that can be printed readily to form pictures of high quality. The constant k usually incorporates a safety factor to assure good exposures even with imprecise exposure calculations.

The present international standard for determining the speeds of black-and-white pictorial films is shown in **Fig. 6**. This standard has been adopted by the American National Standards Institute (ANSI) and the International Organization for Standardization (ISO). Film speeds determined by this method are called ISO speeds and correspond closely to values long known as ASA speeds. Point M is at a density 0.10 above base-plus-fog density. Point N lies 1.3 log H units from

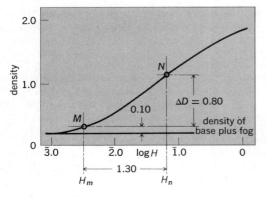

Fig. 6. Determination of the speed of a photographic film according to the 1979 American Standard Method, now adopted internationally. ISO/speed is $0.8/H_M$ when film is processed to achieve gradient defined by points M and N.

M in the direction of increased exposure. Developing time in a standard developer is adjusted so that N also lies at a density 0.8 above the density of M. Then, the exposure H_M (in lux seconds) corresponding to M is the value from which the ISO speed is computed according to $S = 0.8/H_M$. Exposure indexes or speeds for films and papers used in printing and for films and plates used in various technical applications are often determined in the same general manner, but the criteria for required exposure, the processing conditions, and the constant k are usually quite different, reflecting the nature of the application for which the material is intended.

Spectral sensitivity measures the response of photographic emulsions as a function of the exposing wavelength. It is usually determined with a spectrosensitometer, that is, a spectrograph with an optically graded wedge or step tablet over the slit. The contour of the spectrogram for a constant density is related to the spectral response of the material and spectral output of the light source. Ideally, spectral sensitivity curves are adjusted to indicate the response to a source of constant output at all wavelengths. These are known as equal-energy curves.

Reciprocity effects. Defining exposure as the product of illuminance and time implies a simple reciprocal relationship between these factors for any given material and development process. As noted above, the efficiency of latent image formation is, in fact, affected by very long, very short, or intermittent exposures. Variations in the speed and contrast of an emulsion with changes in the rate of exposure are therefore called reciprocity effects. In general, these effects are not dependent upon the exposing wavelength for a single emulsion, but they may vary among the individual emulsions of color materials, thereby affecting color rendition.

Tone reproduction. The relation between the distribution of luminance in a subject and the corresponding distribution of density in a final print is referred to as tone reproduction. It is affected by the sensitometric characteristics of the materials, including their nonlinear aspects, and by subjective factors as well. The exposure and processing of both the negative and print materials are manipulated by skilled photographers to achieve visually appealing results.

Image structure. The granular nature of photographic images manifests itself in a nonhomogeneous grainy appearance which may be visible directly and is always visible under magnification. Granularity, an objective measure of the sensation of graininess, is generally greater in high-speed emulsions and depends also on both the nature of the development and the mean density of the image area being measured. Producing emulsions with increased speed without a corresponding increase in granularity is a constant goal. Graininess in prints increases with increases in magnification, contrast of the print material, and (in black-and-white materials) with mean density of the negative.

The ability of a film to resolve fine detail or to record sharp demarcations in exposure is limited by the turbidity (light-scattering property), grain size, and contrast of the emulsion, and is affected by several processing effects. Resolving power is a measure of the ability to record periodic patterns (bars and spaces of equal width) and is usually expressed as the number of line pairs (bar and space pairs) per millimeter which can just be separated visually when viewed through a microscope. Resolving power values of 100 to 150 line pairs/mm are common but may range from less than 50 to more than 2000.

Sharpness refers to the ability of an emulsion layer to show a sharp line of demarcation between adjacent areas receiving different exposures. Usually such an edge is not a sharp step but graded in density to an extent governed by light scattering (turbidity) and development effects. Subjective assessments of sharpness correlate with a numerical value known as acutance, which is a function of both the rate of change in density and the overall difference in density across a boundary.

A different and generally more useful measure of the ability of a photographic material to reproduce image detail is called the modulation transfer function (MTF). If a sinusoidal distribution of illuminance is applied to the material, the resulting variation in density in the processed image will not, in general, coincide with that predicted by the characteristic curve, again because of localized scattering and development effects. (These effects are excluded in conventional sensitometry.) Modulation transfer is the ratio of the exposure modulation that in an ideal emulsion would have caused the observed density variation to the exposure modulation that actually produced it. This ratio is a function of the spatial frequency of the exposure distribution, expressed in cycles per millimeter. The MTF of a film may be combined with a similar function for a lens to predict the performance of a lens-film combination.

COLOR PHOTOGRAPHY

Color photography is the photographic reproduction of natural colors in terms of two or more spectral components of white light. In 1665 Isaac Newton observed that white light could be separated into a spectrum of colors, which then could be reunited to form white light. In 1861 James Clerk Maxwell prepared positive transparencies representing red-, green-, and blue-filter exposures, projected each through the respective "taking" filter, and superimposed the three images to form a full-color image on a screen.

Maxwell's demonstration was the first example of additive color photography. Red, green, and blue are called additive primary colors because they add to form white light. Their complements are the subtractive primary colors, cyan, magenta, and yellow. A cyan, or minus-red, filter absorbs red light and transmits blue and green; a magenta, or minus-green, filter absorbs green and transmits red and blue; and a yellow, or minus-blue, filter absorbs blue and transmits red and green. Subtractive color processes produce color images in terms of cyan, magenta, and yellow dyes or pigments (**Fig. 7**). Two-color subtractive processes (cyan and red-orange), which provide a less complete color gamut, have also been used. *See* COLOR; COLOR FILTER.

Additive color photography. Additive color is most widely represented by color television, which presents its images as arrays of miniscule red, green, and blue elements.

The screen plate system of additive photography uses a fine array of red, green, and blue filters in contact with a single panchromatic emulsion. Reversal processing produces a positive color transparency comprising an array of positive black-and-white image elements in register with the filter elements. Among the early screen processes were the Lumière Autochrome plate and Dufaycolor film. Polachrome 35-mm transparency film and the earlier Polavision Phototape, an 8-mm instant movie film, each have a hyperfine color stripe screen integral with a silver transfer film.

Lenticular additive photography, last used commercially in 1928 in the 16-mm Kodacolor process, is based on black-and-white film embossed with lenticules. While the camera lens photographs the scene, the lenticules image onto the film a banded tricolor filter placed over the

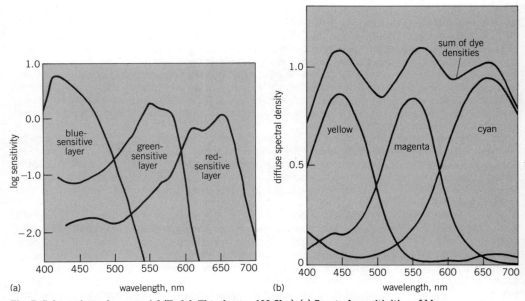

Fig. 7. Subtractive color material (Kodak Ektachrome 100 film). (a) Spectral sensitivities of blue-, green-, and red-sensitive emulsions. (b) Corresponding spectral densities of dyes that form yellow, magenta, and cyan images. Sum of dye densities, corresponding to uniform exposure at all visible wavelengths, is not flat but yields visually neutral density.

camera lens. The resulting composite image is reversal-processed and projected through the same tricolor filter to reconstruct the original scene in color.

Subtractive color photography. Many types of subtractive color negatives, color prints, and positive transparencies are made from multilayer films with separate red-, green-, and blue-sensitive emulsion layers.

In chromogenic or color development processes, dye images are formed by reactions between the oxidation product of a developing agent and color couplers contained either in the developing solution or within layers of the film. For positive color transparencies, exposed silver halide grains undergo nonchromogenic development, and the remaining silver halide is later developed with chromogenic developers to form positive dye images. Residual silver is removed by bleaching and fixing.

Kodachrome was the first commercially successful subtractive color transparency film (**Fig. 8**). The Kodachrome process forms reversal dye images in successive color development steps, using soluble couplers. Following black-and-white development of exposed grains, the remaining red-sensitive silver halide is exposed to red light, then developed to form the positive cyan image. Next, the blue-sensitive layer is exposed to blue light and the yellow image developed, and finally the green-sensitive emulsion is fogged and the magenta image developed.

Simpler reversal processing is afforded by films with incorporated nondiffusing couplers that produce three dye images in a single color development step.

Color negative films with incorporated couplers are developed with a color developer to form cyan, magenta, and yellow negative images for use in printing positive color prints or transparencies. Colored couplers in color negative films provide automatic masking to compensate for the imperfect absorption characteristics of image dyes. For example, a magenta dye, ideally minus-green, absorbs not only green but also some blue light. The magenta-forming coupler is colored yellow, and in undeveloped areas it absorbs as much blue light as the magenta dye absorbs in developed areas, so that blue exposure through the negative is unaffected by the blue absorption of the magenta dye.

Chromogenic films designed for daylight exposure commonly have an ultraviolet-absorbing dye in the protective overcoat to prevent excessive exposure of the blue-sensitive emulsion. (Ultraviolet exposure of the remaining layers is blocked by the yellow filter layer.) Advances in silver halide, dye, and processing chemistry have resulted in chromogenic films with progressive im-

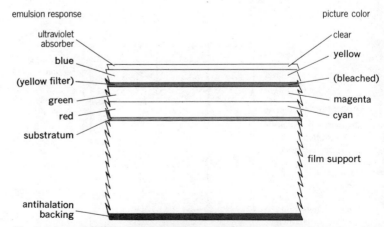

Fig. 8. Schematic cross section of Kodachrome film, showing configuration of three emulsion layers with (left) the spectral response of each emulsion and (right) the color of the dye image formed in each layer through chromogenic development. Protective overcoat contains ultraviolet absorbing dye to prevent overexposure of blue-sensitive emulsion with daylight or electronic flash illuminants.

provements in speed, color rendition, image quality, and dye stability over early products of this type.

One-step color photography. The introduction in 1963 of the single-structure Polacolor film and single-step Polacolor process marked a significant branching in the history and technology of color photography. The availability of instant color photographs directly from the camera provided a useful alternative to conventional color films and print materials, which require a printing step and separate darkroom processing of each material, particularly in applications where enlargement and multiple prints are not primary requirements.

In the Polacolor process and later one-step, or instant, processes, color prints are produced under ambient conditions shortly after exposure of the film. Processing takes place rapidly, using a reagent—sometimes called an activator—provided as part of the film unit. Important concepts include the design of cameras incorporating processing rollers, the use of viscous reagents sealed in foil pods to provide processes that are outwardly dry, and traps to capture and neutralize excess reagent. The viscous reagent makes possible accurately controlled, uniform distribution. The closed system provides fresh reagent for each picture unit and permits use of more reactive chemicals than are practical in conventional darkroom processes.

Rather than forming dyes during processing, the Polacolor process introduced preformed dye developers, compounds that are both image dyes and developers. Development of exposed silver halide crystals causes immobilization of the associated dye developers while dye developers in unexposed regions diffuse to a polymeric image-receiving layer. The positive receiving sheet includes a polymeric acid layer that neutralizes residual alkali. The full-color print is ready for viewing when the negative and positive sheets are stripped apart after 60 s. Polaroid's Colorgraph film, based on similar principles, yields color transparencies suitable for overhead projection. The Polacolor process, like most peel-apart systems, is adaptable to large-format photography.

Polaroid SX-70 Land film, also based on dye developers, was the first fully integral one-step color film (**Fig. 9**). The film unit comprises two polyester sheets, one transparent and one opaque, with all of the image-forming layers between them. The light-sensitive layers are exposed through the transparent sheet, and the final image is viewed through the same sheet against a reflective white-pigmented layer that is formed by the reagent. Cameras for the SX-70 system automatically eject the film immediately, driving the film unit between rollers to spread the reagent within the film. Although still light-sensitive when ejected, the emulsion layers are protected from exposure to ambient light by the combined effects of the white pigment and special opacifying dyes in-

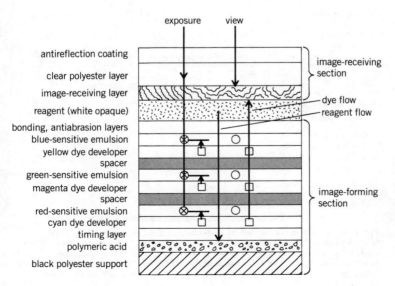

Fig. 9. Schematic cross section of SX-70 Time-Zero Supercolor and "600" integral one-step film units that provide full-color prints within 2 min after exposure. The dye image diffuses from exposed areas to the image-receiving layer and is viewed against the reflective white pigment of the reagent layer.

cluded in the reagent. Timing of the process is self-controlled and there is no separation of the sheets. Images on materials which are exposed and viewed from the same side will be laterally reversed unless an odd number of image inversions is introduced in the exposing process. Hence, cameras intended for use with Polaroid SX-70 film incorporate one or three mirrors. The use of three mirrors permits cameras of compact design.

Kodak instant color films incorporate image-forming and -receiving layers between two polyester sheets, both clear (**Fig. 10**). Direct positive emulsions are selectively exposed through the clear cover sheet and an ultraviolet-absorbing layer. Following exposure, an activator containing potassium hydroxide, a developing agent, and other components is distributed uniformly between the cover sheet and the integral imaging sections as the film unit is ejected from the camera. Carbon black in the activator blocks exposure of the emulsion layers while the activator penetrates them and develops the unexposed silver halide crystals. Oxidized developer resulting from development reacts with dye releasers in an adjacent layer and, in the presence of an alkali, liberates preformed dyes. (Scavenger layers prevent dye release from the wrong layers.) The released dyes migrate through two or more opaque layers to an image-receiving layer. Timing of the process is governed by the neutralizing action of a polymeric acid when the activator penetrates the timing layers.

Because the images formed in Kodamatic films are viewed through the second clear plastic sheet, they are right-reading. Thus, cameras designed to accept these products need contain no mirrors, though two are incorporated in many such cameras to reduce overall camera size. All layers are retained as an integral unit in Kodamatic film, but Kodamatic Trimprint film (Fig. 10) embodies two extra layers and is partially slit during manufacture so that the picture and border section of the image-receiving section may be separated from the unit after processing is complete. The image-forming section, including the activator pod and trap, is discarded, leaving a thin print. Kodak Instagraphic color slide film produces transparencies instead of thin prints. Fol-

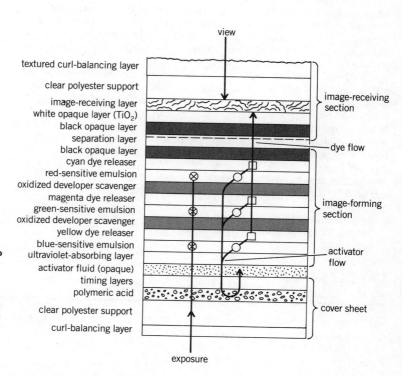

Fig. 10. Schematic cross section of Kodamatic Trimprint integral one-step film. Dyes released from unexposed areas of image flow to image-receiving layer for viewing against white opaque layer. Image-receiving section may then be separated (at broken line) from image-forming section to obtain a thin print.

lowing separation from the image-forming section, the transparencies are trimmed and inserted automatically into 2 by 2 in. (5 by 5 cm) slide mounts with a slide-mounting accessory.

Instant materials manufactured by Fuji also utilize dye-release chemistry and include both integral and peel-apart films.

Color print processes. Color prints from color negatives are made primarily on chromogenic print materials such as Ektacolor paper using simplified processes similar to those that are used for the original negatives. Color transparencies are made from such negatives using special print films which are processed like the original negatives. Chromogenic reversal materials, such as Ektachrome paper and duplicating films, are used to make prints or duplicate transparencies from positive color originals. Alternatively, a color negative intermediate film may be derived from the original positive transparency and used to make multiple prints with negative-working papers.

Color prints and transparencies can be made from original color negatives and transparencies with Ektaflex PCT products in an image transfer system which uses a single, strongly alkaline activator for processing. As appropriate, a negative- or positive-working Ektaflex PCT film is exposed to the original. After 20s immersion in the activator, the film is laminated to a reflective (paper) or transparent (plastic) receiver sheet. The two parts are separated minutes later. Other color print films based on release and diffusion of preformed dyes in the presence of alkaline activators are Agfachrome Speed, a single-sheet material, and Agfa-Gevaert Copycolor, a two-sheet graphic arts product.

Color prints and transparencies are also made from original positive transparencies by the silver dye bleach process, represented by Cibachrome. The process is based on the destruction of dyes rather than their formation or release. Following development of silver images in red-, green-, and blue-sensitive emulsion layers, associated subtractive dyes are selectively destroyed by a bleach reaction catalyzed by the negative silver, leaving a positive color image in place. Residual silver salts are removed by fixing.

Dye transfer prints are made by sequential imbibition printing of cyan, magenta, and yellow images from dyed gelatin relief matrices onto gelatin-coated paper or film (the Kodak Dye Transfer Process). Panchromatic matrix films are exposed from color negatives through appropriate color separation filters; blue-sensitive matrix films are printed from black-and-white separation negatives. The images formed by dye transfer (including the Ektaflex process) and dye bleach (Cibachrome) exhibit good dye stability.

Color motion pictures. Many of the techniques described above have been used in motion picture processes. The Technicolor imbibition transfer process is analogous to dye transfer printing. Since their introduction in the 1950s, chromogenic negative and positive films have become the most widely used motion picture production materials.

Bibliography. A. A. Blaker, *Photography, Art and Technique*, 1980; E. Blickman et al., *Unconventional Imaging Processes*, 1978; B. H. Carroll et al., *Introduction to Photographic Theory: The Silver-Halide Process*, 1980; G. T. Eaton, *Photographic Chemistry*, 1981; H. L. Gibson, *Photography by Infrared: Its Principles and Applications*, 3d ed., 1978; G. Haist, *Modern Photographic Processing*, 2 vols., 1979; T. H. James (ed.), *The Theory of the Photographic Process*, 4th ed., 1977; M. J. Langford, *Basic Photography: A Primer for Professionals*, 4th ed., 1977; R. P. Lovelend, *Photomicrography: A Comprehensive Treatise*, 2 vols., 1981; H. N. Todd and R. D. Zakia, *Photographic Sensitometry: The Study of Tone Reproduction*, 2d ed., 1974; H. J. Walls and G. S. Attridge, *Basic Photo Science: How Photography Works*, 1978.

LATENT IMAGE
R. H. NOBLE

An invisible image produced by a physical or chemical effect of light on the individual crystals (usually silver halide) of photographic emulsions. This image can be rendered visible by the process known as development. For details of latent image formation and development SEE PHOTOGRAPHY.

PHOTOGRAPHIC MATERIALS
ROBERT D. ANWYL

The light-sensitive recording materials of photography, that is, photographic films, plates, and papers. They consist primarily of a support of plastic sheeting, glass, or paper, respectively, and a thin, light-sensitive layer, commonly called the emulsion, in which the image will be formed and stored. The material will usually embody additional layers to enhance its photographic or physical properties.

Supports. Film support, for many years made mostly of flammable cellulose nitrate, is now exclusively of slow-burning "safety" materials, usually cellulose triacetate or polyester terephthalate, which are manufactured to provide thin, flexible, transparent, colorless, optically uniform, tear-resistant sheeting. Polyester supports, which offer added advantages of toughness and dimensional stability, are widely used for films intended for technical applications. Film supports usually range in thickness from 0.0025 to 0.009 in. (0.06 to 0.23 mm) and are made in rolls up to 60 in. (1.5 m) wide and 6000 ft (1800 m) long.

Glass is the predominant substrate for photographic plates, though methacrylate sheet, fused quartz, and other rigid materials are sometimes used. Plate supports are selected for optical clarity and flatness. Thickness, ranging usually from 0.04 to 0.25 in. (1 to 6 mm), is increased with plate size as needed to resist breakage and retain flatness. The edges of some plates are specially ground to facilitate precise registration.

Photographic paper is made from bleached wood pulp of high α-cellulose content, free from ground wood and chemical impurities. It is often coated with a suspension of baryta (barium sulfate) in gelatin for improved reflectance and may be calendered for high smoothness. Fluorescent brighteners may be added to increase the appearance of whiteness. Many paper supports are coated on both sides with water-repellent synthetic polymers to preclude wetting of the paper fibers during processing. This treatment hastens drying after processing and provides improved dimensional stability and flatness.

Before an emulsion is coated on the support, a "sub" (substratum) is applied to ensure good adhesion of the emulsion layer (**Fig. 1**).

Emulsions. Most emulsions are basically a suspension of silver halide crystals in gelatin. The crystals, ranging in size from 2.0 to less than 0.05 micrometers, are formed by precipitation by mixing a solution of silver nitrate with a solution containing one or more soluble halides in the presence of a protective colloid. The salts used in these emulsions are chlorides, bromides, and iodides. During manufacture, the emulsion is ripened to control crystal size and structure. Chemicals are added in small but significant amounts to control speed, image tone, contrast, spectral sensitivity, keeping qualities, fog, and hardness; to facilitate uniform coating; and, in the case of color films and papers, to participate in the eventual formation of dye instead of metallic silver images upon development. The gelatin, sometimes modified by the addition of synthetic polymers, is more than a simple vehicle for the silver halide crystals. It interacts with the silver halide

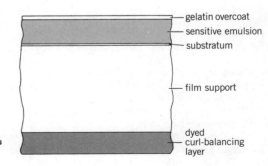

Fig. 1. Schematic cross section of film.

crystals during manufacture, exposure, and processing and contributes to the stability of the latent image.

After being coated on a support, the emulsion is chilled so that it will set, then dried to a specific moisture content. Many films receive more than one light-sensitive coating, with individual layers as thin as 1.0 μm. Overall thickness of the coatings may range from 5 to 25 μm, depending upon the product. Most x-ray films are sensitized on both sides, and some black-and-white films are double-coated on one side. Color films and papers are coated with at least three emulsion layers and sometimes six or more plus filter and barrier layers. A thin, nonsensitized gelatin layer is commonly placed over film emulsions to protect against abrasion during handling. A thicker gelatin layer is coated on the back of most sheet films and some roll films to counteract the tendency to curl, which is caused by the effect of changes in relative humidity on the gelatin emulsion. Certain films are treated to reduce electrification by friction because static discharges can expose the emulsion. The emulsion coatings on photographic papers are generally thinner and more highly hardened than those on film products.

Another class of silver-based emulsions relies on silver-behenate compounds. These materials require roughly 10 times more exposure than silver halide emulsions having comparable image-structure properties (resolving power, granularity); are less versatile in terms of contrast, maximum density, and spectral sensitivity; and are less stable both before exposure and after development. However, they have the distinct advantage of being processed through the application of heat (typically at 240 to 260°F or 116 to 127°C) rather than a sequence of wet chemicals. Hence, products of this type are called Dry Silver films and papers.

Spectral sensitivity. The silver halides (and silver behenates) are normally sensitive only to x-radiation and to ultraviolet, violet, and blue wavelengths, but they can be made sensitive to longer wavelengths by adding special dyes, predominantly polymethines, to the emulsion. The process is known as spectral sensitizing to distinguish it from the chemical sensitizing used to raise the overall or inherent sensitivity of the grains.

Spectrally nonsensitized emulsions are generally termed blue-sensitive or ordinary and exhibit little response beyond 450 nanometers. Emulsions treated with dyes to extend their sensitivity to 500 nm are identified as extended blue, while those with sensitivity extended through the green (550 nm) are termed orthochromatic. Some panchromatic films are sensitized out to 650 nm; others, with extended red sensitivity, respond efficiently out to 700 nm. The upper limit for spectral sensitizing is about 1200 nm, which is in the near-infrared region, but conventional infrared films exhibit little sensitivity beyond 900 nm. Ultraviolet sensitivity is diminished somewhat between 250 and 280 nm and sharply below 250 nm by the strong optical absorption by gelatin. Materials sensitive to much shorter wavelengths can be prepared by several means which result in a high concentration of silver halide crystals at the top surface of the emulsion layer. Films are selected for spectral characteristics to achieve efficient response to various exposing sources and, whenever possible, to provide convenient handling under safelight conditions. Spectral sensitivity is an important consideration in assuring visually satisfying tonal reproduction of colored objects with black-and-white materials or intentionally altering the tonal reproduction to provide greater spectral discrimination than the human eye. The spectral sensitivity of a recording material is altered, in effect, by placing color filters over the camera lens.

Halation. During exposure of a film or plate, part of the radiation entering the emulsion is absorbed; the rest is transmitted into the support. A small fraction of the transmitted light is reflected at the back surface and returned to the emulsion at a point displaced from the incoming radiation, depending upon the angle of incidence and the thickness of the support (**Fig. 2**). This secondary exposure is called halation because a bright point in a scene is recorded as a point surrounded by a halo. Halation can be suppressed to negligible levels by incorporating carbon black or selected dyes in the support, a layer on the back of the support, a layer between the emulsion and the support, or in rare instances, the emulsion. The absorbing layer need not be opaque to be effective since the offending radiation must make two passes through the absorbing layer and is strongly attenuated upon reflection at the back surface. Carbon black is added to some polyester supports to suppress light-piping as well as halation.

Dyes in layers under the emulsion or in curl-balancing layers on the back of some films are bleached or dissolved during processing. The antihalation backings on some plates and films are dissolved or stripped away during processing.

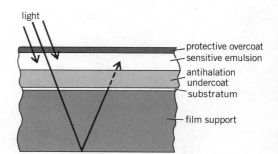

Fig. 2. Suppression of halation by an absorbing medium, shown here as an antihalation undercoat.

Photographic characteristics. For a discussion of characteristics of sensitive materials, as well as the theory of the photographic process, latent image, sensitometry, and image structure, SEE PHOTOGRAPHY.

All of these characteristics are related to the combination of emulsion, developer, and exposure. The structure of the developed image is of importance in determining the quality of the photograph, especially its ability to reproduce fine detail.

Photographic products. Hundreds of types of films, plates, and papers are available in numerous sizes and configurations for a wide range of applications. Each field of use generally requires special properties. Films for amateur, professional, and technical photography typically range in speed from an exposure index (ISO/ASA) of less than 10 to over 1000, and from very fine grain and high sharpness to rather coarse grain with corresponding loss in definition. They can be obtained in a wide range of contrasts and spectral sensitizations, especially for scientific and technical photography. While most materials are negative-working, a number of products provide positive-working emulsions which result in direct positive reproductions with conventional (negative) processing. Materials of extremely high contrast and density are widely used for graphic reproduction and in the printing industry to obtain high-contrast line and halftone images. Many types of x-ray films are made, most sensitized on both sides and used in combination with fluorescent intensifying screens to minimize patient exposure. Several types of color film are available, designed primarily for pictorial photography.

Photographic papers are offered in both contact and enlarging speeds and with a range of contrasts for producing continuous-tone prints from camera negatives. A second class of enlarging papers provides selective control of contrast through the use of filters during exposure of two-component emulsions. One component is blue-sensitive only and has high contrast, while the other has sensitivity extended to the yellow-green and relatively low contrast. Papers are also distinguished by surface texture and tint of the paper support. Developing agents are incorporated in the emulsions of some photographic papers so that internally controlled development can be accomplished quickly by immersing the paper in an alkaline bath. Color printing papers are made for printing from color negative or positive films. High-contrast negative and direct-positive papers are used for photographic reproduction of documents, engineering drawings, or line copy. Several papers are designed for oscillographic recording, photographic typesetting, and other specialized applications.

Image-transfer systems. The image resulting from exposure and processing of most photographic materials is retained in the light-sensitive layer or layers. In another and very useful set of products, the image is transferred by chemical means to a receiver sheet which has been placed in contact with the basic photographic material during part or all of the processing step. Careful design of each component in these image-transfer systems generally results in convenient processing and relatively rapid access to the final image. Most widely known are camera-based systems in which continuous-tone silver or dye images are chemically transferred to a receiver sheet or layer from the light-sensitive and sometimes complex image-forming layers. In the original Polaroid-Land process, unexposed silver halide crystals in the emulsion are transferred by a solvent and reduced to form a positive black-and-white image on the receiver in as little as 10 s (**Fig. 3**). Several newer systems form color images through the transfer of dyes to the receiver

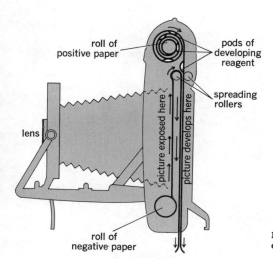

Fig. 3. Polaroid-Land camera with two-part roll-film configuration. (*Polaroid Corp.*)

within 1 to 4 min. These "instant" products are popular for amateur photography but are widely used for professional and technical applications as well, including the photography of cathode-ray-tube images. SEE CAMERA.

Other examples of image-transfer systems include materials for producing large color prints or transparencies from either negative or positive color film originals and several high-contrast black-and-white materials used for preparing short-run printing plates. While the image-forming component of most image-transfer systems is usually discarded after being pulled apart from the receiver, it is sometimes retained as a useful film negative. In some instant products intended primarily for amateur photography, all layers are retained as a unit to avoid litter. These integral film structures, including image-forming layers, an image-receiving layer, and chemicals, are handled in sheet form, while two-part structures are offered in both roll (Fig. 3) and sheet configurations.

Storage of materials. Unexposed films and papers may be damaged by exposure to elevated temperatures and high relative humidities. Protection is provided by packing the materials hermetically at controlled humidities (30–50% relative humidity). Storage at a temperature of 40 to 55°F (4 to 13°C) is sufficient for most materials, but some products require deep-freeze conditions.

The life of processed films and prints depends strongly on processing conditions as well as environmental factors. Thorough fixing and washing are usually a primary requirement. Storage at 30–50% relative humidity and at temperatures not exceeding 70°F (21°C) in carefully selected enclosures is commonly recommended.

Nonsilver processes. Many special-purpose photographic materials use active compounds other than silver salts. In general, they are less sensitive by factors of 100 or more than comparable silver-based products, and they lack the amplification factor associated with the development of silver compounds. Accordingly, they are used mainly in reproduction processes where speed is not essential or in certain recording applications where intense exposures from laser sources are possible. They include inorganic compounds which are photosensitive, especially iron salts and dichromates in colloid layers such as gelatin, glue, albumen, shellac, and polyvinyl alcohol or other synthetic resins which become insoluble on exposure. Other important processes are based on light-sensitive organic compounds, especially diazo compounds; unsaturated compounds such as cinnamic acid derivatives, which are insolubilized by cross-linking on exposure; systems in which polymerization occurs as a result of free-radical formation or exposure; dyes which are formed, bleached, or destroyed directly by exposure to light or electrons; and thermographic systems in which a thermal pattern resulting from intense exposure produces an image by means of melting a composition or initiating a chemical reaction.

Electrophotographic systems rely on the imagewise elimination of a charge by exposure of a uniformly charged photoconductive layer and subsequent development of the remaining charge distribution with charged powders or liquid toners, or by electrolytic deposition of metal on the charge image. Toned images may be fixed by heating so the toner particles are fused either to the photoconductive material or, following physical transfer, to a receiver sheet. Because electrophotographic systems commonly exhibit strong edge effects, they are especially useful in document-copying applications, but continuous-tone systems which rely on careful design of the charging and toning elements have been demonstrated. Many xerographic copiers use coatings of amorphous selenium on rigid supports as the photoconductive layer. The preferred medium in modern copiers of this type consists of an organic photoconductor coated on a flexible support similar to film base.

STROBOSCOPIC PHOTOGRAPHY
Harold E. Edgerton

Stroboscopic or "strobe" photography generally refers to pictures of both single and multiple exposure taken by flashes of light from electrical discharges. Originally the term referred to multiple-exposed photographs made with a stroboscopic disk as a shutter. One essential feature of modern stroboscopic photography is a short exposure time, usually much shorter than can be obtained by a mechanical shutter.

High-speed photography with stroboscopic light has proved to be one of the most powerful research tools for observing fast motions in engineering and in science. Likewise, the electrical system of producing flashes of light in xenon-filled flash lamps is of great utility for studio, candid, and press photography. Spark photography, especially with short air gaps at high voltage, has been used for many years to take short-exposure photographs.

Energy storage. Devices for storing energy are required to produce the high peak power needed for producing pictures with a short exposure time. The electrical capacitor is ideal for this service. The energy stored in a capacitor is given by notation (1), where C is the capacitance in

$$\frac{CE^2}{2} \text{ watt-seconds (or joules)} \tag{1}$$

farads and E is the initial voltage to which the capacitor is charged. The electrolytic type of capacitor has the largest ratio of stored watt-seconds (Ws) per pound, but is limited to about 500 V. Paper capacitors are in widespread use for short-duration flash lights where high voltage is required or where the flash cycle is repeated frequently.

Flash lamps. A gaseous discharge lamp or spark gap is required to convert the stored energy of the capacitor into light. The open spark in air is used where a small volume source of very short duration is required. Otherwise the more efficient xenon-filled flash lamp is in almost universal application since its efficiency is about five times greater than that of the open spark. There is an afterglow in xenon gas of about 10 microseconds or more which is objectionable when fast subjects such as high-speed bullets are photographed.

Flash duration. Light duration from a xenon lamp or a spark gap is influenced by the electrical characteristics of the capacitor, the series inductance of the capacitor and leads, and the resistance of the leads. For air gaps, the effective resistance of the gap is negligible compared to these other factors. However, a long xenon-filled flash lamp of small diameter has a high resistance that limits the current flow. The flash duration of a xenon lamp can be much longer than for an air spark.

Circuit of flash lamp. Figure 1 shows an electronic flash lamp L such as a xenon-filled glass tube with one external and two internal electrodes. The lamp is connected directly across the main flash capacitor C which stores the energy that is to be converted by the lamp into light. Current from the transformer, after being rectified, charges the capacitor C to a voltage equal to the peak voltage of the secondary of the transformer T.

An electrical discharge is started in the flash lamp when switch S is closed. An X contact in a synchronized camera shutter can serve as the switch S. The pulse voltage in the secondary

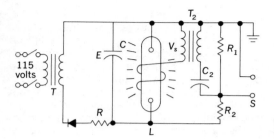

Fig. 1. Elementary circuit of electronic flash lamp L.

of the step-up transformer T_2 is connected to the external electrode on the lamp and starts a current in the tube by condenser action. Once the glow on the tube walls is started, the lamp is lighted briefly but brilliantly by the condenser discharge. There is a transient current that builds up rapidly from zero to a peak and then slowly decays to zero. The transient is a complex one to analyze because the gaseous discharge is nonlinear.

The complete history of a discharge, showing light as well as current and voltage against time, must be measured experimentally for the exact conditions of flash. Data from such tests are available in graphic or tabular form for some of the flash lamps now in use.

Electrical characteristics. As has been described, the current in a long-gap xenon tube rises rapidly from zero after the lamp has been triggered to a peak and then decays in an exponential manner to zero. For many lamps of the xenon-filled types of long arc length, the instantaneous ratio of lamp voltage to current is essentially constant over most of this exponential-like discharge.

Lamp resistance R can be defined, then, as the ratio of the initial anode voltage to the peak current. It follows that the time constant of the current decay is RC seconds and the time constant of the flash duration is $RC/2$ since the light is approximately proportional to the power (current × voltage). These approximate relationships are of great assistance in calculating the expected performance of proposed systems.

As an example, the standard flash lamp type FX-1 has a resistance of $R = 2$ ohms when flashed with 100 microfarads at 2000 V. This lamp has an arc length of 6 in. (15 cm) and a 4-mm (0.158-in.) inside diameter. The resistance of other lamps can be estimated by assuming that a lamp resistance increases directly with length and inversely with the area if the energy density and initial voltage gradient are the same in the two lamps. Thus Eq. (2) is valid, where l is tube

$$\text{Flash duration} = \frac{RC}{2} = 2 \frac{l}{6} \frac{4^2 C}{d^2 2} \text{ seconds} \qquad (2)$$

length in inches and d is inside diameter in millimeters. This approximate equation holds for a flash tube with 8-in. (20-cm) filling of xenon and with an energy density of at least 40 Ws/cm^3 (655 Ws/in.^{3}).

As previously mentioned, spark gaps in air do not follow this rule because their resistance is usually so small that the circuit inductances form the current-limiting inpedances. The discharge current from a capacitor into an air spark gap is oscillatory, at a frequency set by the circuit constants. A pulse of light follows in general the major envelope of the current oscillations. Afterglow in the excited gas in the gap produces light between half cycles of current.

Time between flashes. A flash lamp in a circuit such as that in Fig. 1 cannot produce a second flash of light until the capacitor has been recharged. When a battery or other dc supply is used instead of the rectifier-transformer arrangement of Fig. 1, the charging rate is the straightforward RC time constant of the circuit elements. In the interval comprising four time constants the voltage will be up to 98% of the final value and the light about 96%. For any specific example, the charging time can be made shorter by decreasing the series charging resistance.

As this resistance is reduced for faster charging, a new condition called holdover is encountered in which the flash does not extinguish but is continuously excited by the charging current is about 1–2 A for many practical flash lamps.

Holdover occurs when the end of the discharge current transient and the charging current are equal. The capacitor is not charged during a holdover condition, and the lamp voltage remains at a relatively low value, with a large current from the dc supply and a large loss of energy and heating in the charging circuit resistor.

Holdover has been experienced in 200–1000 Ws flash equipment in the 450- and 900-V types when 10-s charging has been desired. Some of the flash units are equipped with a relay which inserts a large charging resistor for a few seconds at the beginning of the charging cycle to reduce the current so that the lamp will deionize.

Lamps operated with a few watt-seconds per flash have been operated at frequencies up to several hundred flashes per second. With less energy per flash, the lamps appear to be able to operate at a higher frequency without skipping or holdover.

Requirements. For each specific use of stroboscopic photography, the user will want to satisfy three requirements: (1) the flash duration required to "stop" the action, (2) the quantity of light adequate to record the data on the film, and (3) the interval of time between flashes for desired displacement-time information.

The first item, required flash duration, is known once the velocity and the blur definition of the subject are available. For example, suppose one wishes to obtain a clear image of a bullet which has a velocity of 2000 ft/s (600 m/s). Let the definition of the blur on the bullet be less than 0.01 in. (0.25 mm). Now calculate the minimum flash duration from the equation $d = vt$ where $d =$ distance, $v =$ velocity, and $t =$ time. Then, Eq. (3) holds.

$$t = \frac{d}{v} = \frac{.01}{2000 \times 12} = 0.4 \text{ microseconds} \tag{3}$$

The second requirement, quantity of light, can best be estimated by the guide factor method as used by photographers. In its simplest form, the guide factor equation, Eq. (4), is

$$DA = \sqrt{(BCPS)\frac{S}{C}} \tag{4}$$

$DA =$ guide factor
$D =$ distance of the lamp to the subject (must be at least 10 reflector diameters)
$A =$ numerical aperture of the lens

$BCPS =$ the beam-candlepower-second output of the lamp and reflector
$S =$ the ASA exposure index of the film
$C =$ a constant which is 15–25 if D is in feet

limited to the single-lamp, front-lighted case, where the camera lens is at a distance from the subject.

The third item, time between flashes, depends upon the information that is desired. If velocity is desired, only two pictures are needed. The required time interval is calculated from Eq. (3). Again using the bullet as an example, the time for the bullet to travel 2 ft (0.6 m) is 1 millisecond. If one wishes to have a sequence of photographs of a bullet striking an object when the bullet motion between exposures is 0.2 ft (0.06 m), then the interval of time between flashes needs to be 100 μs (at a rate of 10,000 flashes/s).

Conditions for starting. The electrical conditions briefly mentioned before for satisfactory starting of an electronic flash lamp can be presented in chart or curve form such as **Fig. 2**. Below a minimum trigger voltage and a minimum anode voltage the lamp will not flash, as indicated by the no flash area. With ample starting margin as indicated by the point marked rated condition, the lamp will start reliably every time as desired. Now, if the characteristic starting curve engulfs this operating point because of changes during the life of the lamp or circuit, the lamp may begin to misfire at infrequent intervals of time. This condition can be corrected by increasing the size of the trigger capacitor C_2 or by increasing the initial voltage across C_2, since either one of these changes increases the starting voltage V_S.

The methods of overcoming missing require an increase of peak trigger current in the switch S. Should this switch be a delicate X sync contactor in a camera shutter, damage may

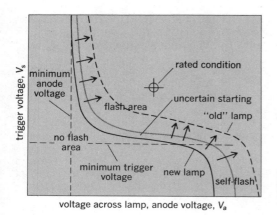

Fig. 2. Idealized starting characteristic of electronic flash lamp showing regions of no flash, flash, and self-start limit as function of anode and trigger voltages.

result to the contacts and the contacts may then fail to close in a positive manner for reliable starting. Furthermore, the contacts usually bounce when they hit, causing sparking at the contacts which reduces the triggering voltage.

A thyratron or a strobotron can be used as a trigger tube in place of the switch S to overcome these difficulties. Then the lamp can be triggered by a microphone, photocell, or other electrical pulse signal.

High-frequency flashes. The frequency limitations discussed previously can be avoided by means of several electrical circuits which do not depend upon the holdover characteristics of the flash lamp.

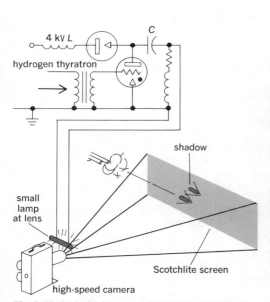

Fig. 3. Method of taking a series of pictures at high frequency of a bullet in flight against Scotchlite screen.

Fig. 4. Photographs of a bullet in flight at a velocity of 2000 ft/s (600 m/s).

LIGHT DETECTION AND PROCESSING

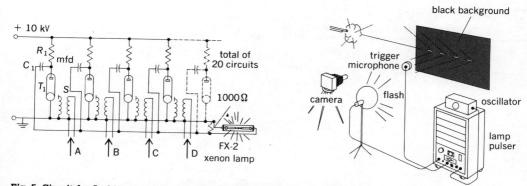

Fig. 5. Circuit for flashing a series of capacitors C into a single flash lamp FX-2 at rates up to 100 kHz, with 1.5 watt-second per flash. A series of triggering pulses is sent through a gating circuit into terminal A, B, C, and so on. The mercury control tube T_1 deionizes rapidly.

1. A control tube, such as a hydrogen thyratron, a hydrogen gap, or a mercury pool tube can be used to switch the current into the flash lamp. These tubes have a rapid deionization of the gas that is used to conduct the pulses of current. Equipment has been developed which operates in short bursts up to many thousands of flashes per second as in **Figs. 3** and **4**. In Fig. 3 a small stroboscope lamp is mounted as near the lens as possible. The discharge capacitor C is charged to 8 kV through the inductance L from the 4-kV power supply. Discharge of the hydrogen thyratron starts the lamp and a pulse of light results. The hydrogen thyratron has a rapid deionization time, and therefore the lamp can be flashed at high frequency such as 10 kHz. To obtain the series of photographs shown in Fig. 4, equipment similar to a Scotchlite screen, which reflects light several hundred times better than a flat white screen, was utilized.

2. A series of separate capacitors which are switched into a flash lamp by means of switching tubes is capable of operating at 10-µs intervals with 1.5 Ws/pulse (**Fig. 5**).

3. Separate flash lamps, each with a complete flashing circuit, can be used. Some arrangement is required to trigger the lamps at the desired intervals of time (**Fig. 6**). This unit was especially developed to measure the early speed of explosion of a dynamite cap with two superimposed photographs spaced about 5 µs apart. SEE PHOTOGRAPHY.

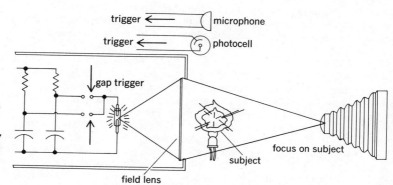

Fig. 6. Circuit diagram of a double flash unit where two discharge capacitors are triggered separately by trigger gap circuits. A field lens is used to collect the light and direct it into the camera lens.

Bibliography. H. Edgerton, *Electronic Flash, Strobe*, 1979; H. E. Edgerton and J. R. Killian, Jr. (eds.), *Moments of Vision: The Stroboscopic Revolution on Photography*, 1979; F. Frungel, *High Speed Pulse Technology*, vols. 1 and 2, 1965, vol. 3, 1976, vol. 4, 1980.

CINEMATOGRAPHY
WALTER GIBBONS-FLY AND E. RUSSEL MCGREGOR

The process of producing the illusion of a moving picture. Cinematography includes two phases: the taking of the picture with a camera and the showing of the picture with a projector. The camera captures the action by taking a series of still pictures at regular intervals; the projector flashes these pictures on a screen at the same frequency, thus producing an image on the screen that appears to move. This illusion is possible only because of the persistence of vision of the human eye. The still pictures appear on the screen many times a second, and although the screen is dark equally as long as it is lighted by the projected image, they do not seem to be a series of pictures but appear to the viewer to be one continuous picture. *See* CAMERA; OPTICAL PROJECTION SYSTEMS; PHOTOGRAPHIC MATERIALS; PHOTOGRAPHY.

Still photographs are taken by motion picture cameras (movie cameras) at the rate of 24 per second, even faster at times. The photographs are sharp and clear, and if they were superimposed on one another would be found to be extremely uniform with respect to position. This last feature is essential in order to have the image appear steady on the screen.

A motion picture camera consists of five basic parts: the lens, the shutter, the gate, the film chamber, and the pull-down mechanism.

Lens. The lens on a motion picture camera collects the light from the scene being photographed and focuses it on the photographic emulsion in the camera aperture. Lenses vary in focal length. The focal lengths most often used on 35-mm cameras are between 30 and 50 mm. Other lenses in common use vary from wide-angle (14.5 to 25 mm), through long-focal-length (60 to 250 mm), to telephoto (300 to 1000 mm). Lenses of shorter focal length have a wider angle of view; those of longer focal length give a larger image size on the film. Lenses for the 16-mm camera which give the same angle of view and the same relative image size are approximately one-half the focal length of those used on a 35-mm camera. Variable-focal-length lenses (zoom lenses) are useful to cinematographers and directors and are used extensively not only for filming sport events but also in making feature films, documentaries, and television films. *See* TELEPHOTO LENS; ZOOM LENS.

To protect the lens from extraneous light and prevent lens flare, a lens shade must be used. On a motion picture camera this shade is called a matte box. It completely surrounds the lens and acts as a lens shade, but it also has slots to hold filters, diffusion, polascreens, fog filters, and optically flat glass for use when needed. These items are used by cinematographers for the many effects required in making a professional motion picture.

Lenses for motion picture cameras have focusing mounts and can be focused on objects as close as 2 or 3 ft (0.6 to 0.9 m). They are fastened to the camera with threaded or bayonet mounts and can be changed when a lens of a different focal length is needed. Although some cameras have turrets on the front of them which hold three or four lenses at one time, making it a simple matter to change from one lens to another, most professional cameras have solid fronts because of the size and weight of the zoom lens, and because with the zoom lens there is little need for frequent lens changes.

The lens has an iris diaphragm which can be opened and closed to control the amount of light reaching the film, thus controlling the exposure. The f-number is a ratio of the focal length of the lens to the diameter of the lens opening, and a lens with a given f-number will always allow the same amount of light per unit area to strike the film regardless of the focal length of the lens. The T-system, used on new lenses, is an ideal f-system and takes into account all of the light losses encountered in a lens, such as the number of reflecting surfaces inside the lens and the various types of coating on the surfaces. T-stops are determined by the amount of light transmitted through the lens, while f-stops are determined by a mathematical formula of the focal length of the lens divided by the diameter of the iris diaphragm opening. Since the T-stops system is more accurate as far as light transmission is concerned, this is usually used for exposure, while the f-number is used to determine the depth of field, that is, a range of subject distances between which all objects will be in sharp focus on the film. *See* LENS.

Shutter. The shutter in a motion picture camera rotates and exposes the film according to its angle of opening. When the shutter is open to its widest angle, the film receives its maximum exposure. Many cameras have variable shutters; that is, the angle of opening can be changed while the camera is running. Such shutter openings usually vary from 170° when the shutter is wide open to 0° when it is closed. Some cameras have a variable shutter which opens to 235° at its widest opening. The shutter in a motion picture camera is almost always located between the lens and the film.

Gate. The aperture called the gate is the passageway through which the film is channeled while it is being exposed. It consists of the aperture plate, which is in front of the film and masks the frame, or picture; the pressure plate, which is in back of the film and holds it firmly against the aperture plate; and the edge guides, which keep the film stable laterally.

Film chamber. This holds the unexposed film at one end and collects the exposed film at the other. Although some motion picture cameras have an interior film chamber, it is usually a separate piece of equipment called the film magazine. Magazines come in 400-ft (120-m) and 1000-ft (300-m) capacities. One thousand feet of film, exposed at the normal speed of 24 frames/s, and with a four-perforation pull-down, lasts slightly more than 11 min. The magazines are loaded with film in a darkroom or in a lightproof changing bag and are fitted on the camera as needed, making camera loading easy and rapid.

Pull-down mechanism. The part of the pull-down mechanism called the intermittent movement pulls the film down through the gate of the camera one frame at a time; in a conventional 35-mm camera, each frame is four perforations high. Some wide-screen cameras have frames five or six perforations high. The pull-down claw engages the perforations and pulls the film down into place to be exposed. At the bottom of its stroke, the claw remains stationary for a moment to position the film and then disengages itself and returns to the upper portion of the stroke to pull another frame into place. During the time the claw is returning to the top of its stroke, the film is stationary and the shutter opens and exposes the film. Some of the more expensive cameras have registration pins which engage the film while the picture is being taken. This ensures that the film is perfectly still and is in exactly the same position as each frame is being exposed.

Film. The film used in the standard motion-picture camera is 35 mm (wide) and is perforated along both edges. There are 64 perforations (or sprocket holes) to each foot of film.

Motion-picture film consists of a cellulose acetate base approximately 0.006 in. (0.15 mm) thick and coated with a light-sensitive emulsion. In color film, several layers of emulsion are applied; each of three of the layers is sensitive to one of the three primary colors of light: blue, green, and red. The film is made in large rolls about 54 in. (1.37 m) wide and as long as several thousand feet. It is slit into 35-mm strips, perforated, and packed in lightproof bags and cans in rolls 100, 200, 400, and 1000 ft (30, 60, 120, and 300 m) in length.

Smaller widths such as 16 and 8 mm are also made from these wide rolls if they are of the same emulsion type. These sizes are seldom used for feature films but are used extensively by professionals in the audiovisual field and for news and low-budget films for television. The kinds of motion picture film are numerous and varied to meet the needs of a many-faceted industry: black and white, color, high and low speed, fine and coarse grain, high and low contrast, and wide and narrow exposure latitude. These film characteristics describe a particular type of film emulsion.

There are two different types of film: negative film, from which a print is made in order to see the original subject in its true likeness (for example, Kodacolor, used in amateur photography), and reversal film, in which a negative is first formed in the original film and from this a positive is formed in the same piece of film (for example, Kodachrome, a film familiar to amateurs). In the negative film two pieces of film are necessary to get a picture that can be projected; in the reversal film only one piece of film is required, making it a much less expensive process. In 16 mm or Super 8, as well as regular 8, the reversal film is generally used because of the cost factor. In 16-mm professional work, the original, that is, the film used in the camera, is never used for projection. A work print is made in order to edit the film and the original is made to conform to the edited work print. The release prints are made from the conformed work print and then the sound is added. The original film is all-important and is kept in a vault at the laboratory; it is used only for making prints.

Aspect ratio. The picture taken in standard 35-mm motion picture cameras is masked by the aperture plate, which varies according to the camera. The silent aperture, that is, the aperture used on cameras that do not leave room on the film for a sound track, is almost the entire width of the film between perforations, slightly less than 1 in. (25 mm) and approximately ¾ in. (19 mm) high. This gives an aspect ratio (the ratio of its width to its height) of 1.33:1 (4/3). Sound apertures are masked down to a width of 0.868 in. (22 mm), to allow room for the sound track, and a height of 0.631 in. (16 mm), which gives an aspect ratio of 1.37:1.

Bibliography. V. Carlson and S. Carlson, *Professional Cameraman's Handbook*, 1981; C. G. Clarke, *American Cinematographer's Manual*, 5th ed., 1980; F. P. Clarke, *Special Effects in Motion Pictures*, 1982; R. Fielding, *The Technique of Special Effects Cinematography*, 4th ed., 1985; R. Fielding, *A Technological History of Motion Pictures and Television*, 1983; J. Mercer, *An Introduction to Cinematography*, 1979; G. Millerson, *Technique of Lighting for Television and Motion Pictures*, 2d ed., 1982; H. M. Souto, *Technique of the Motion Picture Camera*, 4th ed., 1982.

IMAGE PROCESSING
WARREN E. SMITH

Manipulating data in the form of an image through several possible techniques. An image is usually interpreted as a two-dimensional array of brightness values, and is most familiarly represented by such patterns as those of a photographic print, slide, television screen, or movie screen. There are fundamentally two ways to process an image: optically, or digitally with a computer. This article focuses on digital processing of an image to perform one of three separate tasks: enhancement, restoration, or compression.

Digitization. To digitally process an image, it is first necessary to reduce the image to a series of numbers that can be manipulated by the computer. If the image is in the form of a photographic transparency, this digitization can be done by a scanning device that moves a spot of light over the transparency in a rasterlike fashion similar to a television scan. When the light falls on a particular area of the transparency, the amount of light transmitted at that point can be recorded by a photodetector, the signal digitized to, say, eight bits (yielding a possible 2^8, or 256, gray levels), and stored. The spot is then moved a distance equal to its width, and the process is repeated, recording the entire transparency in this way as a long string of numbers. The smaller the spot and the finer the scan (up to the limit of the film grain), the more accurate the description of the transparency will be. Each number representing the brightness value of the image at a particular location is called a picture element, or pixel. A typical digitized image may have 512 × 512 or roughly 250,000 pixels.

Once the image has been digitized, there are three basic operations that can be performed on it in the computer. For a point operation, a pixel value in the output image depends on a single pixel value in the input image. For local operations, several neighboring pixels in the input image determine the value of an output image pixel. In a global operation, all of the input image pixels contribute to an output image pixel value. These operations, taken singly or in combination, are the means by which the image is enhanced, restored, or compressed.

Enhancement. An image is enhanced when it is modified so that the information it contains is more clearly evident, but enhancement can also include making the image more visually appealing. An example is noise smoothing. Given the image of **Fig. 1**a, by replacing the value of randomly chosen pixels by 150 (bright gray) or 0 (black), the noisy image of Fig. 1b results. This kind of noise might be produced, for example, by a defective solid-state TV camera. To smooth the noisy image, median filtering can be applied with a 3 × 3 pixel window. This means that the value of every pixel in the noisy image is recorded, along with the values of its nearest eight neighbors. These nine numbers are then ordered according to size, and the number with as many numbers above it as below it (the median) is selected as the value for the pixel in the new image. The location of this pixel in the new image is the same as the location of the center of the 3 × 3 window in the old image. Eventually, as the window is moved one pixel at a time across the noisy image, the filtered image is formed, as in Fig. 1c. This is an example of a local operation, since the value of a pixel in the new image depends on the values of nine pixels in the old image.

Another example of enhancement is contrast manipulation, where each pixel's value in the new image depends solely on that pixel's value in the old image; in other words, this is a point operation. Contrast manipulation is commonly performed by adjusting the brightness and contrast

Fig. 1. Noise smoothing. (a) Original 256 × 256 pixel image. (b) Original image corrupted by spike noise. (c) Corrupted image filtered by a 3 × 3 pixel median filter.

controls on a television set, or by controlling the exposure and development time in printmaking. Another point operation is that of pseudocoloring a black-and-white image, by assigning arbitrary colors to the gray levels. This technique is popular in thermography (the imaging of heat), where hot objects (high pixel value) are assigned the color red, and cool objects (low pixel value) are assigned blue, with other colors assigned to intermediate values.

Restoration. The aim of restoration is also to improve the image, but unlike enhancement, knowledge of how the image was formed is used in an attempt to retrieve the ideal (uncorrupted) image. Any image-forming system is not perfect, and will introduce artifacts (for example, blurring, aberrations) into the final image that would not be present in an ideal image. To illustrate, the image of **Fig. 2**a is "imaged" on a computer that simulates an out-of-focus lens, producing Fig. 2b. The image of a point object through this system would look like Fig. 2c; this is called the point spread function of the system. Such blurring can be referred to as a local operation, in that the value of a pixel in the blurred image is an average of the values of the pixels around (and including) that pixel location in the original image. The averaging function is the blurring point spread function (Fig. 2c). The blurring process is mathematically referred to as a convolution of the original object with the point spread function.

Another point spread function, called a filter, can be constructed that "undoes" the blurring caused by the point spread function of Fig. 2c. Such a point spread function (not shown) has both positive and negative values, even though the blurring point spread function has only positive values. A general rule is that blurring point spread functions are positive, and sharpening point spread functions have both positive and negative values. By imaging Fig. 2b with the filter point spread function, the restored image of Fig. 2d results. The filter point spread function is spread out more than the blurring point spread function, bringing more pixels into the averaging process. This is an example of a global operation, since perhaps all of the pixels of the blurred image can contribute to the value of a single pixel in the restored image. This type of deblurring (called inverse filtering) will not work well if noise is added to the blurred image before the deblurring operation. A more sophisticated approach, modifying the deblurring filter according to the properties of the noise, is then required.

Fig. 2. Example of restoration. (a) Original 256 × 256 pixel image (*courtesy of Richard Murphy*). (b) Blurred version of the original image. (c) The image of a point subject to the same blurring that produced the blurred version of the original image. (d) The results of the inverse filter applied to the blurred version of the original image.

Fig. 3. Example of compression. (*a*) Original 256 × 256 pixel image, 8 bits per pixel (256 gray levels). (*b*) Original image, reduced to 3 bits per pixel (8 gray levels), for a bit reduction of 0.375. (*c*) Original image, with contrast and geometric transformations applied, followed by a reduction to 3 bits per pixel, for a total bit reduction of 0.188. (*d*) The inverse contrast and geometric transformation applied to *c*, to "decompress" the image. (*From R. N. Strickland and W. E. Smith, Stationary Transform Processing of Digital Images for Data Compression, Appl. Opt.*, 22:2161–2168, July 15, 1983)

There exist mathematical transformations that can be performed on images to facilitate deblurring, as well as other operations. An extremely useful one in image processing is the Fourier transform, which decomposes the image into two-dimensional sine waves of varying frequency, direction, and phase. In fact, convolution of an image with a point spread function (the imaging process) can be carried out in the Fourier domain by simply multiplying the Fourier transform of the image by the Fourier transform of the point spread function, and taking the inverse Fourier transform of this product to get the final image. This process was used in the example of Fig. 2.

Compression. Compression of an image is a way of representing the image by fewer numbers, at the same time minimizing the degradation of the information contained in the image. There are two ways to perform compression: in the image space or in transform space (for example, Fourier). Compression in image space is discussed in the following example.

Figure 3*a*, representing the image to be compressed, contains 256 × 256 pixels, with each pixel having a possible 256 gray levels, so that 8 bits are required on a computer to store each pixel. Thus a total of roughly 500,000 bits is required to store or transmit the image. If the number of bits per pixel is reduced to 3 (8 gray levels), Fig. 3*b* results. The total number of bits have been reduced by 3/8 or 0.375, but at the price of contouring the image. However, if a spatially variable contrast transformation is first applied to Fig. 3*a* with the resulting image then reduced to 3 bits, followed by a geometric distortion that shrinks areas of little variation (the wall and jacket) and leaves areas of large variation (the face) alone, Fig. 3*c* results. The geometric shrinking reduces the number of bits by a factor of 0.500, and the gray-level reduction to 3 bits further reduces the number by 3/8 to a total reduction of 0.188, or roughly 1/5 of the original number of bits required to represent the image. Applying the inverse geometric and contrast transformations to Fig. 3*c* results in Fig. 3*d*, an improvement over Fig. 3*b*, with a greater compression factor than Fig. 3*b* as well.

Applications. Due to its vast scope, image processing is an active area of research in such diverse fields as nuclear medicine, astronomy, electron microscopy, seismology, and many others. The concept of an image is expanding to include three-dimensional data sets (volume "images"), and even four-dimensional volume-time data sets. An example of the latter is a volume image of a beating heart, obtainable with computerized tomography. Advances in computer technology are making efficient processing of the huge data sets required in these problems possible, and such technology promises to bring sophisticated image processing into the home.

Bibliography. H. C. Andrews and B. R. Hunt, *Digital Image Restoration*, 1977; H. H. Barrett and W. Swindell, *Radiological Imaging: The Theory of Image Formation, Detection, and Processing*, 1981; R. C. Gonzalez and P. Wintz, *Digital Image Processing*, 1982; W. K. Pratt, *Digital Image Processing*, 1978.

MEASUREMENT OF LIGHT

Radiometry	342
Exposure meter	343
Radiance	344
Photometry	345
Photometer	347
Illumination	353
Luminous flux	361
Luminous intensity	361
Candlepower	362
Luminous energy	362
Luminance	362
Illuminance	363
Luminous efficacy	363
Luminous efficiency	364

RADIOMETRY
Jon Geist

A branch of science that deals with the measurement or detection of radiant electromagnetic energy. Radiometry is divided according to regions of the spectrum in which the same experimental techniques can be used. Thus, vacuum ultraviolet radiometry, intermediate-infrared radiometry, far-infrared radiometry, and microwave radiometry are considered separate fields, and all of these are to be distinguished from radiometry in the visible spectral region. Curiously, radiometry in the visible is called radiometry, optical radiation measurement science, or photometry, but it is not called visible radiometry. SEE ELECTROMAGNETIC RADIATION; INFRARED RADIATION; LIGHT; ULTRAVIOLET RADIATION.

The use of the word photometry to mean radiometry in the visible portion of the spectrum is misleading. Strictly speaking, photometry is the measurement of electromagnetic radiation according to its ability to produce visual sensation. This, of course, is a subjective attribute of the radiation. Radiometry, on the other hand, is concerned with physical attributes of the radiation such as its energy content and spectral distribution. The reason for the confusion between photometry and radiometry is that the meanings of these words have changed considerably over the last 200 years. SEE PHOTOMETRY.

Photon detectors. These detectors, on the other hand, respond only to photons of electromagnetic radiation that have energies greater than some minimum value determined by the quantum-mechanical properties of the detector material. Since heat radiation from the environment at room temperature consists of infrared photons, photon detectors for use in the visible can be built so that they do not respond to any source of heat except the radiation of interest.

Except for certain measurements of interest in photochemistry and photobiology, chemical reactions that respond to radiation are not very convenient detectors. The key to developing photon detectors that are more convenient than photochemical reactions has been to base them on some electrical principle, so that the most advanced electronic techniques could be used to detect and process the detector signal. So many different approaches have been used that only the most important will be mentioned here.

After vacuum-tube technology was introduced, its techniques were applied to the external photoelectric effect to produce vacuum photodiodes and photomultipliers. For ultraviolet and short-wavelength visible radiation, photomultipliers are the most sensitive detectors available. By using photomultipliers in properly designed experiments, it is possible to record the absorption of a single photon. This mode of operation is called photon counting.

Following the introduction of planar silicon technology for microelectronics, the same tech-

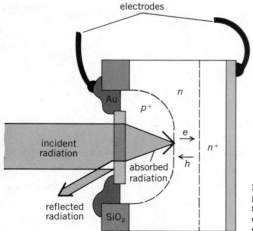

Fig. 1. Silicon photodiode. The separation of a photogenerated electron-hole pair (e-h) by the built-in field surrounding the p^+n junction induces the flow of one electron in an external short circuit across the electrodes.

nology was quickly exploited to make planar photodiodes based on the internal photoelectric effect in silicon. In these devices (**Fig. 1**) the separation of a photogenerated electron-hole pair by the built-in field surrounding the p^+n junction induces the flow of one electron in an external short circuit (such as the inputs to an operational amplifier) across the electrodes. The number of electrons flowing in an external short circuit per absorbed photon is called the quantum efficiency. The use of these diodes has grown to the point where they are the most widely used detector for the visible and nearby spectral regions. Their behavior as a radiation detector in the visible is so nearly ideal that they can be used as a standard, their cost is so low that they can be used for the most mundane of applications, and their sensitivity is so high that they can be used to measure all but the weakest radiation (which requires the most sensitive photomultipliers).

Research efforts have been directed at producing photon detectors based on more exotic semiconductors, and more complicated structures to extend the sensitivity, time response, and spectral coverage of the radiometric techniques that have been developed around the planar silicon photodiode. *See* OPTICAL DETECTORS.

Radiometric sources. Any source of optical radiation can be used as a radiometric source. Lasers have become important radiometric sources. *See* LASER.

Some sources of radiation have properties that have made them useful as standards at various times. Early standard sources include candles of specified composition and construction, gas mantles, and lamps. The best-known standard source is the blackbody, which emits a unique spectral distribution of radiation dependent only upon its temperature. This distribution is given by Planck's law. No real body can behave as a true blackbody, but cavities surrounded by freezing metals can be used as blackbodies to a very high level of approximation. **Figure 2** illustrates how the cavity geometry reduces the reflectance of the cavity opening below that of the surface of the

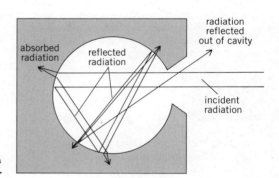

Fig. 2. Cavity approximating a blackbody, that is, an object that absorbs all of the radiation incident on it.

cavity by multiple reflection and absorption of the radiation before it leaves the interior of the cavity. Thus the cavity appears blacker (more absorptive) than a flat surface made from the same material. Other standard sources in use include the radiation from plasma, and synchrotron radiation. These are used as standards primarily in the vacuum ultraviolet.

Bibliography. R. W. Boyd, *Radiometry and the Detection of Optical Radiation*, 1983; W. Budde, *Physical Detectors of Optical Radiation*, vol. 4 of F. Grum and C. J. Bartleson (eds.), *Optical Radiation Measurements*, 1983; F. Grum and R. J. Becherer, *Radiometry*, vol.1 of F. Grum (ed.), *Optical Radiation Measurements*, 1979; C. L. Wyatt, *Radiometric Calibration: Theory and Methods*, 1978.

EXPOSURE METER
FRANK H. ROCKETT

An indicating instrument used in photography to determine lens aperture and shutter speed. An exposure meter may be used either in the darkroom to determine approximate printing time for a contact print or an enlargement or, more usually, with a camera to determine exposure of film.

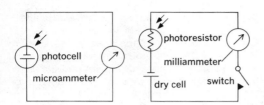

Basic exposure meter circuits. (a) Photovoltaic type circuit. (b) Photoconductive type circuit.

Exposure meters (see **illus**.) are of two basic types, photovoltaic and photoconductive. In photovoltaic meters, a selenium cell converts photons to electrons, producing a current directly proportional to received light. A sensitive microammeter indicates this current to the user. The photocell has rapid response, spectral sensitivity closely approximating that of panchromatic film, and indefinitely long life. Because the power to deflect the microammeter comes entirely from the received light, design of the exposure meter is a compromise between a cell of large area and a meter movement of delicate construction.

In the photoconductive meter, a cadmium sulfide cell changes conductivity in proportion to the received light. A battery (usually a mercury cell) supplies power through the cadmium sulfide cell to the meter movement. Because the received light serves to control the power from the battery, this type of instrument is about three orders of magnitude more power-sensitive than the photovoltaic type. As a consequence, cell area can be small and meter movement can be somewhat more sturdy. Spectral response is greatest in the green and the yellow portion of the spectrum and least in the blue, so a filter is built into the meter. The photoconductive cell recovers slowly from bright light; therefore, it is preferably covered when not in use. Also, a switch opens the electric circuit when the meter is not in use to conserve battery life, nominally about a year.

For the photovoltaic meter, a honeycomb (or flyeye) lens with acceptance angle comparable to that of a camera lens of normal focal length gathers light for the cell. A multi-iris cover may serve to reduce sensitivity for use in bright light, with a mechanically linked change in meter scale. For the photoconductive meter, a small lens controls acceptance angle; or the cell may be mounted inside the camera and thus can use the camera lens, which determines acceptance angle. As an accessory meter separate from the camera, the exposure meter may be fitted with a lens having an acceptance angle of about 3° and a corresponding viewfinder, so that the meter can be used to measure from a distance the brightness of small portions of the subject.

Used at the camera, either type of meter serves as a brightness meter to measure reflected light from the subject. Used at the subject, with a diffuser head over the exposure meter lens, the instrument serves as an illumination meter to indicate incident light. SEE PHOTOMETRY.

To aid in interpreting the indicated light as exposure conditions, the meter may carry a circular slide rule. If built into the camera, the meter movement may either operate an indicator in the field of the viewfinder, or it may control the camera iris to produce an average exposure. In any case, the user presets the rated speed of the emulsion in use into the meter. SEE CAMERA; PHOTOGRAPHY.

Bibliography. J. F. Dunn and G. L. Wakefield, *Exposure Manual*, 4th ed., 1981.

RADIANCE
FRED E. NICODEMUS

The physical quantity that corresponds closely to the visual brightness of a surface. A simple radiometer for measuring the (average) radiance of an incident beam of optical radiation (light, including invisible infrared and ultraviolet radiation) consists of a cylindrical tube, with a hole in each end cap to define the beam cross section there, and with a photocell against one end to measure the total radiated power in the beam of all rays that reach it through both holes (see **illus**.). If A_1 and A_2 are the respective areas of the two holes, D is the length of the tube (distance between holes), and Φ is the radiant flux or power measured by the photocell, then the (average)

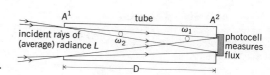

A simple radiometer.

radiance is approximately given by the equation below. In the alternate forms on the second line,

$$L = \Phi/(A_1 \cdot A_2/D^2) \text{ W} \cdot \text{m}^{-2} \cdot \text{sr}^{-1} = \Phi/(A_1 \cdot \omega_2) = \Phi/(A_2 \cdot \omega_1)$$

$\omega_2 = A_2/D^2$ is the solid angle subtended at A_1 by A_2 (at a distance D), and vice versa for ω_1. These approximate relations are good to about 1% or better when D is at least 10 times the widest part of either hole, but the accuracy degenerates rapidly for larger holes or shorter distances between them. A more exact formula, in terms of the ray radiance in each direction at every point across an area through which a beam flows, requires calculus. Then the radiance of a ray in a given direction through a given point is defined as the radiant flux per unit projected area perpendicular to the ray at the point and unit solid angle in the direction of the ray at the point. SEE LIGHT.

Power (flux) is given in SI units of watts (W), areas in square meters (m²), and solid angles in steradians (sr), and so the units of radiance are, as shown, watts per square meter and steradian [$\text{W} \cdot \text{m}^{-2} \cdot \text{sr}^{-1}$ or W/(m² · sr)]. If the flux of visible radiation is given in units weighted for standardized eye response, called lumens (lm), instead of watts, with everything else exactly the same, the corresponding photometric quantity of luminance is obtained in $\text{lm} \cdot \text{m}^{-2} \cdot \text{sr}^{-1}$. When concerned with interactions between radiation and matter, flux can also be measured in numbers of photons or quanta per second ($\text{q} \cdot \text{s}^{-1}$) rather than watts.

A complete specification of a beam of optical radiation requires the distribution of radiance, not only as a function of ray position and direction, but also as a function of wavelength, time, and polarization. A distribution in space and wavelength λ is spectral radiance L_λ in watts per square meter, steradian, and nanometer of wavelength ($\text{W} \cdot \text{m}^{-2} \cdot \text{sr}^{-1} \cdot \text{nm}^{-1}$). Spectral radiance may also be given in terms of wave number σ in reciprocal centimeters (cm^{-1}) or of frequency ν in terahertz (THz). No special names are given to the distributions with respect to time (or frequency of fluctuation or modulation) and polarization, and they are often overlooked even though their effects may be significant.

All of the foregoing assumes that radiant energy is propagated along the rays of geometrical optics that can form sharp shadows and images. The results may be in error when there are significant diffraction or interference effects, as is often the case with coherent radiation such as that from a laser. SEE DIFFRACTION; GEOMETRICAL OPTICS; INTERFERENCE OF WAVES; LASER.

Bibliography. American Association of Physics Teachers, *Radiometry: Selected Reprints*, 1971; R. W. Boyd, *Radiometry and the Detection of Optical Radiation*, 1983; National Bureau of Standards, *A Self-Study Manual on Optical Radiation Measurements*, NBS Tech. Note, 1976; F. E. Nicodemus, Radiance, *Amer. J. Phys.*, 31:368–377, 1963.

PHOTOMETRY
JON GEIST

That branch of science which deals with measurements of light (visible electromagnetic radiation) according to its ability to produce visual sensation. Specifically, photometry deals with the attribute of light that is perceived as intensity, while the related attribute of light that is perceived as color is treated in colorimetry. SEE COLOR; COLORIMETRY.

The purely physical attributes of light such as energy content and spectral distribution are treated in radiometry. Sometimes the word photometry is used to denote measurements that have nothing to do with human vision, but this is a mistake according to modern usage. Such measurements are properly referred to as radiometry, even if they are performed in the visible spectral region. SEE RADIOMETRY.

Relative visibility. The relative visibility of a fixed power level of monochromatic electromagnetic radiation varies with wavelength over the visible spectral region (**Fig. 1**). The relative visibility of radiation also depends upon the illumination level that is being observed. The cone cells in the retina determine the visual response at high levels of illumination, while the rod cells dominate in the dark-adapted eye at very low levels (such as star light). Cone-controlled vision is called photopic, and rod-controlled vision is called scotopic, while the intermediate region where both rods and cones play a role is called mesopic. SEE VISION.

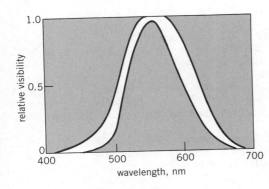

Fig. 1. Range of the relative visibility functions of 125 observers as measured by W. W. Coblentz.

Originally, photometry was carried out by using the human visual sense as the detector of light. As a result, photometric measurements were subjective. That is, two observers with different relative visibility functions, starting with the same standard and following identical procedures, would assign different values to a test radiation having a different spectral distribution than that of the standard.

In order to put photometric measurements on an objective basis, and to allow convenient electronic detectors to replace the eye in photometric measurements, the Commission Internationale de l'Eclairage (CIE; International Commission on Illumination) has adopted two relative visibility functions as standards. These internationally accepted functions are called the spectral luminous efficiency functions for photopic and scotopic vision, and are denoted by $V(\lambda)$ and $V'(\lambda)$, respectively. SEE LUMINOUS EFFICIENCY.

Thus photopic and scotopic (but not mesopic) photometric quantities have objective definitions, just as do the purely physical quantities. However, there is a difference. The purely physical quantities are defined in terms of physical laws, whereas the photometric quantities are defined by convention. In recognition of this difference the photometric quantities are called psychophysical quantities.

Photometric units. According to the International System of Units, SI, the photometric units are related to the purely physical units through a defined constant called the maximum spectral luminous efficacy. This quantity, which is denoted by K_m, is the number of lumens per watt at the maximum of the $V(\lambda)$ function. K_m is defined in SI to be 683 1m/W for monochromatic radiation whose wavelength is 555 nanometers, and this defines the photometric units with which the photometric quantities are to be measured.

At various times, the photometric units have been defined in terms of the light from different standard sources, such as candles made according to specified procedures, and blackbodies at the freezing point of platinum. According to these definitions, K_m was a derived, rather than defined, quantity. SEE ILLUMINATION.

Spectral luminous efficacy. The products of K_m and the spectral luminous efficiencies are called the spectral luminous efficacies denoted by $K(\lambda)$. The spectral luminous efficacy functions for photopic and scotopic vision are shown in **Fig. 2**. Currently, there are no internationally accepted spectral luminous efficiency functions for mesopic vision. Thus there are no objectively defined mesopic quantities in use in photometry. However, research suggests the usefulness o

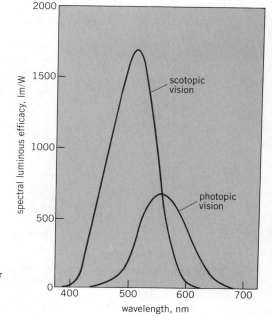

Fig. 2. Internationally accepted standard values for the spectral luminous efficacy of monochromatic radiation for human scotopic (dark-adapted) and photopic (light-adapted) vision.

linear combinations of the curves shown in Fig. 2, where the relative weight assigned to each curve depends upon the level of illumination. SEE LUMINOUS EFFICACY.

Photometric measurements. Photometric measurements are made by a number of different means. The human eye is still used occasionally, particularly in the mesopic region where no other internationally accepted standard exists. It is also common to make photometric measurements with a detector-optical filter combination in which the filter has been chosen to make the relative spectral response of the combination closely resemble the $V(\lambda)$ function. It is sometimes difficult to get a good enough fit of the spectral response function to the $V(\lambda)$ function over a sufficiently broad spectral region to make accurate measurements on exotic light sources. It is becoming accepted practice to measure the spectral distribution of these sources by conventional spectral radiometric techniques, and to calculate the photometric quantities by multiplying the spectral quantities by $K(\lambda)$ and integrating with respect to wavelength. SEE LIGHT.

PHOTOMETER
G. A. HORTON

An instrument used for making measurements of light, or electromagnetic radiation, in the visible range. In general, photometers may be divided into two classifications: laboratory photometers, which are usually fixed in position and yield results of high accuracy; and portable photometers, which are used in the field or outside the laboratory and yield results of lower accuracy. Each class may be subdivided into visual (subjective) photometers and photoelectric (objective or physical) photometers. These in turn may be grouped according to function, such as photometers to measure luminous intensity (candelas or candlepower), luminous flux, illumination (illuminance), luminance (photometric brightness), light distribution, light reflectance and transmittance, color, spectral distribution, and visibility. Since visual photometric methods have largely been supplanted commercially by physical methods, they receive only casual attention here, but because of their simplicity, visual methods are still used in educational laboratories to demonstrate photometric principles. SEE ILLUMINANCE; LUMINANCE; LUMINOUS FLUX; LUMINOUS INTENSITY.

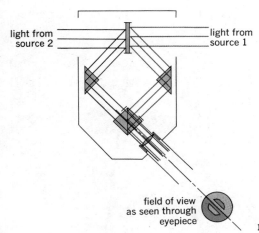

Fig. 1. Lummer-Brodhun contrast sight box.

Visual photometers. These are luminance comparison devices, and most of them utilize the Lummer-Brodhun sight box or some adaptation of its principles. The sight box is an optical device for viewing simultaneously the two sides of a white diffuse plaster screen illuminated by the light sources that are being compared. The general arrangement of the contrast sight box and the field of view as seen through the eyepiece are shown in **Fig. 1**. The contrast sight box gives greater sensitivity of observation than the equality of luminance prism arrangement and appearance of field of view (**Fig. 2**). If there is a color difference in the two light sources being compared, it may be difficult to make a photometric balance. In this case, methods of heterochromatic photometry, for example, the flicker photometer may be used.

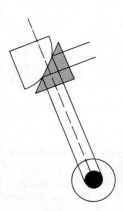

Fig. 2. Diagrammatic sketch of Lummer-Brodhun equality of luminance prism arrangement and appearance of field of view.

Flicker photometer. In this photometer a single field of view is alternately illuminated by the light sources to be compared. The rate of alternation is set at such a speed that the color differences disappear; the photometric balance is obtained by moving the sight box back and forth, the criterion of equality of luminance of the two sources being the disappearance of flicker. Other systems of heterochromatic photometry are the cascade method, the compensation of mixture method, and the use of color-matching or color-equalizing optical filters.

Illuminometer. An illuminometer is a portable visual photometer, as shown in cross section in **Fig. 3**. The light to be measured enters the instrument and is balanced by a Lummer-Brodhun sight box of the equality of luminance type against a comparison lamp, which is moved along the tube. A control box supplies a calibrated current to the comparison lamp, and calibrated optical filters can be placed in the light paths to correct for color differences in the comparison and measured sources and to extend the range of the instrument.

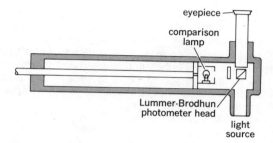

Fig. 3. Essentials of a Macbeth illuminometer.

Luminance meter. This instrument measures photometric brightness and may be of the visual or photoelectric type; it is a major tool for research scientists in visual psychology and illuminating engineering. **Figure 4** shows a photoelectric luminance photometer; a separate unit combines the power supply with the controls and readout meter. The photometer has a telescopic viewing system for imaging the bright surface to be measured on the cathode of a photoemissive tube.

The visual type of illuminometer can also be used to measure the brightness of any surface which is large enough to fill the field of view of the instrument.

Photoelectric photometers. Photometers that use barrier-layer (photovoltaic) photocells and photoemissive electronic tubes as light receptors are classified as photoelectric photometers. Since barrier-layer cells generate their own current when illuminated and therefore do not require a power supply for operation, they may be used either in the laboratory or in the portable instruments. The type generally used consists of an iron plate, often circular, coated with a thin layer of selenium, which in turn is covered with a very thin, transparent film of metal such as gold or

Fig. 4. Spectra Pritchard photoelectric luminance photometer. (*a*) Viewing system. (*b*) Readout meter. (*Photo Research Corp.*)

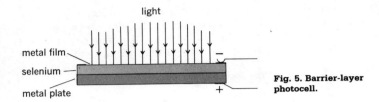

Fig. 5. Barrier-layer photocell.

platinum (**Fig. 5**). A ring of metal is sprayed around the edge of this film; when this metal ring is connected to one terminal of a sensitive microammeter or galvanometer and the other terminal is connected to the iron plate, a current flows through the instrument when light penetrates the metal film to the selenium layer. The photoemissive tube photometers, on the other hand, require an external power supply, and so they are used mainly in laboratory work. Both types of cells should be fitted with optical filters to correct approximately their spectral response to the standard luminosity curve of the eye.

With the advent of semiconductor electronics, it is possible to construct current amplifiers for use with barrier-layer photocells, since both devices have inherent low-resistance characteristics. The linearity of these cells is also a function of the voltage developed across its terminals. To obtain maximum linearity, the load on the cell should approach zero. Semiconductor operational amplifiers have also been found to fulfill this requirement. Transistor amplifiers can be used with barrier-layer cells to drive digital voltmeters, which have made possible the semiautomation of most observational work in photometry.

Photoelectric illuminometer. The measurement of illumination has been revolutionized since 1931 by the introduction of photoelectric portable photometers employing rugged barrier-layer cells. These instruments eliminate the tedious visual comparison required with previous portable visual illuminometers and have considerably simplified illumination measurements. Barrier-layer cells generate a small electric current of about 6 or 7 microamperes/footcandle when light in the visible range falls upon them. The electric instrument used to measure the cell output should have low resistance (100 ohms or less), making the cell current nearly proportional to the luminous flux falling upon it. The photocell should be corrected to correspond to Lambert's cosine law of incidence. SEE PHOTOMETRY.

In this type of cell the current attains its final value after a short time because of the effects of fatigue and temperature. These effects are minimized by low resistance in the instrument circuit.

Barrier-layer cell photometers are also used extensively in photographic light meters to determine proper lighting and camera diaphragm openings. For such applications the instrument can be calibrated on an arbitrary scale. A simple manual computer is often included to convert the meter reading into the proper diaphragm opening and exposure time for a given type of film. SEE PHOTOGRAPHY.

Photoemissive tube photometer. This photometer utilizes a photoelectric tube whose sensitivity is much lower than that of a barrier-layer cell. This device is more accurate than the barrier-layer cell, and the linearity and stability are sufficient for highest-precision photometric work. Since electronic amplification is required to increase the sensitivity to a practicable level, this type of photometer is not usually used outside the laboratory. In calibration, the dark current which flows when the tube is not illuminated is an important factor and should be subtracted or eliminated by circuitry.

Integrating-sphere photometer. Any of the above photometers can be used with an integrating sphere to measure the total luminous flux of a lamp or luminaire. This is ordinarily made as an Uhlbricht sphere (**Fig. 6**), although other geometrical shapes can be used. The inside surface has a diffusely reflecting white finish which integrates the light from the source. A white opaque screen prevents direct light from the source from falling on a diffusing window, which is illuminated only by diffusely reflected light from the sphere walls. The brightness of this window, which is measured by the photometer outside the sphere, is then directly proportional to the flux emitted by the source if the inside finish of the sphere is perfectly diffusing.

Fig. 6. Uhlbricht sphere for measuring luminous flux and efficiency. (*Westinghouse Electric Corp.*)

Reflectometer. This instrument serves the dual purpose of a reflectometer and transmissometer and combines integrating spheres and barrier-layer photocells (**Fig. 7**). The absolute reflectance of test surfaces can be determined by measuring the total reflected light when a beam of light strikes the surface. Transmittance can be measured by placing a flat sample of the material in the opening between the two spheres, with the bottom sphere containing the light source for transmittance measurements, and the upper one the light-measuring cells and a collimated beam of light for reflectance measurements.

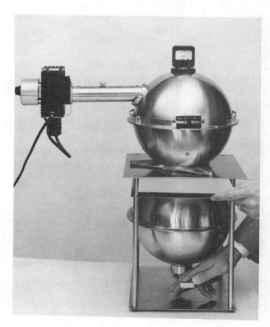

Fig. 7. Dual-purpose reflectometer transmissometer, in which integrating spheres and barrier-layer cell photometers are combined. (*General Electric Co.*)

Light-distribution photometer. One of the most frequently used photometric devices, this measures luminous intensity at various angles from lamps, luminaires, floodlights, searchlights, and the like. Although the most direct method is to tilt the light source so that measurements can be made in various directions, this may be impracticable with sources whose light output is position-sensitive. It is therefore sometimes necessary to use mirror arrangements similar to the one in **Fig. 8**. The source to be measured is placed at a fixed point on the photometric axis (the horizontal line passing through the centers of the source and photocell). The mirror system can then be moved bodily about the axis so that the light emitted by the source in any direction in a vertical plane perpendicular to the photometric axis is reflected to the photocell.

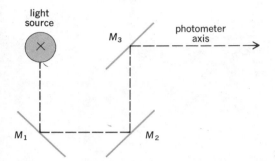

Fig. 8. Mirror system to measure light distribution.

Modern methods of automation have been applied to distribution photometers. They may be controlled by punched cards or tape to give complete programming of the photometer so that no human attention is necessary during the test. The photometric data can be automatically recorded, and curves of luminous intensity and footcandle intensity can be automatically drawn. A modern automated distribution photometer is shown in **Fig. 9**.

Spectrophotometer. Measurements of spectral energy distribution from a light source are made by this device. It measures the energy in small wavelength bands by means of a scanning slit, and the results are presented as a spectral distribution curve.

Fig. 9. Modern automated distribution photometer. (*Westinghouse Electric Corp.*)

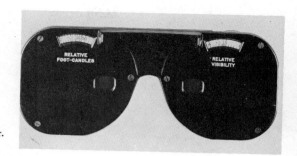

Fig. 10. Luckiesh-Moss visibility meter. (*General Electric Co.*)

Visibility meters. These operate on the principle of artificially reducing the visibility of objects to threshold values (borderline of seeing and not seeing) and measuring the amount of that reduction on an appropriate scale. The Luckiesh-Moss visibility meter is probably the best known and easiest to use. The instrument (**Fig. 10**) consists of two variable-density filters (one for each eye) so adjusted that a visual task seen through them is just barely discernible. Readings are on a scale of relative visibility related to a standard task.

ILLUMINATION
Warren B. Boast

In a general sense, the science of the application of lighting. Radiation in the range of wavelengths of 0.38–0.76 micrometer produces the visual effect commonly called light by the response of the average human eye for normal (photopic) brilliance levels. Illumination engineering pertains to the sources of lighting and the design of lighting systems which distribute light to produce a comfortable and effective environment for seeing. In a specific quantitative sense, illumination is the combination of the spatial density of radiant power received at a surface and the effectiveness of that radiation in producing a visual effect.

Illuminance, the alternate term for illumination, is receiving growing worldwide acceptance, particularly in science, but it is not generally used by illuminating engineers in the United States. *See* Illuminance.

Subjective color. Subjective evaluation of the surfaces of objects as viewed by the human eye has been treated in three classifications. All three are usually grouped by the nonspecialist as "color." *See* Color vision.

One of the classifications is the attribute of brilliance or subjective brightness, in respect to which every subjective evaluation of the seeing process may be classified according to its subjective intensity, ranging from very dim to very bright.

A second classification is hue, which describes color according to the common color names, such as red, yellow, and green.

The third classification is called variously saturation or purity or vividness of hue. Two colors may be subjectively evaluated as having the same brilliance and the same hue, but the first may seem to contain more white than the second. Conversely, one would evaluate the second color as being a more vivid or pure color than the first.

The hue designation of pure (completely saturated) spectral colors may be approximately assigned to regions of the wavelength spectrum.

Spectral response function of human eye. The spectral response of the human eye to radiation at normal (photopic) brilliance levels was tested by K. S. Gibson and E. P. T. Tyndall. The data from these tests were the source from which the Commission Internationale de l'Eclairage in 1924 established by international agreement a standard brilliance response of the human eye to radiant flux. The function was first known as the CIE values of visibility. The letters ICI (International Commission on Illumination) for the English name are also used. Later the function

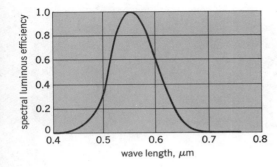

Fig. 1. Spectral response of human eye to radiation. (*After W. B. Boast, Illumination Engineering, 2d ed., McGraw-Hill, 1953*)

was named the relative luminosity function, but is now called the spectral luminous efficiency of radiant flux or simply the spectral luminous efficiency. The ordinates of this function are plotted against the wavelength of radiation in **Fig. 1** and are shown in the **table**. SEE LUMINOUS EFFICIENCY.

At threshold (scotopic) levels of seeing, the spectral response function of the human eye is shifted toward lower wavelengths. This is known as the Purkinje effect.

Photometric definitions. Illumination (or illuminance) is the density of radiant power incident upon a receiving surface evaluated throughout the spectrum in terms of the spectral luminous efficiency. Illumination is symbolized by E and is the density of luminous flux per unit of surface area. The unit of illumination in the International System of Units (SI) is the lux, which is equal to the lumen per square meter (1 lx = 1 lm/m^2). Units of illumination for other area measures are the phot (ph), equal to lumen per square centimeter, and the footcandle (fc), equal to lumen per square foot.

For a line-spectrum component of irradiation, a simple multiplication of the irradiation component G_i and the value of the spectral luminous efficiency v_i taken at the proper value of wavelength i yields the component illumination. For continuous-spectrum irradiation, G_λ is a function of wavelength; therefore, the product of the two functions v_λ and G_λ must be integrated. Equation (1) states mathematically a combination of line spectra and a continuous-spectrum source. Here

$$E = k \sum_{1}^{n} v_i G_i + k \int_{0}^{\infty} v_\lambda G_\lambda \, d\lambda \qquad (1)$$

G_i is the irradiation due to the ith component of the line spectra, v_i is the spectral luminous efficiency at the wavelength of the ith component irradiation, n is the number of the component

Spectral luminous efficiency of radiant flux (value of unity at 0.554 μm wavelength)

Wavelength, μm	Efficiency	Wavelength, μm	Efficiency	Wavelength, μm	Efficiency
0.38	0.00004	0.51	0.503	0.64	0.175
0.39	0.00012	0.52	0.710	0.65	0.107
0.40	0.0004	0.53	0.862	0.66	0.061
0.41	0.0012	0.54	0.954	0.67	0.032
0.42	0.0040	0.55	0.995	0.68	0.017
0.43	0.0116	0.56	0.995	0.69	0.0082
0.44	0.023	0.57	0.952	0.70	0.0041
0.45	0.038	0.58	0.870	0.71	0.0021
0.46	0.060	0.59	0.757	0.72	0.00105
0.47	0.091	0.60	0.631	0.73	0.00052
0.48	0.139	0.61	0.503	0.74	0.00025
0.49	0.208	0.62	0.381	0.75	0.00012
0.50	0.323	0.63	0.265	0.76	0.00006

line-spectra radiations, G_λ is the spectral irradiation of the continuous spectrum at λ wavelength, v_λ is the spectral luminous efficiency at the corresponding wavelengths, and k is a constant equal to 683 lm/W.

Colorimetry. The experiments from which the curve of Fig. 1 was obtained involved matching the brilliance of radiant flux from narrow spectral bands of adjacent wavelengths by 52 observers. Many scientists had concluded that by adding the radiations from three distinctly different spectral sources properly distributed in the visible spectrum, combinations could be obtained that would uniquely define measurements for all colors. The science of such measurements is called colorimetry. SEE COLOR; COLORIMETRY.

Luminous flux. As previously implied, the lumen is the SI unit of luminous flux that results when radiant power is evaluated through the spectral luminous efficiency function. Related to illumination (or illuminance), the luminous flux Φ received over an area A is given by Eq. (2).

$$\Phi = \int_A E\, dA \tag{2}$$

The lumen is the only presently used unit of luminous flux and is the SI standard. In Eq. (2) the units of E and A must be consistent combinations between lux and square meter, phot and square centimeter, or footcandle and square foot. If E is uniform over the area A, then $\Phi = EA$. SEE LUMINOUS FLUX.

Luminous intensity. The luminous intensity I of a source is the radiant intensity J of that source evaluated throughout the spectrum in terms of the spectral luminous efficiency (v_i or v_λ or both) in the same manner as illumination was derived from radiant power density upon a receiving surface in Eq. (1).

The concepts of radiant intensity and luminous intensity apply strictly only to a point source. In practice, however, flux (radiant or luminous) emanating from a source whose dimensions are negligibly small in comparison with distances from which it is observed may be considered as coming from a point.

The luminous intensity I of a small source, when viewed in a particular direction, is related to the differential of luminous flux $d\Phi$ emitted in that direction in a differential solid angle $d\omega$ by Eq. (3). The candela, the SI unit of luminous intensity, is related to the lumen as 1 candela equals

$$I = d\Phi/d\omega \tag{3}$$

1 lumen per unit solid angle (often called steradian).

Quantitative standardization in the field of illumination has been established in various manners by international agreement over the years. At a Paris meeting in October 1979, the General Conference on Weights and Measures redefined the base SI unit candela as the luminous intensity, in a given direction, of a source that emits monochromatic radiation of frequency 540×10^{12} Hz and of which the radiant intensity in that direction is 1/683 watt per steradian (W/sr).

The frequency of 540×10^{12} Hz corresponds to a wavelength of 0.555 μm in free space where the velocity of light is essentially 3×10^8 m/s (1.86×10^5 mi/s). This frequency and wavelength is essentially that where the peak of unity occurs for the spectral luminous efficiency function of Fig. 1. Thus Eq. (4) relates J_i in W/sr for the monochromatic radiation to the resulting

$$I_i = k v_i J_i \tag{4}$$

monochromatic intensity I_i in candela. The constant k is 683 lm/W as before in Eq. (1). For the source defining the candela, v_i is unity at $f = 540 \times 10^{12}$ Hz, and J_i is 1/683 W/sr, giving Eq. (5).

$$I_i = k v_i J_i = \left(683\, \frac{\text{lm}}{\text{W}}\right)(1.000)\left(\frac{1}{683}\, \frac{\text{W}}{\text{sr}}\right) = 1\, \frac{\text{lm}}{\text{sr}} = 1\ \text{candela} \tag{5}$$

Consistency thus exists among Eqs. (3) and (4) and the constant k from Eq. (1). SEE LUMINOUS INTENSITY.

Luminance. Many sources of luminous flux are not small relative to the distance between the luminous source and points on receiving surfaces where illumination results. Consequently, another concept called luminance, or photometric brightness, is very convenient.

The luminance at a point on a radiating surface, acting either from self-emission or by reflection of luminous flux, can be related to a differential segment of the radiating surface dA and the differential luminous intensity dI when the differential area is observed from a particular direction by Eq. (6). The angle α is that between the particular direction of observation and the normal

$$L = \frac{dI}{\cos \alpha \, dA} \qquad (6)$$

to the differential surface area dA. The denominator $\cos \alpha \, dA$ is the projected area of the differential element from the viewed direction. The equation states that the luminance L is the ratio of the differential luminous intensity to the projected differential area from the direction of observation.

The metric unit of candela per square meter (cd/m^2) has been named the nit (nt) and is recognized as the SI unit of luminance. The fractional value (also metric) of cd/cm^2 is called the stilb (sb). Additional units involving a factor π are the apostilb (asb = $cd/\pi m^2$), lambert (L = $cd/\pi cm^2$), and the footlambert (ftL = $cd/\pi ft^2$). The reason for the π factor in these units is discussed below. *See* LUMINANCE.

Inverse-square law. Consider a differential area dA of a luminous surface possessing a differential intensity dI in a specific direction. At a distance D in that specific direction, the differential illumination dE_P upon a receiving surface at a point P is given by Eq. (7), where β is the angle between the normal to the receiving surface and D. Equation (7) indicates that the differential illumination varies inversely as the square of the distance between source and receiving surface. Substituting dI from Eq. (6) into Eq. (7) yields Eq. (8). The total illumination from an

$$dE_P = \frac{dI}{D^2} \cos \beta \qquad (7) \qquad\qquad dE_P = \frac{L \cos \alpha \cos \beta \, dA}{D^2} \qquad (8)$$

extended surface may be obtained by integrating the right-hand side of Eq. (8) over the surface.

If the total luminous source is negligibly small relative to D, the inverse square law is given by Eq. (9), where I is the intensity in the direction of D toward P. *See* INVERSE-SQUARE LAW.

$$E_P = \frac{I}{D^2} \cos \beta \qquad (9)$$

Light sources. Nature's source of radiation, the Sun, produces radiant power on the Earth extending from wavelengths below 0.3 μm in the ultraviolet region to well over 3 μm in the infrared region of the spectrum. The Sun's spectral radiation per unit of wavelength is greatest in the region per unit of wavelength is greatest in the region of 0.4–0.9 μm. The response of the human eye is well matched to this range of wavelengths. *See* VISION.

Oil-flame and gas lighting were used before the advent of the electric light, but since about 1900 electric energy has been the source of essentially all modern lighting devices.

Incandescent lamps. These electric lamps operate by virtue of the incandescence of a filament heated by electric current. The filament usually is composed of tungsten and in the conventional lamp is contained in either a vacuum or inert-gas-filled glass bulb. More recent designs use a halogen regenerative cycle. Presently iodine or bromide vapors are used in a quartz bulb, which in some designs is internal to an outer, conventional inert-gas-filled bulb. The evaporated tungsten reacts chemically to form a halogen compound, which returns the tungsten to the filament. Bulb blackening is prevented, higher-temperature filaments with increased luminous efficacy result, and higher luminous output can be achieved in relatively small bulbs. Much more efficient optical designs can be achieved for projection equipment, where the source size of light should be small for efficient utilization of the luminous flux.

Vapor lamps. These operate by the passage of an electric current through a gas or vapor. In some lamps the light may be produced by incandescence of one or both electrodes. In others the radiation results from luminescent phenomena in the electrically excited gas itself.

Fluorescent lamps. These are usually in the form of a glass tube, either straight or curved, coated internally with one or more fluorescent powders called phosphors. Electrodes are located at each end of the tube. The lamp is filled to a low pressure with an inert gas, and a small amount of mercury is added. The electric current passing through the gas and vapor generates ultraviolet radiation, which in turn excites the phosphors to emit light. If the emission of light continues only during the excitation, the process is called fluorescence. If the materials continue to emit light after the source of excitation energy is removed, the process is called phosphorescence. Phosphors for fluorescent lamps are chosen to accentuate the fluorescent action.

Reflection and transmission. The control of light is of primary importance in illumination engineering because light sources rarely have inherent characteristics of distribution, brightness, or color desirable for direct application. Modification of light may be provided in a number of ways, all of which may be grouped under the general topics of reflection and transmission. Reflection from a surface and transmission through it each may be classified according to their spatial and spectral characteristics.

Spatial characteristics. Spatially a surface may exhibit reflection conditions ranging from a regular, or specular, reflection to an ideally diffuse characteristic. Similarly, transmission may range spatially from complete transparency to an idealized diffuse transmission.

Figure 2 illustrates the extremes of specular reflection such as would be obtained from polished metal or silvered glass, and of an ideal mat-finished surface possessing microscopic roughness of minute crystals or pigment particles. The specular reflector gives a direct image of the source, with the angle of reflection equal to the angle of incidence (Fig. 2a). The plot shown for the diffuse reflector (Fig. 2b) illustrates the differential luminous intensity dI of a differential area dA as observed from different directions in the plane of the incident ray. For perfect diffusion, where L is constant for all angles of observation, it follows from Eq. (6) that dI must possess a spatial distribution with α which contains the $\cos \alpha$ function. The ideal diffuser illustrated in Fig. 2b has such a distribution. The dI distribution with α for a perfect diffuser is the same cosine distribution for all planes passing through the normal to the reflecting surface element dA.

The differential flux density, $d\Phi_E/dA$, emitted by any reflecting surface is symbolized by M and called luminous exitance (luminous emittance is used also but is deprecated). For the perfectly diffusing surface of constant luminance L (called a Lambert surface), differential luminous intensity can be integrated over α from 0 to $\pi/2$ and shown to produce a differential flux $d\Phi_E$ of $\pi L/dA$. The ratio of $d\Phi_E/dA$ is the luminous exitance M. Thus for the perfectly diffusing surface, Eq. (10) is valid.

$$M = \frac{d\Phi_E}{dA} = \pi L \qquad (10)$$

The luminance units apostilb, lambert, and footlambert, which contain the factor π in their denominators, are sometimes called rationalized units because they are convenient for expressing the luminance of a perfectly diffusing surface. It is unfortunate that the use of units of luminance established to match for a special situation of perfect diffusion has persisted.

Practical surfaces possess spatial-reflection characteristics intermediate between the ideals of diffuse and specular reflection, particularly at the diffuse end of the range. Typical distribution curves are shown in **Fig. 3**. In all of these illustrations the light source is small. Figure 3a demonstrates that even with the best practical mat surfaces, such as dull-finished metals or those

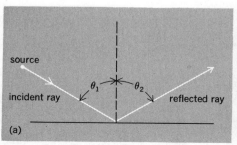
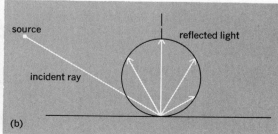

Fig. 2. Ideal reflection. (a) Specular, such as from a polished surface. (b) Diffuse, such as from a mat surface of microscopic roughness. (*After W. B. Boast, Illumination Engineering, 2d ed., McGraw-Hill, 1953*)

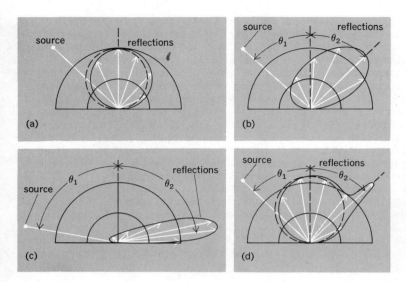

Fig. 3. Typical reflection characteristics. (a) Practical mat surface. (b) Semimat surface. (c) Semimat surface at large angle of incidence. (d) Porcelain-enameled steel. Broken-line curves in a and d illustrate ideal cosine distribution. (*After W. B. Boast, Illumination Engineering, 2d ed., McGraw-Hill, 1953*)

painted with flat paint, the location of the source of light has some slight influence upon the intensity distribution as evidenced by an irregularity of the cosine distribution of intensity of the small area of the surface. Figure 3b and c demonstrates the pronounced influence of the location of the light source upon the intensity distribution for surfaces covered with semimat materials, such as satin-finish paint. Figure 3d illustrates a mixed-reflection phenomenon resulting when a diffusing surface is overlaid with a surface possessing sheen.

Spatial characteristics of transmission are illustrated in **Fig. 4**. Figure 4a shows regular or transparent transmission, such as occurs with clear glass or plastic. Figure 4b shows the diffuse transmission that would result with an idealized diffusing material. In Fig. 4a a refraction of the transparent image occurs, but angles θ_1 and θ_2 are equal.

Practical transmission materials possess spatial transmission characteristics intermediate between these ideals and exhibit intensity distribution curves illustrated by **Fig. 5**. Figure 5a demonstrates that a perfect cosine intensity distribution is not obtained even with a solid opal- or milk-glass material. A more direct transmission is shown in Fig. 5b for a flashed opal-glass medium, in which the majority of the base material is clear glass. Sandblasted or frosted-glass material is much less diffusing, and the location of the light source greatly influences the intensity distribution curves of Fig. 5c and d.

Spectral characteristics. If a homogeneous radiation of wavelength λ impinges upon a surface, the reflection characteristic or the transmission characteristic will result geometrically

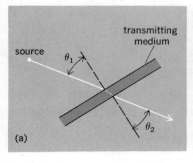
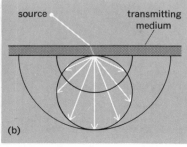

Fig. 4. Spatial characteristics of ideal transmission media. (a) Specular. (b) Diffuse. (*After W. B. Boast, Illumination Engineering, 2d ed., McGraw-Hill, 1953*)

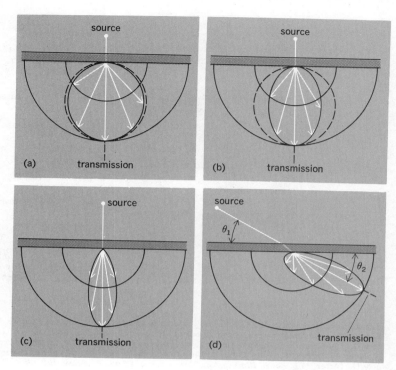

Fig. 5. Transmission characteristics. (a) Solid opal glass. (b) Flashed opal glass. (c) Sand-blasted glass. (d) Sand-blasted glass, incident light at 30° from plane at surface. (*After W. B. Boast, Illumination Engineering, 2d ed., McGraw-Hill, 1953*)

according to the spatial considerations of the preceding section. The magnitude of total power density reflected, or transmitted, to that received is defined as the spectral reflectance ρ_λ or the spectral transmittance τ_λ respectively. Thus $\rho_\lambda = J_\lambda/G_\lambda$ and $\tau_\lambda = J'_\lambda/G_\lambda$, where J_λ is the reflected spectral emission at wavelength λ, J'_λ is the transmitted spectral emission at wavelength λ, and G_λ is the spectral irradiation at wavelength λ.

A device for measuring ρ_λ or τ_λ is called a spectrophotometer. Comparison of J_λ or J'_λ with G_λ (or its essential equivalence) may be accomplished by visual or photoelectric methods of photometry.

Either continuous- or line-spectra sources of irradiation may constitute a practical source of radiant-power density. The resulting secondary source of radiant-power density J_λ or J'_λ will be modified, compared with the original, depending on spectral reflection or transmission of the reflecting or transmitting material.

An evaluation of J_λ or J'_λ using the spectral luminous efficiency ν_λ may be effected for the secondary sources of radiant-power density in the same manner as is done for incident radiant-power density conditions. Both line- and continuous-spectrum effects may be included as demonstrated for the evaluation of the illumination. The evaluation of the reflected or transmitted luminous flux density yields the luminous exitance, symbolized by M for the reflecting surface and M' for the transmitting surface. The functional operations are shown graphically in the series of curves of **Fig. 6** for the reflection resulting from a continuous-spectrum source. The ratio of M or M' to the illumination E results in $\rho = M/E$ and $\tau = M'/E$, where ρ is the reflectance of the surface and τ is the transmittance of the material. The reflectance and transmittance are dependent upon the source of illumination as well as the spectral characteristics of the surface or the transmitting medium. Reflectances and transmittances of surfaces and media are frequently published for the case of a tungsten-filament light source (often without reference to this source) and are useful as guides and indications of magnitude if the sources of illumination do not deviate too greatly from this spectral distribution.

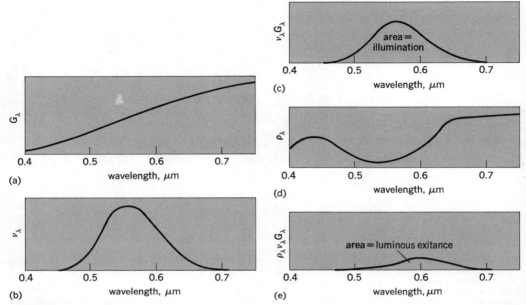

Fig. 6. Process of evaluating the luminous exitance of a reflecting surface after evaluating the illumination. (a) Spectral irradiation G_λ curve. (b) Spectral luminous efficiency v_λ curve. (c) Curve of $v_\lambda G_\lambda$. Illumination is the area under this curve, the integration of $v_\lambda G_\lambda$. (d) Spectral reflectance ρ_λ curve. (e) Curve of $\rho_\lambda v_\lambda G_\lambda$. Luminous exitance is the area under this curve, the integration of $\rho_\lambda v_\lambda G_\lambda$. (*After W. B. Boast, Illumination Engineering, 2d ed., McGraw-Hill, 1953*)

Light-control methods. Either a control of the character of the reflecting surface or transmitting material or a control of the primary source of luminous flux determines the degree of light control possible. If a light ray is reflected or transmitted diffusely, the shape of the reflecting or transmitting surface is of little importance, although its size may be of great importance. A larger diffuse surface reduces the brightness of the primary source. If a light ray is reflected from a specular surface, more accurate control of the light can be effected.

Parabolic reflectors. These are probably the most useful of all reflector forms. The rays emerge essentially parallel from such a reflector if the light source is placed at its focal point. Searchlights use parabolic mirror reflectors.

Elliptic reflectors. A relatively large amount of light may be made to pass through a small opening before spreading out by using an elliptic reflector. A light source is placed at one of the focal points of a nearly complete elliptical reflector. A small opening is placed at the second focal point where the light is focused. Such a design is used for the pinhole spotlight.

Lens control. Refraction of light rays at the boundary between glass and air may be used to control transmitted light in a manner similar to reflector control. Simple thick lenses may be cut away if the cutaway piece is duplicated in the new equivalent surface. Slight irregularities in the field of the light beam occur. These may be of little consequence for searchlights or floodlights, but they would be undesirable for high-quality picture projection. SEE LENS.

Polarization of light. Light rays emitted by common sources may be considered as waves vibrating at all angles in planes at right angles to the direction of the ray. As they pass through some substances or are reflected from specular surfaces, particularly at certain angles, wave components in some directions are absorbed more than those in other directions. Such light is said to be polarized.

Polarized light may be controlled by a transmitting section of a polarizing material. When the absorbing section is oriented to permit passage of the polarized light, minimum control is exercised. If the absorbing section is rotated 90°, the polarized light is essentially completely absorbed. Because sunlight reflected from water or specular surfaces is highly polarized, glare from such sources is effectively controlled by polarized sunglasses or camera filters. Polarized light is used also for detection of strains and defects in glass or plastics. The strained areas appear as color fringes in the material. SEE POLARIZED LIGHT.

Control of primary source. Control of the source of luminous flux can be achieved by controlling the voltage supplied and thus controlling the magnitude of current for some types of light sources or by controlling the time at which the current is permitted to begin on each half cycle of the electric supply for other types of light sources.

For simple incandescent lamps the control of the magnitude of alternating voltage and current can be controlled by variable transformer-type dimmers. Very small silicon controlled rectifiers (SCRs), a modern solid-state development, can accomplish the same control function with equipment which is so small that a unit can be installed in the same space as would be occupied by a conventional, residential "on-off" switch and with very small inherent heat loss caused by the control itself. SCRs are also used in larger capacities for control of large lighting installations.

The dimming control of fluorescent lamps or other sources employing a gas or vapor discharge is more involved because of the relatively high striking voltage of the vapor discharge. To achieve a wide range of dimming control, the timing on each half-cycle at which current is permitted to begin can be controlled through electronic means and thus produce dimming to low levels. The complexity and cost of such control equipment, however, is greater than the control equipment needed for incandescent lamp dimming.

Bibliography. C. L. Amick, *Fluorescent Lighting Manual*, 2d ed., 1947; W. B. Boast, *Illumination Engineering*, 2d ed., 1953; Bureau Central de la Commission Internationale de l'Eclairage, *International Lighting Vocabulary*, 3d ed., 1970; D. G. Fink and H. W. Beaty (eds.), *Standard Handbook for Electrical Engineers*, 12th ed., 1987; K. S. Gibson and E. P. T. Tyndall, *Visibility of Radiant Energy*, Nat. Bur. Stand. Sci. Pap. 19, 1923; Illuminating Engineering Society, *IES Lighting Handbook*, 6th ed., 1981.

LUMINOUS FLUX
RUSSELL C. PUTNAM

The time rate of flow of light. It is radiant flux in the form of electromagnetic waves which affects the eye or, more strictly, the time rate of flow of radiant energy evaluated according to its capacity to produce visual sensation. The visible spectrum is ordinarily considered to extend from 380 to 760 nanometers in wavelength; therefore, luminous flux is radiant flux in that region of the electromagnetic spectrum. The unit of measure of luminous flux is the lumen. SEE PHOTOMETRY.

LUMINOUS INTENSITY
RUSSELL C. PUTNAM

The solid angular luminous flux density in a given direction from a light source. It may be considered as the luminous flux on a small surface normal to the given direction, divided by the solid angle (in steradians) which the surface subtends at the source of light. Since the apex of a solid angle is a point, this concept applies exactly onto to a point source. The size of the source, however, is often extremely small when compared with the distance from which it is observed, so in practice the luminous flux coming from such a source may be taken as coming from a point. For accuracy, the ratio of the diameter of the light source to the measuring distance should be about 1:10, although in practice ratios as large as 1:5 have been used without excessive error.

Mathematically, luminous intensity I is given by the equation below, where ω is the solid

$$I = dF/d\omega$$

angle through which the flux from the point source is radiated, and F is the luminous flux. The luminous intensity is often expressed as candlepower. SEE CANDLEPOWER; PHOTOMETRY.

CANDLEPOWER
RUSSELL C. PUTNAM

Luminous intensity expressed in candelas. The term refers only to the intensity in a particular direction and by itself does not give an indication of the total light emitted. The candlepower in a given direction from a light source is equal to the illumination in footcandles falling on a surface normal to that direction, multiplied by the square of the distance from the light source in feet. The candlepower is also equal to the illumination of metercandles (lux) multiplied by the square of the distance in meters.

The apparent candlepower is the candlepower of a point source which will produce the same illumination at a given distance as produced by a given light source.

The mean horizontal candlepower is the average candlepower of a light source in the horizontal plane passing through the luminous center of the light source.

The mean spherical candlepower is the average candlepower in all directions from a light source as a center. Since there is a total solid angle of 4π (steradians) emanating from a point, the mean spherical candlepower is equal to the total luminous flux (in lumens) of a light source divided by 4π (steradians). SEE LUMINOUS INTENSITY; PHOTOMETRY.

LUMINOUS ENERGY
RUSSELL C. PUTNAM

The radiant energy in the visible region or quantity of light. It is in the form of electromagnetic waves, and since the visible region is commonly taken as extending 380–760 nanometers in wavelength, the luminous energy is contained within that region. It is equal to the time integral of the production of the luminous flux. SEE PHOTOMETRY.

LUMINANCE
RUSSELL C. PUTNAM

The luminous intensity of any surface in a given direction per unit of projected area of the surface viewed from that direction. The International Commission on Illumination defines it as the quotient of the luminous intensity in the given direction of an infinitesimal element of the surface containing the point under consideration, by the orthogonally projected area of the element on a plane perpendicular to the given direction. Simply, it is the luminous intensity per unit area. Luminance is also called photometric brightness.

Since the candela is the unit of luminous intensity, the luminance, or photometric brightness, of a surface may be expressed in candelas/cm^2, candelas/in.^{2}, and so forth.

Mathematically, luminance L may be found from the equation below, where θ is the angle

$$L = dI/(dA \cos \theta)$$

between the line of sight and the normal to the surface area A considered and I is the luminous intensity.

The stilb is a unit of luminance (photometric brightness) equal to 1 candela/cm^2. It is often used in Europe, but the practice in America is to use the term candela/cm^2 in its place.

The apostilb is another unit of luminance sometimes used in Europe. It is equal to the luminance of a perfectly diffusing surface emitting or diffusing light at the rate of 1 lumen/m^2. SEE LUMINOUS INTENSITY; PHOTOMETRY.

ILLUMINANCE
Russell C. Putnam

A term expressing the density of luminous flux incident on a surface. This word has been proposed by the Colorimetry Committee of the Optical Society of America to replace the term illumination. The definitions are the same. The symbol of illumination is E, and the equation is $E = dF/dA$, where A is the area of the illuminated surface and F is the luminous flux. See Illumination; Luminous flux; Photometry.

LUMINOUS EFFICACY
G. A. Horton

There are three ways this term can be used: (1) The luminous efficacy of a source of light is the quotient of the total luminous flux emitted divided by the total lamp power input. Light is visually evaluated radiant energy. Luminous flux is the time rate of flow of light. Luminous efficacy is expressed in lumens per watt. (2) The luminous efficacy of radiant power is the quotient of the total luminous flux emitted divided by the total radiant power emitted. This is always somewhat larger for a particular lamp than the previous measure, since not all the input power is transformed into radiant power. (3) The spectral luminous efficacy of radiant power is the quotient of the luminous flux at a given wavelength of light divided by the radiant power at that wavelength. A plot of this quotient versus wavelength displays the spectral response of the human visual system. It is, of course, zero for all wavelengths outside the range from 380 to 760 nanometers. It rises to a maximum near the center of this range. Both the value and the wavelength of this maximum depend on the degree of dark adaptation present. However, an accepted value of 683 lumens per watt maximum at 555 nm represents a standard observer in a light-adapted condition. The reciprocal of this maximum spectral luminous efficacy of radiant power is sometimes referred to as the mechanical equivalent of light, with a probable value of 0.00147 W/lm.

If the spectral luminous efficacy values at all wavelengths are each divided by the value at the maximum (683 lm/W), the spectral luminous efficiency of radiant power is obtained. It is dimensionless.

For purposes of illuminating engineering, light is radiant energy which is evaluated in terms of what is now simply and descriptively the spectral luminous efficiency curve, V_λ. In this case it is the intent that an illuminating engineer have a measure for the capacity or capability for the production of luminous flux from radiant flux, or that radiant flux has efficacy for the production of luminous flux.

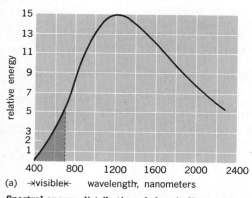 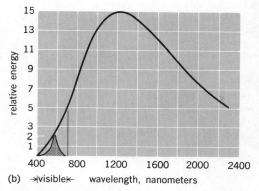

Spectral energy distribution of electric filament lamp. (*a*) Colored area is measure of power of visible radiation. (*b*) Colored area is measure of light output.

Since the human eye is sensitive only to radiations of wavelengths between about 400 and 700 nm, it follows that the sensation of sight evoked by a radiator is due only to the radiation within this limited wavelength band. If all the wavelengths within this band were equally effective for the purpose of vision, then the area under the graph of radiant power distribution would be a measure of the light output.

The **illustration** shows the spectral distribution curve for an electric tungsten filament lamp. The total area under the curve between the wavelength limits $\lambda = 0$ and $\lambda = \infty$ is proportional to the power of the total radiation. In illus. *a* the shaded area between the limits of $\lambda = 400$ and $\lambda = 700$ nm is a measure of the power of visible radiation. However, this is not the same thing as luminous efficacy because of the variable sensitivity of the eye with respect to wavelength. If each ordinate, say at 10-nm wavelength intervals, is multiplied by the appropriate value of V_λ a new distribution is obtained, as shown in the shaded area of illus. *b*. This area is a measure of the light output of the source. The ratio of the shaded area of illus. *b* to the area under the larger curve is the luminous efficacy in terms of lumens per watt, and this is the method of determination that is generally used. The name luminous efficacy expresses exactly what is meant by the effectiveness of 1 watt of radiant power in producing luminous flux. SEE ILLUMINATION; LUMINOUS EFFICIENCY; LUMINOUS FLUX; PHOTOMETRY.

LUMINOUS EFFICIENCY
G. A. HORTON

Visual efficacy of visible radiation, a function of the spectral distribution of the source radiation in accordance with the "spectral luminous efficiency curve," usually for the light-adapted eye or photopic vision, or in some instances for the dark-adapted eye or scotopic vision. Account is taken of this fact in photometry, or the measurement of light. Actually, instead of indicating directly the physical radiant power of light, photometers are designed to indicate the corresponding luminous flux of the visually effective value of the radiation.

The spectral luminous efficiency of radiant flux is the ratio of luminous efficacy for a given wavelength to the value of maximum luminous efficacy. It is a dimensionless ratio. Values of spectral luminous efficiency for photopic vision V_λ at 10-nanometer-wavelength intervals were provisionally adopted by the International Commission on Illumination in 1924 and by the International Committee on Weights and Measures in 1933, as a basis for the establishment of photometric standards of types of sources differing from the primary standard in spectral distribution of radiant flux. These standard values of spectral luminous efficiency were determined by visual photometric observations with a 2° photometric field having a moderately high luminance (photometric brightness); photometric evaluations based upon them, consequently, do not agree exactly with other conditions of observations, such as in physical radiometry. Watts weighted in accordance with these standard values are often referred to as light-watts.

Values of spectral luminous efficiency for scotopic vision V'_λ at 10-nm intervals were provisionally adopted by the International Commission on Illumination in 1951. These values of spectral luminous efficiency were determined by visual observations by young, dark-adapted observers, using extra-foveal vision at near-threshold luminance.

The main difference between the terms luminous efficacy and luminous efficiency is that the latter came into use a century or so before the former. It should also be noted that in engineering, general values of efficiency never exceed 1.0, and illuminating engineers were the only ones who had values of efficiency up to 683, but this has now been changed by the usage of the term luminous efficacy. In view of the above discussion, the term luminous efficiency is evidently a misnomer. SEE ILLUMINATION; LUMINOUS EFFICACY; PHOTOMETRY.

HUMAN PERCEPTION OF LIGHT

Vision	366
Color	370
Colorimetry	372
Color vision	374
Stereoscopy	378
Moiré pattern	379

VISION
Lorrin A. Riggs

The sense of sight, which perceives the form, color, size, movement, and distance of objects. Of all the senses, vision provides the most detailed and extensive information about the environment. Conversely, blindness is recognized as more disabling than deafness or any other sensory handicap. In the higher animals, especially the birds and primates, the eyes and the visual areas of the central nervous system have developed a size and complexity far beyond the other sensory systems.

Visual stimuli. These are typically rays of light entering the eyes and forming images on the retina at the back of the eyeball (**Fig. 1**). The intensity and wavelength characteristics of the light vary according to the light source and the object from which they are reflected. Human vision is most sensitive for light comprising the visible spectrum in the range 380–720 nanometers in wavelength. Sunlight and common sources of artificial light contain substantially all wavelengths in this range but each source has a characteristic spectral energy distribution. In general, light stimuli can be measured by physical means with respect to their energy, dominant wavelength, and spectral purity. These three physical aspects of the light are closely related to the perceived brightness, hue, and saturation, respectively. SEE COLOR; LIGHT.

Atypical (sometimes called inadequate) stimuli for vision include momentary pressure on the eyeball, electric current through the eyes or head, a sudden blow on the back of the head, or disturbances of the central nervous system, caused by drugs, fatigue, or disease. Any of these may yield visual experiences not aroused by light. They are of interest because they show that the essentially visual character of the sensory experience is determined by the region stimulated (eyes, visual tracts) rather than by the nature of the stimulus. Indeed any observant person can detect swirling clouds or spots of "light" in total darkness or while looking at a homogeneous field such as a bright blue sky. These phenomena illustrate the spontaneous activity that is characteristic of the nervous system in general. They show that the visual system is continuously active, which in turn means that the effect of a stimulus is to modify existing activity and not merely to initiate new activity.

Anatomical basis for vision. The anatomical structures include the eyes, optic nerves and tracts, optic thalamus, and visual cortex. The eyes are motor organs as well as sensory; that is, each eye can turn directly toward an object to inspect it. The two eyes are coordinated in their inspection of objects, and they are able to converge for near objects and diverge for far ones. Each eye can also regulate the shape of its crystalline lens to focus the rays from the object and to form a sharp image on the retina. Furthermore, the eyes can regulate the amount of light reaching the sensitive cells on the retina by contracting and expanding the pupil of the iris. These motor responses of the eyes are examples of involuntary action that is controlled by various reflex pathways within the brain.

The process of seeing begins when light passes through the eye and is absorbed by the sensitive cells of the retina. These cells are activated by the light in such a way that electrical potentials are generated. These potentials are probably responsible for many features of the electroretinogram, an electrical response wave that can be detected by means of electrodes attached

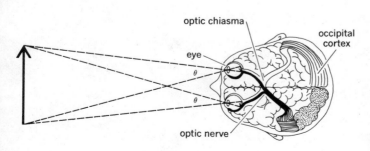

Fig. 1. Diagram showing the eyes and visual projection system. The visual angle θ is measured in degrees.

to the outside of the eye. Some of these potentials are the signs of electrochemical activity that serves to generate nerve responses in various successive neural cells (bipolars, ganglion cells, and others) in the vicinity of excitation. Finally, impulses emerge from the eye in the form of repetitive discharges in the fibers of the optic nerve. It must be emphasized, however, that the optic nerve impulses do not mirror exactly the excitation of the sensory cells by light. Complex interactions within the retina serve to enhance certain responses and to suppress others. Furthermore it is a fact that each eye contains more than a hundred times as many sensory cells as optic nerve fibers. Thus it would appear that much of the integrative action of the visual system has already occurred within the retina before the brain has had a chance to act.

The optic nerves from the two eyes traverse the optic chiasma. Figure 1 shows that the fibers from the inner (nasal) half of each retina cross over to the opposite side, while those from the outer (temporal) half do not cross over but remain on the same side. The effect of this arrangement is that the right visual field, which stimulates the left half of each retina, activates the left half of the thalamus and visual cortex. Conversely the left visual field affects the right half of the brain. This situation is therefore similar to that of other sensory and motor projection systems in which the left side of the body is represented by the right side of the brain and vice versa.

The visual cortex includes a projection area in the occipital lobe of each hemisphere. Here there appears to be a point-for-point correspondence between the retina of each eye and the cortex. Thus the cortex contains a "map" or projection area, each point of which represents a point on the retina and therefore a point in visual space as seen by each eye. But the map is much too simple a model for cortical function. Vision tells much more than the location at which an object is seen. Visual tests show that other important features of an object such as its color, motion, orientation, and shape are simultaneously perceived. In monkey experiments, neurophysiologists have identified many types of cortical cell, each one responding selectively to these critical features. The two retinal maps are merged to form the cortical projection area. This merger allows the separate images from the two eyes to interact with each other in stereoscopic vision, binocular color mixture, and other phenomena. In addition to the projection areas on the right and left halves of the cortex, there are visual association areas and other brain regions that are involved in vision. Complex visual acts, such as form recognition, movement perception, and reading, are believed to depend on widespread cortical activity beyond that of the projection areas.

Night animals, such as the cat or the owl, have eyes that are specialized for seeing with a minimum of light. This type of vision is called scotopic. Day animals such as the horned toad, ground squirrel, or pigeon have predominantly photopic vision. They require much more light for seeing, but their daytime vision is specialized for quick and accurate perception of fine details of color, form, and texture, and location of objects. Color vision, when it is present, is also a property of the photopic system. Human vision is duplex; humans are in the fortunate position of having both photopic and scotopic vision. Some of the chief characteristics of human scotopic and photopic vision are enumerated in the **table**.

Characteristics of human vision		
Characteristic	Scotopic vision	Photopic vision
Photochemical substance	Rhodopsin	Cone pigments
Receptor cells	Rods	Cones
Speed of adaptation	Slow (30 min or more)	Rapid (8 min or less)
Color discrimination	No	Yes
Region of retina	Periphery	Center
Spatial summation	Much	Little
Visual acuity	Low	High
Number of receptors per eye	120,000,000	7,000,000
Cortical representation	Small	Large
Spectral sensitivity peak	505 nm	555 nm

Scotopic vision. This occurs when the rod receptors of the eye are stimulated by light. The outer limbs of the rods contain a photosensitive substance known as visual purple or rhodopsin. This substance is bleached away by the action of strong light so that the scotopic system is virtually blind in the daytime. Weak light causes little bleaching but generates neural inhibitory signals that lower the overall sensitivity of the eye. In darkness, however, the rhodopsin is regenerated by restorative reactions based on the transport of vitamin A to the retina by the blood. One experiences a temporary blindness upon walking indoors on a bright day, especially into a dark room or dimly lighted theater. As the eyes become accustomed to the dim light the scotopic system gradually begins to function. This process is known as dark adaptation. Complete dark adaptation is a slow process during which the rhodopsin is restored in the rod receptors of the retina. Faulty dark adaptation or night blindness is found in persons who lack rod receptors or have a dietary deficiency in vitamin A. These rare persons are unable to find their way about at night without the aid of strong artificial illumination.

Dark adaptation is measured by an adaptometer, a device for presenting test flashes of light after various periods of time spent in the dark. The intensity of flash is varied to determine the momentary threshold for vision as dark adaptation proceeds. A 10,000-fold increase in sensitivity (that is, a reduction of 10,000 to 1 in the threshold intensity of flash) is often found to occur during a half-hour period of dark adaptation. By this time some of the rod receptors are so sensitive that only one elementary quantum (photon) of light is necessary to trigger each rod into action. A person can detect the presence of a flash of light that simultaneously affects only a few of the millions of rod receptors. Thus the scotopic sensitivity of the human eye approaches the ideal case of a receiver that is capable of responding to a single quantum of energy.

The variation of the scotopic threshold with wavelength of light is shown in the rod curve of **Fig. 2**. In spite of the variations in wavelength, the subject does not see any color when the intensity of the light is low enough to fall in the rod portion of the diagram. This scotopic vision is colorless or achromatic, in agreement with the saying that in the night all cats appear gray.

Normal photopic vision. Normal photopic vision has the characteristics enumerated in the table. Emphasis is placed on the fovea centralis, a small region at the very center of the retina of each eye.

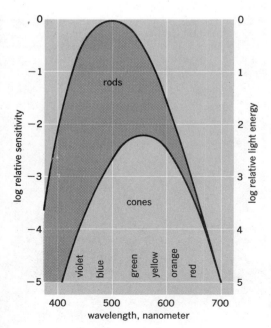

Fig. 2. Spectral sensitivity curves for human vision. The rod curve shows that scotopic vision, based on rhodopsin, is most sensitive to light of about 505 nm. The cone curve shows that photopic vision is generally less sensitive than scotopic vision, except for light at the red end of the spectrum.

Foveal vision. This is achieved by looking directly at objects in the daytime. In **Fig. 3** the image of a small object at F falls within a region almost exclusively populated by cone receptors. These are so closely packed together in the central fovea that their density is about 100,000 per square millimeter. Furthermore, each of the cones in the fovea is provided with a series of specialized nerve cells that process the incoming pattern of stimulation and convey it to the cortical projection area. In this way the cortex is supplied with superbly detailed information about any pattern of light that falls within the fovea centralis.

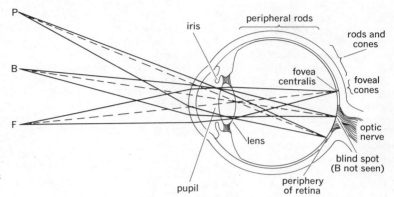

Fig. 3. Foveal and peripheral vision. Looking directly at F, the eye focuses the light from F on the central fovea, the region of clearest vision. P is seen poorly, and B not at all.

Peripheral vision. This is vision that takes place outside the fovea centralis (Fig. 3). As an example, look directly at a single letter at the center of a printed page. This letter, and a few letters immediately adjacent to it, appear clear and black because they are seen with foveal vision. The rest of the page is a blur in which the lines of print are seen as gray streaks. This is an example of peripheral vision. Vision actually extends out to more than 90° from center, so that one can detect moving objects approaching from either side. This extreme peripheral vision is comparable to night vision in that it is devoid of sharpness and color.

There is a simple anatomical explanation for the clarity of foveal vision as compared to peripheral vision. The cone receptors become less and less numerous in the retinal zones that are more and more remote from the fovea. In the extreme periphery there are scarcely any cones, and even the rod receptors are more sparsely distributed. Furthermore, the plentiful neural connections from the foveal cones are replaced in the periphery by network connections in which hundreds of receptors may activate a single optic nerve fiber. This mass action is favorable for the detection of large or dim stimuli in the periphery or at night, but it is unfavorable for visual acuity or color vision, both of which require the brain to differentiate between signals arriving from closely adjacent cone receptors.

Visual acuity. Visual acuity is defined as the ability to see fine details of an object. In Fig. 1 the small arrow at the back of each eye shows the image of the test object that is focused on the retina. Standard visual acuity is defined as the ability to see an object so small that the angle θ subtended at the eye is only 1 minute of arc, or $1/60$ of a degree. At 20 ft (0.6 m) the size of such a test object is therefore only about 0.07 in., or 1.75 mm. The image of the object on the fovea (neglecting diffraction and optical aberrations that deteriorate the image) has a length of only 0.005 mm. Small as this is, it is twice the diameter of the smallest foveal cone receptor. One therefore comes to the conclusion that normal visual acuity approaches the limit imposed by diffraction and by the optical aberrations in the eye.

The specific forms of test object used for determining visual acuity yield different results. The angle θ in Fig. 1 can be made so small that it represents the size of test object that can barely be seen by the normal eye. This angular size can be infinitely small in the case of a bright point seen against a dark background. The stars at night provide a good example of this, since they are so distant that their angular subtense at the eye is practically zero. A long dark line can be detected against a bright field (for example a flagpole seen against the sky) if it subtends 1 second of arc (1/60 of a minute) at the eye. On the other hand, letters of the alphabet, as used in optometric wall charts, need to be composed of black lines 1 minute thick in order to be recognizable to the normal eye. A similar value holds for the lines of a grating (with parallel black and white lines of equal width) when viewed under good illumination.

The apparent discrepancy between acuity for single points or lines on the one hand, and for more complex forms on the other hand, can be explained by the effects of optical diffraction. This is a phenomenon resulting from the wave nature of light. It means that no optical image can ever be completely sharp and clear. The retinal image of a star is not a point but a blurred circle of light. The diameter of this blurred image is never less than 1 minute of arc, no matter how small an angle the star itself may subtend. Thus the star is seen, provided that the blurred image is noticeably brighter than the surrounding field. In the case of a grating (black and white stripe pattern) the image of each line is blurred also. If the lines are too close (less than 1 minute) together, the blurred image of one overlaps the blurred image of the next and the separate lines cannot be resolved by the eye. One comes to the conclusion, then, that the chief factors limiting the visual acuity of the fovea are (1) optical diffraction, (2) the ability to discriminate relative brightnesses within the images blurred by diffraction, and (3) the compactness of the pattern of cone receptors.

Bibliography. H. Davson, *The Physiology of the Visual System*, 3d ed., 1972; C. H. Graham (ed.), *Vision and Visual Perception*, 1965; R. Held, H. W. Leibowitz, and H. L. Teuber (eds.), *Perception*, vol. 8 of *Handbook of Sensory Physiology*, 1978; D. Jameson and L. M. Hurvich (eds.), *Visual Psychophysics*, pt. 4 of vol. 7, *Handbook of Sensory Physiology*, 1972; R. Jung (ed.), *Central Processing of Visual Information*, pt. 3 of vol. 7, *Handbook of Sensory Physiology*, 1973; D. C. Marr, *Vision*, 1982; H. Obstfeld, *Optics in Vision*, 1982.

COLOR
ROBERT M. BOYNTON

That aspect of visual sensation enabling a human observer to distinguish differences between two structure-free fields of light having the same size, shape, and duration. Although luminance differences alone permit such discriminations to be made, the term color is usually restricted to a class of differences still perceived at equal luminance. These depend upon physical differences in the spectral compositions of the two fields, usually revealed to the observer as differences of hue or saturation.

Photoreceptors. Color discriminations are possible because the human eye contains three classes of cone photoreceptors that differ in the photopigments they contain and in their neural connections. Two of these, the R and G cones, are sensitive to all wavelengths of the visible spectrum from 380 to 700 nanometers. (Even longer or shorter wavelengths may be effective if sufficient energy is available.) R cones are maximally sensitive at about 570 nm, G cones at about 540 nm. The ratio R/G of cone sensitivities is minimal at 465 nm and increases monotonically for wavelengths both shorter and longer than this. This ratio is independent of intensity, and the red-green dimension of color variation is encoded in terms of it. The B cones, whose sensitivity peaks at about 440 nm, are not appreciably excited by wavelengths longer than 540 nm. The perception of blueness and yellowness depends upon the level of excitation of B cones in relation to that of R and G cones. No two wavelengths of light can produce equal excitations in all three kinds of cones. It follows that, provided they are sufficiently different to be discriminable, no two wavelengths can give rise to identical sensations.

The foregoing is not true for the comparison of two different complex spectral distributions. These usually, but not always, look different. Suitable amounts of short-, middle-, and long-wavelength lights, if additively mixed, can for example excite the R, G, and B cones exactly as does a light containing equal energy at all wavelengths. As a result, both stimuli look the same. This is an extreme example of the subjective identity of physically different stimuli known as chromatic metamerism. Additive mixture is achievable by optical superposition, rapid alternation at frequencies too high for the visual system to follow, or (as in color television) by the juxtaposition of very small elements which make up a field structure so fine as to exceed the limits of visual acuity. The integration of light takes place within each receptor, where photons are individually absorbed by single photopigment molecules, leading to receptor potentials that carry no information about the wavelength of the absorbed photons. SEE LIGHT.

Colorimetry. Although colors are often defined by appeal to standard samples, the trivariant nature of color vision permits their specification in terms of three values. Ideally these might be the relative excitations of the R, G, and B cones. Because too little was known about cone action spectra in 1931, the International Commission on Illumination (CIE) adopted at that time a different but related system for the prediction of metamers (the CIE system of colorimetry). This widely used system permits the specification of tristimulus values X, Y, and Z, which make almost the same predictions about color matches as do calculations based upon cone action spectra. If, for fields 1 and 2, $X_1 = X_2$, $Y_1 = Y_2$, and $Z_1 = Z_2$, then the two stimuli are said to match (and therefore have the same color) whether they are physically the same (isometric) or different (metameric). SEE COLORIMETRY.

The use of the CIE system may be illustrated by a sample problem. Suppose it is necessary to describe quantitatively the color of a certain paint when viewed under illumination by a tungsten lamp of known color temperature. The first step is to measure the reflectance of the paint continuously across the visible spectrum with a spectrophotometer. The reflectance at a given wavelength is symbolized as R_λ. The next step is to multiply R_λ by the relative amount of light E_λ emitted by the lamp at the same wavelength. The product $E_\lambda R_\lambda$ describes the amount of light reflected from the paint at wavelength λ. Next, $E_\lambda R_\lambda$ is multiplied by a value $\bar{x}_\lambda$, which is taken from a table of X tristimulus values for an equal-energy spectrum. The integral $\int E_\lambda R_\lambda \bar{x}_\lambda d\lambda$ gives the tristimulus value X for all of the light reflected from the paint. Similar computations using $\bar{y}_\lambda$ and $\bar{z}_\lambda$ yield tristimulus values Y and Z.

Tables of $\bar{x}_\lambda$, $\bar{y}_\lambda$, and $\bar{z}_\lambda$ are by convention carried to more decimal places than are warranted by the precision of the color matching data upon which they are based. As a result, colorimetric calculations of the type just described will almost never yield identical values, even for two physically different fields that are identical in appearance. For this and other reasons it is necessary to specify tolerances for color differences. Such differential colorimetry is primarily based upon experiments in which observers attempted color matches repeatedly, with the standard deviations of many such matches being taken as the discrimination unit.

Chromaticity diagram. Colors are often specified in a two-dimensional chart known as the CIE chromaticity diagram, which shows the relations among tristimulus values independently of luminance. In this plane, y is by convention plotted as a function of x, where $y = Y/(X + Y + Z)$ and $x = X/(X + Y + Z)$. [The value $z = Z/(X + Y + Z)$ also equals $1 - (x + y)$ and therefore carries no additional information.] Such a diagram is shown in the **illustration**, in which the continuous locus of spectrum colors is represented by the outermost contour. All nonspectral colors are contained within an area defined by this boundary and a straight line running from red to violet. The diagram also shows discrimination data for 25 regions, which plot as ellipses represented at 10 times their actual size. A discrimination unit is one-tenth the distance from the center of an ellipse to its perimeter. Predictive schemes for interpolation to other regions of the CIE diagram have been worked out.

If discrimination ellipses were all circles of equal size, then a discrimination unit would be represented by the same distance in any direction anywhere in the chart. Because this is dramatically untrue, other chromaticity diagrams have been developed as linear projections of the CIE chart. These represent discrimination units in a relatively more uniform way, but never perfectly so.

A chromaticity diagram has some very convenient properties. Chief among them is the fact that additive mixtures of colors plot along straight lines connecting the chromaticities of the colors being mixed. Although it is sometimes convenient to visualize colors in terms of the chromaticity chart, it is important to realize that this is not a psychological color diagram. Rather, the chromaticity diagram makes a statement about the results of metameric color matches, in the sense that a given point on the diagram represents the locus of all possible metamers plotting at chromaticity coordinates x, y. However, this does not specify the appearance of the color, which be dramatically altered by preexposing the eye to colored lights (chromatic adaptation) or, in the complex scenes of real life, by other colors present in nearby or remote areas (color contrast and

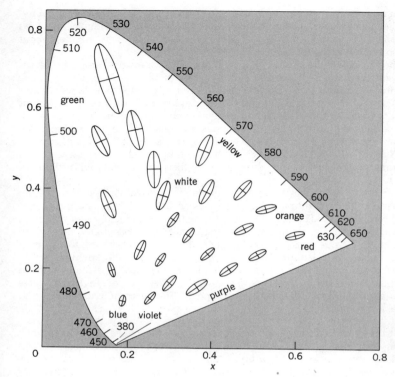

The 1931 CIE chromaticity diagram showing MacAdam's ellipses 10 times enlarged. (*After G. S. Wyszecki and W. S. Stiles, Color Science, John Wiley and Sons, 1967*)

assimilation). Nevertheless, within limits, metamers whose color appearance is thereby changed continue to match.

For simple, directly fixated, and unstructured fields presented in an otherwise dark environment, there are consistent relations between the chromaticity coordinates of a color and the color sensations that are elicited. Therefore, regions of the chromaticity diagram are often denoted by color names, as shown in the illustration.

Although the CIE system works rather well in practice, there are important limitations. Normal human observers do not agree exactly about their color matches, chiefly because of the differential absorption of light by inert pigments in front of the photoreceptors. Much larger individual differences exist for differential colorimetry, and the system is overall inappropriate for the 4% of the population (mostly males) whose color vision is abnormal. The system works only for an intermediate range of luminances, below which rods (the receptors of night vision) intrude, and above which the bleaching of visual photopigments significantly alters the absorption spectra of the cones. *See* COLOR VISION.

Bibliography. R. M. Boynton, *Human Color Vision*, 1979; R. W. Burnham, R. M. Hanes, and C. J. Bartleson, *Color: A Guide to Basic Facts and Concepts*, 1963; D. B. Judd and G. W. Wyszecki, *Color in Business, Science, and Industry*, 3d ed., 1975; R. Overheim and D. L. Wagner, *Light and Color*, 1982; G. W. Wyszecki and W. S. Stiles, *Color Science*, 2d ed., 1982.

COLORIMETRY
Deane B. Judd and Jack L. Lambert

Any technique by which an unknown color is evaluated in terms of known colors. Colorimetry may be visual, photoelectric, or indirect by means of spectrophotometry. These techniques are widely used in scientific studies involving the appearance of objects and lights, but are of greatest importance in the color specification of the raw materials and finished products of industry.

Visual colorimetry. In this type, the unknown color is presented beside a comparison field into which may be introduced any one or a range of known colors from which the operator chooses the one matching the unknown. To be generally applicable, the comparison field must not only cover a sufficient color range but must also be continuously adjustable in color in three independent ways: (1) by superposition of light primaries (for example, red, green, and blue) in any required proportions (additive colorimeters, see **illus**.); (2) by successive transmission through primary filters (for example, yellow, cyan, and magenta) of adjustable thickness (subtractive colorimeters); or (3) by a rotating sectored disk whose four differently colored sectors are adjustable in relative area (disk colorimeters). Many colorimeters, however, provide but a single series of comparison colors and are limited to specimens known in advance to have colors identical or nearly identical to one or another of this single series of colors.

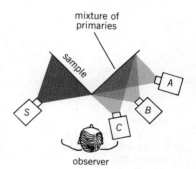

The tricolorimeter, in which the given color in S is matched by the addition of any three primaries in A, B, and C.

Indirect colorimetry. In this type, the light leaving the unknown specimen is split into its component spectral parts by means of a prism or diffraction grating, and the amount of each component part is separately measured by a photometer. The quantity evaluated is spectral radiance of a light source, spectral transmittance of a filter (glass, plastic, gelatin, or liquid), or spectral reflectance of an opaque body.

From spectrophotometric data such as these, it is possible to predict the amounts of known light primaries (red, green, and blue) required to produce the color of the specimen by superposition. This prediction is made by using the color-mixture functions (CIE standard observer) recommended for this purpose in 1931 by the International Commission on Illumination (CIE). These functions give the amounts of the light primaries required by the standard observer to match unit radiance of each part of the spectrum in turn. The prediction is found by adding up for the whole visible spectrum the amounts of red primary that are appropriate to the unknown specimen, and then doing the same thing for the appropriate amounts of green and blue primaries. *See* Color.

Photoelectric colorimetry. In this type, the light leaving the specimen is measured separately by three photocells. The spectral sensitivity of these photocells is adjusted, usually by color filters, to conform as closely as possible to the three color-mixture functions for the average normal human eye (CIE standard observer). The responses of the photocells give directly the amounts of red, green, and blue primaries required to produce the color of the unknown specimen for the kind of vision represented by the three photocells. Photoelectric colorimetry is used primarily for the control of the colors of manufactured articles to a preestablished tolerance.

Metamers. If two objects have the same color because the light leaving one of them toward the eye is spectrally identical to that leaving the other, any type of colorimetry serves reliably to establish the fact of color match. If, however, the two lights are spectrally dissimilar, they may still color-match for any one observer; such pairs of lights are called metamers. Normal color vision differs sufficiently from person to person so that a metameric color match for one observer may be seriously mismatched for another. On this account, the question of color match of spectrally dissimilar lights can be reliably settled only by the indirect method which uses spectrophotometry combined with a precisely defined standard observer. Examples are the color matching of a fluorescent lamp to an incandescent lamp or the matching of a specimen colored by one set of dyes or pigments to a specimen colored by a different set. SEE COLOR FILTER; COLOR VISION.

Colorimetric analysis. By strict definition, the term colorimetric analysis is limited to the techniques for visual identification and comparison of colored solutions. By common usage among analytical chemists, it has become a generic term for all types of analysis involving colored solutions.

Identification of a substance by the hue of its solution is termed qualitative analysis. Determination of its concentration in a solution by comparison of the intensity of its color to color intensity standards is termed quantitative analysis. When the human eye is used as the detector, quantitative colorimetric methods have relatively poor precision. Moreover, the precision varies with the color because of the varying response of the eye to different colors.

Most visual color-comparison methods utilize incident white light containing all wavelengths in the visible region. Filters or monochromators are seldom used, because colorimetric methods are advantageous principally for their simplicity, speed, and low cost, and for their modest demands for skill and training of the operator. Color intensities of solutions are usually compared to a set of permanent color standards. These standards may be solutions containing the same substance of similar hue, or colored glass.

One technique for the comparison of sample solutions to standard solutions is the use of long test tubes with flat bottoms called Nessler tubes, which may contain as much as 100 ml of solution. These tubes permit comparisons through depths of as much as 300 mm of solution, which increases the color intensities of dilute solutions. The sample solution is matched to one, or placed between two, of a series of standard solutions of graduated concentrations. A more sophisticated technique, now largely obsolete due to the availability of spectrophotometers, is based on the Duboscq colorimeter, in which one standard solution is used. With this instrument, the depths of the sample and standard solution are made continuously variable by means of transparent glass plungers. The color intensity of the sample solution is matched to that of the standard solution through the use of a split-field optical system, and the sample concentration is then determined by calculations based on the Beer-Lambert law.

A simple device used in water analysis for pH, chlorine, ammonia, iron, and phosphate is the Hellige comparator, which utilizes a series of colored glass standards mounted on a rotatable disk. The sample solution, after appropriate treatment with a reagent, is matched to a colored glass standard. With one device, the sample solution is viewed through a long Nessler tube; with a pocket-type comparator, the solution is viewed transversely in a square glass cell.

Bibliography. G. Charlot, *Colorimetric Determination of Elements: Principles and Methods*, 1964; F. W. Clulow, *Color: Its Principles and Their Applications*, 1972; *Industrial Color Technology*, American Chemical Society Advances in Chemistry Series 107, 1972; D. L. MacAdam, *Color Measurement*, 2d ed., 1985; F. D. Snell and C. T. Snell, *Colorimetric Methods of Analysis*, 3d ed., 1972–1976; L. C. Thomas and G. J. Chamberlin, *Colorimetric Chemical Analytical Methods*, 9th ed., 1980.

COLOR VISION
LORRIN A. RIGGS

The ability to discriminate light on the basis of wavelength composition. It is found in humans, in other primates, and in certain species of birds, fishes, reptiles, and insects. These animals have visual receptors that respond differentially to the various wavelengths of visible light. Each type of receptor is especially sensitive to light of a particular wavelength composition. Evidence indicates that primates, including humans, possess three types of cone receptor, and that the cones of each type possess a pigment that selectively absorbs light from a particular region of the visible spectrum.

If the wavelength composition of the light is known, its color can be specified. However, the reverse statement cannot be made. A given color may usually be produced by any one of an infinite number of combinations of wavelength. This supports the conclusion that there are not many different types of color receptor. Each type is capable of being stimulated by light from a considerable region of the spectrum, rather than being narrowly tuned to a single wavelength of light. The trichromatic system of colorimetry, using only three primary colors, is based on the concept of cone receptors with sensitivities having their peaks, respectively, in the long, middle, and short wavelengths of the spectrum. The number of such curves and their possible shapes have long been subjects for study, but not until the 1960s were direct measurements made of the spectral sensitivities of individual cone receptors in humans and various animals. SEE COLOR; CO-LORIMETRY.

Color recognition. Color is usually presented to the individual by the surfaces of objects on which a more or less white light is falling. A red surface, for example, is one that absorbs most of the short-wave light and reflects the long-wave light to the eye. The surface colors are easily described by reference to the color solid shown in **Fig. 1**. The central axis defines the lightness or darkness of the surface as determined by its overall reflectance of white light, the lowest reflectance being called black and the highest, white. The circumference denotes hue, related primarily to the selective reflectance of the surface for particular wavelengths of light. The color solid is pointed at its top and bottom to represent the fact that as colors become whiter or blacker they lose hue. The distance from central axis to periphery indicates saturation, a characteristic that depends chiefly on the narrowness, or purity, of the band of wavelengths reflected by the surface. At the center of the figure is a medium-gray surface, one which exhibits a moderate amount of reflectance for all wavelengths of light. Colors that are opposite one another are called complementaries, for example, yellow and blue, red and blue-green, purple and green, and white and black.

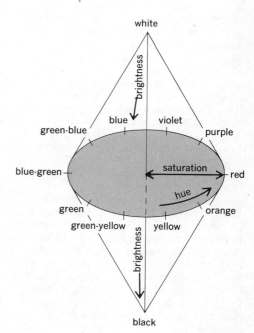

Fig. 1. Diagram of a color solid for illustrating the appearance of surface colors. (After N. L. Munn, Psychology, 2d ed., Houghton Mifflin, 1951)

Color mixture and contrast. Two complementaries, when added together in the proper proportions, can oppose or neutralize one another to produce a colorless white or gray. Various contrast effects also attest to the opposition of the complementaries. Staring at a bright-red surface against a gray background results in the appearance of a blue-green border around the red; this is an example of simultaneous contrast. Similarly, when a red light is turned off, there is frequently a negative afterimage that appears to have the opposite color, blue-green.

A set of primary colors can be chosen so that any other color can be produced from additive mixtures of the primaries in the proper proportions. Thus, red, green, and blue lights can be added together in various proportions to produce white, purple, yellow, or any of the various intermediate colors. Three-color printing, color photography, and color television are examples of the use of primaries to produce plausible imitations of colors of the original objects. SEE PHOTOGRAPHY.

Achromatic colors. Colors lying along a continuum from white to black are known as the gray, or achromatic, colors. They have no particular hue, and are therefore represented by the central axis of the color diagram in Fig. 1. White is shown at the top of the diagram since it represents the high-brightness extreme of the series of achromatic colors. With respect to the surface of an object, the diffuse, uniform reflectance of all wavelengths characterizes this series from black through the grays to white in order of increasing reflectance. Whiteness is a relative term; white paper, paint, and snow reflect some 80% or more of the light of all visible wavelengths, while black surfaces typically reflect less than 10% of the light. The term white is also applied to a luminous object, such as a gas or solid, at a temperature high enough to emit fairly uniformly light of all visible wavelengths. In the same connotation, a sound is described as white noise if its energy is nearly the same at all audible frequencies.

Gray. Gray is the term applied to all the intermediate colors in the series of achromatic colors. Gray may result from a mixture of two complementary colors, from a mixture of all primary colors, or from a fairly uniform mixture of lights of all wavelengths throughout the visible spectrum. Grayness is relative; a light gray is an achromatic color that is lighter than its surroundings, while a dark gray is so called because it is darker than its surroundings. Thus, the same surface may be called light or dark gray when carried from one situation to another. A gray color is one of minimum saturation; it corresponds to zero on a scale of excitation purity. At the other extreme on this scale is the color evoked by pure monochromatic light.

Black. The opposite extreme from white in the series of achromatic colors is black. Blackness is a relative term applied to surfaces that uniformly absorb large percentages of light of all visible wavelengths. A black object in sunlight absorbs a large percentage of the light, but it may reflect a larger absolute quantity of light to the eye than does a white object in the shade. Black may also be used to refer to invisible light; ultraviolet rays, for example, may be called black light if they fall on fluorescent materials that thereby emit visible light.

Color blindness. Color blindness is a condition of faulty color vision. It appears to be the normal state of animals that are active only at night. It is also characteristic of human vision when the level of illumination is quite low or when objects are seen only at the periphery of the retina. Under these conditions, vision is mediated not by cone receptors but by rods, which respond to low intensities of light. In rare individuals, known as monochromats, there is total color blindness even at high light levels. Such persons are typically deficient or lacking in cone receptors, so that their form vision is also poor.

Dichromats are partially color-blind individuals whose vision appears to be based on two primaries rather than the normal three. Dichromatism occurs more often in men than in women because it is a sex-linked, recessive hereditary condition. One form of dichromatism is protanopia, in which there appears to be a lack of normal red-sensitive receptors. Red lights appear dim to protanopes and cannot be distinguished from dim yellow or green lights. A second form is deuteranopia, in which there is no marked reduction in the brightness of any color, but again there is a confusion of the colors normally described as red, yellow, and green. A third and much rarer form is tritanopia, which involves a confusion among the greens and blues.

Many so-called color-blind individuals might better be called color-weak. They are classified as anomalous trichromats because they have trichromatic vision of a sort, but fail to agree with normal subjects with respect to color matching or discrimination tests. Protanomaly is a case of this type, in which there is subnormal discrimination of red from green, with some darkening of the red end of the spectrum. Deuteranomaly is a mild form of red-green confusion with no marked brightness loss. Nearly 8% of human males have some degree of either anomalous trichromatism or dichromatism as a result of hereditary factors; less than 1% of females are color-defective. A few forms of color defect result from abnormal conditions of the visual system brought on by poisoning, drugs, or disease.

Color blindness is most commonly tested by the use of color plates in which various dots of color define a figure against a background of other dots. The normal eye readily distinguishes the figure, but the colors are so chosen that even the milder forms of color anomaly cause the figure to be indistinguishable from its background. Other tests involve the ability to mix or distinguish colored lights, or the ability to sort colored objects according to hue.

Theories. Theories of color vision are faced with the task of accounting for the facts of color mixture, contrast, and color blindness. The schema shown in **Fig. 2** may be useful in considering the various theories of color vision. It has no necessary resemblance to the structures or functions that are actually present in the visual system.

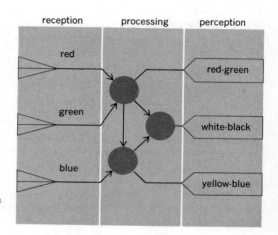

Fig. 2. Schematic representation of the three phases of color vision which can be generally used to describe the various theories of color vision.

Young-Helmholtz theory. Thomas Young, early in the nineteenth century, realized that color is not merely a property of objects or surfaces. Rather, it involves a sensory experience, the characteristics of which are determined by the nature of the visual receptors and the way in which the signals that they generate are processed within the nervous system. Young was thus led to a trichromatic theory of color vision, because he found that only three types of excitation were needed to account for the range of perceived colors. Hermann von Helmholtz clarified the theory and provided it with the idea of differential absorption of light of various wavelengths by each of three types of receptor. Hence, the theory, commonly known as the Young-Helmholtz theory, is concerned with the reception phase in the color vision schema shown. This theory assumes the existence of three primary types of color receptor that respond to short, medium, and long waves of visible light, respectively. Primary colors are those that stimulate most successfully the three types of receptor; a mixture of all three primaries is seen as white or gray. A mixture of two complementary colors also stimulates all three. Blue light, for example, stimulates the blue-sensitive receptors; hence, its complementary color is yellow, which stimulates both the red receptors and the green. Similarly, the complement of red is blue-green and the complement of green is purple (blue and red). Protanopia is explained as a condition in which no long-wave sensitive receptors are present. Deuteranopia results from loss of the green receptor.

Hering theory. The theory formulated by Ewald Hering is an opponent-colors theory. In its simplest form, it states that three qualitatively different processes are present in the visual system, and that each of the three is capable of responding in two opposite ways. Thus, this theory is concerned with the processing phase (Fig. 2), in which there is a yellow-blue process, a red-green process, and a black-white process. The theory obviously was inspired by the opposition of colors that is exhibited during simultaneous contrast, afterimages, and complementary colors. It is a theory best related to the phenomenal and the neural-processing aspects of color rather than to the action of light on photopigments. This theory gains credibility when it is realized that opponent responses are common enough in the central nervous system. Well-known examples are polarization and depolarization of cell membranes, excitation and inhibition of nerve impulses, and reciprocal innervation of antagonistic response mechanism. The theory also maintains that there are four primary colors, since yellow appears to be just as unique perceptually, as red, green, or blue. However, it does not necessarily assume the existence of a separate yellow receptor.

Other theories. Various forms of stage theory or zone theory of color vision have been proposed. One form combines the principles of the Young-Helmholtz and Hering theories. It proposes that (1) there are three types of cone receptor; (2) the responses of red-sensitive, green-sensitive, and blue-sensitive receptors (and possibly others) are conducted to the higher visual centers; (3) at the same time, interactions are occurring at some stage along these separate conducting paths so that strong activity in a red response path inhibits, for example, the activity of the other response paths; and (4) this inhibiting effect is specific to the time and place of the strong red activity. The last point means that in neighboring regions and at subsequent times the blue and green mechanisms are less inhibited, so that blue-green afterimages and borders are commonly experienced as an opponent reaction to strong red stimulation. SEE VISION.

Bibliography. R. M. Boynton, *Human Color Vision*, 1979; C. H. Graham (ed.), *Vision and Visual Perception*, 1965; L. M. Hurvich, *Color Vision*, 1981; G. Jacobs (ed.), *Comparative Color Vision*, 1981; J. Mollion and L. T. Sharpe (eds.), *Colour Vision: Physiology and Psychophysics*, 1983; W. S. Stiles, *Mechanisms of Color Vision*, 1978.

STEREOSCOPY
KENNETH N. OGLE

The phenomenon of simultaneous vision with two eyes, producing a visual experience of the third dimension, that is, a vivid perception of the relative distances of objects in space. In this experience the observer seems to see the space between the objects located at different distances from the eyes. The stereoscopic effect is so unique that it cannot be easily described to one who does not possess it. Stereopsis, or stereoscopic vision, provides the individual with the most acute sense of relative depth and is of vital importance in visual tasks requiring the precise location of objects.

Stereopsis is believed to have an innate origin in the anatomic and physiologic structures of the retinas of the eyes and the visual cortex. It is present in normal binocular vision because the two eyes view objects in space from two points, so that the retinal image patterns of the same object points in space are slightly different in the two eyes. The stereoscope, with which different pictures can be presented to each eye, demonstrates the fundamental difference between stereoscopic perception of depth and the conception of depth and distance from the monocular view. In the **illustration**, each of the two eyes views a pair of vertical lines A and B drawn on cards. The separation of these lines for the right eye is greater than that for the left eye. If the difference in separation is not too great, the images of the lines fuse when the two targets are observed by

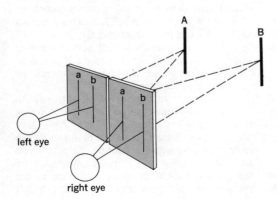

Diagram of stereoscopic vision.

the two eyes. There is almost an immediate stereoscopic spatial experience that the two lines are located in space, line B being definitely more distant than line A. No hint of this spatial experience occurs in the observation of each target alone. This difference between the images in the two eyes provides the stimulus for emergence of the stereoscopic experience.

Stereoscopic acuity and the presence of stereopsis may be tested by several methods or instruments such as the ordinary hand stereoscope and similar devices, vectograph targets, the Howard-Dolman peg test, the Verhoeff stereopter, the Hering falling bead test, pins with colored heads stuck in a board to test for near vision, and afocal meridional magnifying lenses placed before one eye while the subject views a special field (leaf-room or tilting table). SEE VISION.

Bibliography. C. H. Graham (ed.), *Vision and Visual Perception*, 1965; W. L. Gulick and R. B. Lawson, *Human Stereopsis: A Psychological Approach*, 1976; E. Hering, *The Theory of Binocular Vision*, 1977; K. N. Ogle, *Optics: An Introduction for Ophthalmologists*, 2d ed., 1979; K. N. Ogle, *Research in Binocular Vision*, 1950, reprint 1964; D. Pickwell, *Binocular Vision Anomalies: Investigation and Treatment*, 1984; R. W. Reading, *Binocular Vision: Foundations and Applications*, 1983; H. Solomon, *Binocular Vision*, 1980.

MOIRÉ PATTERN
GERALD OSTER

When one family of curves is superposed on another family of curves, a new family called the moiré pattern appears. A familiar example of moiré is the pattern one sees on looking through the folds of a nylon curtain. When the curtain is moved slightly, the moiré pattern is observed to move wildly about. This effect raises at once the possibility of using the moiré effect to measure minute displacements. Moreover, the moiré patterns in the curtain are reminiscent of the complex patterns of waves seen at the shore of a lake. This suggests again that moiré might be used to describe in graphical form certain physical phenomena.

To produce moiré patterns, the lines of the overlapping figures must cross at an angle of less than about 45°. The moiré lines are then the locus of points of intersection. **Figure 1** illustrates the case of two identical figures of simple gratings of alternate black and white bars of equal spacing. When the figures are crossed at 90°, a checkerboard pattern with no moiré effect is seen. At crossing angles of less than 45°, however, one sees a moiré pattern of equispaced lines, the moiré fringes. The spacing of the fringes increases with decreasing crossing angle. This provides one with a simple method for measuring extremely small angles (down to 1 second of

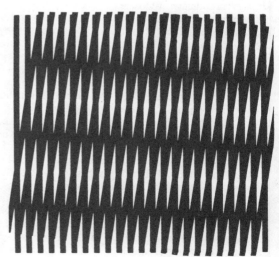

Fig. 1. Two simple gratings crossed at a small angle.

arc). As the angle of crossing approaches zero, the moiré fringes approach 90° with respect to the original figures.

Even when the spacing of the original figures are far below the resolution of the eye, the moiré fringes will still be readily seen. When two diffraction gratings of the transmitting type are superimposed and a distant light source viewed through them, moiré fringes can be seen by the unaided eye. This experiment, first performed by Lord Rayleigh in 1874, provides a means of checking the fidelity of a replica of a diffraction grating. If the repeat spacing of one grating differs slightly from that of the other, a beat pattern will be observed when the lines of the two figures are crossed at zero angle. The beat spacing is inversely proportional to the difference in spacings of the two gratings. Thus with nearly identical gratings a slight relative displacement will cause many fringes to pass by. This beat effect is the same as that seen when one is moving past two repetitive structures, such as the railings of a bridge; the separation between the posts of the railing closer to the observer appears to be slightly greater than that of the farther railing. As a consequence, one sees a beat when a post of the nearer railing gets in step (or is in phase) with an open space of the farther railing. The moiré-beat method, using diffraction gratings, has been applied industrially to automotive precision machinery, the fringes being detected by a photocell and the electric pulses so produced fed into a computer. SEE DIFFRACTION GRATING.

Industrial applications. Moiré techniques are used in the stress analysis of metals. One procedure is to produce photographically a copy of a grating on the surface of the material under examination and to view it through the master grating. In the unstressed condition no pattern is observed (that is, the moiré fringes are at infinity). When the sample is subjected to elongation or shear, a moiré pattern appears that can be readily interpreted in terms of the strain distribution. SEE PHOTOELASTICITY.

The degree of flatness of a surface can be determined by the moiré technique. The shadow of a grating on the surface serves as the second grating. If the surface is not perfectly flat, the moiré pattern observed no longer consists of parallel lines, and from the pattern a contour map of the surface is obtained. In this manner the examination of large optical surfaces can be carried out.

Any lens alters the apparent spacings of a grating. If now this altered image is viewed through another grating, the moiré pattern obtained is determined by the focal length of the lens. The aberrations of the lens become apparent, since the fringes are no longer straight and equidistant. When a periodic grating is viewed through a refractive index gradient (for example, sugar molecules diffusing into water), the spacings of the grating are modified. On overlaying this figure with another periodic grating, a moiré pattern is obtained which gives directly the refractive index gradient curve. SEE ABERRATION.

Phenomena in physics. Many concepts in physics may be demonstrated with a moiré kit, which consists of families of curves printed on transparent plastic. A simple grating of black and white bars can be regarded as the representation of a plane periodic wave, whose wavelength is given by the length of the periodic element of the grating. The moiré pattern of Fig. 1 represents the interference (or superposition) of two plane monochromatic waves. This occurs, for example, in x-ray diffraction; the interfringe distance corresponds to the reflecting planes of atoms, and Bragg's law is satisfied. The figure also represents other situations in which two plane waves are crossing, such as can occur with reflection at a seawall, in holography, or in a microwave cavity. The rays (perpendicular to the wavefront) satisfy the condition of reflection, namely, that the angle of incidence equals the angle of reflection.

Cylindrical or spherical waves are represented by a figure consisting of equispaced concentric circles. Superposing two such figures gives a moiré pattern (**Fig. 2**) consisting of hyperbolas, and in the central portion a family of ellipses. The hyperbolas give the location of the fringes seen in Thomas Young's famous experiment, in which he illuminated two fine slits with a single source of light. The family of ellipses represents the standing-wave modes in an elliptical cavity. It would represent, for example, the sound patterns in a so-called "whispering gallery," where one speaks at one focus of the room and the voice is heard clearly at the other focus.

Diffraction of light by a straightedge can be represented by the moiré pattern produced by the superposition of a straight-line grating and a circular grating; each point of the straightedge becomes a source of cylindrical waves. The moiré pattern is the interference pattern, or hologram, of a point scatterer (a speck of dust, for example). Indeed, the hologram obtained for more com-

Fig. 2. Superposition of two families of equispaced concentric circles.

plicated objects may be regarded as the moiré pattern produced by the superposition of many such simple figures. SEE HOLOGRAPHY.

The moiré technique can also be applied to problems of fields and flows. Thus in hydrodynamics the flow of liquid from a source may be represented by a figure of radiating lines; the greater the source strength, the smaller will be the angular openings. As no positive or negative radial direction is indicated, the figure can also be used to represent a sink. Two such figures when superposed give a moiré pattern, which is the resultant flow of liquid for a source near a sink. This is also the field for an electric dipole, a positive charge close to a negative charge. When a simple parallel-line grating is superposed on a radial figure, one obtains the flow pattern for a source of liquid in a moving stream. This result, which is ordinarily difficult to calculate, is readily obtained and in graphic form by the moiré method.

The vortex, or circular motion, of fluids is represented by circles in which the spacing is greater as one goes out from the center. Superposition of two such figures gives the resulting flow for two vortex centers of opposite sense (**Fig. 3**). The moiré pattern also represents the lines of the same voltage (the equipotentials) looking down the ends of two wires which are oppositely charged. It should be noticed that the pattern is turned 90° from that for the radial figures. By superposing this circular figure with other figures, one can obtain the flow of air about an airplane wing, for example.

Fig. 3. Moiré pattern produced by superposition of two figures representing vortex or circular motion.

CONTRIBUTORS

CONTRIBUTORS

Allen, Dr. Richard G. Steward Observatory, University of Arizona.
Anwyl, Robert D. Retired; formerly, Photographic Consultant, Eastman Kodak Company, Rochester, New York.

Barer, Prof. Robert. Department of Human Biology and Anatomy, University of Sheffield.
Beaglehole, Prof. David. Department of Physics, Victoria University of Wellington.
Billings, Dr. Bruce H. Special Assistant to the Ambassador for Science and Technology, Embassy of the United States of America, Taipei, Taiwan.
Billmeyer, Dr. Fred W. Retired; formerly, Department of Chemistry, Rensselaer Polytechnic Institute.
Boast, Dr. Warren B. Anson Marston Distinguished Professor Emeritus of Electrical Engineering, Iowa State University.
Boynton, Dr. Robert M. Department of Psychology, University of California, San Diego.
Brienza, Dr. Michael J. Retired; formerly, Norden Division, United Technologies Corp., Norwalk, Connecticut.

Casasent, Prof. David. Department of Electrical and Computer Engineering, Carnegie-Mellon University.

Chapman, Dr. Robert D. Section Head, Laboratory for Solar Physics and Astrophysics, NASA Goddard Space Flight Center, Beltsville, Maryland.
Chu, Prof. Benjamin. Department of Chemistry, State University of New York, Stony Brook.

Dieke, Prof. G. H. Deceased; formerly, Chairperson, Department of Physics, Johns Hopkins University.
Duguay, Dr. Michael A. Bell Laboratories, Holmdel, New Jersey.

Economou, George A. Group Vice President—Optical Systems, Contraves Goerz Corporation, Pittsburgh, Pennsylvania.
Edgerton, Dr. Harold E. Institute Professor, Emeritus, Massachusetts Institute of Technology.
Evenson, Dr. Kenneth M. Time and Frequency Division, National Bureau of Standards, Boulder, Colorado.

Feldman, Dr. Barry J. Naval Research Laboratories, Washington, D.C.
Fisher, Dr. Robert A. Los Alamos National Laboratory, Los Alamos, New Mexico.

Galt, Dr. J. K. Vice President, Sandia Laboratories, Albuquerque, New Mexico.
Geist, Jon. Radiometric Physics Division, National Bureau of Standards, Washington, D.C.
Gibbons-Fly, Walter. CBS Television Network, Hollywood, California.
Gibbs, Prof. Hyatt M. Optical Science Center, University of Arizona.
Goodman, Dr. Joseph W. Stanford Electronics Laboratory, Stanford University.
Greenler, Dr. Robert. Department of Physics, University of Wisconsin.
Grey, Dr. David S. Formerly, Aerospace Corporation, El Secundo, California.

Harris, Lawrence A. Oak Ridge National Laboratory, Oak Ridge, Tennessee.
Harrison, George R. Deceased; formerly, Dean Emeritus, School of Science, Massachusetts Institute of Technology.
Herzberger, Max J. Deceased; formerly, Consulting Professor, Department of Physics, Louisiana State University.
Hill, Armin J. Dean, College of Physical and Engineering Science, Brigham Young University.
Horton, G. A. Manager, Engineering Laboratories, Westinghouse Electric Corporation, Cleveland, Ohio.
Hubbard, Dr. W. M. Central Services Organization, Bell Communications Research, Holmdel, New Jersey.
Hull, Dr. McAllister H., Jr. Department of Physics, University of New Mexico.

Ippen, Prof. Erich P. Department of Electrical Engineering and Computer Science, Massachusetts Institute of Technology.

Jacobs, Prof. Stephen F. Optical Science Center, University of Arizona.
Jenkins, Prof. Francis A. Deceased; formerly, Department of Physics, University of California, Berkeley.
Judd, Dr. Deane B. Deceased; formerly, Office of Colorimetry, National Bureau of Standards.

Kaminow, Ivan P. Bell Telephone Laboratories, Holmdel, New Jersey.
Kaprelian, Edward K. Vice President and Technical Director, Keuffel and Esser Company, Morristown, New Jersey.

Klock, Dr. Benny L. Plans and Requirements, Advanced Weapons Technology Division, Defense Mapping Agency, Washington, D.C.
Kogelnik, Dr. Herwig. Crawford Hill Laboratory, Bell Laboratories, Holmdel, New Jersey.

Lambert, Dr. Jack L. Department of Chemistry, Kansas State University.
Lord, Prof. Richard C. Department of Chemistry, Massachusetts Institute of Technology.

McGregor, Prof. E. Russel. Department of Cinema/Television, University of Southern California.
Madison, Dr. Vincent. Department of Medicinal Chemistry, School of Pharmacy, University of Illinois.
Mellon, Prof. M. G. Department of Chemistry, Purdue University.
Moscowitz, Dr. Albert. Department of Chemistry, University of Minnesota.
Mottet, Dr. N. Karle. Professor of Pathology and Director of Hospital Pathology, University Hospital, University of Washington.

Nagel, Suzanne R. Bell Laboratories, Murray Hill, New Jersey.
Nicodemus, Dr. Fred E. Physicist, Optical Radiation Section, Heat Division, Institute for Basic Standards, National Bureau of Standards.
Noble, R. H. Optical Division, Institute Nacional Astrofis, Pueblo, Mexico.

Ogle, Dr. Kenneth N. Deceased; formerly, Section of Biophysics, Mayo Clinic, Rochester, Minnesota.
O'Neill, Jerome P., Jr. Rumrill-Hoyt, Inc., Rochester, New York.
Oster, Prof. Gerald. Mount Sinai School of Medicine, City University of New York.

Pasachoff, Dr. Jay M. Hopkins Observatory, Williams College.
Pecora, Prof. Robert. Department of Chemistry, Stanford University.
Peyghambarian, Nasser. Optical Sciences Center and Optical Circuitry Cooperative, University of Arizona.
Putnam, Prof. Russell C. Professor Emeritus, Case Institute of Technology.

Reintjes, John F. Naval Research Laboratory, Washington, D.C.

Richards, Dr. Oscar W. College of Optometry, Pacific University.

Riggs, Dr. Lorrin A. Department of Psychology, Brown University.

Rockett, Frank H. Engineering Consultant, Charlottesville, Virginia.

Rosenberger, Harold E. Scientific Division, Bausch and Lomb, Inc., Rochester, New York.

Russell, Dr. Howard W. Deceased; formerly, Technical Director, Battelle Memorial Institute, Columbus, Ohio.

Sargent, Dr. Murray, III. Max-Planck Institut für Festkörperforschung, Stuttgart, West Germany.

Schawlow, Prof. Arthur L. Department of Physics, Stanford University.

Scott, Prof. J. F. Department of Physics, University of Colorado.

Shack, Prof. Roland V. Optics Science Center, University of Arizona.

Shannon, Prof. Robert R. Optical Sciences Center, University of Arizona.

Siegman, Dr. A. E. Stanford, California.

Smith, Dr. Warren E. Optical Sciences Center, University of Arizona.

Smythe, Dr. William R. Department of Physics, California Institute of Technology.

Stewart, Dr. John W. Department of Physics, University of Virginia.

Streifer, Dr. William. Palo Alto Research Center, Xerox Corp., Palo Alto, California.

Stroke, Dr. George W. Department of Electrical Sciences, and Head, Electro-Optical Sciences Center, State University of New York, Stony Brook.

Stuart, Dr. Edward G. Deceased; formerly, West Virginia School of Medicine.

Valdmanis, Janis A. Bell Laboratories, Murray Hill, New Jersey.

Van Winkle, Prof. Quentin. Department of Chemistry, Ohio State University.

Ververka, Prof. Joseph. Laboratory for Planetary Studies, Cornell University.

Walworth, Vivian K. Polaroid Corporation, Cambridge, Massachusetts.

Watson, Dr. W. W. Professor Emeritus of Physics, Yale University.

West, Dr. William. Eastman Kodak Company, Rochester, New York.

Winter, Prof. Rolf G. Department of Physics, College of William and Mary.

Wolfe, Dr. William L. Optical Sciences Center, University of Arizona.

Wyant, Prof. James C. Optical Sciences Center, University of Arizona.

Zuk, Dr. William. School of Architecture, University of Virginia.

INDEX

INDEX

Asterisks indicate page references to article titles.

Abbe, E. 12, 78, 92
Aberration 11, 13-20*
  astigmatism 18
  caustics 14-15
  chromatic *see* Chromatic aberration
  coma 17
  distortion 16
  field curvature 15
  of geometry 15-16
  higher-order 18-19
  measures 14
  orders of 15-19
  origins of 19-20
  of point images 16-17
  spherical 16-17
Absorption filter *see* Color filter
Absorption of electromagnetic radiation 185-194*
  absorption and emission coefficients 189-190
  atmospheric 243
  color filter 195-196*
  dichroism 220-221*
  dispersion 190-194
  fluorescence 189
  laws of absorption 185-187
  luminescence 189
  measurement of absorption 187

Absorption of electromagentic radiation (*cont.*):
  phosphorescence 189
  physical nature of absorption 187-189
  scattering 187
  selective reflection 194
Acoustooptics 252-253*
  applications 252-253
  magnitude of effect 252
  optical modulators 254-255
  production of sound 253
  types of effect 252
Adams, W.S. 116
Adaptive optics 102-106*
  basic techniques 102-105
  multidither COAT system 102-103
  nonlinear phase conjugation 105
  operating systems 105-106
  phase-conjugate system 103-104
  sharpness-function technique 103
  wavefront correctors 104
Afocal system: geometry 7-8
Airy, G. 114
Albedo 180-182*
Andrews, T. 224
Aperture: geometrical optics 9

Apodization 159
Astigmatism 18
  eyeglasses 42
Astronomical transit instrument 53-59
  applications 54-55
  broken-back transit 59
  corrections 55
  meridian circle 55-57
  photographic zenith tube 57-59

Babcock, H.W. 102
Babinet's principle 161
Beer's law 186-187
Bessel, F.W. 116
Billet split lens 138
Binoculars 67*
  prisms 27-28
Birefringence 220*
  birefringent crystals and polarized light 167-168
  electrooptics 246
  Kerr effect 247-248*
  use in photoelasticity 257
  visible light 199
Blackbody: incandescence 128-129*
Blackett, P.M.S. 117
Bond albedo 180-181

Bouguer's law *see* Lambert's law
Bradley, J. 114
Brillouin scattering 222-223
  optical phase conjugation 299

Camera 73-74*
  ballistic 70-71
  finders 74
  focusing 73-74
  rangefinder 64
  shutters 74
Candlepower 362*
Caustics: aberration 14-15
Cemented lenses 33
Chadwick, J. 117
Chemical laser 265
Chevalier, C. 36
Chief ray 9
Chromatic aberration 20-23*
  color correction 21-23
  dispersion formula 20-21
Chromatic resolving power 36-37
Cinematography 336-338*
Circular dichroism 207
  Cotton effect 210-213*
Coherence 270-276*
  examples 275-276
  quantitative definitions 274-275
  single beam 273-274
  two beams 270-273
Color 370-372*
Color filter 195-196*
Color photography 322-324
Color vision 374-378*
Colorimetry 355, 372-374*
  colorimetric analysis 374
  indirect 373
  metamers 373-374
  photoelectric 373
  visual 373
Coma 17
Comparison microscope 82-83
Compound lens 32
Compound microscope 76
Compton, A.H. 226

Compton effect 113
Concave diffration grating 164
Conic mirror 29-30
Conics of revolution 25
Cornu's spiral 161
Cotton effect 210-213*
  measurements 211
  molecular structure 211-213
Cotton-Mouton effect 249-250
Critical opalescence 224, 225
Crystal optics 213-220*
  biaxial crystals 218-220
  dichroism 220-221*
  index ellipsoid 213-214
  interference in polarized light 216-218
  Kerr effect 247-248*
  optical parametric oscillators 294
  pleochroism 220*
  Pockels cell 246
  ray ellipsoid 214-216
  rotatory dispersion 210
  trichroism 221-222*
Curie, I. 117

de Broglie, L. 113
Debye, P. 224
Dichroism: circular *see* Circular dichroism
  dichroic crystals and polarized light 167
Differential interference contrast microscope 98-99
Diffraction 153-162*
  atmospheric 243
  Fraunhofer 154-159
  Fresnel 159-162
  fringe 142*
  light 109
  microwaves 162
Diffraction grating 157, 162-165*
  concave grating 164
  echelette grating 164
  echelle grating 164
  grating spectroscopes 163-164
  mountings 164-165

Diffraction grating (*cont*.):
  production 162-163
  properties 163
  resolving power 37-38
Diode laser 289
Diopter 13*
Dirac, P.A.M. 113
Dispersion (radiation) 4, 200*
  anomalous 191
  dispersing prisms 30-31
  formula for chromatic aberration 20-21
  light 112-113
  by a prism 190-191
  relation to absorption 191-194
Dissecting microscope 83
Distortion (aberration) 16
Double refraction *see* Birefringence
Dyson microscope 97

Eccentric mirror 26
Echelette grating 164
Echelle grating 164
Einstein, A. 223
Electromagnetic radiation 118-121*
  absorption *see* Absorption of electromagnetic radiation
  damped waves 119-120
  detection 118-119
  electric dipole 121
  free-space waves 119
  infrared *see* Infrared radiation
  plane waves 119
  radiation pressure 122*
  radiometry 342-343*
  reflection *see* Reflection of electromagnetic radiation
  spherical waves 119
  ultraviolet *see* Ultraviolet radiation
  wave impedance 120-121
  see also Light

INDEX **393**

Electromagnetic wave 117-118*
Electrooptics 246-247*
  electrooptic sampling 246-247
  integrated optical circuit switching and modulation 289-290
  Kerr effect 247-248*
  optical modulators 253-254
  Pockels cell 246
Ellipsometry 182-184*
  applications 183-184
  principles 182-183
Etalon devices 295-296
Exposure meter 343-344*
Eyeglasses 41-43*
  diopter 13
Eyepiece 59-60*

Fabry-Perot etalon 295-296
Fabry-Perot interferometer 148-149
  light scattering 234
  optical bistability 302-303
Faraday, M. 250
Faraday effect 248, 250-251*
Fermat's principle 4
Field curvature 15
Filter: color filter 195-196*
  interference filter 151-153*
Fizeau interferometer 146
Flare (ghost image) 24
  absorption of radiation 189
Fluorescence microscope 101*
Focal length 13*
  geometry 6-7
Fraunhofer diffraction 153, 154-159
  applications 158-159
  determination of resolving power 157-158
  diffraction grating 157
  vibration curve 155-157
Free-electron laser 266
Fresnel, A. 114, 133, 153
Fresnel biprism 136-138

Fresnel diffraction 153, 159-161
  Babinet's principle 161
  Cornu's spiral 161
  zone plate 160-161
Fresnel double mirror 136
Fresnel rhomb: polarized light production 171
Fringe 142*

Gabor, D. 276
Gans, R. 223
Gas-discharge laser 264-265
Gas-dynamic laser 266
Geometrical optics 4-10*
  aberrations of geometry 15-16
  afocal systems 7-8
  apertures 9
  basic concepts 4-5
  focal systems 6-7
  ideal image formation 6
  Lagrange invariant 9-10
  marginal ray and chief ray 9
  paraxial optics 8-9
  ray paths 4-5
  rays 4
  refractive index and dispersion 4
  sources and wavefronts 4
Geometrical wavefront 4
Ghost image 23-24*
  flare 24
  lens coatings 24
Grating spectroscope 163-164
Gunsights 62-63*
  aircraft 63
  artillery 63
  rifles 62-63

Heisenberg, W. 113
Hertz, H. 118
High-power short-pulse laser 266-268
Holographic interferometry 149-150
Holography 276-279*
  applications 277-279

Holography (*cont.*):
  display of images 279
  fundamentals of the technique 276-277
  in interferometry 278
  in microscopy 278
  optical elements 279
  in optical memories 278-279
Huygens, C. 133
Huygens eyepiece 59
  light microscope 80
Huygens-Fresnel principle 133
Huygens' principle 133*
Hyperopia: eyeglasses 42
Hyper-Raman effect 230-231

Illuminance 363*
Illumination 353-361*
  inverse-square law 356
  light-control methods 360-361
  light sources 356-357
  photometric definitions 354-356
  reflection and transmission 357-359
  spectral response function of human eye 353-354
  subjective color 353
Illuminometer 349
  photoelectric 350
Image *see* Optical image
Image processing 338-340*
  applications 340
  compression 340
  digitization 338
  enhancement 338-339
  optical information systems 321-322
  restoration 339-340
Image tube (astronomy) 309-313*
  image-intensifier cameras 312-313
  image-intensifier tubes 309-312
  photocathode 309
Incandescence 128-129*

Information systems, optical 291-294
Infrared photography 314
Infrared radiation 122-126*
Integrated optical circuit 287
Integrated optics 287-291*
　coupling external light beams 289
　guided waves 288
　lasers and distributed feedback 289
　materials and fabrication 288
　spectrum analyzer 290-291
　switches and modulators 289-290
Integrating-sphere photometer 350
Intensity fluctuation spectroscopy 233-234
Interference filter 151-153*
　basic properties 151
　frustrated reflection 152
　multilayer types 151-152
Interference microscope 96-101*
　biological applications 99-101
　comparison with phase-contrast type 99
　differential interference contrast microscope 98-99
　measurement of phase changes 99
　nonbiological applications 101
　practical realization 96-98
　theory 96
Interference of waves 133-142*
　amplitude splitting 138-140
　crystal interference in polarized light 216-218
　diffraction grating 162-165*
　fringe 142*
　holography 276-277
　interference filter 151-153*
　light 110
　multiple-beam interference 140-142

Interference of waves (cont.):
　splitting of light sources 134-138
　theory of interference microscope 96
　two-beam interference 134
Interferometer: resolving power 38
Interferometry 142-151*
　basic classes of interferometers 143
　Fabry-Perot interferometer 148-149
　Fizeau interferometer 146
　holographic 149-150
　Mach-Zehnder interferometer 146-147
　Michelson interferometer 143-144
　Michelson stellar interferometer 147-148
　phase-shifting 150-151
　shearing type 147
　speckle 150
　Twyman-Green interferometer 144-146
　use of holography 278
Inverse-square law 132*
　illumination 356
Inverted microscope 82

Joliot, F. 117
Jones, R.C. 172
Jones calculus 172-173
Jordan, P. 113

Kerr effect: Kerr constant 247
　Kerr shutter 248
　magnetooptic 250
　optical modulators 253
　optical phase conjugation 298-299
Kirchhoff, G. 133
Krishnan, K.S. 227

Lagrange invariant 9-10
Lambert's law 186

Land, E.H. 168
Landsberg, G. 227
Laser 262-268*
　applications 268
　chemical 265
　coherence 276
　comparison with other sources 262-263
　free-electron 266
　gas-discharge 264-265
　gas-dynamic 266
　high-power short-pulse 266-268
　Kerr effect 248
　optical communications transmitters 286-287
　optical pumping 263-264, 268-269*
　photodissociation 265-266
　pulsed gas 265
　quantum electronics 269-270*
　semiconductor 266
　speckle 279-282*
　stabilized 109
　Twyman-Green interferometer 144-146
Latent image 326*
Law of Malus 167
Lens 31-36*
　cemented lenses 33
　chromatic aberration 20-23*
　compound lenses 32
　diopter 13*
　enlarger lenses 35-36
　eyeglasses see Eyeglasses
　focal length 13*
　ghost image 23-24*
　magnifiers 35-36
　meniscus 34
　optical microscopes 77-79
　photographic objectives 33-35
　projection lens 45
　single lenses 32-33
　systems 33-36
　telephoto lens 74*
　telescope systems 33
　types 32-33
　zoom lens, 74–75*

INDEX

Light 108-117*
 blackbody radiation 113
 Compton effect 113
 corpuscular phenomena 113
 diffraction 109
 dispersion 112-113
 electromagnetic wave propagation 112
 general relativity results involving light 116-117
 gravitational and cosmological redshifts 116
 historical approach 110-111
 illumination 353-361*
 incandescence 128-129*
 interference 110
 matter and radiation 117
 Michelson-Morley experiment 114-115
 photoelectric effect 113
 photometer 347-353*
 photometry 345-347*
 photon 129*
 polarization 110
 polarized see Polarized light
 principal effects 108-111
 quantum theories 113-114
 quasielastic light scattering 232-235*
 reflection 109-110
 refraction 110
 refraction of visible light 198-200
 refractive index 112
 relativistic aberration 114
 relativistic effects 114-117
 speed of 108-109
 stabilized lasers 109
 theory 111-117
 velocity in moving media 114
 wave optics 132*
 wave phenomena 111-113
Light see also Electromagnetic radiation
Light amplification by stimulated emission of radiation see Laser

Light-distribution photometer 352
Light microscope 80-82
 fluorescence microscope 101*
Light ray 4
Linear polarizer 167-168
Lloyd's mirror 138
Luminance 355-356, 362*
Luminescence: absorption of radiation 189
Luminous efficacy 363-364*
Luminous efficiency 364*
Luminous energy 362*
Luminous flux 355, 361*
Luminous intensity 355, 361-362*
 candlepower 362*

Mach-Zehnder interferometer 146-147
Magnetooptics 248-250*
 Cotton-Mouton effect 249-250
 Faraday effect 248, 250-251*
 Kerr effect 250
 Majorana effect 250
 Voigt effect 248-249
 Zeeman effect 248
Magnification 12-13*
 magnifying lenses 35-36
Magnifying lenses 35-36
Magnifying power 12
 microscope lens 77
Majorana effect 250
Malus, law of 167
Mandelstam, L. 227
Marginal ray 9
Maser: coherence 276
Maxwell, J.C. 111, 118
Meniscus lens 34
Metallurgical microscope 84
Meteorological optics 241-243*
 diffraction 243
 emission and absorption 243
 light scattering 241-242

Meteorological optics (cont.):
 reflection 242
 refraction 242-243
Michelson, A.A. 158, 276
Michelson interferometer 143-144
Michelson-Morley experiment 114-115
Michelson stellar interferometer 147-148
 coherence 276
 Fraunhofer diffraction 158
Microscope 75-76*
 see also Fluorescence microscope; Interference microscope; Optical microscope; Phase-contrast microscope; Reflecting microscope
Microwave: diffraction 162
Mie, G. 224
Mie scattering 224
Mirror: conic 29-30
 eccentric 26
 plane 26-27
 spherical 28-29
Mirror optics 26-30*
 conic mirrors 29-30
 Lloyd's mirror 138
 mirror coatings 30
 plane mirrors 26-27
 prisms 27-28, 30
 spherical mirrors 28-29
Moiré pattern 379-381*
 industrial applications 380
 phenomena in physics 380-381
Mueller matrices 174
Myopia: eyeglasses 42

Near-infrared microscopy 89-90
Newton, I. 279
Nonlinear optical devices 294-297*
 etalon devices 295-296
 four-wave mixing beam deflectors 297

Nonlinear optical devices (*cont.*):
frequency generation 294-295
optical signal processing 295-297
waveguide devices 297
Nonlinear optics 236-241*, 270
coherent effects 240-241
devices *see* Nonlinear optical devices
inelastic scattering 240
nonlinear materials 236
nonlinear spectroscopy 239-240
optical phase conjugation 297-300*
second-order effects 236-237
self-action and related effects 238-239
third-order effects 237-238
time measurements 240
Nonlinear spectroscopy 239-240

Occhialini, G.P. 117
Ocular *see* Eyepiece
Opalescence 225-226*
critical 224, 225
time dependency 225-226
Opaque medium 195*
Optical activity 206-209*
correlation with molecular structure 207-209
Cotton effect 211-212
methods of measurement 206-207
rotatory dispersion 209-210*
Optical bistability 300-304*
all-optical systems 301-302
bistable optical devices 302-303
fundamental studies 304
Optical communications 284-287*
atmospheric 285
free-space 285

Optical communcations (*cont.*):
optical fiber communications 285-286
optical receivers 287
optical transmitters 286-287
Optical detectors 308*
communications receivers 287
photon detectors 342-343
Optical fibers 282-284*
attenuation of light 283-284
designs 282-283
use in communications 285-286
Optical glass 201-204
available types and limitations 204
colored glass 203-204
effect of absorption 202-203
glass types 201-202
Optical image 10-12*
aberration *see* Aberration analysis 12
eyepiece 59-60*
formation in optical microscopes 78-79
ghost image 23-24*
ideal image formation 6
mirror optics 26-30*
processing *see* Image processing
resolution 11-12
resolving power 36-38*
Optical information systems 291-294*
image processing 291-292
optical computing 293
signal processing 292-293
Optical interferometer *see* Interferometry
Optical materials 200-206*
available glasses and limitations 204
cautions 206
crystalline materials 204
for infrared and ultraviolet 205-206
nonlinear 236
optical fibers 282-284*

Optical materials (*cont.*);
optical glass 201-204
plastics 205
Optical microscope 77-90*
calibration and measurement of specimen 87-89
catadioptric lens systems 78
color correction 78
comparison microscope 82-83
condenser aperture 79
condenser system 79
condensers 79-80
critical illumination 85
dark-field condensers 79-80
dark-field illumination 86-87
dissecting microscope 83
Fraunhofer diffraction 158-159
illumination 85-86
illumination modification by filters 87
image formation 78-79
inverted microscope 82
Köhler's method of illumination 85-86
lenses 77-79
light microscope 80-82
magnifying power 77
metallurgical microscope 84
microscopy 85-89
near-infrared microscopy 89-90
objectives 77
oblique illumination 86
resolving power 38
Shillaber's type 3 illumination 86
types 80-84
vertical illumination 86
Optical modulators 253-256*
acoustooptic modulation and deflection 254-255
electrooptic effect 253-254
electrooptic intensity modulation 254
optical waveguide devices 255-256

INDEX **397**

Optical phase conjugation 297-300*
  fundamental properties 297-298
  practical applications 300
  techniques 298-300
Optical prism 30-31*
  diopter 13*
  dispersion 30-31, 190-191
  mirror optics 27-28
  reflecting 30
  resolving power 37
  Wollaston prism 167-168
Optical projection systems 43-46*
  condenser 44
  light source 43-44
  object holder 44-45
  projection lens 45
  projection screen 45-46
Optical pulses 304-305*
Optical pumping 268-269*
Optical rotation 206-207
Optical signal processing: information systems 292-293
  nonlinear optical devices 295-297
Optical surfaces 24-26*
  conics of revolution 25
  eccentric mirrors 26
  general aspherics of revolution 26
  mirror coatings 30
  nonrotationally symmetric surfaces 26
  spherical and aspherical 24-25
Optical telescope 46-59*
  afocal system geometry 8
  associated tools 49-50
  astronomical transit instruments 53-59
  conic mirrors 29-30
  future developments 53
  lens system 33
  limitations 50-51
  notable telescopes 51-53
  reflecting telescopes 47-48
  refracting telescopes 46-47

Optical telescope (*cont.*):
  resolving power 38
  Schmidt camera 48-49, 60-62*
  site selection 49
  solar telescopes 49
  tracking telescopes 71
  types 46-49
Optical tracking instruments 68-73*
  ballistic cameras 70-71
  satellite optical tracking 71-73
  spatial position determination 68-70
  tracking telescopes 71
Optically pumped laser 263-264
Ornstein, L.S. 224
Orthoscopic eyepiece 59

Paraxial optics 8-9
Pauli, W. 113
Periscope 40-41*
  miscellaneous types 41
  plane mirrors 26-27
  submarine type 40-41
  tank type 40
Petzval objective 35
Phase-contrast microscope 92-96*
  biological applications 94-95
  comparison with interference type 99
  image interpretation 94
  nonbiological applications 95-96
  practical realization 93-94
  theory 92-93
Phase-shifting interferometry 150-151
  absorption of radiation 189
Photodissociation laser 165-166
Photoelasticity 257-259*
  determination of principal stresses 257-259
  measurements on actual objects 259

Photoelasticity (*cont.*):
  polariscope 257
  three-dimensional measurements 259
  use of birefringent phenomenon 257
Photoelectric effect 113
Photoelectric illuminometer 350
Photoelectric photometer 349-350
Photoemissive tube photometer 350
Photographic materials 327-331*
  emulsions 327-328
  halation 328
  image-transfer systems 329-330
  nonsilver processes 330-331
  photographic products 329
  spectral sensitivity 328
  storage 330
  supports 327
Photographic objectives 33-35
Photography 313-326*
  color 322-326
  exposure meter 343-344*
  infrared 314
  latent image 326*
  materials *see* Photographic materials
  radiometry 315
  sensitometry and image structure 318-321
  stroboscopic photography 331-335*
  theory of the process 315-318
  ultraviolet 314-315
Photometer 347-353*
  illuminometer 349
  integrating-sphere 350
  light-distribution 352
  luminance meter 349
  photoelectric 349-350
  photoelectric illuminometer 350
  photoemissive tube 350
  reflectometer 351

Photometer (*cont.*):
  spectrophotometer 352
  visibility meters 353
  visual 348
Photometry 345-347*
  photometer 347-353*
  photometric measurements 347
  photometric units 346
  relative visibility 346
  spectral luminous efficacy 346-347
Photon 129*
Physical optics 177*
Plane mirror 26-27
Pleochroism 220*
Pockels cell 246
Point image: aberrations of 16-18
Point source 4, 132*
  inverse-square law 132*
Polariscope 257
Polarization of waves 165-166*
Polarized light 110, 166-174*
  analyzing devices 171-172
  dichroism 220-221*
  Faraday effect 250-251*
  law of Malus 167
  linear polarizing devices 167-168
  optical activity 206-209*
  pleochroism 220*
  polarization by scattering 168
  production 169-171
  retardation theory 172-174
  rotatory dispersion 209-210*
  types 169
Polaroid sheet polarizer 168
Pound, R.V. 116
Poynting's vector 121-122*
Presbyopia: eyeglasses 42
Prism *see* Optical prism
Pulsed gas laser 165

Quantum electronics 269-270*
  applications of lasers 270
  nonlinear optical phenomena 270

Quantum electronics (*cont.*):
  stimulated emission and amplification 269-270
Quasielastic light scattering 232-235*
  applications 235
  Fabry-Perot interferometry 234
  intensity fluctuation spectroscopy 233-234
  rotational diffusion coefficients 235
  static light scattering 232-233
  translational diffusion coefficients 234-235

Radiance 344-345*
Radiation pressure 122*
Radiometry 342-343*
  photon detectors 342-343
  radiometric sources 343
Raman, C.V. 222, 227
Raman effect 222, 226-232*
  discovery 226-227
  hyper-Raman effect 230-231
  optical phase conjugation 299
  resonance Raman effect 230
  special forms 230-232
  spectroscopy 227-228
  stimulated Raman effect 231-232
  theory 228-230
Raman spectroscopy 227-228, 232
Ramsden eyepiece 59
Rangefinder 63-66*
  cameras 64
  coincidence 64-65
  depression 66
  errors 66
  heightfinders 66
  military types 64-65
  stadia methods of rangefinding 66
  stereoscopic 65-66
Ray paths 4-5

Rayleigh-Gans-Debye theory 224
Rayleigh scattering 222-223
Real image 10
Rebka, G.A. 116
Reflecting microscope 90-92*
  optical system 90-91
  uses 91-92
Reflection and transmission coefficients 184-185*
Reflection of electromagnetic radiation 5, 177-180*
  albedo 180-182*
  antireflection coatings 180
  atmospheric 242
  from dielectrics 178-180
  ellipsometry 182-184*
  ghost image 23-24*
  high-reflectivity multilayers 180
  light 109-110
  from metals 178
  mirror optics 26-30*
  optical surfaces 24-26*
  reflecting microscope 90-92*
  reflecting prisms 30
  reflecting telescopes 47-48
  reflection angle 177
  reflection coefficients 184-185
  reflectivity 177
  selective 194
Reflectometer 351
Refraction of waves 4, 196-200*
  atmospheric 199-200, 242-243
  birefringence *see* Birefringence
  double refraction 199
  electromagnetic waves 200
  eyeglasses 41-43*
  light 110
  optical surfaces 24-26*
  refracting telescopes 46-47
  refractometry 199
  Snell's law 196-198
  visible light 198-200
Refractometry 199

Resolving power 12, 36-38*
  chromatic 36-37
  Fraunhofer diffraction 157-158
  gratings 37-38
  interferometers 38
  microscopes 38
  prisms 37
  telescopes 38
Resonance Raman effect 230
Rotatory dispersion 209-210*
  Cotton effect 210-213*
  optically active materials 210
  reasons for variation 210

Scattering of electromagnetic radiation 222-224*
  atmospheric 241-242
  Brillouin and Rayleigh scattering 222-223
  critical opalescence 224
  distinguished from absorption 187
  inelastic scattering 222
  large particles 223-224
  light polarization 168
  nonlinear 224, 240
  opalescence see Opalescence
  quasielastic light scattering 232-235*
  Raman effect see Raman effect
  Tyndall effect 226*
Schmidt camera 48-49, 60-62*
  related systems 62
  Schmidt-Cassegrain design 62
  sky surveys 61-62

Schmidt telescope see Schmidt camera
Schrödinger, E. 113
Semiconductor laser 266
  optical communications transmitter 286
Shadow 194-195*
Shearing interferometer 147
Sheet polarizer 168
Single lens 32-33
Smith-Baker microscope 98
Smoluchowski, M. 223
Snell's law 196-198
Speckle 279-282*
  applications 281-282
  basic phenomenon 280
Speckle interferometry 150
Spectrophotometer 352
Speed of light 108-109
Spherical aberration 16-17
Spherical mirror 28-29
Spherochromatism 19
Stabilized lasers 109
Stellar speckle interferometry 282
Stereoscopy 378-379*
Stimulated Raman effect 231-232
Stroboscopic photography 331-335*
  conditions for starting 333-334
  energy storage 331
  flash lamps 331-333
  high-frequency flashes 334-335
  requirements 333

Telephoto lens 35, 74*
Telescope see Optical telescope

Translucent medium 195*
Transparent medium 195*
Trichroism 221-222*
Twyman-Green interferometer 144-146
Tyndall effect 223, 226*

Ultraviolet photography 314-315
Ultraviolet radiation 126-128*

Virtual image 10
Visibility meter 353
Vision 366-370*
  color vision 374-378*
  eyeglasses 41-43*
  stereoscopy 378-379*
Visual photometer 348
Voigt effect 248-249

Wave optics 132*
Wave plate: polarized light production 169-171
Waveguide: integrated optics 288
  nonlinear devices 297
Wollaston, W.H. 167
Wollaston prism 167-168

Young, T. 111, 134
Young's two-slit experiment 134-136
  coherence phenomena 271

Zeeman effect 248
Zernike, F. 93, 224
Zoom lens 74-75*